Atomic Physics of
Highly Ionized Atoms

NATO Advanced Science Institutes Series

A series of edited volumes comprising multifaceted studies of contemporary scientific issues by some of the best scientific minds in the world, assembled in cooperation with NATO Scientific Affairs Division.

This series is published by an international board of publishers in conjunction with NATO Scientific Affairs Division

A	**Life Sciences**	Plenum Publishing Corporation
B	**Physics**	New York and London
C	**Mathematical and Physical Sciences**	D. Reidel Publishing Company Dordrecht, Boston, and London
D	**Behavioral and Social Sciences**	Martinus Nijhoff Publishers The Hague, Boston, and London
E	**Applied Sciences**	
F	**Computer and Systems Sciences**	Springer Verlag Heidelberg, Berlin, and New York
G	**Ecological Sciences**	

Atomic Physics of Highly Ionized Atoms

Edited by

Richard Marrus

University of California, Berkeley
Berkeley, California

Plenum Press
New York and London
Published in cooperation with NATO Scientific Affairs Division

Proceedings of a NATO Advanced Study Institute on
Atomic Physics of Highly Ionized Atoms,
held June 7—18, 1982,
in Cargèse, Corsica, France

Library of Congress Cataloging in Publication Data

NATO Advanced Study Institute on Atomic Physics of Highly Ionized Atoms (1982:
Cargèse, Corsica)
  Atomic physics of highly ionized atoms.
  (NATO advanced science institutes series. Series B, Physics; v. 96)
  "Published in cooperation with NATO Scientific Affairs Division."
  "Proceedings of a NATO Advanced Study Institute on Atomic Physics of Highly Ionized
Atoms, held June 7—18, 1982, in Cargese, Corsica"—T.p. verso.
  Includes bibliographical references and index.
  1. Ions—Congresses. 2. Atoms—Congresses. 3. Collisions (Nuclear physics)—Con-
gresses. I. Marrus, Richard. II. North Atlantic Treaty Organization. Scientific Affairs Divi-
sion. III. Title. IV. Title: Highly ionized atoms. V. Series.
QC702.N37 1982   539.7   83-8081
ISBN-13: 978-1-4613-3720-1   e-ISBN-13: 978-1-4613-3718-8
DOI: 10.1007/978-1-4613-3718-8

© 1983 Plenum Press, New York
Softcover reprint of the hardcover 1st edition 1983
A Division of Plenum Publishing Corporation
233 Spring Street, New York, N.Y. 10013

PREFACE

The last decade has seen dramatic progress in the development of devices for producing multicharged ions. Indeed it is now possible to produce any charge state of any ion right up through fully-stripped uranium (U^{92+}). Equally dramatic progress has been achieved in the energy range of the available ions. As an example, fully-stripped neon ions have been produced in useable quantities with kinetic energies ranging from a few ev to more than 20 Gev.

Interest in the atomic physics of multicharged ions has grown apace. In the fusion program, the spectra of these ions is an important diagnostic tool. Moreover the presence of multicharged ions presents a serious energy loss mechanism in fusion devices. This fact has motivated a program to study the collision mechanisms involved. In another area, multicharged ions are present in the solar corona and the interstellar medium and knowledge of their collision properties and spectra is essential to understanding the astrophysics. Other possible applications are to x-ray lasers and heavy ion inertial fusion. On a more fundamental level, new possibilities for testing quantum electrodynamics with multicharged ions have emerged.

As a result, there has been a large world-wide effort to study the atomic physics of these ions. The goal of the organizers of the school was to put together a set of courses which would survey some of the broad areas of study. Theoretical and experimental efforts both in spectroscopy and collision physics were included, and are summarized in the chapters contained herein. In addition, courses were given in related topics such as ion source development and higly-ionized atoms in Tokamak discharges. Several seminars were given, and there was also a symposium on low-energy electron capture but these are not included in the manuscripts.

Many people and institutions contributed to the success of the school. Thanks are due to the Scientific Affairs Division of NATO and the CNRS of France for their support. Marie France Hanseler and Annie Vernhet made possible the wonderful arrangement in Cargese and most of all, I thank my colleague Professor Jean-Pierre Briand for taking the burden of so many of the tasks on his shoulders.

Richard Marrus
Berkeley, California

CONTENTS

EXPERIMENTAL INVESTIGATIONS OF THE
STRUCTURE OF HIGHLY IONIZED ATOMS

Indrek Martinson

Department of Physics, University of Lund

S-223 62 Lund, Sweden

INTRODUCTION

Contemporary atomic physics has been revitalized by several important experimental developments. The structures of neutral and singly ionized atoms can nowadays be very accurately determined by exciting the atoms using e.g. tunable dye lasers or synchrotron radiation. Another important line of research concerns the different atomic structure problems occurring in highly ionized atoms. Powerful light sources such as laser-produced plasmas or excited fast ions from particle accelerators may now yield more than 40 times ionized atoms. The spectroscopy of such ions reveals the importance of several effects, due to relativity, nuclear structure or quantum electrodynamics (QED), that are either absent or insignificant in the case of neutral or mildly ionized atoms.

Experimental studies of highly ionized atoms are also motivated by problems and data needs in astrophysics and plasma physics including fusion research.

Some Important Developments

Investigations of the spectra of highly ionized atoms have been carried out for many years. More than 50 years ago Bowen and Millikan (1925) observed two spectral lines in Cl VII (six times ionized Cl). A decisive experimental breakthrough was in the 1930's made in M. Siegbahn's laboratory at Uppsala where a powerful spectrograph for the far ultraviolet (UV) region was combined with an efficient light source (vacuum spark).

Using this equipment Edlén, Tyrén and their collaborators

1

were able to produce and investigate the properties of highly ionized atoms. For example, the Na-like spectra of K IX – Cu XIX (Edlén
1936) and Ne-like spectra of Cr XV – Co XVIII (Tyrén 1938) were studied. More than 20 times ionized atoms were often produced (e.g.
Sb XXIV), a record in ionization that stood unsurpassed for about
35 years. That careful work had several astrophysical applications.
Thus Bowen and Edlén (1939) identified some strong lines in Nova
Pictoris as "forbidden" transitions in Fe VII and some years later
Edlén (1942) identified many lines in the solar corona as forbidden
transitions in 9 – 15 times ionized Ca, Fe and Ni. After World War II
it has been possible to investigate the solar spectrum at wavelengths
shorter than 3000 Å, initially with a rocket-borne spectrograph
(Tousey et al., 1947). Spectroscopic studies of the sun and the
stars using rockets or satellites now form a very productive and
important area in astrophysics and a wealth of data about highly
ionized atoms has thereby been obtained (Fawcett 1974, 1981, Feldman 1981).

Work in plasma physics towards controlled thermonuclear fusion is also dependent upon atomic parameters, particularly regarding highly ionized atoms. Studies of e.g. Tokamak discharges have
shown the presence of impurity atoms in the hot plasmas. These
atoms e.g. Cr, Fe, Ni, Mo, Ta and W in high ionization degrees)
which enter the plasma as the result of sputtering processes, give
rise to substantial energy losses through electromagnetic radiation
(Hinnov 1976, Drawin 1981).

There has been much further work to develop efficient laboratory light sources that produce highly ionized atoms. A number of
very efficient light sources such as low-inductance vacuum spark,
exploding wire, theta-pinch, plasma focus, laser-produced plasma, and
foil-excited ion beam were developed in the 1960's and early 1970's.
Parallel to this experimental work there have been several developments have been partially analyzed, which should be compared with a
and atomic many-body effects. Several advanced computer programs
are now available which provide valuable theoretical explanations
of the experimental material.

Despite the continuous efforts to produce new data about multiply ionized atoms the present information is far from complete.
Thus Cowan (1981) points out that 1019 spectra of 99 chemical elements have been partially analyzed, which should be compared with a
total of 5460 spectra for 104 elements. Also, even for spectra where
the analyses have been considered as reasonably complete much new
information can be found. Thus Johansson (1982) identified over
300 new levels in Cr II, a spectrum earlier labeled as "complete".

In Fig. 1 we show the number of LS-terms presently known
in all 26 spectra of Fe (Fe I - Fe XXVI). Although this is one
of the most thoroughly studied elements the information is still

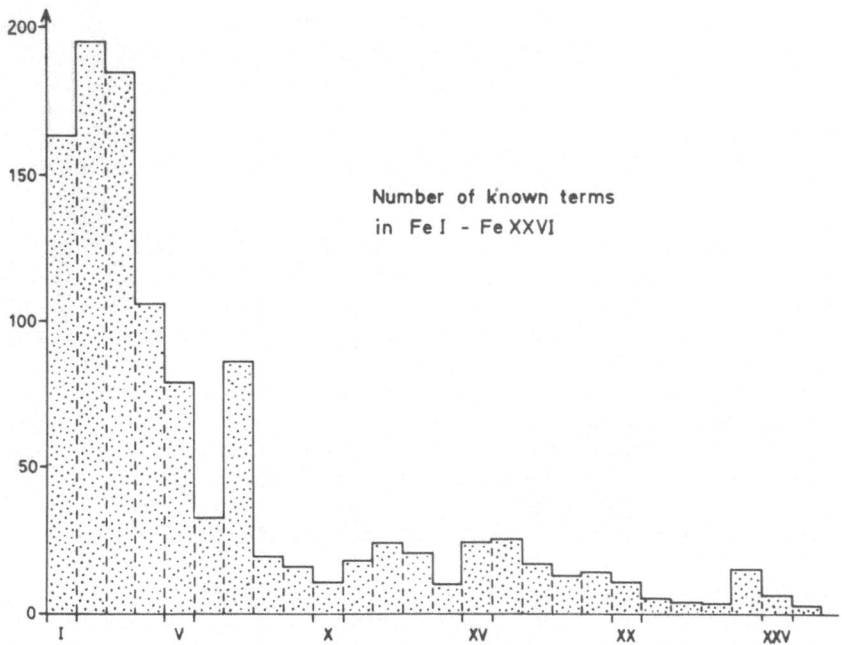

Fig. 1. The number of experimentally found LS-terms in Fe I - Fe XXVI

fairly limited for ionization degrees higher than Fe VIII. There exists much less data about lifetimes and transition probabilities in several times ionized Fe, only a few limited measurements have been reported so far.

Scope of this Article

The experimental data considered in the present review include wavelengths, excitation and ionization energies, fine structure and hyperfine structure, lifetimes of excited levels in ions, transition probabilities for decay by photon or electron emission and QED effects. It is not possible to give more than a sketchy review of all available material. Many references to original work will therefore be included here. There are also several thorough review articles and books.

Among the latter the recent monograph by Cowan (1981a) contains much information about experimental results, besides giving an extensive survey of atomic theory. The structure of atomic spectra has been systematically reviewed by Edlén (1964,1976). Much interesting material can also be found in the reviews by Fawcett (1974, 1981), Kononov (1978), Feldman (1981) and Drawin (1981). These

works which essentially deal with energy-level data can be complemented with reviews on lifetimes (Wiese 1979, Imhof and Read 1977).
There are also some reviews that discuss energy levels as well as
transition probabilities for highly ionized atoms (Martinson 1980,
1981, Drawin 1982).

LIGHT SOURCES FOR HIGHLY IONIZED ATOMS

Several light sources can be used for the production of highly ionized atoms (Minnhagen 1964, Fawcett 1974, 1981, Martinson
1980). One group of light sources, characterized by rather high
electron densities ($10^{18} - 10^{21} cm^{-3}$) includes vacuum sparks, exploding wires and laser-produced plasmas. Some plasma sources such as
the theta pinch have electron densities of the order of 10^{16} cm^{-3}.
In Tokamak plasmas the electron densities are rather low, typically
10^{13} cm^{-3}. Similar values may exist in solar flares whereas the
electron density is much lower in the solar corona. In the foil-
excited beams the particle density is typically 10^6 cm^{-3} and the
electron densities may be an order of magnitude higher.

The laser-plasma light source has developed largely as a result of the research towards laser fusion, which has motivated the
construction of high-power lasers. Similarly, the foil excitation
method for atomic physics studies has benefitted from the design of
powerful heavy-ion accelerators largely intended for research in
nuclear physics. These two light sources will be discussed at some
length while the other important sources for highly ionized atoms
will be more briefly mentioned.

Vacuum Spark

This light source was used in the early studies of highly
ionized atoms, see e.g. Edlén (1936). A capacitor of about 0.3 -
0.5 µF is connected to two electrodes which are in vacuum a few mm
apart. One of the electrodes is often a carbon rod while the other
consists of the element to be studied. The capacitor is charged to
high voltage, typically 70 - 80 kV, until breakdown occurs. The inductance in the circuit must be very low if extreme ionization stages are investigated.

This has been realized in the low-inductance vacuum spark
(Feldman et al. 1967). Here the voltage is rather low, 10 - 20 kV
whereas the capacity is comparatively high, 15 - 30 µF. The inductance is typically 2 - 100 nH. With such a light source Beier and
Kunze (1978) observed radiation from He-like and Li-like Mo (Mo XLI
and Mo XL).

The vacuum spark is a relatively simple and inexpensive light
source for highly ionized atoms. The main disadvantage is that the
spectra obtained with it can be rather complicated because many

ionization degrees can be produced during a discharge. The separation of these ionization degrees may present considerable problems in the case of complicated spectra. e.g. those of the iron-group elements. The reproducibility of the discharges may also cause problems. A detailed review of the results obtained with low-inductance vacuum sparks has been given by Feldman and Doschek (1977). The plasma-physics properties of this source have also been discussed by several authors, e.g. Cilliers et al (1975).

Exploding Wires

Thin wires (10 - 100 μm) explode when they undergo very sudden electric heating whereby light is emitted. Burkhalter et al. (1977, 1978) have applied this technique for the spectroscopy of highly ionized atoms. The explosions were obtained with 10^3 GW discharges of relativistic electrons. The plasmas produced exhibited transitions in He-like Ti and Fe. The Ni-like spectra of W XLII, Pt LI and Au LII were also observed. These belong to the highest ionization stages ever obtained in the laboratory.

Laser-produced Plasmas

For about 15 years high power lasers (ruby or neodymium lasers) have been used for the spectroscopy of highly ionized atoms. By focusing the light from the laser on solid target a plasma is generated. With the so-called Q-switching technique very short and energetic laser pulses are obtained. In one of the earliest atomic physics experiments of this kind Fawcett et al. (1966) used a ruby laser (0.694 μm wavelength), which gave pulses of 8 J energy and 15 ns length. The power density on the target was about 10^{12} Wcm^{-2}. The spectra were very clean, showing lines from only a few ionization stages, e.g. Fe XV, Fe XVI and Ni XVII, Ni XVIII.

This method has subsequently developed into one of the most efficient techniques for producing highly ionized atoms. Very high-powered lasers (several GW) have been employed while the focal spots have been reduced to about 10 μm. The power densities on target have thus been very high which leads to extreme ionization stages. Thus in the Ni I isoelectronic sequence up to 50 times ionized atoms have been produced (Zigler et al. 1980).

A typical experimental setup is shown in Fig. 2, (from Reader et al. 1979).

The plasma-physics processes occurring in laser-produced plasmas have been discussed in several reviews (Peacock 1976, Key and Hutcheon 1980, Carroll and Kennedy 1981). The laser power required for producing various ions has been investigated by many authors. Fawcett (1974) points out that only 20 MW are needed for studying the spectra of C IV whereas the production of Fe XV and Fe XVI demands

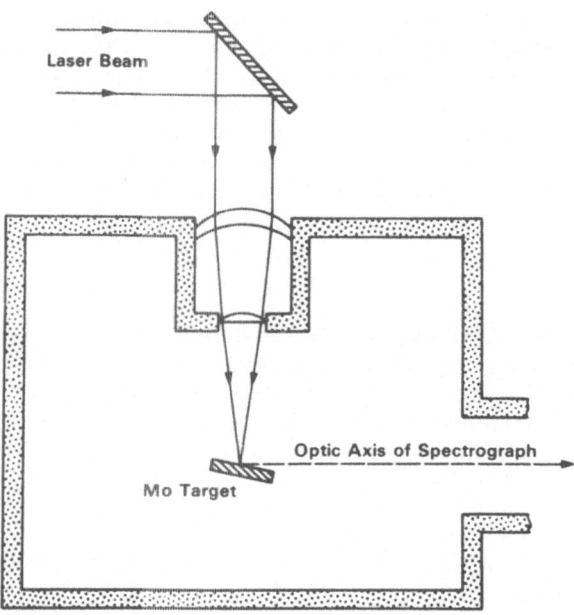

Fig. 2.   Experimental arrangement for an atomic-physics study of a
          laser-produced palsma (Reader et al., 1979)

about 400 MW. The optical focusing of the laser light on target
plays a very important role (Fawcett 1974).

  As part of a systematic study of laser-produced plasma, Boland et al. (1968) investigated transitions in C I - C VI from a
laser-produced plasma and determined electron temperatures, electron densities, ion velocities and energy deposition. One of the interesting results was that most of the laser energy is transformed
into kinetic energy of the ions. The typical ion velocities 10^4 -
10^6 ms^{-1} result in Doppler broadening and asymmetries on spectral
lines (Feldman 1976).

  In Fig. 3 two spectra of Fe are shown, one from a spark light
source, the other from a laser-produced plasma (Feldman 1976). The
latter is much cleaner, showing transitions from only a few ionization stages.

  The ionization stages can be varied e.g. by changing the laser power.

  Table I gives some examples of the lasers used in atomic
spectroscopy and the spectra studied in this way. A very detailed
report which summarizes the data obtained at the Lebedev Institute,
Moscow, has been written by Boiko et al. (1978).

Table I. Examples of Lasers Used in Atomic Spectroscopy

Laser	Energy (J)	Pulse Length (ns)	Spectra Studied	Reference
Nd	10	2	Hf XLV - Re XLVIII	Zigler et al. (1980)
Nd	15	10	Mo XIV	Reader et al. (1979)
Nd	24	6	K XVI - V XX	Fawcett et al. (1980)
Nd	160	0.07	Xe XLVI - Xe XLVIII	Conturie et al. (1982)

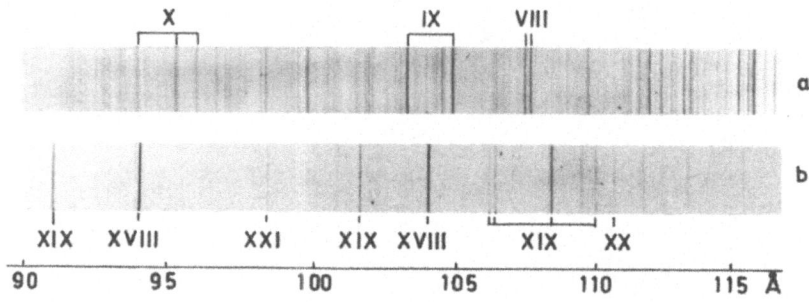

Figure 3.   Spectrum of Fe (9- - 115 Å), from
             a vacuum spark (a) and a Laser-
             produced plasma (b).   The low
             ionization states are absent in
             the latter case (Feldman 1976).

Plasma Light Sources

The Tokamak is one of the most interesting light sources for the spectroscopy of highly ionized atoms (Hinnov 1976, Drawin 1978, 1981). The electron and ion temperatures in Tokamaks can reach several keV whereas the densities are orders of magnitude lower than in the light sources discussed above. One of the most important properties of the Tokamak light source is that forbidden transitions (e.g. of the M1 or E2 type) can be observed. In denser light sources the collisional processes usually dominate over radiative decays. Much of our knowledge about forbidden lines originates from Tokamak studies (Suckewer and Hinnov 1982).

Recent examples of the application of the Tokamak to problems in basic atomic physics can be found in the paper by Stamp et al. (1981) which describes a study of the structure of helium-like ions, performed at the DITE machine in Culham.

There are other plasma light sources, such as the theta pinch and the plasma focus device, that provide excellent spectra of ions (Fawcett 1974). In particular the theta pinch is very valuable in connection with studies of the spectra of gaseous elements.

Astrophysical Light Sources

The explanation of the origin of the unknown lines in the solar corona (Edlén 1942) showed that the temperature in the corona is $2 \cdot 10^6$ K. The electron density varies between 10^{11} and 10^8 cm^{-3}. The light elements are totally ionized in the solar corona and Fe is up to 15 times ionized. Even higher ionization states are present in ı solar flares where the temperature is approximately $2 \cdot 10^7$ K. The electron density is here 10^{13} cm^{-3} and elements up to Fe and Ni may be fully ionized (Dupree 1978, Jordan 1979, Feldman 1981).

More than 70 elements have been identified in the solar spectrum (including the photosphere, chromosphere and corona), see e.g. Engvold (1977).

The distribution of ionization stages observed for the light elements H - Ni (Z = 1 - 28) can be seen in Fig. 4. In several cases the spectra are not sufficiently well known to permit identification of lines in the solar spectrum.

The sun is thus a very important light source for the spectroscopy of highly ionized atoms. Much important data has been obtained from the Skylab Mission (Feldman 1981) as well as from the Solar Maxium Mission (Feldman 1981). One X-ray spectrum, from the latter experiment is shown in Fig. 5, (Fawcett 1981).

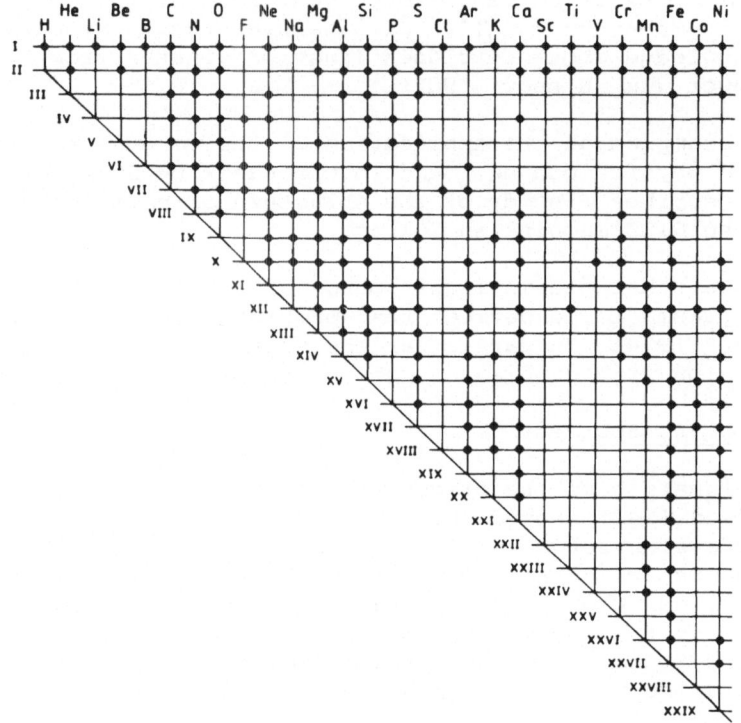

Figure 4.   The distribution of ionization stages
           observed in the sun, (Dupree 1978).

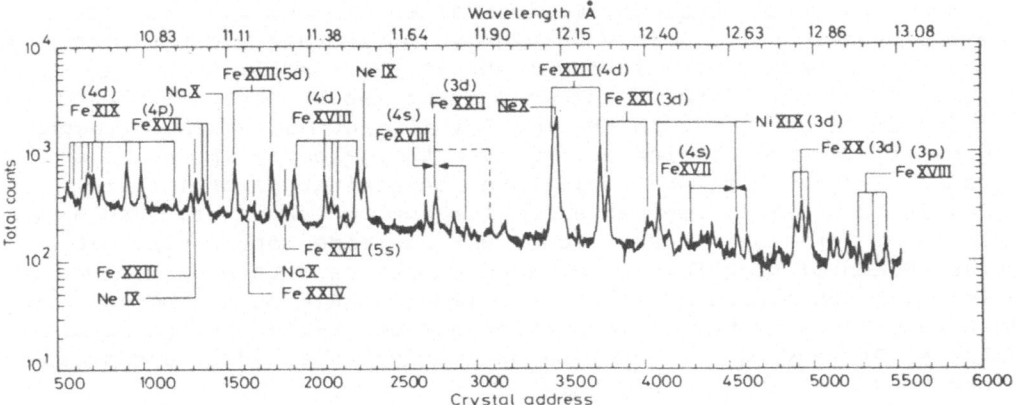

Figure 5. Spectrum of solar flare, (Fawcett 1981).

Accelerated Ions Excited by Collisions

Beams of fast ions from heavy-ion accelerators have for nearly 20 years been used for research in atomic physics. The ions are excited by collisions with atoms in a solid (e.g. a thin foil of carbon) or in a target gas. The most common method, beam-foil spectroscopy (BFS) was introduced by Kay (1963) and Bashkin (1964), who found that ions from Van de Graaff accelerators were strongly excited after passing the foil.

The principle of BFS is illustrated in Fig. 6, which also indicates a few typical experiments. The ions from an accelerator (e.g. a heavy ion linear accelerator, or a Van de Graaff generator) are magnetically analyzed (to ensure a chemically and isotopically pure beam) and then directed through the thin foil (typical thickness 100 nm) in the target chamber. Because of collisions with the foil atoms the beam ions may emerge in highly ionized and excited states which decay in vacuum by emitting light in the visible or UV regions, X-rays or electrons. Perhaps the most important property of the foil excited beam of fast ions is the spatial resolution, which corresponds to an excellent time resolution, because the velocity of the ions is accurately known. This fact facilitates determination of lifetimes for levels in multiply ionized atoms (see below). However, even as a light source for atomic spectroscopy BFS has some interesting properties. Beams of practically any element can be obtained from modern accelerators and indeed more than 70 elements have been studied with BFS. A very large number of ionization and excitation states is also available. By varying the accelerator energy the yield of a desired ionization state can be optimized.

The wavelength resolution in BFS has been inferior to that of other light sources (e.g. the spark or laser-produced plasma). Because of Doppler effects (the light is emitted by fast ions with v/c reaching several per cent) spectral line broadenings are difficult to avoid in BFS. Furthermore, the light source is a rather weak one, and this necessitates the use of comparatively small optical instruments with high transmission. However, much progress has been made in recent years and methods have thus been found which largely reduce the Doppler broadening effects in beam-foil spectra. An example of a modern beam-foil spectrum is shown in Fig. 7.

An interesting aspect of the BFS method is that multiply excited states in atoms and ions are profusely populated (Berry 1975). Such states can be studied both by photon spectroscopy and by electron spectroscopy. Another important class of levels that are prominent in beam-foil spectra are those with high n and ℓ quantum numbers (Rydberg orbits) in several times ionized atoms. Such levels have very large radii and they are populated when the ions leave the foil.

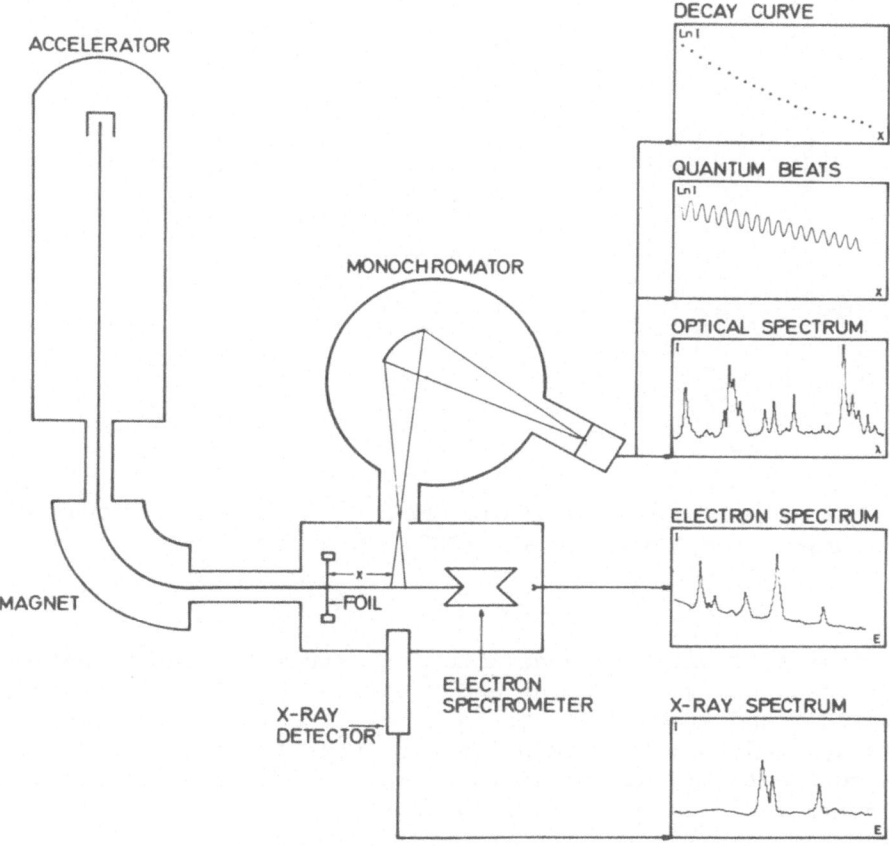

Fig. 6.   Experimental arrangement for atomic physics experiments
with fast ions from an accelerator. The fast ions are
mass-analyzed and sent through a thin excited foil. The
radiation emitted by the ions on the downstream side is
analyzed with a monochromator, or X-ray detector. The deexcitation by electron emission can also be investigated.
Decay curve measurements yield lifetimes of excited levels.
In certain cases also quantum beats can be investigated.

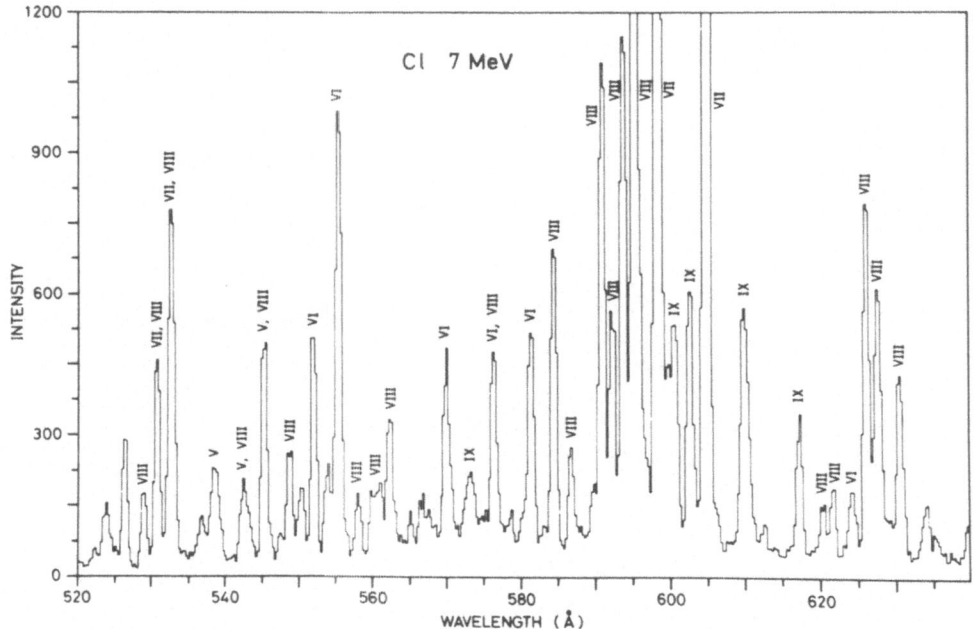

Fig. 7.   Beam-foil spectrum of Cl, observed with a 1 m vacuum mono-
          chromator (Jupén et al. 1982, Martinson 1982)

        Beam-foil data are available for extremely highly ionized
atoms such as Ar XVIII (Marrus and Schmieder, 1972), Fe XXV (Buchet
et al. 1981) and Kr XXXV (Dietrich et al. 1980). However, BFS can
also be applied to a few times ionized atoms as well as neutrals.
Many experiments are discussed in recent review articles (Cocke
1976, Berry 1977, Andrä 1979, Pegg 1980).

        When very energetic and highly charged ions from an accele-
rator collide with atoms in a gaseous target the latter may become
highly ionized - often in inner-shell excited states - but their
kinetic energies are usually quite low, typically only a few eV
(Beyer et al. 1980). This fact clearly facilitates precision studies
Using projectiles such as Xe^{24+} and U^{40+} Folkmann et al. (1981)
observed transitions in multiply ionized Ne. This method is also
of importance for atomic collision studies.

        Optical spectra of highly ionized atoms can further be obser-
ved when fast ions collide with atoms in a gas target under single-
collision conditions. An example is shown in Fig. 8, from the
work of Hvelplund et al. (1981). As the result of electron capture
by multiply charged ions highly excited Rydberg levels are populated.

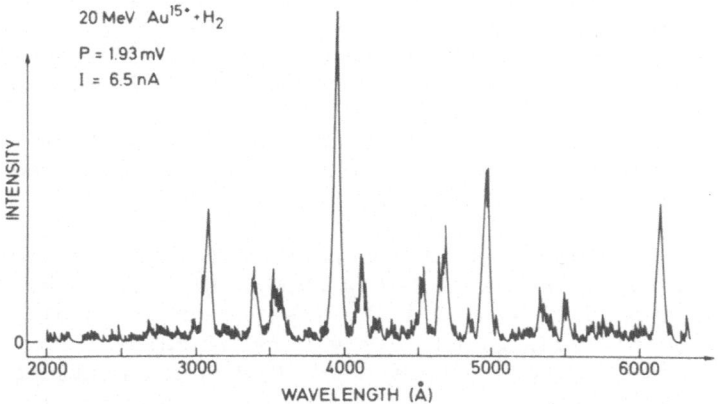

Fig. 8.   A spectrum of highly ionized Au (Hvelplund et al. 1982).
          The observed lines are from $\Delta n = 1$ transitions with
          $n = 12, 13, 14$ and 15 in Au XV.

ATOMIC ENERGY LEVELS

Experimental Aspects

The light emitted by the various sources discussed above must
be first dispersed with a suitable optical instrument. Highly ionized atoms may emit strong transitions in the X-ray region and a
variety of X-ray instruments have therefore been used for such
studies. Low-resolution spectra can be recorded with solid-state
detectors such as the Si(Li) detector. The efficiency is quite high
while the resolution is typically 150 eV at 5 - 6 keV. An interesting
development in X-ray spectroscopy consists of the Doppler-tuned
X-ray spectrometer (Schmieder and Marrus 1973) which is applicable
to fast, excited ions. The Doppler-shifted emission from the beam is
matched with a known absorption feature in a thin filter between
the source and the detector. High resolving powers together with
favorable detection efficiency are thereby possible. In a recent
experiment Mowat et al. (1979) applied this method to transitions
in Li-like Mg. Very high resolving power can be obtained with crystal spectrometers but here the detection efficiency is usually
quite low.

Good examples of high-resolution X-ray spectroscopy can be
found in the work of Bitter et al. (1979) who used a Bragg crystal
spectrometer, bent to a radius of curvature of 333 cm, to record
high-resolution satellite spectra of Fe XXV (around 1.85 Å) from
the Princeton Large Torus (PLT) device. The resolving power was
150,000 at 1.85 Å. The spectra of the Tokamak plasma resemble those
observed in solar flares (Presnyakov and Urnov 1979, Feldman 1981).

Beam-foil spectra in the X-ray region (4 - 7 Å) have been investigated by Träbert et al. (1979) who used a 10 cm curved-crystal
spectrometer. The resolving power was about 2000.

For spectral studies at higher wavelength optical spectrometers or spectrographs are used. Below 2000 Å they are operated in
vacuum. In the region 20 Å - 600 Å the instrument is usually of the
"grazing-incidence" type. In the early work of Edlén (1936) a 5 m
grazing incidence spectrograph was used. The instrument, equipped
with a ruled grating with about 600 lines/mm, could resolve lines
less than 0.02 Å apart in the first order (Edlén 1966). Only with
the 10.7 m grazing incidence spectrograph at the National Bureau
of Standards has it been possible to obtain significantly higher
resolution.

When very highly charged ions are investigated the conditions
in the light source are usually such that significant Doppler effects may broaden the spectral lines. Extremely high spectral resolution can therefore not be achieved. Thus grating spectrographs
or spectrometers with 1 - 2 m radius may be quite sufficient. This
is also the case when the light source is rather weak, e.g. the
foil-excited ion beam. Using a commercial 2.2 m instrument Bruch et
al. (1981) have been able to obtain good-quality beam-foil spectra
with linewidths of 0.035 Å around 170 Å.

For wavelengths from about 500 Å and higher instruments of
the so-called Seya-Namioka design (Namioka 1959) or of the normal-
incidence type (see e.g. Edlén 1963, Samson 1976) are advantageous.
For high-resolution work very large instruments have been employed,
e.g. a 10.7 m spectrograph at the National Bureau of Standards. This
instrument uses the so-called Eagle mounting (Klinkenberg
1976). More detailed information about spectrometers as well as
about detectors for photon spectroscopy is found in the reviews by
Edlén (1963), Klinkenberg (1976) and Samson (1976). It should also
be noted that there are now very rapid developments in these areas,
largely motivated by research in astrophysics, plasma physics and
experiments using synchrotron radiation. Important progress has e.g.
taken place in grating design; ruled gratings are now being replaced
by holographic ones. There is also significant progress in the case
of detectors. Photographic plates are now frequently replaced by
electronic multichannel arrangements.

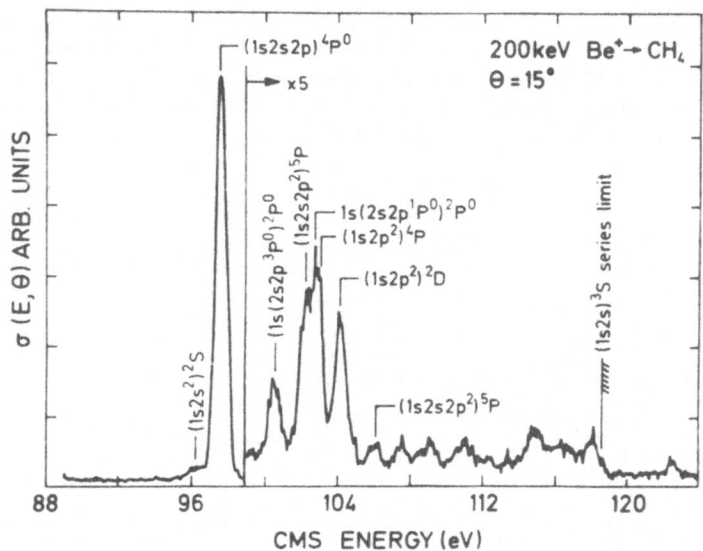

Fig. 9.   Electron spectrum observed by exciting Be$^+$ ions in a target
gas (Bruch et al. 1977). The transitions are from multiply
excited, autoionizing states in Be I and Be II.

So far we have only discussed the spectroscopy of photons.
However, very valuable experimental data can also be obtained by
studying the energies and intensities of the electrons often emitted by excited ions, e.g. from beam-foil or beam-gas experiments.
Such ions are often in multiply-excited states which decay by the
emission of electrons (Sellin 1976). The energies of these electrons
range typically from tens of eV to several keV. An electron spectrum
is shown in Fig. 9. (from the work of Bruch et al. 1977). The resolution is here not as good as in most photon spectroscopy, largely
because of substantial Doppler broadening. However, Bachmann et al.
(1982) have recently demonstrated that energy resolution as low as
0.01% can be achieved in electron spectroscopy on fast ions.

In all atomic spectroscopy it is important to obtain as high
wavelength (energy) resolution as possible. Otherwise line blending
will cause several problems. For example, Edlén (1966) shows a copper spectrum that contains more than 60 lines between 140 and 145 Å.
Without good resolution only a continuum would have been obtained

here. High accuracy is particularly important in the far UV (Edlén
1963). Thus the relation $\Delta\sigma = \Delta\lambda/\lambda^2$ (where $\Delta\sigma$ and $\Delta\lambda$ are the uncertainties in wavenumber and wavelength, respectively) shows that
even small wavelength errors will here lead to substantial energy
uncertainties. The wavelengths are usually measured with respect to
so-called "reference lines", i.e. spectral lines with accurately
known wavelengths (Kaufman and Edlén 1974).

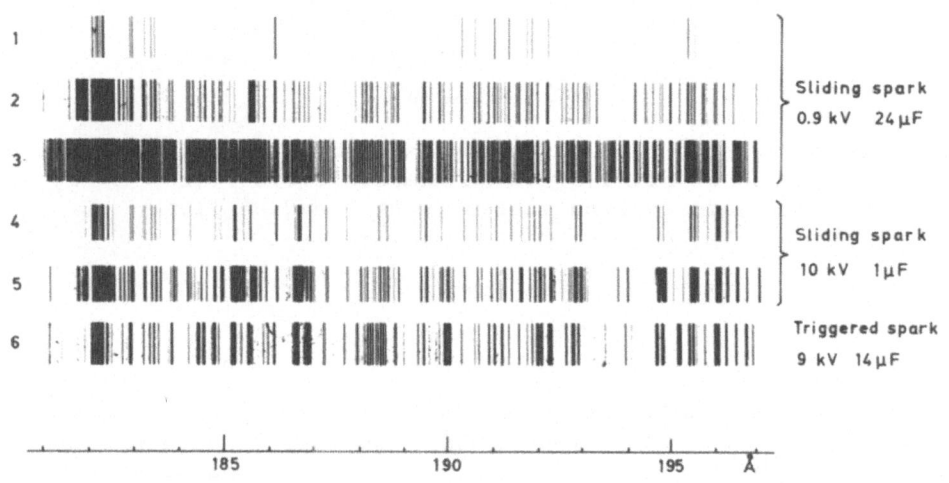

Fig. 10. Spectrum of Fe (180 - 200 Å) recorded with different spark
light sources. For a given source the inductance has also
been varied (Ekberg 1974).

It is also very important to determine the ionization stages
of the observed spectral lines. When a spark light source is used
the inductance of the circuit can be varied - the lower the inductance the higher charge states are produced in the plasma. Examples
of an iron spectrum recorded under various light source conditions
are shown in Fig. 10 (Ekberg 1974).

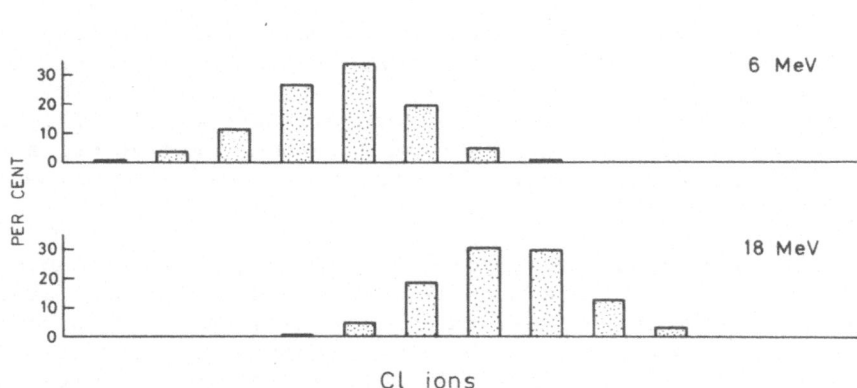

Fig. 11. Equilibrium charge distributions for 6 and 18 MeV chlorine
         ions (Wittkower and Betz 1973).

When laser-produced plasmas are investigated different charge states
are generated by varying the laser power. In Tokamaks and theta-pinch
plasmas the lines belonging to different ionization stages have different time evolutions as well as spatial locations in the plasma.
Finally, in beam-foil experiments it is often necessary to record
spectra using different ion energies. The charge distributions as
a function of ion energy are frequently known (Wittkower and Betz
1973). An example of this is shown in Fig.11.

        The next step (after wavelength/energy measurement and charge-
state determination) consists of the classification, i.e. the assignment of the observed transitions to energy levels in the ions. This
is often accomplished by means of well-known relations in atomic
spectroscopy such as isoelectronic regularities, Ritz formulae,
quantum-defect analyses, polarization formulae, etc. (Edlén 1964).

        Theoretical calculations of atomic structure are frequently of
great help in connection with the analyses of experimental data.
Several powerful computer programs have been developed for this
purpose. Much important information can be found in the monograph
by Cowan (1981a).

        In the following review we will present some data for atomic
energy levels, obtained from experimental studies. It is only possible to give a rather sketchy overview of this extensive field.
References to several more comprehensive reivews will therefore be
included.

Helium-like Ions

Although they have been studied for many years the spectra of helium-like ions continue to present interesting experimental and theoretical problems. Here the Schrödinger equation cannot be exactly solved. Various approximation methods (Accad et al. 1971, Drake 1979, Safronova 1981) have resulted in very high theoretical accuracies which can only be equalled by the most careful experimental investigations, however.

An extensive amount of experimental material is available for He I – C V while the data are much less complete for systems with higher Z (Martin 1981). Besides the resonance lines $1s^2$ $^1S - 1snp$ ^1P, ^3P and some other lines in the X-ray region, only the $1s2s$ $^3S -$ $1s2p$ ^3P multiplet has been carefully studied in highly ionized two-electron systems. Comparisons between theoretical and experimental energies for the triplet levels provide valuable information about electron correlation as well as the effects on relativity and QED on atomic structure. For several ions from O VII to Fe XXV the

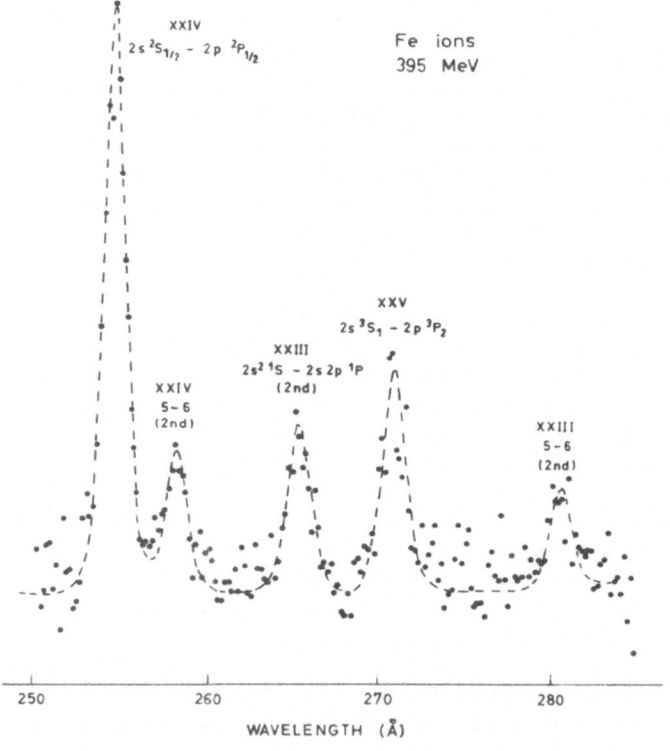

Fig.12. Beam-foil spectrum of highly ionized Fe (Buchet et al. 1981)

1s2s ^{3}S - 1s2p ^{3}P transition wavelengths have been measured using
light sources such as the theta pinch (Engelhardt and Sommer 1971),
Tokamak (Stamp et al. 1981) or the beam-foil method (Davis and
Marrus 1977, O'Brien et al. 1979, DeSerio et al. 1981). A recent
beam-foil result for Fe XXV (Buchet et al. 1981) is shown in
Fig.12.

Here the authors determine the 1s2s ^{3}S$_1$ - 1s2p ^{3}P$_2$ transition
wavelength to be 271.02±0.09 Å which was compared with the theoretically calculated value of 267.9 Å. The difference is due to QED
effects which could be determined to 3% accuracy. Thorough discussions of the measurements and underlying theories can be found in
several of the papers mentioned above and in the work of Ermolaev
(1979), Peacock (1980), Berry et al. (1980), and Livingston (1982).

Ions with 3-9 Electrons

The spectra belonging to the Li I - F I isoelectronic sequences
have been very extensively investigated in recent years. The number
of electrons is sufficiently low to permit quite accurate calculations of transition energies and lifetimes. Extensive theoretical
data, using relativistic wavefunctions have recently been provided
by Cheng et al. (1979). The transitions in highly ionized members
of these sequences are further of considerable importance to astrophysics and plasma spectroscopy. Eldén (1979, 1981) has made
extensive analyses of the level structures of these ions, using
both theoretical and experimental data. The latter originate both
from laboratory measurements (including Tokamak spectra) and from
astrophysical observations, the spectra of the solar corona and solar flares. The latter data go up to Ni (Z=28) while the following
elements have too low abundances. Systems with higher Z can be studied in the laboratory, however. For example, data for the F I isoelectronic sequence are available for Z ≤ 42, largely from the
spectroscopy of laser-produced plasmas (Boiko et al., 1978).

The work of Edlén has resulted in semiempirical energy expressions for many important levels in the Li I-F I like systems.
The relations obtained are valuable for interpolation and extrapolation work whenever the experimental energies are not accurately
known. One result, for B-like ions, is shown in Fig.13.

The level structure of extremely highly ionized atoms has
been calculated by Kononov and Safronova (1977). The results are in
good agreement with the semi-empirical work of Edlén (1979,
1981) and experiments. However, the theoretical data go as high as
Z=100. The LS-coupling which is valid for low Z   in the isoelectronic sequences is according to theory gradually changing into jj
coupling as Z increases. There are good reasons for extending experimental work to very highly charged systems.

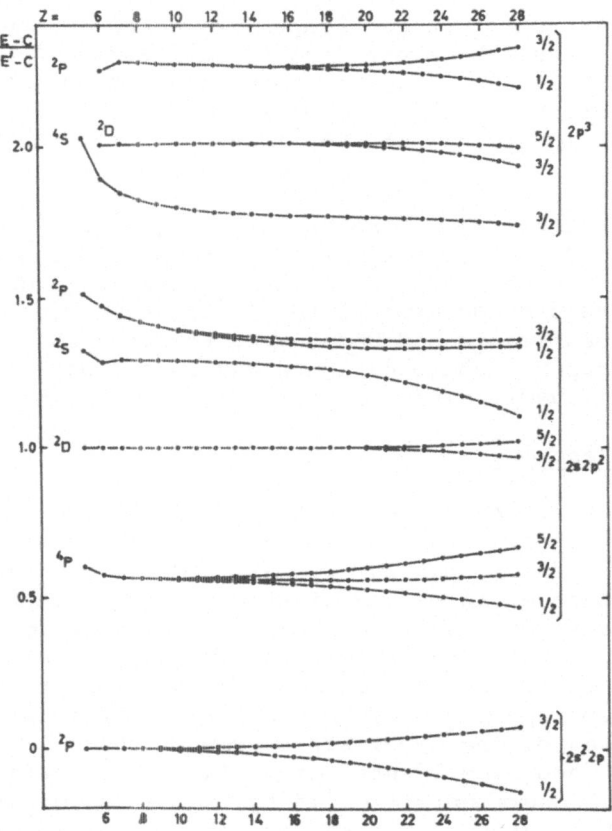

Fig. 13. Energy level structure of n = 2 levels in the B I isoelectronic sequence (Edlén 1981)

Heavier Systems with One Valence Electron

The Na I-like spectra have been investigated for ionization
stages as high as Mo XXII (Mansfield et al. 1978, Edlén 1978). Here
the structure is comparatively simple and the spectra can usually
be interpreted without too much effort. Very accurate expressions
for level energies, ionization potentials and transition wavelengths
in this sequence are now available (Edlén 1978) and these semi-empirical data can be reliably extrapolated to Z > 42 where no experimental
results have yet been reported.

An electron structure that is rather simple for low Z in an
isoelectronic sequence may become much more complicated as Z in-

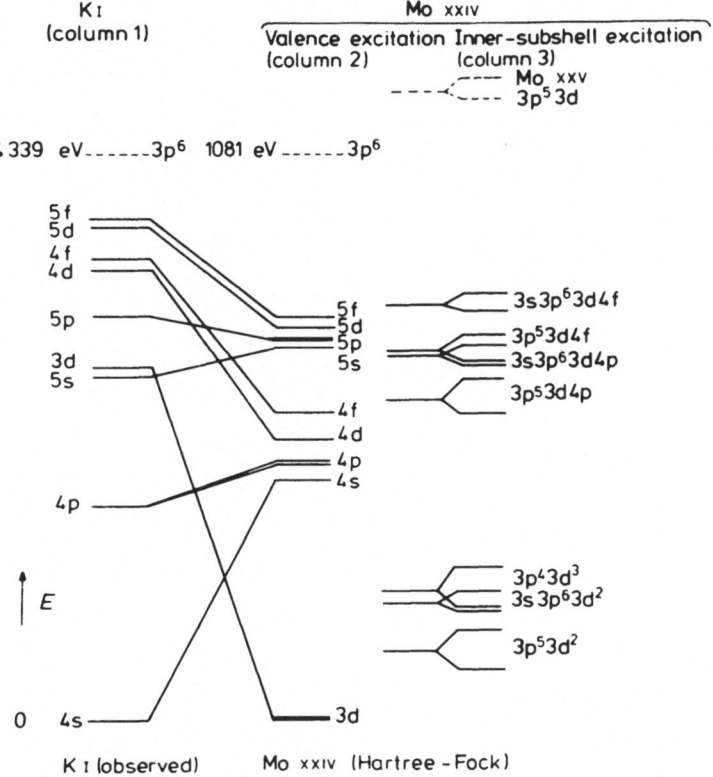

Fig. 14. Comparison of the energy level structure for K I and Mo XXIV (Mansfield et al. 1978).

creases. Empty orbitals, e.g. 3d, may become partially filled for higher Z. An interesting example of this is shown in Fig. 14, from the work of Mansfield et al. (1978). To the alkali-like structure $3p^6nl$ ($n \geq 4$) of K I are in Mo XXIV added several configurations which involve excitation of 3s or 3p electrons to the 3d shell (see Fig.14). The spectrum is thus more difficult to analyze than the comparatively simple system of K I.

Interesting data have also been found for the Cu I isoelectronic sequence. Here the 3d shell is filled and the ground configuration is $3d^{10}4s$ ^2S while the excited levels are of the type $3d^{10}$ nl ^2L. As can be seen from the work of Reader and Luther (1981) this sequence has been followed to W LXVI. The spectra of the lower members such as Sr X - Mo XIV were obtained using a low-inductance spark light source (Reader and Acquista 1977) whereas the plasmas produced with a powerful laser (30 J, 0.3 ns Nd:glass laser) provided transitions in Ba XXVIII - W XLVI. Some of the results are shown in Fig. 15.

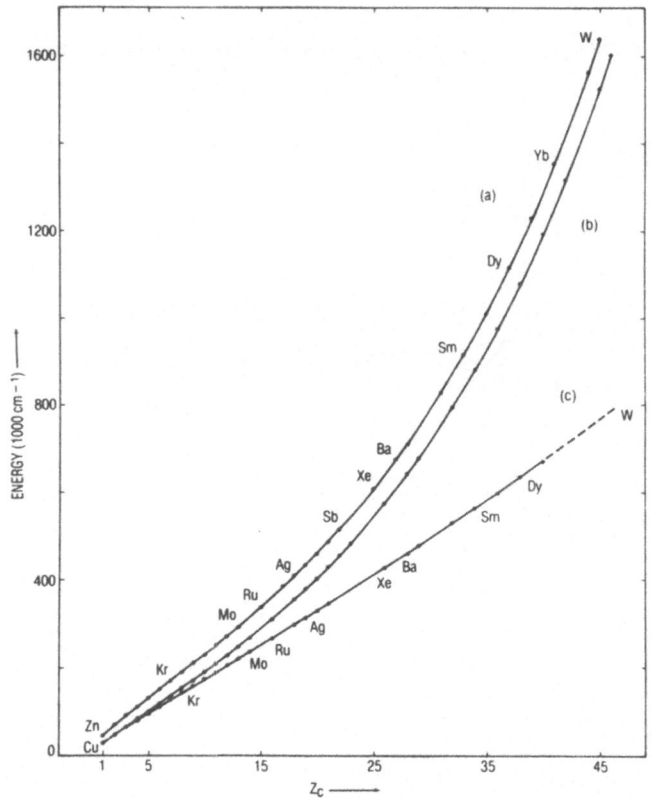

Fig. 15. Energies of the resonance transitions $4s^2\ {}^1S - 4s4p\ {}^1P$ in Zn-
like ions (a) and $4s\ {}^2S_{1/2} - 4p\ {}^2P_{3/2}$ (b) and $4s\ {}^2S_{1/2} - 4p$
$^2P_{1/2}$ in Cu-like ions. (Reader and Luther 1981).

This seems to be one of the most thoroughly investigated isoelectronic sequences, ranging from the neutral atom to a 45+ ion.

    Experimental investigations of the systems described here are
largely motivated by needs in the spectroscopy of hot plasmas. The
ions with a few valence electrons emit strong resonance lines that
are useful for diagnostic purposes (Hinnov 1976). It should also be
mentioned that the ions discussed here, e.g. those belonging to the
Cu I sequence have been the subjects of several theoretical analyses
(Cheng et al. 1978, Curtis 1977, 1979).

Complex Spectra

    We have already seen how a relatively simple structure becomes
more complex when Z increases (K I - Mo XXIV, discussed above). In
general the complexity increases with the number of valence electrons

or whenever there are partially filled shells, e.g. 3d or 4f.

Already for Mg-like ions (with two valence electrons outside the $2p^6$ shell) complications may arise when configurations of the type 3dnl start to interact with levels of the 3snl and 3pnl configurations. The $3d^2$ levels have been localized already for P IV (Zetterberg and Magnusson 1977). Indeed the Mg I sequence is accurately known only for Mg I - S V.

The spectra of the iron-group elements (Sc - Ni) often show great complexity, because of the partially filled 3d shell. Here the lower ionization stages usually have the ground configuration $3s^2 3p^6 3d^k$ (k=1-10) and higher configurations arise by promoting a 3p or a 3d electron. In many cases it is practically impossible to perform a complete analysis of anything but the lowest configurations, $3d^k$, $3d^{k-1}$4s and $3d^{k-1}$4p. As an example of such a study Fig. 16 shows the energy levels of Fe V, k=4 (Ekberg 1975). These three configurations give rise to about 180 energy levels, practically all of which were established by classifying about 1000 spectral lines.

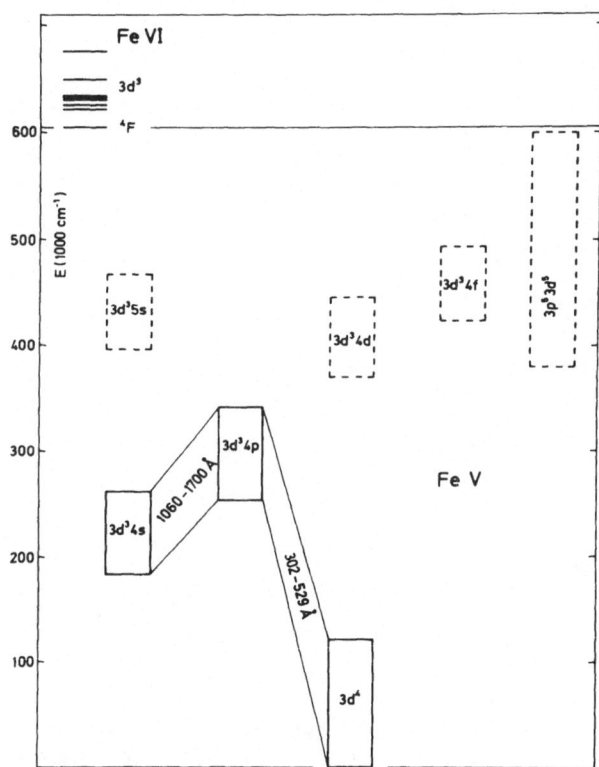

Fig. 16. Energy level structure of Fe V (Ekberg 1975).

Very compicated spectra are observed in the case of highly
ionized heavy elements. Interesting examples are e.g. provided by
highly stripped W which has been observed as an impurity in the Oak
Ridge tokamak (Isler et al. 1977). Recent analyses seem to indicate
that the measured spectra - which show pronounced band structure -
should be largely due to W XXVIII which belongs to the Ag I isoelectronic sequence (Sugar and Kaufman 1980). Such spectra can at present
only be analyzed by combining the experimental data with theoretical
analyses (Cowan 1981 a).

Recently, Carroll and O'Sullivan (1982) investigated the ground
state configurations and the 4d - 4f transitions in the spectra of
the rare-earth elements using a laser-plasma light source. When the
4f shell is partially filled the spectra will be extremely complex.
For example, there are 83 024 predicted spectral lines for the
$4d^{10}4f^6 - 4d^9 4f^7$ transition in LS coupling! Detailed analyses are
clearly ruled out and the experimental data are usually treated by
means of recently developed statistical methods whereby unresolved
bands of spectral lines can be successfully investigated (Bauche-
Arnoult et al., 1979).

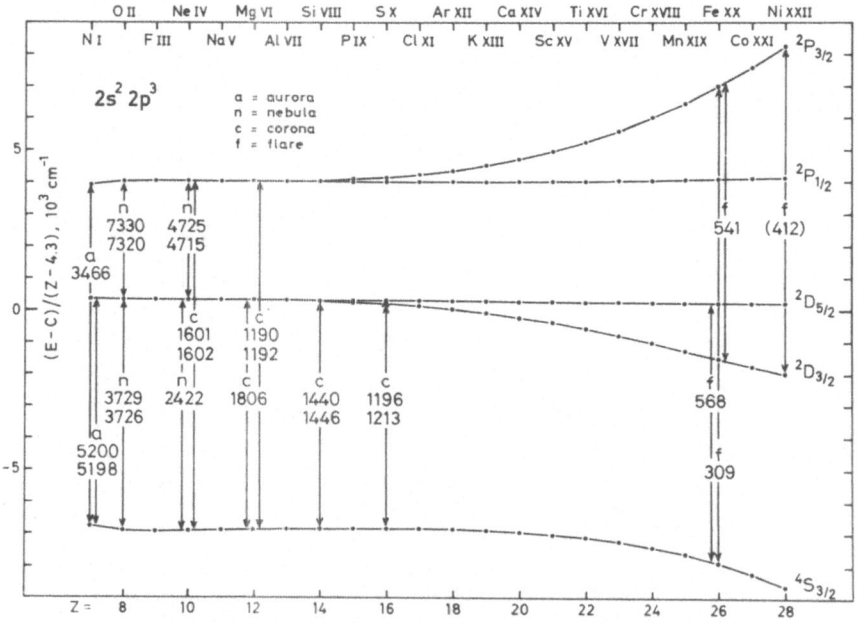

Fig. 17. Fine structure of the $2s^2 2p^3$ configurations in the N I
         isoelectronic sequence (Edlén 1982). The wavelengths of
         the forbidden lines, as observed in celestial light sources,
         are indicated.

Forbidden Transitions

The forbidden lines in astrophysics (mentioned in the introduction) are due to magnetic-dipole (M1) transitions between fine-structure levels of a term. For example, several of the corona lines (Edlén 1942) arise from transitions within the ground configurations $2s^2 2p^k$ of Ca XI-XV and $3s^2 3p^k$ of Fe X - XIV and Ni XII - XVI. There have subsequently been many additional classifications of the corona lines, see e.g. Edlén and Smitt (1978) and Jordan (1979).

The forbidden transitions have also been observed in Tokamak discharges (Suckewer and Hinnov 1979, Suckewer 1981) where they play an important role in plasma diagnostics, e.g. in connection with the determination of ion temperatures.

An example of forbidden lines, Fig. 17 shows the transitions within the $2s^2 2p^3$ ground configuration in N-like ions as observed in various celestial light sources (Edlén 1982).

Hydrogen-like States

In many light sources it is difficult to excite states with high n,l quantum numbers in multiply ionized atoms. The beam-foil method seems here to be an exception and transitions between such

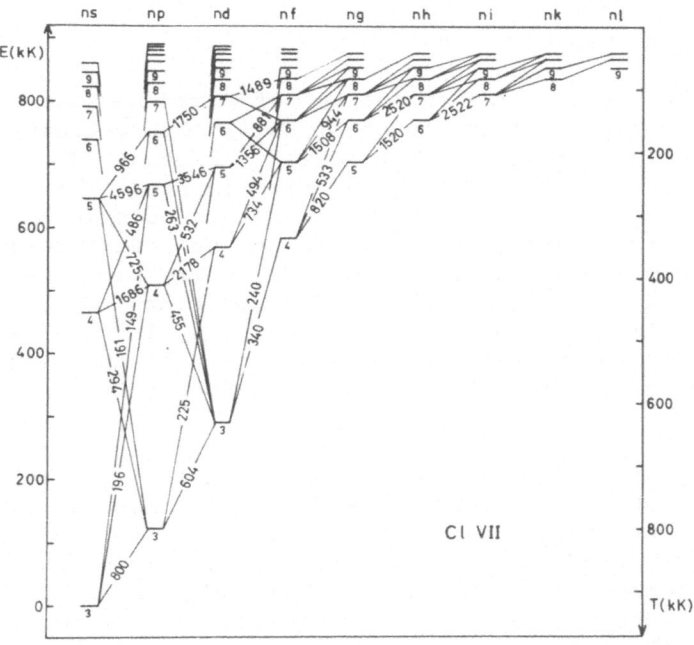

Fig. 18. Energy level diagram of Cl VII (Jupén et al. 1982)

states appear very strong in beam-foil spectra. The term values of
such states can be expressed as

$$T = T_H + \Delta_p \tag{1}$$

where T_H is the hydrogenic value (corrected for relativistic effects)
and Δ_p the polarization energy, usually expressed as

$$\Delta_p = \alpha_d \, R \, <r^{-4}> + \alpha_q \, R \, <r^{-6}> \tag{2}$$

Here the α_d and α_q are the dipole and quadrupole polarizabilities of
the core, R the Rydberg constant and $<r^{-4}>$ and $<r^{-6}>$ the expectation
values of the radial wave functions. There is a great interest in
obtaining values for α_d and α_q in highly ionized atoms. Accurate
wavelength meausrements using beam-foil spectra have here yielded
much data on these quantities. As an example, Fig. [18] shows the
energy level diagram for Cl VII (Jupén et al. 1982) where most results
for high n,l levels originate from fairly accurate beam-foil experiments.

Recently, Curtis (1981,1981 a) has reported detailed semi-empirical investigations of the polarization formula. Of particular interest are here the effects of penetration and configuration interaction.
Interesting cases can also be found when radiative lifetimes for high
n,l states are modified because of core-polarization effects.

Multiply Excited States

States that involve the excitation of more than one electron may
be populated in several light sources. Early observations of the UV-
spectrum of helium (Kruger 1930) revealed a line at 320.3 Å which
arises from the $1s2p \, ^3P - 2p^2 \, ^3P$ transition in He I. The $2p^2 \, ^3P$ level
lies 35 eV above the He II ground state but it does not autoionize by
the electrostatic interaction for which the selection rules are
$\Delta J = \Delta L = \Delta S = 0$ and no parity change. In their spark spectra of
carbon Edlén and Tyrén (1939) observed transitions of the type $1s^2 nl -$
$1s2pnl$ as satellites to the C V resonance line, $1s^2 \, ^1S - 1s2p \, ^1P$.
Such satellites are of fundamental importance for the diagnostics
of laboratory and astrophysical plasmas. (Dubau and Volonté (1980),
Gabriel (1982)).

Beginning with the first beam-foil studies of multiply-excited
states (Bickel at el. 1969, Buchet et al. 1969) much attention has
been paid to non-autoionizing or weakly autoionizing multiply excited
states in light atoms and ions. Several new doubly-excited levels in
He-like Be III have recently been established by means of electron
spectroscopy (Rødbro et al. 1979) and optical spectroscopy (Andersen
et al. 1980). These data nicely complement each other.

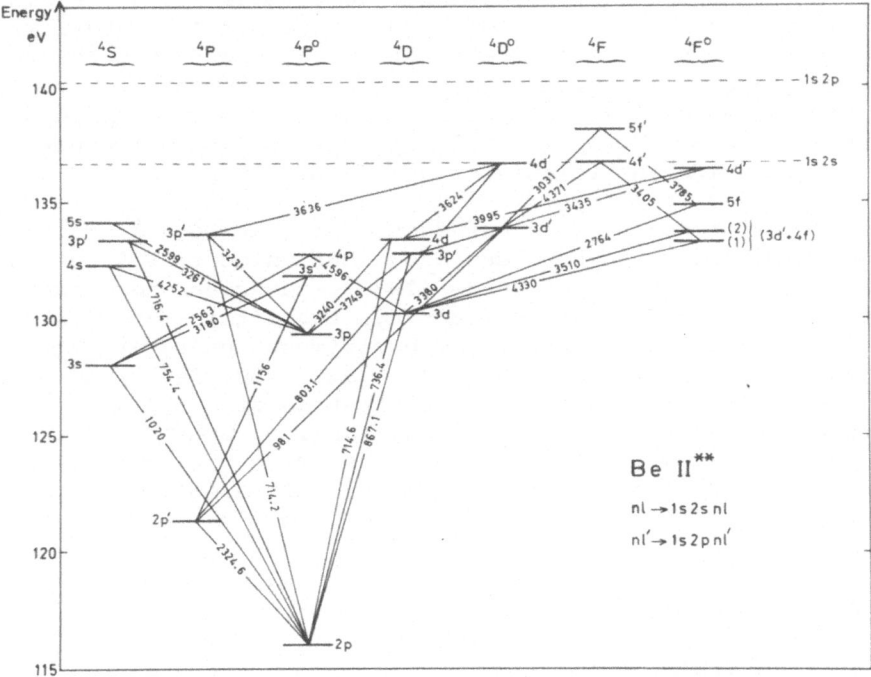

Fig. 19. Experimentally found transitions in the quartet system of
         Be II (Bentzen et al. 1981, 1982, Mannervik et al. 1981)

However, the majority of the work performed so far concerns the
1s2snl and 1s2pnl ^{4}L levels in Li-like ions. These multiply excited
levels often show strong effects of electron correlation which
clearly complicate theoretical analyses. For example odd ^{4}P terms
in Li-like ions may arise from several spectral series (1s2snp,
1s2pns and 1s2pnd ^{4}P).

An example of recent beam-foil results is shown in Fig. 19.
The experimental material for doubly-excited Be II was obtained by
Bentzén et al. (1981, 1982) and Mannervik et al. (1981). The spectral
identifications were facilitated by accurate theoretical calculations
(Larsson et al. 1979, Galan and Bunge 1981, Laughlin 1982, Froese-
Fischer 1982). The Be II quartet levels lie about 100-120 eV above
the Be III ground state. None of the transitions shown in Fig. 19
has been observed with other light sources.

These quartet states in Li I - Ar XVI have been investigated with photon or electron spectroscopy of fast ions (Berry 1975, Sellin 1976), as well as with the recoil-ion technique (Beyer et al. 1980). It has also been possible to make accurate determinations of the 1s2s2p $^4P_{1/2,3/2,5/2}$ and 1s2p^2 $^4P_{1/2,3/2,5/2}$ fine structures in C IV - O VI (Livingston and Berry 1978). The experimental data were in good agreement with advanced theoretical calculations which included both electron correlation and relativistic effects (Cheng et al., 1978a).

Much interest has been focussed on a line at 3489 Å in the beam-foil spectrum of Li which has been known but unidentified for many years. Recent calculations of Bunge (1980) made clear that this line is due to the 1s2s2p^2 $^5P_{1,2,3}$ - 1s2p^3 $^5S_2$ transition in the negative Li-ion. This result was supported by experiments using electrostatic fields (Mannervik et al. 1980, Brooks et al. 1980, Denis and Deses-quelles 1981). This is the first demonstration of a radiative transition between two bound states in a negative ion. The ^5P - ^5S transition has now been observed in Li$^-$ - F VI and the $^5P_{1,2,3}$ fine structure has also been resolved. The results are graphically displayed in Fig.20.

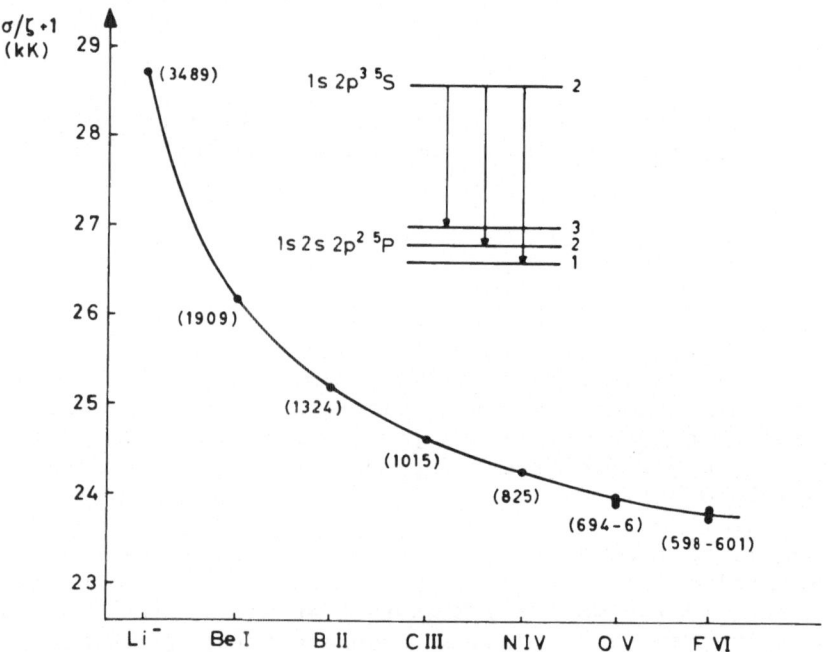

Fig. 20. Isoelectronic comparison of the 1s2s2p ^5P - 1s2p^3 ^5S transition in Li$^-$ - F VI. (Brooks et al.1980,Livingston and Hinterlong 1980, Martinson 1982).

LIFETIMES AND TRANSITION PROBABILITIES

The following well-known relation holds in the case of the electric dipole (E1) approximation

$$\tau_i^{-1} = A_{if} = (4\alpha \omega^3 /3c^2) |<\phi_i |Q| \phi_f>|^2 \qquad (3)$$

Here τ_i is the lifetime of the excited level, A_{if} the probability for spontaneous decay, $\hbar\omega$ the photon energy, α the fine-structure constant, c the speed of light, ϕ_i and ϕ_f the wavefunctions of the levels involved and $Q = \Sigma r_j$ the dipole operator. A convenient quantity is the absorption oscillator strength (or f-value), numerically related to A_{if} according to

$$f = 1.499 \cdot 10^{-16} \lambda^2 \cdot A_{if} \cdot g_i/g_f \qquad (4)$$

The wavelength λ is given in Å while g_i and g_f are the statistical weights of the levels.

Experimental transition probabilities provide useful information about the merits of various quantum mechanical calculations of atomic properties. Very accurate theoretical methods have now been developed for light atoms. However, with increasing number of electrons such calculations become quite difficult, largely because of correlation effects between valence electrons. In heavy atoms as well as in highly ionized systems relativistic effects also play an important role (Desclaux 1982, Drake 1982). Theoretical studies of atomic structure have been reviewed in several articles, e.g. by Weiss (1973) and Hibbert (1975, 1977).

A number of experimental techniques are available for the determination of atomic lifetimes and transition probabilities. The methods include measurements of spectral line intensities in emission or absorption, the determination of refractive index as a function of wavelength ("hook" method), the excitation of atoms with short pulses (or sinusoidally modulated waves) of photons (e.g. from a laser or a synchrotron light facility) or electrons and - finally - techniques based on level-crossing and resonance phenomena, e.g. the Hanle-effect method. Detailed information about these and additional experimental approaches can be found in reviews by Corney (1968), Imhof and Read (1977), Huber (1977), Martinson (1978), Penkin (1979) and Wiese (1979).

All these methods work quite well for neutral atoms and they can usually also be applied to singly-ionized species. However, extensions to multiply ionized atoms are only possible in exceptional cases. Here the beam-foil method is the only available technique that can be routinely used. We will therefore limit further discussions to BFS and some modifications of it.

Lifetime Measurements Using BFS

The excellent spatial – and thus temporal – resolution of the beam-foil light source (see e.g. Fig. 6) and the absence of secondary collisions makes it easy to obtain meaningful decay curves of spectral-line intensities. In the ideal case a lifetime τ can then be determined from the simple relation

$$I(x) = I(0) \exp(-x/v\tau) \qquad\qquad (5)$$

where $I(0)$ and $I(x)$ are the photon (or electron) counting rates close to the foil and at a variable distance x along the beam, respectively, and v the mean velocity of the excited ions.

The lifetimes of excited levels in atoms and ions are typically between 10^{-8} and 10^{-12} s and such short times can be measured by BFS in a straight-forward way. The range of lifetimes that can be conveniently studied has been estimated by Bromander (1973) and Andrä (1979). A graphical result, from the latter work, is shown in Fig. 21 .

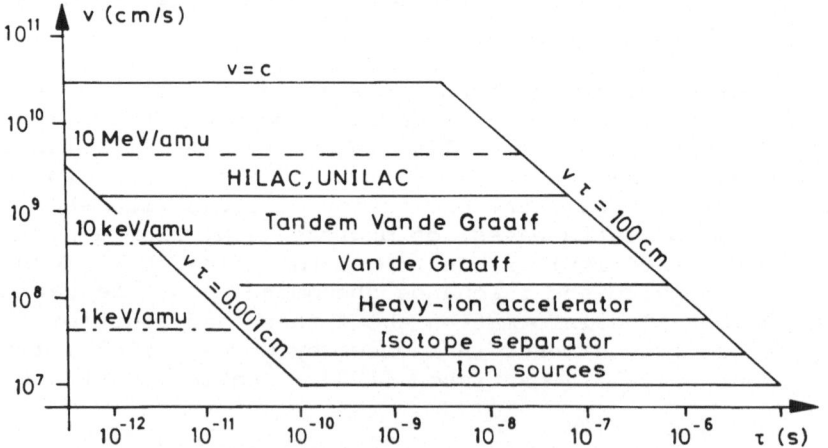

Fig. 21. Estimated range of atomic lifetimes that can be studied by beam-foil decay measurements (Bromander 1973, Andrä 1979). The two limits of decay lengths have been assumed to be 0.01 mm (spatial resolution) an 1 m (target chamber size).

An interesting example of a successful measurement of a very short lifeitme, 1.2 ps, is shown in Fig. 22 (Knystautas and Drouin 1976). Special methods have also been developed whereby even shorter lifetimes can be determined. Thus Betz et al. (1974) found a way of measuring lifetimes in the 10^{-14} s range. The method, later also employed by Panke et al. (1975) and Varghese et al. (1976), is based on the determination of X-ray yields as a function of target thickness when fast, heavy ions (e.g. 90 MeV S-ions) traverse the target. The lifetimes of excited states in projectiles enter as parameters in the X-ray yields.

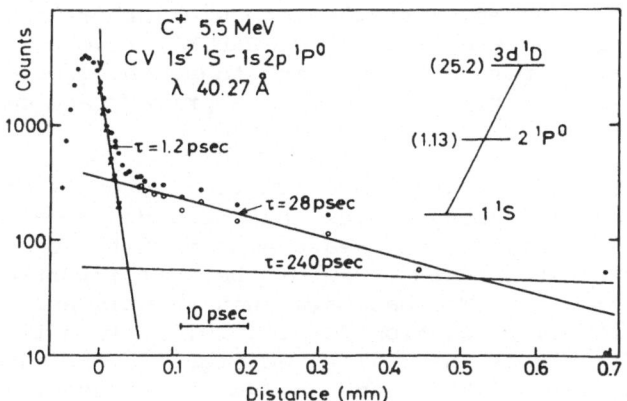

Fig. 22. Decay measurement for a very short-lived level, 1s2p ^{1}P in C V (Knystautas and Drouin 1976). A careful experimental design resulted in excellent spatial resolution along the beam. The decay curve also shows the effects of cascading from higher levels.

The BFS method can be applied to very many levels in atoms and ions. Lifetime data have so far been obtained for many species ranging from neutrals to more than 30 times ionized atoms. There should be no substantial difficulties in extending the work to 40 - 50 times ionized systems.

Experimental Problems

A number of problems must be mastered before reliable data can be extracted from the decay curves obtained in BFS. Cascading processes belong to the most obvious sources of experimental error. Additional problems may be caused by the ion velocity uncertainty and spread after the foil and the modest spectral resolution - mostly used so far - which may lead to line blends.

The ion-foil interaction results in the population of several levels in a spectrum. The simple relation (Eq. 5) must then be replaced by a sum of exponentials

$$I_{if}(x) = I_{if}(0) \exp (-x/v\tau_i) + \sum_k I_{ik}(0) \exp (-x/v\tau_k) \qquad (6)$$

and the experimental data should be analyzed using non-linear least-squares fitting procedures. Efficient computer programs such as HOMER (Irwin and Livingston 1974) or DISCRETE (Provencher 1976) are available for this purpose but even with these the problems may be very difficult, in particular whenever the primary and cascade lifetimes are similar.

A powerful method for cascade correction, usually called ANDC (arbitrarily normalized decay curves) has been developed by Curtis et al. (1971, 1976). It includes measurements of the decay curves of the levels of interest (i) and those of the most important levels (k) that directly feed i. The data are then jointly analyzed using a simple population rate equation (Curtis 1976). In a lifetime study for the 3p ^{2}P level in Cl VII (see Fig. 18) the decay curves of 3p ^{2}P and those of 4s, 5s ^{2}S and 3d, 4d, 5d ^{2}D should be measured and incorporated in the lifetime analysis. Information about indirect cascading to 3p ^{2}P (e.g. from ^{2}P, ^{2}F and ^{2}G levels) is then automatically included. Note also that the relative intensities of the various transitions fortunately need not be determined but these numbers are treated as parameters in the analysis.

The ANDC method has been employed in several cases with great success. Measurements of the 4p ^{2}P lifetime in Kr VIII (which belongs to the Cu I isoelectronic sequence), by Pinnington et al. (1979) and Livingston et al. (1980), have yielded a value that differs by about 40% from the extracted from curve-fitting. However, the ANDC result is in excellent agreement with advanced theoretical analyses (Cheng et al. 1978).

In connection with a study of the 2s2p ^{1}P lifetimes in Be-like N IV and O V Engström et al. (1981) also found that curve-fitting and ANDC analyses gave quite different results despite the presence of experimental data of high quality. One of their decay curves is displayed in Fig. 23 .

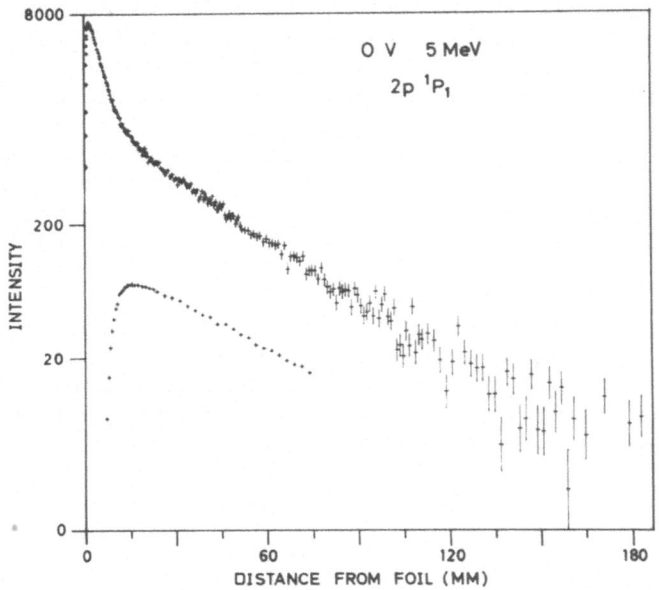

Fig. 23. Decay curve for the 2s2p ^{1}P level in O V (Engström et al. 1981)

Here, also, the ANDC results confirmed theoretical calculations.

The conclusion of these and similar other results is that ANDC must be applied whenever cascading is suspected. It is then also necessary to know the energy level structure with sufficient accuracy - otherwise the cascading transitions cannot be identified.

Cascading was often underestimated in early BFS studies - many decay curves may appear deceptively simple - and those results thus contained too optimistic error estimates. This has been criticized by several authors (Wiese 1970, Younger and Wiese 1978 and others). However, modern data can be considered as highly reliable.

The energy loss in the foil plays an important role when slow heavy ions (with 100 - 1000 keV energies) are used. However, with energies of several MeV - which is the case when lifetimes for highly ionized atoms are measured - this problem is rather less significant and the ion velocities are quite accurately known. Line blending may appear as a trivial source of uncertainty but it must be carefully examined, especially when the spectra are complex.

A number of methods have been developed to eliminate the problems discussed above, in particular cascading. The most powerful remedy is the beam-laser method, introduced by Andrä et al. (1973).

The ions are now excited with monochromatic laser light, e.g. from a
dye laser, into on particular state. The decay curve thereby obtained
is easy to analyze correctly. This method has yielded very accurate
lifetimes for Li I, Na I (Gaupp 1979), Ne I (Kandela and Schmoranzer
1981), Rb II (Ceyzeriat et al., 1980) and several other neutral and
singly-ionized atoms. Extensions to multiply ionized atoms are not
easy, however, one of the drawbacks being the spectral limitations
of the presently available lasers. The resonance lines of multiply
ionized atoms are often in the far UV range, an inconvenient region
for lasers.

Coincidence methods which are well known in nuclear spectroscopy
have also been applied to atomic lifetime measurements using fast
beams. In the first study of this kind (Masterson and Stoner 1973)
the accuracy was limited by very low coincidence counting rates.
New experimental developments seem to be possible, however (Becker
et al., 1982).

Results

Atomic lifetimes and f-values show interesting regularities and
systematic trends, for example with respect to isoelectronic or homo-
logous sequences (Wiese and Weiss 1968). The well-known Z-expansion
method yields the following expression in the non-relativistic case

$$f = f_o + f_o Z^{-1} + f_1 Z^{-2} + \ldots \ldots \tag{7}$$

where f_o is the hydrogenic f-value ($f_O = 0$ for $\Delta n = 0$ transitions) and
f_1 and f_2 are constants. More complicated expressions involving a
double expansion in Z^{-1} and $(\alpha Z)^2$ can be derived in the relativistic
case.

One of the systems most frequently studied by experimentalists
is the 2s $^2S_{1/2}$ - 2p $^2P_{1/2,3/2}$ resonance doublet in Li-like ions.
Experimental results for Fe XXIV (an ion of great importance in
plasma physics and astrophysics), as obtained by Dietrich et al.
(1978) are shown in Fig. 24. Note that these decay curves show no
effects of cascading. The dominant cascades are from $\Delta n = 1$ transitions the probability of which scales as Z^4, to be compared with the
2s - 2p rate which is proportional to Z. The n = 3 and n = 4 levels have
thus very short lifetimes ($\sim$1ps in Fe XXIV) and their decay does not
influence the present lifetime measurement.

Some f-values for the 2s - 2p transition in Li-like ions are displayed in Fig. 25. The experimental data for Fe XXIV and Kr XXXIV
were obtained at the SUPER-HILAC accelerator at Berkeley (Dietrich
et al. 1978, 1980) whereas the Ti XX f-values were determined using
the dual tandem accelerator at Brookhaven (Johnson et al. 1981). Note
that these experimental data confirm the relativistic calculations
of Kim and Desclaux (1976).

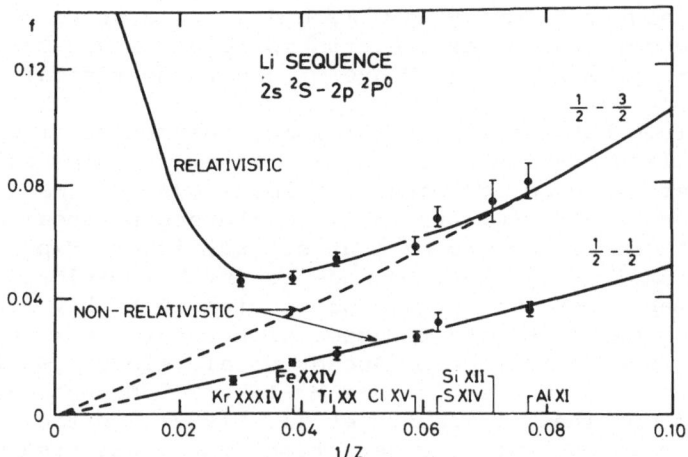

Fig. 24. Lifetime measurement of the 2p $^2P_{1/2}$ and $^2P_{3/2}$ levels in Fe XXIV (Dietrich et al. 1978)

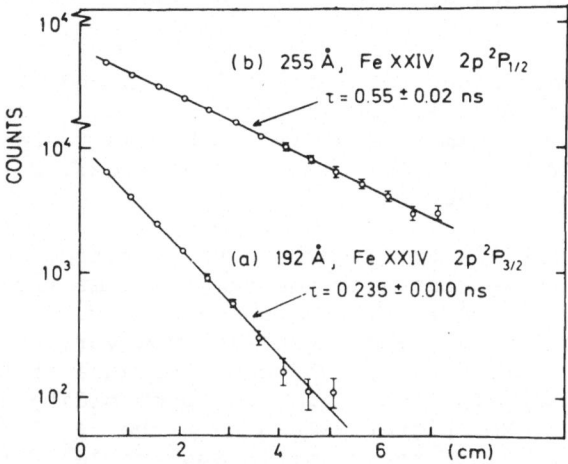

Fig. 25. Oscillator strengths for the $^2S - ^2P$ resonance lines in highly ionized Li-like ions.

Similar data are also available for the $2s^2\ ^1S - 2s2p\ ^1P$ resonance line in Be-like ions but here the highest system experimentally
investigated seems to be Cl XIV (Pegg 1980, Martinson 1982). Cascading
from $2p^2\ ^1S$ and $2p^2\ ^1D$ complicates experimental analyses. However,
extensions to higher Z, using ANDC methods, would be worth undertaking.
Because of the combination of electron correlation and relativistic
effects Be-like ions with high Z are of great theoretical interest.

A comparison between experimental and theoretical f-values for
the resonance lines in the Na I and Mg I isoelectronic sequences
shows that there is good agreement for low Z systems (Na I, Mg II,
Mg I, Al II) but thereafter systematic differences appear and the
experimental f-values tend to be consistently lower. Experimental
studies using ANDC should here be made to further explore the reason
of these discrepancies. Recently, two studies of highly ionized Fe
have been reported which tend to agree with theory, however. Thus
using careful cascade correction Buchet et al. (1980) confirm the
theoretical f-values for Na-like and Mg-like Fe. Essentially similar
data were also obtained by Träbert et al. (1982). In the latter beam-
foil study a position-sensitive detection system was used which
allowed the simultaneous registration of decay curves for many excited
levels.

Recent theoretical analyses of Al-like ions using a relativistic
parametric potential method (Farrag et al. 1982) have shown the
existence of several irregularities for f-values as Z increases.
There is little experimental material available to check these theoretical predictions; new and careful measurements would be of great
interest.

The work of Marrus and Schmieder (1972) showed that accurate
decay probabilities for forbidden transitions can be obtained using
highly ionized atoms. Summaries of all such results can be found in
the literature (Marrus and Mohr 1978, Marrus 1982). Only a few examples will therefore be given here.

In the He I isoelectronic sequence the lifetimes of the $1s2p$
$^3P_{0,1,2}$ levels have been measured for many systems from He I to
Kr XXXV. A summary of the results is shown in Fig. 26. For low values
of Z all ^3P levels have similar lifetimes. However, when Z increases
the $^3P_1$ lifetime is drastically shortened, because of the $1s^2\ ^1S_0 -$
$1s2p\ ^3P_1$ intercombination transition the probability for which is
proportional to Z^{10}. For higher values of Z the $^3P_2$ lifetime will
also be reduced because of the magnetic quadrupole (M2) transitions
to the ground state. Whenever the nuclear spin differs from zero
there is also a shortening of the ^3P lifetimes because of hyperfine-
induced transitions to the ground state. This was demonstrated for
the $^3P_2$ lifetime in V XXII by Gould et al. (1974). Subsequently,
Mohr (1976) calculated the rate of the nuclear-spin-induced decay
of the $^3P_0$ state. These calculations have been verified for F VIII

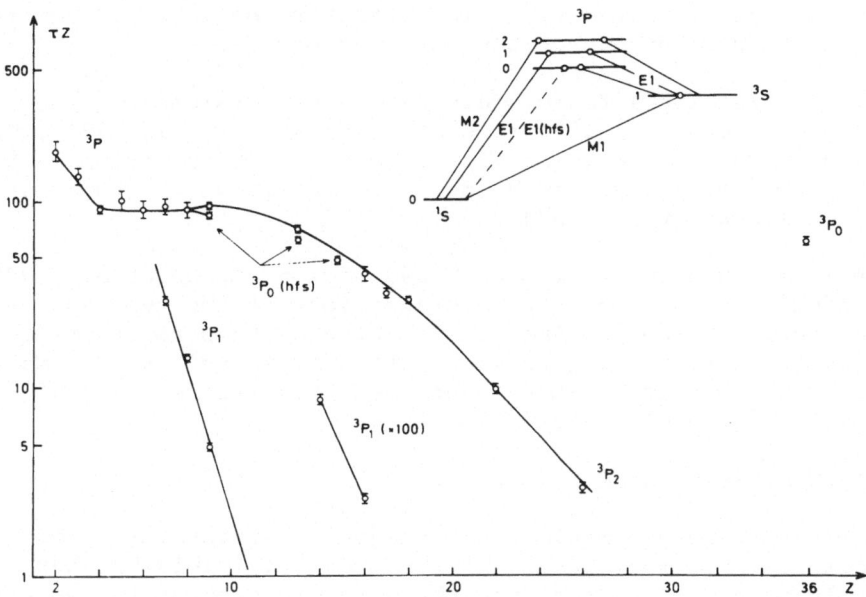

Fig. 26. Lifetimes of the 1s2p ^{3}P levels in two-electron systems

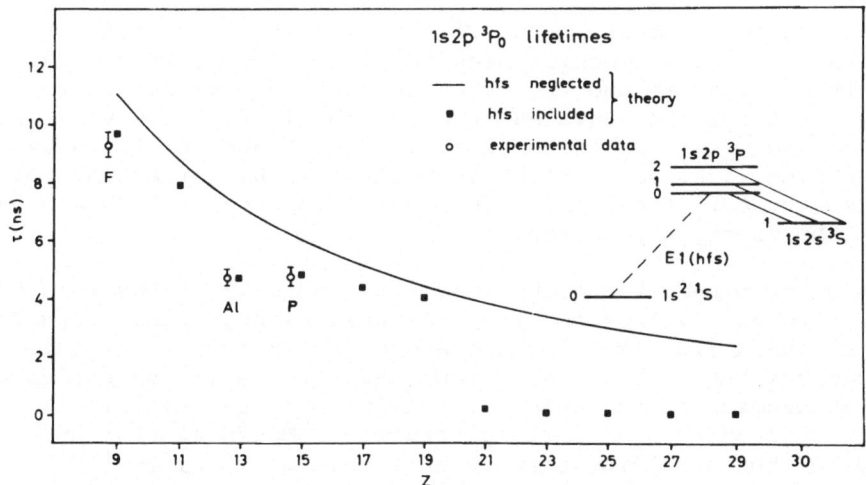

Fig. 27. Theoretical and experimental data for the 1s2p ^{3}P$_0$ lifetime for two-electron ions with non-zero nuclear spin.

(Engström et al. 1980), Al XII (Denne et al. 1980) and P XIV (Hinterlong and Livingston 1982), as shown in Fig.27. Theory (Mohr 1976) further predicts the hyperfine effect to be very substantial for several higher members of the He I sequence.

Very interesting lifetime data have been obtained for the $1s2s2p\ ^4P_{5/2}$ level in Li-like ions up to Ar XVI (Sellin 1976). This level autoionizes by the relativistic spin-spin interaction mechanism. For high values of Z also the M2 decay to the $1s^2 2s\ ^2S$ ground state is possible (Cocke et al. 1974).

Beam-foil experiments in the X-ray region (Cocke et al. 1975) have yielded probabilities for E2 transitions of the type $3d^{10}$ - $3d^9 4s$ in highly ionized iodine, I XXV which belongs to the Ni I sequence. It is interesting to note that this transition, in Mo XV, has been observed from Tokamak impurities (Klapisch et al. 1978 , Mansfield et al. 1978).

FINE-STRUCTURE MEASUREMENTS

Fine-structure separations are frequently determined from wavelength measurements for spectrally resolved multiplets. Higher accuracy is usually obtained from the determination of the wavelengths of M1 transitions within spectral terms. Such studies of forbidden lines, discussed above, have resulted in quite precise data on fine-structure intervals. This material is of considerable theoretical interest, largely because ab-initio calculations of fine-structure separations are usually quite difficult (Cheng et al. 1981a).

A very accurate measurement of the $1s2p\ ^3P_2$ - $^3P_1$ fine structure in He-like F VIII has recently been completed (Myers et al. 1981). Foil-excited F ions in the $^3P_2$ state interacted with light from a powerful CO_2 laser, the wavelength of which (10.6 μm) is very close to that of the $^3P_2$ - $^3P_1$ transition. The resonance was found by Doppler-tuning, i.e. varying the ion velocity. By correcting for hyperfine structure the authors obtained a value of 957.88 ± 0.03 cm^{-1} for the energy separation.

The quantum-beat technique (see e.g. Andrä 1979) has resulted in rather accurate values for fine- as well as hyperfine- structure in neutral and a few times ionized atoms. Extensions to high ionization degress have not yet been reported. However, a direct wavelength measurement of the $1s2s\ ^3S_1$ - $1s2p\ ^3P_{0,1,2}$ multiplet in F VIII by "ordinary" beam-foil spectroscopy (Myers et al. 1982) has resolved the hyperfine structure of the $^3S_1$ level in ^{19}F (I = ½), as shown in Fig. 28 .

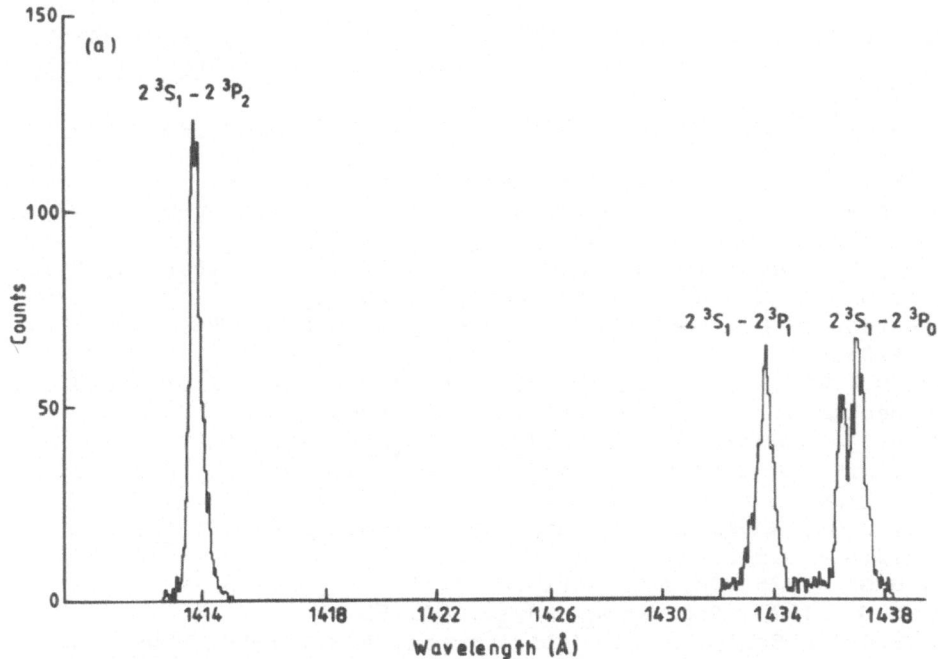

Fig. 28. The 1s2s $^3S_1$ - 1s2p $^3P_{0,1,2}$ multiplet in beam-foil excited
F VIII. The $F = 3/2 - F = 1/2$ separation of the $^3S_1$ level,
760 GHz (or 25 cm^{-1}) is clearly resolved in the case of
the $^3S_1$ - $^3P_0$ transition (Myers et al. 1982).

ACKNOWLEDGEMENTS

    I am grateful to Drs. H.G. Berry, R.D. Cowan, J.O. Ekberg,
Se. Johansson, and U. Litzén for enlightening discussions and to
Ms. H. Forsberg and Ms. S. Mickelsson for assistance in preparing
the manuscript.

    This work was supported by the Swedish Natural Science Research
Council (NFR) and the National Swedish Board for Energy Source
Devlopment.

REFERENCES

Accad, Y., Pekeris, C.L., and Schiff, B., 1971, Phys. Rev, A4:516
Andersen, T., Bentzen, S.M., and Poulsen, O., 1980, Physica Scripta,
    22:119
Andrä, H.J., Gaupp, A., and Wittmann, W., 1973, Phys. Rev. Lett.,
    31:501
Andrä, H.J., 1979, Fast-Beam (Beam-Foil) Spectroscopy, in: "Progress
    in Atomic Spectroscopy". W. Hanle and H. Kleinpoppen, eds.,
    Plenum, New York (p.829)
Bachmann, P., Eberlein, A., and Bruch, R., 1982, J. Phys. E, 15:207
Bashkin, S., 1964, Nucl. Instr. Methods, 28:88
Bauche-Arnoult, C., Bauche, J., and Klapisch, M., 1979, Phys. Rev.
    A20:2424
Becker, B., Winter, H., Zimny, R., Schirmacher, A., Andrä, H.J., and
    Heckmann, P.H., 1982, to be published.
Beier, R., and Kunze, H.-J., 1978, Z. Physik, A285:347
Bentzen, S.M., Andersen, T., and Poulsen, O., 1981, J. Phys. B,
    14:1345
Bentzen, S.M., Andersen, T., and Poulsen, O., 1982, J. Phys. B,
    15:L71 (1982)
Berry, H.G., 1975, Physica Scripta, 12:5
Berry, H.G., 1977, Rep. Prog. Phys., 40:155
Berry, H.G., DeSerio. R., and Livingston, A.E., 1980, Phys. Rev.,
    A22:998
Betz, H.D., Bell, F., Panke, H., Kalkoffen, G., Welz, M., and Evers,
    D., 1974, Phys. Rev. Lett., 33:807
Beyer, H.F., Schartner, K.-H., and Folkmann, F., 1980, J. Phys. B,
    13:2459
Bickel, W.S., Bergström, I., Buchta, R., Lundin, L., and Martinson,
    I., 1969, Phys. Rev., 178:118
Bitter, M., Hill, K.W., Santhoff, N.R., Efthiminion, P.C., Mesewey,
    E., Roney, W., von Goeler, S., Horton, R., Goldman, M., and
    Stodiek, W., 1979, Phys. Rev. Lett., 43:129
Boiko. V.A., Faenov, A.Ya., and Pikuz, S.A., 1978, J. Quant. Spec-
    trosc. Radiat. Transfer, 19:11
Boland. B.C., Irons, R.E., and McWhirter. R.W.P., 1968, J. Phys. B,
    1:1180
Bowen, I.S., and Millikan, R.A., 1925, Phys. Rev., 25:295
Bowen, I.S., and Edlén, B., 1939, Nature, 143:374
Bromander, J., 1973, Nucl. Instr. Methods, 110:11
Brooks, R., Hardis, J.E., Berry, H.G., Curtis, L.J., Cheng, K.T.,
    and Ray. W., 1980, Phys. Rev. Lett., 45:1318
Bruch, R., Rødbro. M., Bisgaard, P., and Dahl. P., 1977, Phys. Rev.
    Lett., 39:801
Bruch, R., Heckmann, P.H., Müller, H.R., Raith, B., Sander, U.,
    Schneider, G., and Träbert, E., 1981, in: "European Conference
    on Atomic Physics, Book of Abstracts", J. Kowalski, G. zu
    Putlitz and H.G. Weber, eds., EPS, Geneva (p.271)

Buchet. J.P., Denis, A., Desesquelles, J., and Dufay, M., 1969, Phys. Lett., 28A:529

Buchet, J.P., Buchet-Poulizac, M.C., Denis, A., Désesquelles, J., and Druetta, M., 1980, Phys. Rev., A22:2061

Buchet, J.P., Buchet-Poulizac, M.C., Denis, A., Désesquelles, J., Druetta, M., Grandin, J.P., and Husson, X., 1981, Phys. Rev., A23:3354

Burkhalter, P.G., Dozier, C.M., and Nagel, D.J., 1977, Phys. Rev., A15:700

Burkhalter, P.G., Schnwider, R., Dozier, C.M., and Cowan, R.D., 1978, Phys. Rev., A18:718

Bunge, C.F., 1980, Phys. Rev. Lett., 44:1450

Carroll, P.K., and Kennedy, E.T., 1981, Contemp. Phys., 22:61

Carroll, P.K., and O'Sullivan, G., 1982, Phys. Rev., A25:275

Ceyzeriat, P., Pegg. D.J. Carré, M., Druetta, M., and Gaillard, M.L., 1980, J. Opt. Soc. Am., 70:901

Cheng, K.T., Desclaux, J.P., and Kim. Y.-K., 1978, At. Data. Nucl. Data Tables, 22:547

Cheng, K.T., Desclaux, J.P., and Kim. Y.-K., 1978a, J. Phys. B, 11:L359

Cheng, K.T., Kim, Y.-K., and Desclaux, J.P., 1979, At. Data Nucl. Data Tables, 24:111

Cilliers, W.A., Datla, R.U., and Griem, H.R., 1975, Phys. Rev., A12:1408

Cocke, C.L., Curnutte, B., and Randall, R., 1974, Phys. Rev., A9:1832

Cocke, C.L., Varghese, S.L., Bednar, J.A., Bhalla, C.P., Curnutte, B., Kauffman, R., Randall, R., Richard, P., Woods, C., and Scofield, J.H., 1975, Phys. Rev., A12:2413

Cocke, C.L., 1976, Beam-Foil Spectrosocopy, in: "Methods of Experimental Physics, 13B", D. Williams, ed., Academic Press, New York (p.213)

Conturie, Y., Yaakobi, B., Feldman, U., Doschek, G.A., and Cowan, R.D., 1981, J. Opt. Soc. Am., 71:1309

Corney, A., 1969, Adv. Electron. Electron. Phys., 29:115

Cowan, R.D., 1981, Physica Scripta, 24:615

Cowan, R.D., 1981a, "The Theory of Atomic Structure and Spectra", University of California Press, Berkeley

Curtis, L.J., Berry, H.G., and Bromander, J., 1979, Phys. Lett., 34A:169

Curtis, L.J., 1976, Lifetime Measurements, in: "Beam-Foil Spectroscopy", S. Bashkin, ed., Springer, Berlin (p.63)

Curtis, L.J., 1977, Phys. Lett., 64A:43

Curtis, L.J., 1979, Phys. Lett., 72A:427

Curtis, L.J., 1981, Phys. Rev., A23:362

Curtis, L.J., 1981a, J. Phys. B, 14:1373

Davis, W.A., and Marrus, R., 1977, Phys. Rev., A15:1963

Denis, A., and Desesquelles, J., 1981, J. Physique, 42:L-59

Denne, B., Huldt, S., Pihl. J., and Hallin, R., 1980, Physica Scripta, 22:45

DeSerio, P., Berry, H.G., Brooks, R.L., Hardis, J., Livingston, A.E., and Hinterlong, S.J,. 1981, Phys. Rev., A24:1872

Desclaux, J.P., 1982, these proceedings
Dietrich, D.D., Leavitt, J.A., Bashkin, S., Convay, J.G., Gould, H.,
     MacDonald, D., Marrus, R., Johnson, B.M., and Pegg, D.J.,
     1978, Phys. Rev., A18:208
Dietrich, D.D., Leavitt, J.A., Gould, H., and Marrus, R., 1980, Phys.
     Rev. A22:1109
Drake, G.W.F., 1979, Phys. Rev., A19:1387
Drake, G.W.F., 1982, these proceedings
Drawin, H.W., 1978, Phys. Reports, 37:125
Drawin, H.W., 1981, Physica Scripta, 24:622
Drawin, H.W., 1982, Ann. Physique, (in press)
Dubau, J., and Volonté, S., 1980, Rep. Prog. Phys., 43:199
Dupree, A.K., 1978, Adv. Atom. Molec. Phys., 14:393
Edlén, B., 1936, Z. Physik, 100;621
Edlén, B., and Tyrén, F., 1939, Nature, 143:940
Edlén, B., 1942, Z. Astrophysik, 22:30
Edlén, B., 1963, Rep. Progr. Phys., 26:181
Edlén, B., 1964, Atomic Spectra, in: "Handbuch der Physik, Vol. 27",
     S. Flügge, ed., Springer, Berlin (p.80)
Edlén, B., 1966, J. Opt. Soc. Am., 56:1285
Edlén, B., 1976, The Term Analysis of Atomic Spectra, in: "Beam-Foil
     Spectroscopy", I.A. Sellin and D.J. Pegg, eds., Plenum, New
     York (p.1)
Edlén, B., and Smitt, R., 1978, Solar Physics, 57:329
Edlén, B., 1978, Physica Scripta, 17:565
Edlén, B., 1979, Physica Scripta, 19:255; 20:129
Edlén, B., 1981, Physica Scripta, 22:593; 23:1079
Edlén, B., 1982, private communication
Ekberg, J.O., 1974, unpublished material
Ekberg, J.O., 1975, Physica Scripta, 12:42
Engelhardt, W., and Sommer, J., 1971, Astrophys. J., 167:201
Engvold, O., 1977, Physica Scripta, 16:48
Engström, L., Jupén, C., Denne, B., Huldt, S., Weng Tai Meng, Kaijser,
     P., Litzén, U., and Martinson, I., 1980, J. Phys. B, 13:L143
Engström, L., Denne, B., Ekberg, J.O., Jones, K.W., Jupén, C.,
     Litzén, U., Weng Tai Meng, Trigueiros, A., and Martinson, I.,
     1981, Physica Scripta, 24:551
Ermolaev, A.M., 1979, Quantum Electrodynamical Effects in Atomic
     Spectra, in: "Progress in Atomic Spectroscopy", W. Hanle
     and H. Kleinpoppen, eds., Plenum, New York (p.149)
Farrag, A., Luc-Koenig, E., and Sinzelle, J., 1982 (to be published)
Fawcett, B.C., Gabriel, A.H., Irons, F.E., Peacock, N.J., and
     Saunders, P.A.H., 1966, Proc. Phys. Soc., 88:1051
Fawcett, B.C., 1974, Adv. Atom. Molec. Phys., 10:223
Fawcett, B.C., Ridgeley, A., and Hatter, A.T., 1980, J. Opt. Soc. Am.,
     70:1349 (1980)
Fawcett, B.C., 1981, Physica Scripta, 24:663
Feldman, U., Swartz, M., and Cohen, L., 1967, Rev. Sci. Instr., 38:
     1372
Feldman, U., 1976, Astrophys. Space Sci., 41:155

Feldman, U., and Doschek, G.A., 1976, Spectroscopy of Highly-Ionized
        Atoms Produced by a Low-Inductance Vacuum Spark, in: "Atomic
        Physics 5", R. Marrus, M. Prior, and H. Shugart, eds., Plenum,
        New York (p.473)
Feldman, U., 1981, Physica Scripta, 24:681
Folkmann, F., Beyer, H.F., Mann, R., and Schartner, K.-H., 1981,
        Nucl. Instr. Methods, 181:99
Froese Fischer, C. 1982 to be published
Gabriel, A.H., 1982, these proceedings
Galan, M., and Bunge, C.F., 1981, Phys. Rev., A23:1624
Gaupp, A., 1979, Dissertation, Free University, Berlin
Gould, H., Marrus, R., and Mohr, P.J., 1974, Phys. Rev. Lett.,
        33:676
Hibbert, A., 1975, Rep. Progr. Phys.
Hibbert, A., 1977, Physica Scripta, 16:7
Hinnov, E., 1976, Phys. Rev., A14:1533
Hinterlong, S.J., and Livingston, A.E., 1972, to be published
Huber, M.C.E., 1977, Physica Scripta, 16:16
Hvelplund, P., Haugen, H.K., and Knudsen, H., 1981, Physica Scripta,
        23:193
Imhof, R.E., and Read, F.H., 1977, Rep. Prog. Phys., 40:1
Irwin, D.J.G., and Livingston, A.E., 1974, Comp. Phys. Commun., 7:95
Isler, R.C., Neidigh, R.V., and Cowan, R.D., 1977, Phys. Letters,
        63A:295
Johansson, Se., (1982), to be published
Johnson, B.M., Jones, K.W., and Gregory, D.C., 1981, IEEE Trans.
        Nucl. Sc., NS28:1350
Jordan, C., 1979, Applications of Atomic Physics to Astrophysical
        Plasmas, in: "Progress in Atomic Spectroscopy", W. Hanle
        and H. Kleinpoppen, eds., Plenum, New York (p.1453)
Jupén, C., Fremberg. J., and Fawcett, B.C.,1982, to be published
Kandela, S., and Schmoranzer, H., 1981, Phys. Letters, A86:101
Kaufman, V., and Edlén, B., (1974), J. Phys. Chem. Ref. Data, 3:825
Kay, L., 1963, Phys. Letters, 5:36
Key, M.H., and Hutcheon, R.J., 1980, Adv. Atom. Molec. Phys., 16:202
Kim, Y.-K., and Desclaux, J.P., 1976, Phys. Rev. Lett., 36:139
Klapisch, M., Schwob, J.L., Finkenthal. M., Fraenkel, B.S., Egert,
        S., Bar-Shalom, A., Breton, C., DeMichelis, C., and Mattioli,
        M., 1978, Phys. Rev. Lett. 41:403
Klinkenberg. P.F.A., 1976, Optical Region in: "Methods of Experimental
        Physics 13A", D. Williams, ed., Academic Press, New York,
        (p.253)
Knystautas, E.J., and Drouin, R., 1976, in: "Beam-Foil Spectroscopy",
        I.A. Sellin and D.J. Pegg. eds., Plenum, New York (p.377)
Kononov, E.Ya., 1978, Physica Scripta, 17:425
Kruger, P.G., 1930, Phys. Rev., 36:855
Larsson, S., Crossley, R., and Ahlenius, T., 1979, J. Physique, 40:C1-6
Laughlin, C., 1982, J. Phys. B, 15:L67
Livingston, A.E., and Berry, H.G., 1978, Phys. Rev., A17:1966
Livingston, A.E., and Hinterlong, S.J., 1980, Phys. Lett., 80A:372

Livingston, A.E., Curtis, L.J., Schectman, R.M., and Berry, H.G.,
        1980, Phys. Rev., A21:771
Livingston, A.E., 1982, these proceedings
Mannervik, S., Astner, G., and Kisielinski, M., 1980, J. Phys. B,
        13:L441
Mannervik, S., Martinson, I., and Jelenkovic, B., 1981, J. Phys. B,
        14:L275
Mansfield, M.W.D., Peacock, N.J., Smith, C.C., Hobby, M.G., and
        Cowan, R.D., 1978, J. Phys. B, 11:1521
Marrus, R., and Schmieder, R.W., 1972, Phys. Rev., A5:1160
Marrus, R., and Mohr, P.J., 1978, Adv. Atom. Molec. Phys., 14:182
Marrus, R., 1982, these proceedings
Martin, W.C., 1981, Physica Scripta, 24:725
Martinson, I., 1978, Experimental Studies of Atomic and Molecular
        Lifetimes, in: "Excited States in Quantum Chemistry", C.A.
        Nicolaides and D.R. Beck, eds., Reidel, Dordrecht (p.1)
Martinson, I., 1980, Experimental Studies of Energy Levels and
        Oscillator Strengths of Highly Ionized Atoms, in: "Atomic and
        Molecular Processes in Controlled Thermonuclear Fusion",
        M.R.C. McDowell and A.M. Ferendici, eds., Plenum, New York
        (p.391)
Martinson, I., 1981, Physica Scripta, 23:126
Martinson, I., 1982, to be published
Masterson, K.D., and Stoner, Jr., J.O., 1973, Nucl. Instr. Methods,
        110:441
Minnhagen, L., 1964, J. Res. NBS, 68C:327
Mohr, P.J., 1976, in: "Beam-Foil Spectroscopy", I.A. Sellin and
        D.J. Pegg, eds., Plenum, New York (p.57)
Mowat, R., Jones, K.W., and Johsnon, B.M., 1979, Phys. Rev., A20:1972
Myers, E.G., Kuske, P., Andrä, H.J., Armour, I.A., Jelley, N.A.,
        Klein, H.A., Silver, J.D., and Träbert, E., 1981, Phys. Rev.
        Lett., 47:81
Myers, E.G., Klein, H.A., Silver, J.D., and Träbert, E., 1982, un-
        published material
Namioka, T., 1959, J. Opt. Soc. Am., 49:951
O'Brien, R., Silver, J.D., Jelley, N.A., Bashkin, S., Träbert, E.,
        and Heckmann, P.H., 1979, J. Phys. B, 12:L41
Panke, H., Bell, F., Betz, H.-D., Stehling, W., Spindler, E., and
        Laubert, R., 1975, Phys. Lett., 53A:457
Peacock, N.J., 1976, Spectroscopy of Highly-Stripped Ions in Laser
        Induced Plasmas, in: "Beam-Foil Spectroscopy", I.A. Sellin and
        D.J. Pegg, eds., Plenum, New York, (p.925)
Peacock, N.J., 1980, Optical Spectroscopic Features of the Emission
        from Highly-Ionised Atoms in Tokamak Discharges, in: "The
        Physics of Ionized Gases, SPIG-80", M. Matic, ed., Boris
        Kidric Institute, Beograd (p.687)
Pegg, D.J., 1980, Radiative and Auger Beam-Foil Measurements, in:
        "Methods of Experimental Physics, Vol. 17", P. Richard, ed.,
        Academic Press, Now York (p.529)

Penkin, N.P., 1979, Experimental Determination of Electronic Transi-
    tion Probabilities and the Lifetimes of the Excited Atomic
    and Ionic States, in: "Atomic Physics 6", A.M. Prokhorov,
    R. Damburg and O. Kuklane, eds., Plenum, New York (p.33)
Pinnington, E.H., Gosselin, R.N., O'Neill, J.A., Kernahan, J.A.,
    Donnelly, K.E., and Brooks, R.L., 1979, Physica Scripta,
    20:151
Presnyakov, L.P., and Urnov, A.M., 1979, J. Physique, 40:C7-279
Provencher, S.W., 1976, J. Chem. Phys., 64:2772
Reader, J., and Acquista, N., 1977, Phys. Rev. Lett., 39:184
Reader, J., Luther, G., and Acquista, N., 1979, J. Opt. Soc. Am.,
    69:144
Reader, J., and Luther, G., 1981, Physica Scripta, 24:732
Rødbro, M., Bruch, R., and Bisgård, P., 1979, J. Phys. B, 12:2413
Safronova, U.I., 1981, Physica Scripta, 23:253
Samson, J.A., 1976, Far Ultraviolet Region, in: "Methods of Experi-
    mental Physics 13A", D. Williams, ed., Academic Press, New
    York (p.204)
Schmieder, R.W., and Marrus, R., 1973, Nucl. Instr. Methods, 110:459
Sellin, I.A., 1976, Adv. Atom. Molec. Phys., 12:215
Stamp, M.F., Armour, I.A., Peacock, N.J., and Silver, J.D., 1981,
    J. Phys. B, 14:3551
Suckewer, S., and Hinnov, E., 1979, Phys. Rev., A20:578
Suckewer, S., 1981, Physica Scripta, 23:72
Suckewer, S., and Hinnov, E., 1982, Atomic Processes for Diagnostic
    of Magnetically Confined Plasmas, in: "Physics of Electronic
    and Atomic Collisions", S. Datz, ed., North-Holland, Amsterdam
    (p.783)
Sugar, J., and Kaufman, V., 1981, Physica Scripta, 24:742
Tousey, R., Strain, C.V., Johnson, F.S., and Oberly, J.J., 1947,
    Astrophys. J., 52:158(A)
Träbert, E., Armour, I.A., Bashkin, S., Jelley, N.A., O'Brien, R.,
    and Silver, J.D., 1979, J. Phys. B, 12:1665
Träbert, E., Jones, K.W., Johnson, B.M., Gregory, D.C., and Kruse,
    T.H., 1982, Phys. Letters,
Tyrén, F., 1938, Z. Physik, 111:314
Varghese, S.;., Cocke, C.L., Curnutte, B., and Seaman, G., 1976,
    J. Phys. B, 13:L387
Weiss, A.W., 1973, Adv. Atom. Molec. Phys., 9:1
Wiese, W.L., and Weiss, A.W., 1968, Phys. Rev., 175:50
Wiese, W.L., 1970, Nucl. Instr. Methods, 90:25
Wiese, W.L., 1979, Atomic Transitions Probabilities and Lifetimes,
    in: "Progress in Atomic Spectroscopy", W. Hanle and H.
    Kleinpoppen, eds., Plenum, New York (p.1101)
Wittkower, A.B., and Betz, H.D., 1973, Atomic Data, 5:113
Younger, S.M., and Wiese, W.L., 1978, Phys. Rev. A17:1944
Zetterberg, P.O., and Magnusson, C.E., 1977, Physica Scripta, 15:189
Zigler, A., Zmora, H., Spector, N., Klapisch, M., Schwob, J.L. and
Bar-Shalom, A., 1980, J. Opt. Soc. Am., 70:129

HIGHLY IONIZED ATOMS IN TOKAMAK DISCHARGES

Einar Hinnov

Princeton University, Plasma Physics Laboratory
Princeton, New Jersey, USA

INTRODUCTION

Tokamaks, stellarators and related plasma confinement devices have existed in a fruitful and yet usually precarious symbiosis with atomic physics. In principle, the only atoms involved in such plasmas are those of hydrogen isotopes, and perhaps eventually helium, but in practice this has not been the case. Other elements have intruded in a variety of ways mostly unexpected, and they have been blamed, usually unjustly, for practically every ailment besetting the plasmas.

As the devices have increased in size and temperature, the interest of atomic physics has shifted toward heavier elements, because the lighter elements become completely stripped of their electrons and hence practically unobservable, except at the outer periphery of the plasma. The preoccupation with these heavier elements is not just, or even primarily, the radiative energy loss that they cause, but rather the localized information that they are capable of providing in the hot interior of the plasma. It is this aspect, the diagnostic use of appropriate elements in the plasma, that is the topic of the present work.

We shall not discuss the plasma physics aspects of tokamaks, either theoretical or observational. Only such description of plasmas will be included that is essential for the atomic physics problems. The best general reference for tokamaks is still the review article by Furth,[1] although a huge and growing literature exists on the subject – an indication that the behavior of these plasmas is not adequately understood, either theoretically or empirically.

Various diagnostic methods of tokamak interest have been described in the Varenna School of Plasma Physics procedings[2] and a recent review by de Michelis and Mattioli.[3] Radiative efficiency of elements in high-temperature plasma[4,5] and radiative energy losses in tokamak plasmas have been discussed abundantly in recent literature.[6-8] These references are only samples, and they emphasize strongly the work done at Princeton. On work closely related to the present topic we would like to mention two papers by the author,[9-10] with further, more specialized references given below.

Up to the present time (1982), most of the tokamaks have been producing plasmas with peak electron temperatures 1.0-1.5 keV, with a few, notably the PLT tokamak in Princeton, capable of 2-3 keV range. However, within the next few years a new generation of considerably larger tokamaks is scheduled to become operational, with prospective peak temperatures in the 5-10 keV range. The underlying leitmotif of the present discussion is to consider what plasma spectroscopic diagnostics would be required or desirable for the study of the inner workings of such unprecedented plasma regimes.

We shall first describe some essential features of tokamak plasmas, then review some of the important aspects of atomic physics in relation to such plasmas, and finally present a scheme of necessary data base appropriate for spectroscopic diagnostics of tokamak plasmas in the present and near future.

DESCRIPTION OF TOKAMAK PLASMAS

As an example of presently operating tokamaks, Fig. 1 shows a schematic view of the PLT tokamak. [The letters stand for Princeton Large Torus, a name somewhat anachronistic partly because our concept of tokamak size has evolved, but partly also because the early concept, 1970-71, visualized a considerably larger device than the one actually built.] It consists of an externally imposed toroidal magnetic field $B_\phi \lesssim 32$ kG, a transformer to produce a toroidal current of typically $I_\theta \approx 500$ kAmps, which induces a poloidal magnetic field B_θ, and auxiliary "shaping" field coils, which produce an essentially vertical magnetic field for the purpose of maintaining the current channel (by $\vec{j} \times \vec{B}$ forces) near the center of the (stainless steel) vacuum vessel. At one toroidal location there is a massive rectangular current-aperture limiter (usually graphite, but stainless steel and tungsten have been used at times), the purpose of which is to reduce the intensity of interaction between the plasma and the vacuum vessel, or any apparatus e.g., rf antennae near the vacuum walls. The limiter aperture is variable, but normally set at a = 40 cm, with the vacuum vessel minor radius about 49 cm. Titanium is often evaporated on the vacuum walls between discharges, in order to

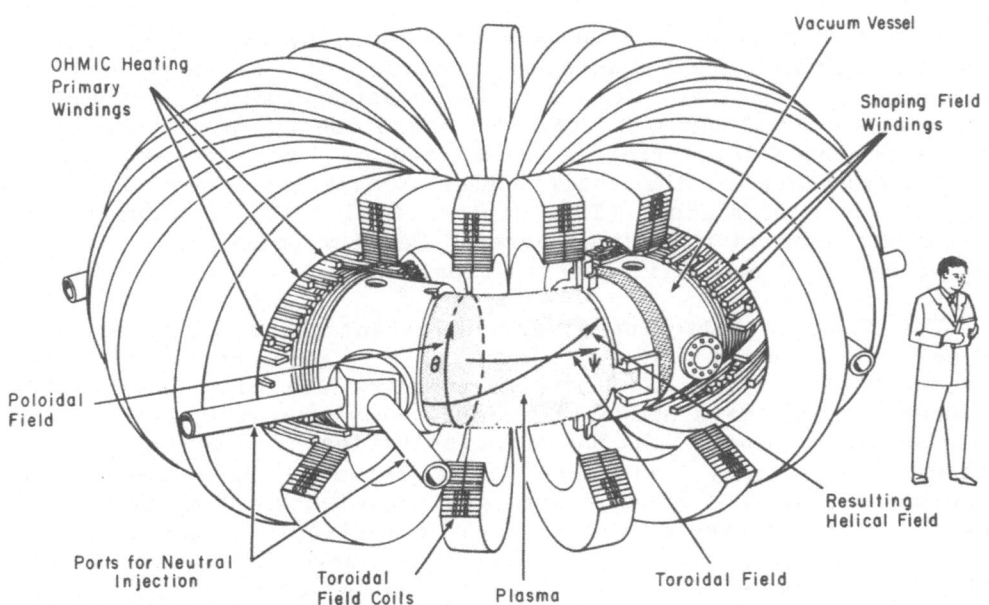

OHMIC Heating
Primary
Windings

Vacuum Vessel

Shaping Field
Windings

Poloidal
Field

Resulting
Helical Field

Ports for Neutral
Injection

Toroidal
Field Coils

Plasma

Toroidal Field

Fig. 1   A schematic view of the PLT tokamak showing the various
field-coils, vacuum vessel and access ports, and the directions of
the magnetic field resulting from the imposed toroidal field B_Φ and
current-generated poloidal field B_θ.

reduce the importance of the walls as sources of hydrogen, oxygen
and carbon atoms, and thereby acquire better control over the rate
of gas admission and plasma composition.   Without the titanium
evaporation considerable quantities of hydrogen, as well as oxygen
and carbon compounds adhere to the walls, and are released in an
essentially uncontrollable fashion during the next discharge.

     The discharge produced by the toroidal current, called ohmic-
heating discharge lasts (in PLT) typically 0.7-0.8 seconds, during
most of which time (i.e., ~ 0.5 sec.) the plasma is in a
quasisteady state.   This means that gross plasma characteristics
such as current, power input, average density and temperature, etc.
do not vary appreciably on a time-scale shorter than the plasma
duration.   However, as the current amplitude, gas inlet rate, the
shaping fields, etc., are externally variable to some extent, even
during the discharge, the details of the plasma characteristics are

quite varied, and depend significantly on the skill and experience of the operator.

Radial profiles of the electron density and temperature, measured during the quasisteady phase of a representative discharge, are shown in Fig. 2. The measured points are derived from single-shot Thomson scattering measurements of a ruby laser beam, with the density profile normalized to 2 mm microwave phase-delay measurement of the total number of electrons per unit area along the line of sight. (This number of electrons divided by the limiter diameter is the "line-average" density, equal to 2/3 of the peak value for a parabolic distribution.)

The Thomson scattering gives an instantaneous radial profile; for time-development it is necessary to rely on shot-to-shot reproducibility. The electron temperatures are also continuously measurable from the slope of the soft X-ray energy spectrum[12] and the electron cyclotron frequency emissivities.[13] Generally, good internal consistency has been found between these independent measurements, so in tokamak discharges the electron temperature and density, indispensable for any quantitative analysis of spectroscopic data, may be assumed to be known (i.e., measurable) with an accuracy better than 10% in favorable cases. [The accuracy depends on the state of the apparatus, and tends to deteriorate at low densities ($\lesssim 10^{13}$ cm^{-3}) and both very low (< 100 eV) and very high (> several keV) temperatures.]

The continuous curves in Fig. 2 are not drawn to the measured points, but analytical expressions—a parabola and the square of a parabola, that have been usually found to represent the electron density and temperature profiles, respectively, fairly well. It is interesting to note that the plasma conductivity (aside from relatively minor effects caused by toroidal curvature) is expected to be proportional to $T_e(r)^{3/2}$ and, in the quasisteady state, this would also be the radial distribution of the current density, and of ohmic power input. Thus, the power input occurs primarily near the center of the discharge and the periphery is presumably heated mostly by kinetic energy transport outward, by conduction or convection. However, if the analytic relationships in Fig. 2 adequately describe $n_e(r)$ and $T_e(r)$ or more generally if $T_e(r) \propto n_e^2(r)$ then the local electron energy is proportional to the local power input. This curious relationship has been noted long ago,[14] and it seems to hold quite well in most measured radial $T_e(r)$, $n_e(r)$ profiles.

The ohmic power input in PLT discharges is in the neighborhood of 0.5 MW, which, distributed as indicated above, implies a near-central power input density typically 500-1000 mW/cm^3. The minimum radiative energy loss -- bremsstrahlung by a pure hydrogen discharge -- would be about 5-10 mW/cm^3. This range of power

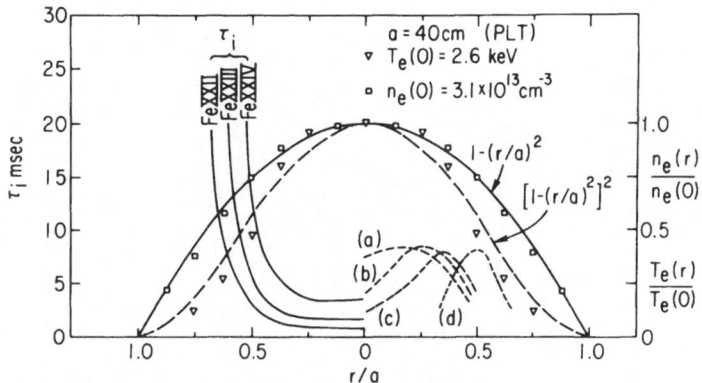

Fig. 2 Representative radial electron density and temperature distributions in the PLT tokamak discharge, together with measured ion density distributions of (a) Fe XXIV, (b) Fe XXIII, (c) Fe XXII, (d) Ne X, and collisional ionization times for the iron ions.

densities ~ 10-1000 mW/cm^3 is then the range of interest for electron energy losses in the interior of tokamak plasmas, and this will probably be not very different in the next generation tokamaks or even in the tokamak-based fusion reactors. Thus, a radiation level of ~ 100 mW/cm^3 may be considered an approximate tolerance limit for restricting the concentration of heavier elements in the discharge.

Besides hydrogen isotopes, the discharge generally contains carbon and oxygen[15,16] at concentrations of the order of 10^{-2} n_e, and wall materials[17,18] (i.e., Fe, Cr, Ni, Ti) of the order of $10^{-3} n_e$. Other elements may be deliberately added at predetermined times[19] and quantities in order to provide the means of spectroscopic plasma diagnostics and to study the effect of such added elements on the plasma behavior.

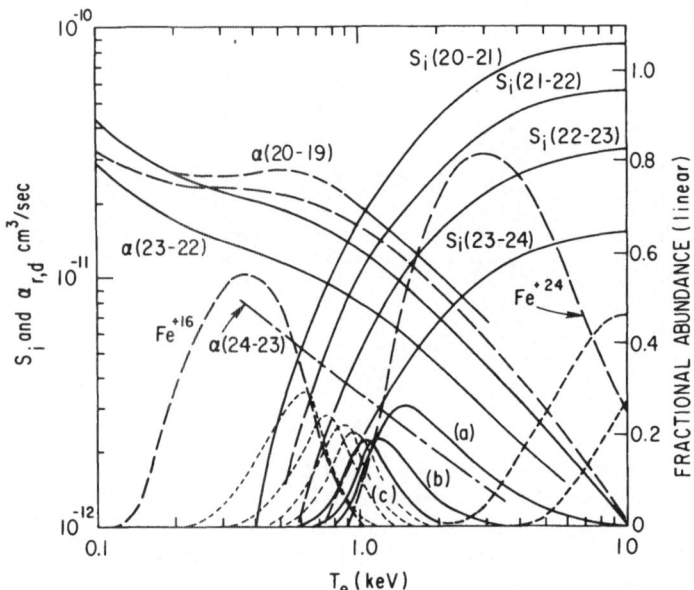

Fig. 3 Ionization (S_i) and recombination (α) rate coefficients for several iron ions, and the fractional abundances (scale right) at coronal ionization equilibrium. Curves marked (a), (b), and (c) refer to the ions shown in Fig. 2. Numbers in parentheses are the net charges on the ions.

        The curves labelled (a), (b), and (c) in Fig. 2 show the measured radial distributions of the three iron ions -- the lithiumlike FeXXIV (ionization potential E_i = 2.0 keV), berylliumlike FeXXIII (E_i = 1.9 keV), and boronlike FeXXII (E_i = 1.8 keV). Like the electron temperature and density profiles, these ion density profiles are quasisteady: their shapes do not change appreciably in time intervals of the order of 100 msec, yet, the collisional ionization times, as indicated on the left of the figure, are only of the order of 1 msec near the center, and volume recombination times (radiative and dielectronic recombination) are also only a few milliseconds. Thus, coronal ionization equilibrium

-- collisional ionization balanced by radiative-dielectronic recombinations -- should be well established within 10-20 msec.

Figure 3 shows the ionization[20,21] and recombination[22] rate coefficients for the various iron ions, and the fractional abundances of these ions expected at coronal equilibrium.[4] A comparison of the curves labelled (a), (b), and (c) at the appropriate temperatures readily shows a substantial discrepancy in the sense that the observed ionization stages occur at significantly higher temperatures (i.e., smaller radii) than expected for coronal equilibrium. This implies that the observed quasisteady state is a dynamic rather than a static phenomenon i.e., that the ion radial motions are sufficiently large to be competitive with the atomic (ionization or recombination) rates, or, in other words, that during an ionization time an ion can move to a radius of significantly different temperature, a distance of typically several centimeters across the magnetic field.[9]

In coronal equilibrium, the maximum abundance of a given ionization stage occurs roughly at $T_e \approx 1/2\ E_i$, at which temperature the relevant ionization and recombination times are about equal. At higher temperatures (as ionization rates increase and recombination rates decrease), recombination rapidly becomes negligible and the actual radial (hence temperature) distribution of the ion in the discharge must be determined by ionization rate and radial motion. Conversely, at lower temperature ionization rate becomes negligible and the profile is determined by radial motion and recombination. The observed profiles of the ions indicated in Fig. 2, with abundance peaks roughly at $T_e(r) \approx E_i$ imply a preferentially inward motion of the ions, with a velocity of the order of a few cm/msec.

It is also evident from Fig. 3 that the heliumlike FeXXV (Fe^{24+}) ion, with its small ionization and recombination rates is quite inert and can spread over a large radial (i.e., temperature) range. It is furthermore practically undetectable except at quite high temperatures. The apparent paradox of steady state of the plasma combined with the continued rapid inward motion of the observed ions is probably resolved by assigning a large amount of the compensatory outward flux to the heliumlike ion. There is indeed some experimental evidence[23] from measured spectra of argon (added for diagnostic purposes to the discharge) for such outward flux. In the same category are also the recent measurements of the radial profiles of fully stripped oxygen and carbon ions in the in the PDX tokamak discharges.[24] The characteristics of relative inertness are also evident in the neonlike ions e.g., Fe XVII (Fe^{16+}) although in a less pronounced fashion.

The question of radial transport of particles and kinetic energy is the most important problem in toroidal confinement devices, and it is here that the physics of highly-ionized atoms will have to play a major role. An important advance in the experimental technique in this regard is the laser blow-off method[25],[19] of introducing predetermined amounts of various elements into the discharge. This allows in principle the study of both inward and outward transport of ions of elements that may be particularly suitable for given plasma conditions.[26-28] Gaseous elements, e.g., noble gases may be mixed with the hydrogen, or added as a short puff during the discharge. Curve (d) in Fig. 2 shows the measured[29] radial distribution of the hydrogenlike neon, introduced in this manner. For the present we only note that this ion shares the characteristics of iron ions -- i.e., it is radiating only over a limited radial range in a quasisteady state, and it is significantly displaced from coronal equilbrium distribution. Lower ionization states are similarly present at larger radii, whereas further inward neon would be fully stripped and hence not observable spectroscopically.

During the quasisteady phase of the discharge additional heating is commonly employed to raise the temperature of the plasma. In PLT, two methods have been extensively used, with power levels larger than ohmic power: 40 keV neutral hydrogen or deuterium beams,[30] and radio-frequency power near the ion cyclotron frequency.[31] Both methods have achieved power input of about 3 MW, for 0.1-0.2 second, and both methods heat preferentially the ions [up to $T_i(o) \approx$ 6-7 keV], although some significant electron heating [to $T_e(o) \approx$ 3-4 kev] and other changes in plasma conditions can also occur. However, with regard to the physics of highly-ionized atoms the effect is mainly quantitative -- with one known exception: the neutral beam injection increases substantially the neutral hydrogen density in the hot interior of the plasma (to levels of $10^8 - 10^9$ cm^{-3} or $\sim 10^{-5}$ n_e). This level of H` is sufficient to cause significant recombination of highly-ionized atoms through charge-exchange, especially in ionization states where dielectronic recombination rate is small, with the consequence that the ionization balance is lowered during the injection. The modification of the ionization balance has been observed[32] in highly-ionized iron and titanium. It has implications on the radiation efficiency of a given element and the allowable impurity concentrations in tokamak plasmas,[5] and it provides plasma diagnostic possibilities, including the quantitative spectroscopic detection of fully stripped atoms.[33],[34],[24] However, we shall not discuss these aspects of the atomic physics at present, beyond noting that the above-mentioned observed deviations from coronal ionization equilibrium appear not to be caused primarily by charge-exchange recombination with H`, although this process needs to be quantitatively considered also in ohmic heating discharges.

Some observational aspects also need to be mentioned. We are

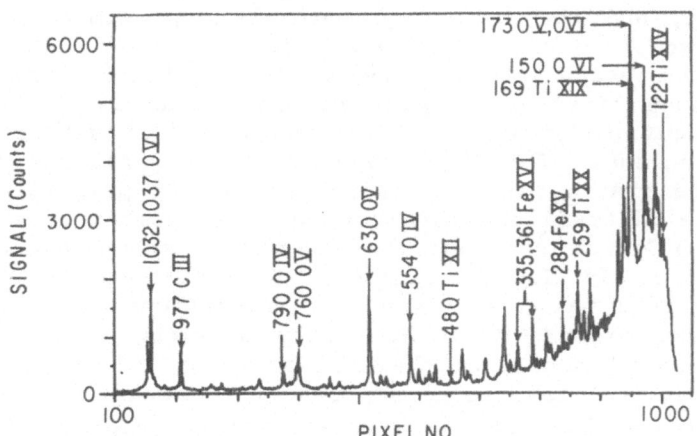

Fig. 4 A representative spectrum of a PDX tokamak discharge during the quasisteady phase, with peak temperature 0.8 keV, and integration time 20 msec. The sensitivity is approximately uniform for wavelengths above 150 Å.

considering either space and time resolved absolute line-emissivity measurements for the purpose of determining local ion densities, or line spectral profile measurements for the purpose of determining ion temperatures or drift-velocities from the Doppler effect. In either case, strong isolated lines are desirable for accurate measurements, and the experimental difficulties increase toward shorter wavelengths. Natural boundaries for instrumental problems occur at ∼ 2000 Å where air (oxygen) becomes opaque, at ∼ 1100-1200 Å beyond which there are no practical windows, and more vaguely around 500 Å where essentially normal-incidence mirrors become inoperable, and around 10-20 Å beyond which crystal spectrometers or extreme grazing incidence gratings must be used. For simplicity we shall call these wavelength regions visible, Schumann range, vacuum uv, extreme uv, and soft X-ray, respectively.

Figure 4 shows the spectrum of a tokamak discharge taken with
a SPRED spectrometer[35] on the PDX tokamak.[36]   Although this is a
relatively low-temperature discharge, with $T_e(0) \sim 0.8$ keV, it shows
many of the characteristic features of tokamak spectra.   The
spectral resolution is about 1 Å throughout, the time integration
about 20 msec, and the intensity scale is approximately constant
and linear, except near the low wavelength cutoff ($\lambda \lesssim 120$ Å) where
it is not well established.   The spectrum clearly exhibits the
strong resonance lines of oxygen, carbon, titanium and iron ions
and a continuous background that increases rather strongly toward
shorter wavelengths.   This background consists partly of scattered
light from the grating surface (which can be in principle reduced
by appropriate filters) and partly of a large number of weak lines
from $n = 2 \rightarrow n > 2$ transitions in oxygen and carbon, and $n = 3 \rightarrow n
> 3$ transitions in iron, chromium, titanium etc.   A similar
concentration from the $n = 2 \rightarrow n > 2$ transitions from the latter
elements occurs around 10-15 Å.   Only a small fraction of the
observed background occurs at these wavelengths. Clearly, only very
strong line emissivities can be separately measured in such crowded
spectral regions, but the prospects improve rapidly at longer
wavelengths — except of course in the immediate neighborhoods of
the strong oxygen and carbon lines.

In order to deduce the radial distribution of the
emissivities, it is necessary to measure the line brightness at
different chords of the quasicylindrical plasma.   Even a cursory
look at the maze of coils on Fig. 1 makes it clear that this is
only feasible, over the entire width of the plasma, by means of
rotating mirrors, or, possibly by an array of identical instruments
viewing different chords.   The latter method is expensive, not just
because of the cost of instrumentation but also of data handling
capability and the manpower needed to keep the instruments in
proper working condition.   Nevertheless, this will probably be the
preferred method in the large tokamaks of the future, with severe
limitations on the number of discharges, and the access to the
device.

To summarize the description of tokamak discharges, we have a
toroidal plasma with electron densities roughly parabolic within
the limiter radius and peak densities about 5×10^{13} cm^{-3} within a
factor $\sim 2-3$.   The electron temperature distribution is narrower
(i.e., there is a hot plasma core surrounded by plasma with
slightly lower density but substantially lower temperature), with
central values ranging from 1-3 keV in the presently operating
devices and 5-10 keV anticipated in the near future.   The plasma is
mostly hydrogen, but contains small ($\sim 10^{-2}$) quantities of oxygen
and carbon and ($\sim 10^{-3}$) wall materials ($Z \sim 25$).   Practically any
other element may be temporarily added for diagnostic purposes, but
only by small quantities with the tolerated amount decreasing with

atomic number Z. The ions of the added element move fairly rapidly (several cm/msec) in radius, resulting in a more-or-less homogeneous radial distribution, with each ionization state appearing roughly at radii where $T_e(r) \approx E_i$. The electron temperature and density distributions are presumed known and steady over times large compared to ion radial transport time and relevant ionization and recombination times.

The problem is to measure the radial distributions of the various ionization stages of the appropriate element, and any other local plasma parameter, such as ion Doppler temperature, plasma rotation velocities, local density or temperature fluctuations etc., in the presence of both intrinsic plasma radiation and any additional radiation generated by the added diagnostic element itself.

CHARACTERISTICS OF HEAVIER ATOMS IN TOKAMAK PLASMAS

In this section we consider some important aspects of the physics of highly ionized atoms in the environment of tokamak plasmas. It is quite impossible to aim for anything resembling comprehensive treatment, so we present only some representative examples, for which we have chosen iron and zirconium. This choice is rather arbitrary, but is motivated by the facts that iron (Z = 26) has been more thoroughly studied than any neighboring element, because of its astrophysical importance, and zirconium (Z = 40) is near the upper end of the elements that are likely to be useful for tokamak diagnostics.

Figure 5 shows a partial energy level diagram of the doublet system of boronlike FeXXII (Z = 26, E_i = 1.8 keV), with the two lowest configurations (including the lowest quartet term) given on a ten times expanded scale at right. The known energy levels, and the corresponding electric dipole radiative transition probabilities (for Fe, Co, and Ni) have been recently compiled by Fuhr et al.[37] A very important feature of the energy levels, characteristic of the nearly-coulomb field of highly-ionized atoms, is the substantial energy gap between the near-ground n = 2 levels and the high n $\geq$ 3 levels. This implies that the strongest lines emitted by such ions tend to fall to two district wavelength regions — in the present case in the neighborhood of 10 Å, and upward of 100 Å, respectively. Furthermore, these wavelength ranges change only quite slowly from element to element along isoelectronic sequences, or from one ionization state to another within a given n-shell of the same element, in the sense that the relatively crowded wavelength regions of several neighboring ion species tend to overlap. A second important feature that will be extensively discussed below is the level structure of the ground configuration, which in the present example consists of only one doublet term, $2s^2 2p \ ^2P$, with a separation corresponding to 845.6 Å.

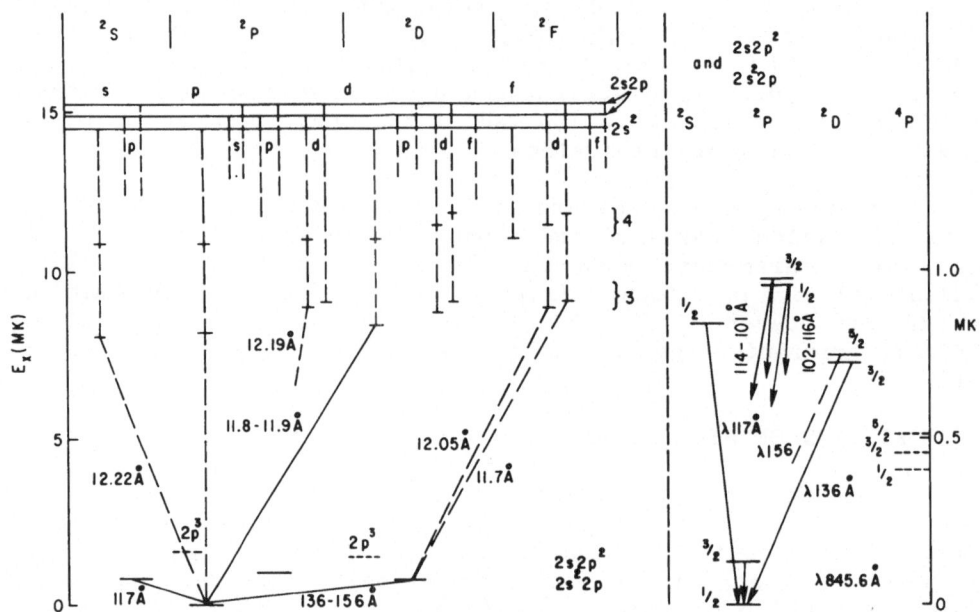

Fig. 5 Partial energy level diagram of the doublet system of Fe XXII, with the two lowest configurations on expanded scale at right, and representative wavelengths in Angstroms. The population of the $^2P_{3/2}$ level in the ground term proceeds to a large extent through cascading via the ^2P term of $2s2p^2$ configuration.

In Fig. 6 there are several representative electron collisional excitation rate coefficients (for the transitions indicated by heavy solid lines in Fig. 5) deduced from calculated cross-sections,[38-41] together with the ionization and recombination rates as in Fig. 3. The electron temperature range of principal interest is in the neighborhood of the ionization potential as indicated, and we assume an electron density of 5×10^{13} cm^{-3} for illustration — this implies an ionization rate ($n_e S_i$) about 10^3/sec and a recombination rate about a factor 2 smaller.

Of the two $2s^2 2p \; ^2P_{1/2} \rightarrow 2s^2 3d \; ^2D_{3/2,5/2}$ transitions shown, the first is representative of the strongest electric dipole transitions to the n = 3 levels, and is roughly comparable to the ionization rate. The other, involving $\Delta J = 2$ transition, is about 20 times weaker. The excitation rates to the n = 3 levels are thus in the range 10^2 to a few times 10^3 sec^{-1}.

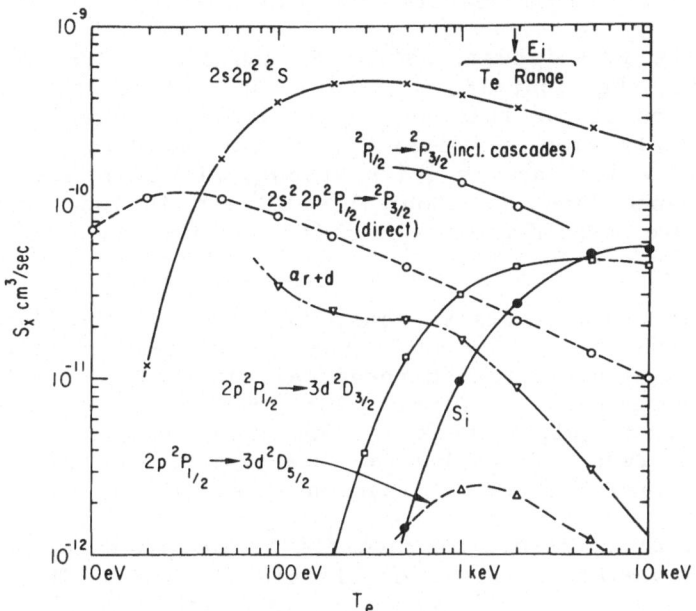

Fig. 6 Representative excitation rate coefficients for the $\Delta n = 1$, and $\Delta n = 0$ levels from the ground level of Fe XXII, compared to the ionization and recombination rates.

The excitation rates scale roughly as the (radiative) oscillator strengths, $f/E_x^{3/2}$, where E_x is the excitation potential. The rates to the $n = 4$ levels are thus about a factor 10 smaller than to $n = 3$, and the rates drop rapidly with increasing n. Excitation through recombination, which is necessarily smaller than the total recombination rate, clearly does not alter the excitation rate substantially, although it may make quantitative changes in the higher-n populations.

The excitation rate of the $2s^2 2p \rightarrow 2s 2p^2$ transitions is evidently much larger, or roughly in the 10^4–10^5/sec range. We also note two other features of these excitation rates: since $T_e \approx E_i \gg E_x$ the rates are relatively insensitive to the near-threshold uncertainties of the calculated cross-sections (unlike the $\Delta n > 1$ where $E_x \approx E_i$), and the rates are rather insensitive to T_e in the range of interest. It is thus these, strong $\Delta n = 0$ type transitions, that are particularly suitable for determining the local ion densities from measured emissivities.

Within the ground term, $2s^2 2p\ ^2P$, the excitation rate is also comparable to the ionization rate, but unlike the previously considered transitions, this one is strongly affected by radiative cascading (mostly from the $2s2p^2\ ^2P$ levels in this case) so that the total excitation rate is quite strong, within a factor 3 of the allowed resonance lines or about 5×10^3/sec in the present case. The temperature dependence is similar to the $\Delta n = 0$ resonance transitions.

Two other types of transitions, not shown on the figures, need to be mentioned. The $1s^2 \rightarrow 1snl$ type inner-shell excitation rate, because of its high excitation potential, is small, ~ 1-10/sec, and strongly temperature sensitive. The transitions to the quartet system are also fairly weak. The most important of these, $2s^2 2p\ ^2P_{1/2} \rightarrow 2s2p^2\ ^4P_{1/2}$ has a rate about 5×10^2/sec, i.e., comparable or less than the ionization rate.

The radiative transition rates[37,42] are generally much larger and have more variation than the collisional rates. Thus, from the $n = 3,4$ levels the typical radiative rates are 10^{11}-10^{13}/sec: from the excited $n = 2$ levels about 10^{10}/sec for the allowed electric dipole transitions and $\sim 10^7$-10^8/sec from the $2s2p^2\ ^4P$ levels. Thus, relative to collisional excitation rates, there are no "metastable" states above the ground configuration. Collisional excitation is immediately followed by radiative transitions. Doubly-excited states – even the $2p^3$ in the present case – are very weakly populated, unlike higher-density sources such as laser-plasmas or sparks. The excitation pattern is heavily dominated by direct single-electron excitations from the ground level, and precise radiative transition probabilities are generally of interest only for branching ratios in cascading. However, the levels of the ground configuration require detailed consideration. In the present case the $^2P_{3/2}$ level has a magnetic dipole radiative transition rate of 1.5×10^4 sec^{-1}, which is not very far from the estimated excitation rate of about 5×10^3 sec^{-1}.

We want to emphasize that the general qualitative features of the relative transition rates and the characteristics of emitted radiation do not change very markedly from one ionization stage to the next in a given shell, or between neighboring elements, i.e., comparing iron with chromium and nickel, or even titanium and copper, etc. The reason is that the energy levels scale roughly with the ionization potential, and so does the temperature at which the ion appears, and hence the ionization rates etc. There are of course important quantitative differences of detail, which we shall consider below. However, these general features of the emitted radiation for FeXXII (and neighboring ions) are as follows. There are a considerable number of relatively weak lines from the $n \geq 3$ levels with photon energies within about a factor 2 of the ionization potential, or about 8-13 Å, in the soft X-ray range. This is a very difficult spectral region for quantitative high-resolution measurements. Furthermore, the excitation rates are subject to near-threshold uncertainties in calculated excitation cross-sections, and (especially near the lower end of the relevant temperature range) the rates are fairly strongly temperature sensitive. These lines are therefore not, in general, very suitable for quantitative spectroscopic plasma diagnostics.

The near-ground, $\Delta n = 0$ transitions give rise to a much smaller number of strong resonance lines, in principle well suited for quantitative ion density determination from measured local emissivities. However, there are still considerable measurement difficulties. It needs to be borne in mind that the spectrometer looks through the entire plasma, and consequently all the ionization states simultaneously. Although the different ionization states differ somewhat in their space and time behavior, the distinction between neighboring states is not very great. Furthermore, the plasma generally contains more than one neighboring element (e.g., Ti, Cr, Fe) and the lines of different ionization states of comparable ionization potentials are quite similar in their space and time dependence and spectral region, as mentioned above. The allowable amount of addition of any element is severely limited, especially for heavier elements. And the relevant spectral region is still the technically difficult extreme ultraviolet. In practice it has been found that the very strong resonance lines of ns (i.e., Li, Na, Cu sequences) and ns^2 (Be, Mg, Zn sequences) configurations are easily detectable and very useful, that some of the stronger lines in ns^2np^x configurations are usable (esp. for x = 1 and 5) but anything more complicated is rather hopeless.

Finally, a few lines resulting from magnetic dipole transitions within the configuration appear at still longer wavelengths. Although these are weaker than the lines in the second group they appear in the relatively uncrowded spectral region, and the generally higher detection efficiency of the normal-incidence spectrometers more than compensates for this. These lines are therefore suitable for high-resolution measurements of Doppler widths and shifts of line profiles, or rapid optical scanning of spatial brightness distribution. However, their quantitative interpretation in terms of ion densities is more difficult than in the case of the $\Delta n = 0$ resonance lines, because the excitation process usually involves considerable cascading contributions. Also, when the radiative transition rates are comparable to collisional rates the emissivity is not a linear function of the local plasma density. Of course when the radiative transition rate is so small that (local) Boltzmann distribution is established by collisions, the interpretation is very simple, but then the emissivity is not very large.

Thus both the $\Delta n = 0$ resonance lines and the ground-configuration magnetic dipole lines are of great interest in the plasma diagnostics, and their characteristics are to a considerable extent complementary.

We now consider the diagnostic applicability of different elements to particular tokamak discharge conditions. We use for this purpose the observed criterion that the density of a given ion peaks at a radius where $T_e(r) \cong E_i$. (However, it is possible that in the large future tokamaks at higher temperature, hence larger electrical conductivity, the radial ion transport rate slows down. In this case the ion density peaks would appear at somewhat lower temperatures, closer to coronal ionization equilibrium.)

Figure 7 shows a hypothetical $T_e(r)$ profile with central temperature 6 keV, such as might be expected in the TFTR or JET tokamaks, and the location of the various ionization stages of iron and of zirconium according to the above criterion. Clearly zirconium, or neighboring elements would be appropriate for such plasma, since elements as light as iron would be stripped to at least that heliumlike state. Of course, iron in the heliumlike and hydrogenlike states at such temperatures would also offer interesting diagnostic possibilities, but they are quite different from the kind considered here.

In the gap between the 2p and 3s, zirconium would be mostly in the neonlike state, Zr XXXI, which is also a special case, as it has no low-lying energy levels. For the type of diagnostics under consideration, an intermediate element such as germanium or selenium would be appropriate at these radii.

A survey of the important energy levels of zirconium ions is given in Fig. 8. The ionization potentials are calculated values,[43] the excitation potentials are mostly extrapolated or interpolated from calculated values at the higher ionization stages, and from a few measurements at the lower ionization stages. Only a few points represent actual measurements, with references given below. Areas surrounded by dashed curves show probable locations of indicated levels. At right is the corresponding wavelength scale for transitions to the ground level [except the crosses, (x), which give wavelengths between some ground-configuration levels other than the ground level]. M, W, and A indicate wavelengths above which mirrors and windows are usable, and air becomes transparent, respectively.

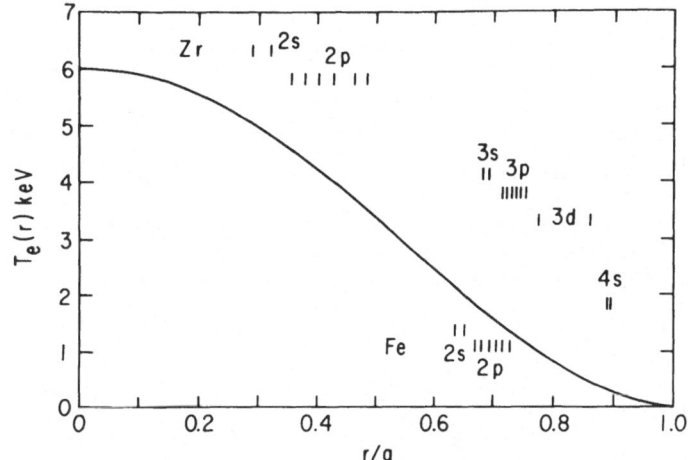

Fig. 7   Approximate radial locations of the peak densities of several zirconium and iron ions in a $T_e(0) = 6$ keV tokamak plasma, assigned on the basis of $E_i \approx T_e(r)$.

The principal features discussed above in connection with the FeXXII ion are clearly evident. The transitions with $\Delta n = 1$ from the ground level follow the ionization potentials within about a factor 2, and there is a substantial energy gap between these and the $\Delta n = 0$ levels (an exception is the nearly-filled 3d subshell, where 3p → 3d and 3d → 4p, 4f energies are comparable). And there is another group of transitions at considerably longer wavelengths between levels of the ground configurations, some of which are still well in the normal-incidence spectrometer range even in the 2p shell (of zirconium and neighboring elements).

As implied above, very little of the information in Fig. 8 is sufficiently quantitative for immediate use in tokamak diagnostics, and this applies to all elements heavier than nickel or copper. The first task for high-temperature diagnostics is thus to establish the wavelengths and energy levels, especially in the $n = 2$ shell, and then the appropriate transition probabilities, both collisional and radiative to allow interpretation of measured emissivities.

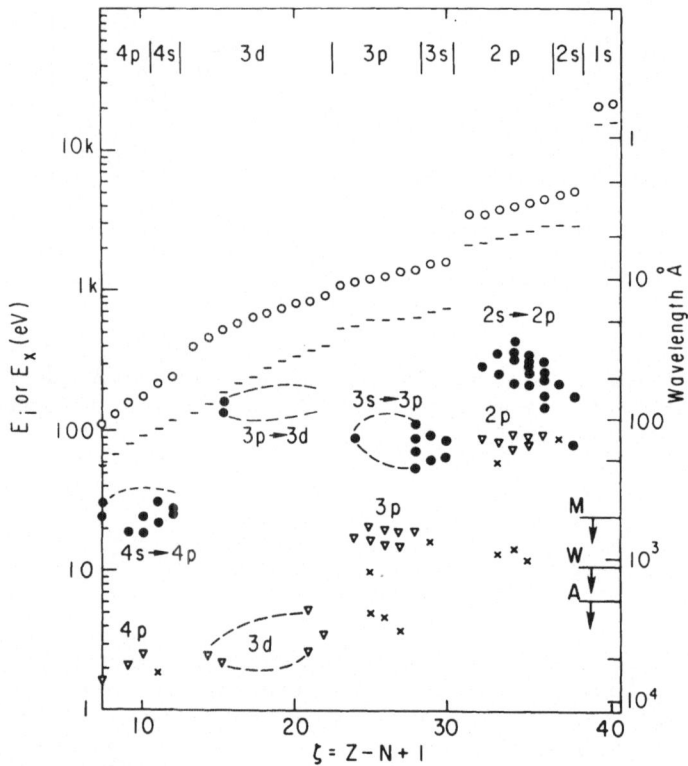

Fig. 8 Energy level scheme of zirconium ions, with ionization
potentials (O) indicating the approximate electron temperature
where the ions will be located. Note the substantial gap between
the lowest $\Delta n = 1$ levels (−), the $\Delta n = 0$ resonance transitions (●)
and the transitions within the ground configuration (∇). Crosses
(X) indicate photon energies of some transitions in the ground
configuration, although not to the lowest level, and M, W, A the
short wavelength limits of mirrors, windows, and air transmission.

Since the radiative efficiency of atoms in a high-temperature plasma increases fairly rapidly with Z, and hence their allowable concentration in tokamak discharges decreases, it seems reasonable for the present to regard molydenum (Z = 42) as a probable limit for likely plasma diagnostic use in the immediate future. This rather arbitrary choice is promoted on the one hand from the fact that molybdenum has been used as construction material in tokamaks, and as a result its spectra have already received considerable attention, and on the other hand because the vaunted long-wavelength advantages of near-ground transitions are gradually being lost in the n = 2 shell, as the fine-structure separations grow rather rapidly with Z. As a consequence, the wavelength range of the ground configuration magnetic dipole lines for Z > 40 begin to overlap the range of the $\Delta n = 0$ resonance lines, and of course, both groups move deeper into the extreme ultraviolet, with increasing Z.

AVAILABLE DATA AND FUTURE TASKS

We now consider the status of available atomic physics data concerning the low configurations of ions of elements up to molybdenum (Z = 42).

There are several recent comprehensive review papers by Edlen on the lithiumlike,[44] and berylliumlike[45] ions, and the $2s^2 2p^x$ configurations,[46] based on observations, and as already mentioned[42] a systematic multiconfiguration Dirac-Fock calculation of both energy levels and radiative transition probabilities. The latter is not sufficiently precise for wavelength determination, but it is very valuable for establishing trends and variations along sequences and, of course, for the transition probabilities.

The observational base in Edlén's papers is fairly adequate up to copper (Z = 29), although a number of gaps and inconsistencies still exist. Beyond that experimental data are very sparse, but from the established trends the data are extrapolated to krypton (Z = 36). In Fig. 9 we have further extrapolated the data to zirconium, using Cheng et al.[42] as a guide.

There are three predicted lines at wavelengths >500 Å (those are the transitions indicated by crosses in Fig. 8), that would be appropriate e.g., for Doppler temperature measurements. However, the 842 Å line of Zr XXXIV and especially the 917 Å line of Zr XXXIII will probably be too weak for this at allowed zirconium densities in tokamaks because the expected radiative branchings favor strongly the short wavelength component. Nevertheless, they may be adequate for radial distribution measurements, and a measured branching ratio would be also useful for intensity calibration of extreme uv spectrometers. The 1052 Å line of Zr XXXV on the other hand has no competing radiative transition and

should be highly useful for more detailed plasma diagnostics.
These qualitative comments of course apply to corresponding
transitions in other elements along the isoelectronic sequences.[10]

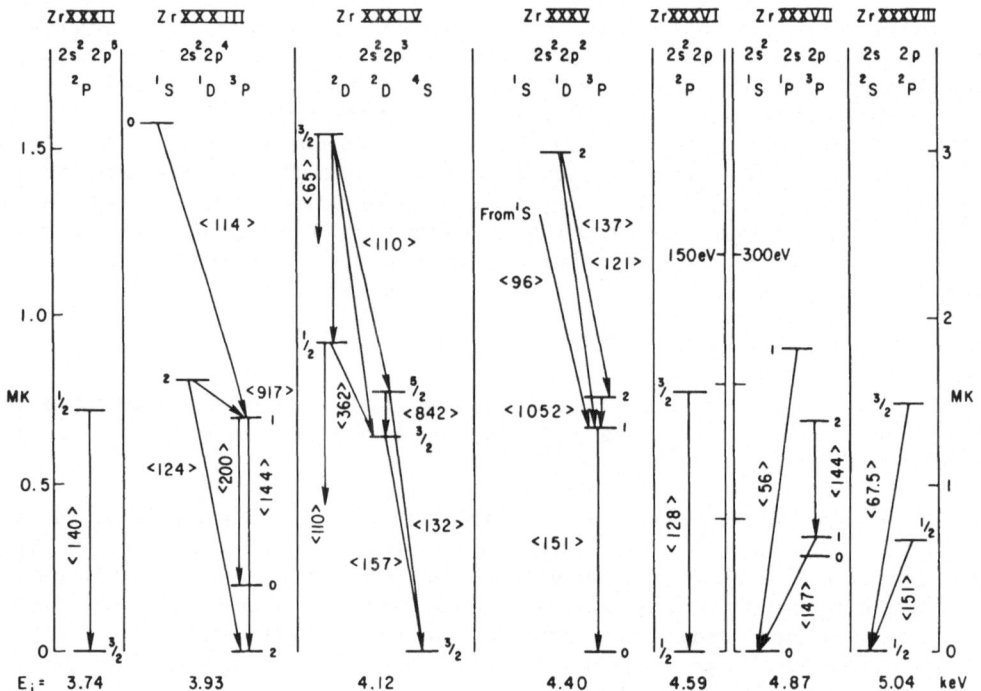

Fig. 9 Energy levels of the lowest configurations of n = 2 shell
in zirconium ions, with predicted wavelengths in Ångstroms.

A striking feature of the wavelengths in Fig. 9 is that many
of the lines occur in a quite narrow wavelength range, i.e., ∼ 140–
160 Å. (These lines would also be roughly comparable in intensity,
provided that the electron temperature is sufficient to reach the
lithiumlike states.)   This would make it worthwhile to consider
developing a specialized grazing-incidence spectrometer in such a
small wavelength range, emphasizing high resolution and spectral
purity, for the purpose of line profile measurements in the
presence of potentially interfering lines due to other elements in
the plasma.   The line profiles would presumably be measured
simultaneously, with a multichannel detector, yielding a radial ion
temperature measurement over a range such as that shown in
Fig. 7.   This feature, mostly due to the 2p spin-orbit interaction,
is not unique to zirconium, but recurs, with quantitative changes,
in other elements.

There is no immediate prospect of establishing any of these wavelengths experimentally, as the electron temperatures required to reach these ionization states are not yet available in tokamaks. However, such ionization states can be reached[47,48] by means of plasmas produced by powerful lasers. At the high densities and short duration of the laser produced plasmas the magnetic dipole lines are not directly observable, but the $\Delta n = 0$ type electric dipole resonance lines – which, as discussed above, are also of direct tokamak interest on their own – allow the determination of the energy levels, and hence more reliable predictions of the expected wavelengths. This type of experiment should be strongly encouraged, because the discharge of a giant tokamak, whatever its other virtues may be, is not a particularly adept light source for the search of an unknown line, even after such discharges will become available.

However, electron temperatures in the 2–3 keV range can be produced in the presently operating tokamaks such as the PLT, and this will allow at least partial investigation of the $n = 2$ shell wavelengths in the range $Z \lesssim 36$, which measurements can then be used for improving the isoelectronic extrapolations. Likewise, the $n = 3$ and higher-n ionization stages of heavier elements are readily produced in such discharges. These ionization states are useful not only for the lower-temperature regions of the future tokamaks as pointed out above, but also for the diagnostics of the present tokamaks. It may be of great interest to study the ion transport of elements of significantly different mass in the same plasma location, e.g. the $n = 2$ shells of scandium or titanium, and the $n = 3$ shells of zirconium or molybdenum. And of course the heavier elements may have particularly appropriate wavelengths for the available equipment.

Figure 10 shows the zirconium energy levels of the $n = 3$ shell corresponding to the $n = 2$ levels of Fig. 9. The ionization potentials indicate that these ions are of diagnostic interest in the 1–2 keV range. The wavelengths in boxes have been observed in the PLT discharges, the others are again predictions based on isoelectronic extrapolations. The assignment of the 742 Å line to Zr XXIX is regarded as tentative, because the corresponding transition in any neighboring element has not yet been observed.

Some of the 3d and $n = 4$ shell wavelengths of actual or potential diagnostic interest of the lower temperature periphery of tokamak discharges are given in Fig. 11, again with the PLT observations shown in boxes. Of particular interest is the appearance of visible lines, i.e., lines transmitted by air, of several of these ions (and corresponding transitions in molybdenum[28]). In spite of their small transition probabilities their emissivities are readily measurable, except in the case of accidental proximity to some oxygen or carbon lines. The low

(100–300 sec^{-1}) radiative transition probabilities imply that the relative populations of these levels are close to their statistical weights, so the measured emissivities are also interpretable as ion densities without very detailed knowledge of all the collisional rate coefficients populating the levels.

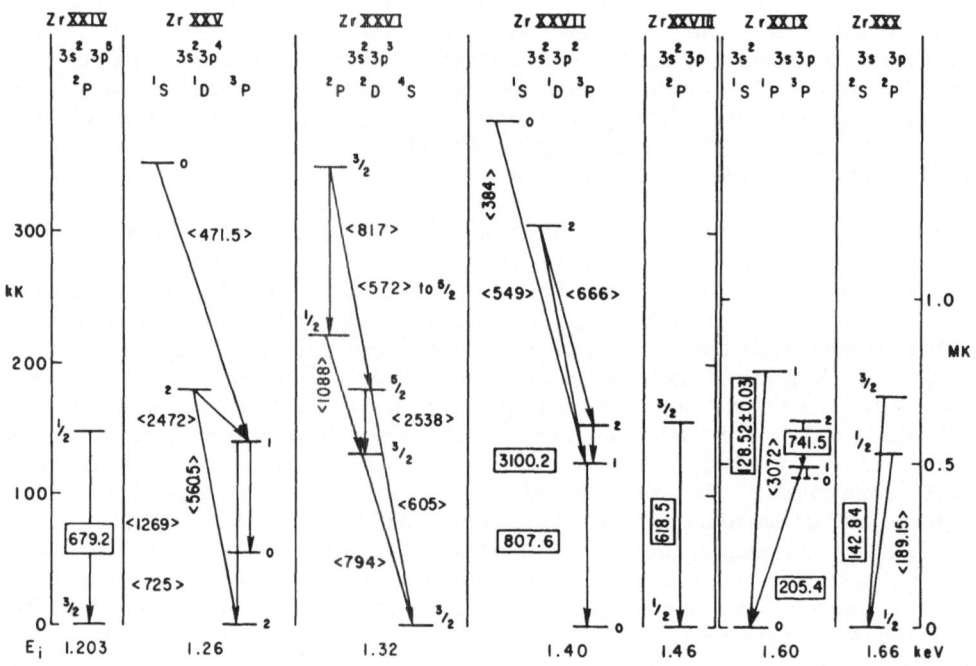

Fig. 10   Energy levels of the lowest configurations of the 3s and 3p subshells of zirconium ions with predicted and observed (in boxes) wavelengths.

The states Zr XVI through Zr XX with the ground configurations $3d^7$ through $3d^3$ have more intricate near-ground level structure, and the prospect of finding suitably strong lines there appears more remote. There do not seem to be any experimental data for these configurations at present, in the Z-range of interest, and at lower-Z elements the radiative transition probabilities are too weak for adequate emissivities.

It is very unlikely that anything approaching adequate collisional excitation rate-coefficient structure could be established for the $3p^6 3d^x$ configurations or even most of the $3s^2 3p^x$. Fortunately this is not absolutely essential for significant interpretation of measured emissivities. If the absolute ion densities can be established for some of the states, such as for example the copper and zinc sequence (4s and $4s^2$) and perhaps some adjacent states such as $4s^2 4p$, and again in the sodium and magnesium sequences (3s and $3s^2$) and the adjacent $3s^2 3p$ and perhaps $2s^2 2p^5$ – all of which do not appear too forbidding – then measured relative (radial) profiles of the intervening states should provide sufficent constraints to reconstruct the entire distribution, with the help of some modelling calculations. Such a procedure would hopefully result in a empirical system of effective line-emission coefficents, presumably as a function of plasma density, and scaling smoothly along isoelectronic sequences for corresponding transitions.

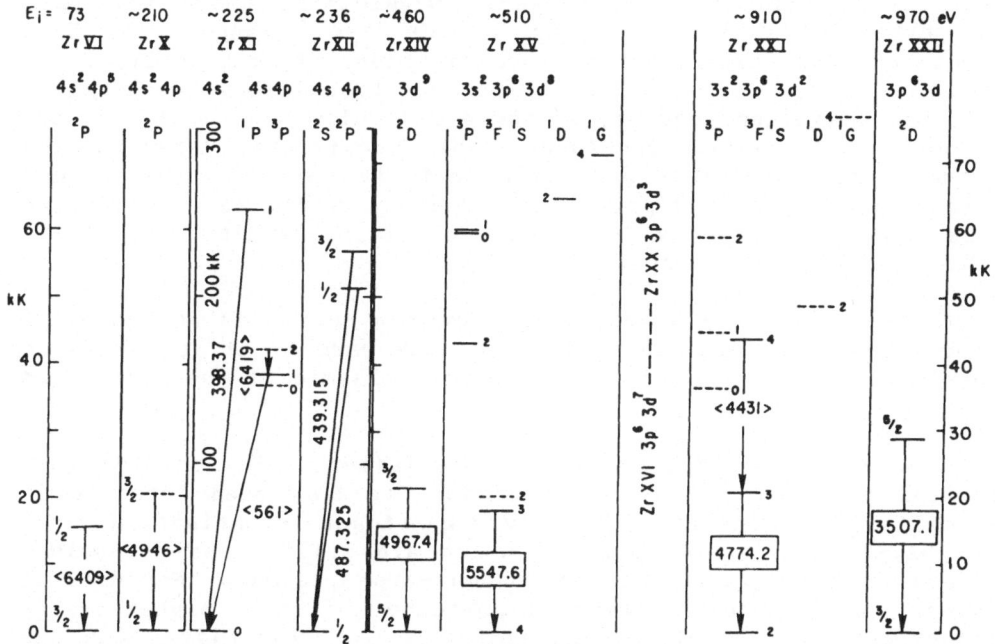

Fig. 11 Energy levels and wavelengths of some 3d, 4s, and 4p subshells of zirconium ions, that are useful for diagnostics in the low-temperature periphery of tokamak plasmas.

For radiative transitions the important range at tokamak densities is about 10^3–10^5 sec^{-1}. Anything slower than this would be negligible for affecting population distributions, and anything much faster is effectively instantaneous. Only branching ratios which may affect cascading contributions significantly - need to be known accurately for the rapid transitions, and the approximate magnitude - which determines the brightness of the line for transitions between the levels at or near statistical (Boltzmann) distributions - for the slow transitions. In this regard, magnetic dipole transitions in the $3d^x$ configurations become too weak for elements with $Z \lesssim 36$, and in the $3p^x$ configurations for elements with $Z \lesssim 30$.

The essential directions of research for the immediate future thus appear as follows: 1) Experimental determination of the ground-configuration energy levels and wavelengths for any of the transitions shown in Figs. 9-11 for all elements up to $Z \approx 42$. 2) Investigation of the $3d^3-3d^7$ ground configurations for suitable magnetic dipole lines in elements with $Z \gtrsim 36$. 3) Theoretical calculations of the corresponding radiative transition probabilities (aided by experimental wavelengths where feasible) with particular emphasis in the 10^3-10^5 sec^{-1} range. 4) Both theoretical and experimental investigation of the collisional excitation cross-sections of the resonance transitions in the zinc, magnesium, beryllium, copper, sodium and lithium sequences, and the neighboring np, $(n-1)p^5$ sequences. 5) Experimental determination of the relationship between the emissivities of the ground configuration magnetic dipole lines, and any of the $\Delta n = 0$ electric dipole resonance lines, as a function of plasma density, especially in the $10^{12}-10^{15}$ cm^{-3} range.   .

This is of course a very large program and it is not going to be completed any time soon. However, any part of it is likely to be useful in the diagnosis of the tokamak plasmas of the future, as well as present.

At the end we must include a disclaimer. The outline just presented does not in any way diminish the continued interest in the collisional ionization, and any significant recombination rates of highly-ionized atoms, both of which are still clearly unsolved problems of major importance in tokamak-type plasmas. Nor does it infringe on the prospective importance of the higher-energy diagnostics, such as the ls-2l excitation of heliumlike ions and inner-shell transitions - even if they propose to measure the same plasma property, e.g. ion transport rate or Doppler temperature. Because of the inability of existing plasma theory to describe or predict the behavior of actual tokamak plasmas, any redundancy in localized spectroscopic measurements of plasma parameters must be considered highly desirable.

ACKNOWLEDGMENTS

For the experimental data presented in this paper the author is indebted to the PLT operators, the PPL surface physics group, particularly Dr. S. Cohen, and members and visitors of the spectroscopy and atomic physics group, particularly Drs. S. Suckewer, R. Fonck, K. Sato, M. Finkenthal, and B. Denne.

This work has been supported by United States DOE Contract No. DE-AC02-76-CHO-3073

REFERENCES

1. H.P. Furth, Nucl. Fusion 15, 487 (1975).
2. Diagnostics for Fusion Experiments, E. Sindoni and C. Wharton, editors, Pergamon Press, Oxford, 1979.
3. C. deMichelis and M. Mattioli, Nuclear Fusion 21 677 (1981).
4. D.E. Post, R.V. Jensen, C.B. Tarter, W.H. Grasberger, and W.A. Lokke, Atomic Data and Nuclear Data Tables 20 397 (1977).
5. R.A. Hulse, D.E. Post, and D.R. Mikkelsen, J. Phys. B 13 3895 (1980).
6. S. Suckewer, E. Hinnov, D. Hwang, J. Schivell, G.L. Schmidt et al., Nuclear Fusion 21 981 (1981).
7. E. Hinnov, J. Hosea, H. Hsuan, F. Jobes, E. Meservey, G. Schmidt, and S. Suckewer, Nuclear Fusion 22, 325 (1982).
8. R. Isler, E.C. Crume, and H.C. Howe, Nuclear Fusion 19 787 (1979).
9. E. Hinnov, in Atomic and Molecular Processes in Controlled Thermonuclear Fusion, M.R.C. McDowell and A.M. Ferendeci, editors, Plenum Publishing Corp., New York, 1980 pp 449-470.
10. E. Hinnov, Nuclear Instruments and Methods (1982) to be published. (Proc. VI Intern. Conf. on Fast Ion Beam Spectroscopy, Quebec, Aug. 1981.)
11. D. Dimock, ref. 2 p.271.
12. S. von Goeler, ref. 2, p.79.
13. P.C. Efthimion, V. Arunasalam, and J.C. Hosea Phys. Rev. Lett. 44 396 (1980).
14. D. Dimock, H. Eubank, E. Hinnov, L. Johnson, and E. Meservey Nucl. Fusion 13 271 (1973).
15. E. Hinnov, S. Suckewer, K. Bol, R. Hawryluk, J. Hosea, and E. Meservey, Plasma Physics 20 723 (1978).
16. Equipe TFR, Nucl. Fusion 15 1053 (1975).
17. E. Hinnov, ref. 2, p.139.
18. J.L. Terry, K.C. Chen, H.W. Moos, and E.S. Marmar, Nucl. Fusion 18 485 (1978).
19. S. Cohen et al., J. Vac. Sci. Technol. 20 1226 (1982).
20. L.B. Golden and D.H. Sampson, J. Phys. B 10 2229 (1977).
21. S.M. Younger, J. Quant. Spectr. Rad. Transfer 27 541 (1982).
22. A.L. Merts, R.D. Cowan, and N.H. Magee Jr., LASL Report LA-6220-MS, Los Alamos N.M. (1976).
23. K. Brau, S. von Goeler, M. Bitter et al., Phys. Rev. A 22 2769 (1980).

24. R. Fonck, M. Finkenthal, R. Goldston, D. Herndon, R. Hulse,
     R. Kaita, and D. Meyerhofer, subm. Phys. Rev. Lett., 1982.
25. E.S. Marmar, J.L. Cecchi, and S.A. Cohen Rev. Sci. Instr. 46
     1149 (1975).
26. E.S. Marmar, J.E. Rice, and S.L. Allen, Phys. Rev. Lett. 45
     2025 (1980).
27. S. Suckewer, J. Cecchi, S. Cohen, R. Fonck, and E. Hinnov,
     Physics Letters 80A, 259 (1980).
28. S. Suckewer, E. Hinnov, S. Cohen, M. Finkenthal, and K. Sato,
     Phys. Rev. 1982 in press.
29. S. von Goeler, private communication (1982).
30. H. Eubank et al., Phys. Rev. Lett. 43 270 (1979).
31. J. Hosea et al., Plasma Physics and Controlled Nuclear Fusion
     Research (Proc. 8th International Conf. Brussels 1980) IAEA
     Viena 1981, Vol. II, p.95.
32. E. Hinnov, S. Suckewer, M. Bitter, R. Hulse, and D. Post,
     Phys. Rev. A22 725 (1980).
33. R. Isler, Phys. Rev. Lett. 38 1359 (1977).
34. V.V. Afrosimov, Yu. S. Gordeev, A.N. Zinovev, and A.A.
     Korotkov Sv. J. Plasma Phys. 5 551 (1979).
35. R.J. Fonck, A.T. Ramsey, and R.V. Yelle, Appl. Optics, 1982
     in press.
36. D. Meade et al., Plasma Physics and Controlled Nuclear Fusion
     Research (Proc. 8th Intern. Conf., Brussels 1980) IAEA,
     Vienna 1981 Vol. I., p.665.
37. J.R. Fuhr, G.A. Martin, W.L. Wiese, and S.M. Younger, J.
     Phys. Chem. Reference Data 10 305-565 (1981).
38. H.E. Mason and P.J. Storey, M.N. R.A.S. 191 631 (1981).
39. U. Feldman, G.A. Doschek, C.C. Cheng, and A.K. Bhatia, J.
     Appl. Phys. 51 190 (1980).
40. A.L. Merts, J.B. Mann, W.D. Robb, and N. H. Magee, Jr.
     "Electron Excitation Collision Strengths: A Collection of
     theoretical Data," Los Alamos Scientific Laboratory report
     LA-8267-MS (1980).
41. R.J.W. Henry, Phys. Rep. 68 3, (1981).
42. K.T. Cheng, Y.K. Kim, and J.P. Desclaux, Atomic Data and
     Nuclear Data Tables 24, 111 (1979).
43. T.A. Carlson, C.W. Nestor, N. Wasserman, and J.D. McDowell,
     Atomic Data 2 63 (1970).
44. B. Edlén, Physica Scripta 19 255 (1979).
45. B. Edlén, Physica Scripta 20 129 (1979).
46. B. Edlen, Physica Scripta 1982, to be published, also Physica
     Scripta 22 593 (1980).
47. J. Reader and G. Luther, Physica Scripta 27 732 (1981).
48. J. Reader, $2s^2 2p^5 - 2s 2p^6$ Transitions in Sr^{29+} and Y^{30+} to be
     published.

RELATIVITY IN ELECTRONIC STRUCTURE : HOW AND WHY?

J.P. Desclaux

Centre d'Etudes Nucléaires de Grenoble
DRF/Laboratoire d'Interactions Hyperfines
85X - 38041 Grenoble Cedex, France

INTRODUCTION

Interest in theoretical studies of highly ionized atoms has increased in the last few years mainly because of the importance of such ions in plasma diagnostics, astrophysics and laboratory beam-foil measurements. We will hear about many such topics during this institute. On the theoretical side the strong interest in these systems originates from the growing importance of relativistic effects with the increase of the nuclear charge.

The incidence of relativistic corrections is easily predictable if we start with the following simple remarks : the eigenvalues of the Schrödinger equation are directly proportional to the mass of the particle. The spatial extent of the wave function as measured by the mean value of the position operator r is inversely proportional to the mass. The mass of an electron moving at velocity v is increased with respect to its rest mass m_0 to the value

$$m = m_0 / \left[1 - (v/c)^2 \right]^{1/2} \tag{1}$$

where c is the speed of light. We can now immediately deduce that the introduction of the relativistic kinematics will result in an increase of the electron binding energies along with a contraction of the charge distributions towards the nucleus. Obviously the size of these corrections will depend upon the value of $(v/c)^2$ and if we adopt $(v/c)^2 \geq 0.1$ as the criterion for significant relativistic effects, it is trivial, with the help of the virial theorem, to show that :

$$\langle v^2/c^2 \rangle = -2E_{nr}/c^2 = (Z/nc)^2 \qquad\qquad (2)$$

where E_{nr} is the non relativistic energy and n the principal quantum number.(Note that we shall use atomic units throughout, so that
m=ƛ=e=1, $c=\alpha^{-1}$=137. The unit of energy is then 27.2eV and that of
length is the hydrogen Bohr radius). Thus our criterion tells us
that relativistic effects will be important if $Z/n \geqslant 40$. Let us
immediately point out that more localized effects are expected when
the electron comes close to the nucleus where, because of the strong
attractive potential, it will acquire a large velocity. These effects
cannot be deduced from the above average considerations. If we
neglect the electron energy compared to the nuclear energy potential,
our criterion gives approximately $10^{-3}Z$ (a.u.) as the critical
distance of approach. Since low angular momentum orbitals are more
penetrating, these effects will exhibit a strong dependence on the
value of this angular momentum.

In addition to these two direct effects there exists a third
indirect one for a N electron system : when performing a self-
consistent field (SCF) calculation, the contraction of the inner
shell orbitals will screen more effectively the nuclear charge for
the outer electrons. As a consequence the valence electrons will
tend to be less localized. This relativistic SCF expansion will
compete with the direct effects and the net result cannot be predicted a priori. It turns out that a strong dependence on the angular momentum is found again. To extract the relative contributions
of the direct and indirect effects one can proceed in the following
way[1] : a full non-relativistic SCF calculation is first performed.
Then the core electron charge distributions are frozen and relativistic dynamics is used to recalculate the valence electrons in the
field of the frozen non-relativistic core. The inverse process can
be done by starting first with a relativistic SCF. Table 1 illustrates the results reported in Ref. 1. Comparison of columns
1 (R,R) and 2 (NR,NR) gives the sum of the direct and indirect
effects while their individual contributions are obtained as the
difference between columns 2 and 4 (NR,R) and between columns 2 and
3 (R,NR) respectively. For the 6s valence orbital of gold the two
calculations with relativistic kinematics (columns 1 and 4) are quite
similar and the SCF expansion is of minor importance as shown by
the two calculations with non-relativistic kinematics (columns 2
and 3). Thus in the present case the behaviour of the 6s orbital is
dominated by the direct effect. The opposite situation is found for
the 5d orbital of lutetium. For this non-penetrating orbital the
relativistic contraction of the core is far more important than the
direct effect even if this latter one is not completely negligible.

These subtleties in the modifications induced by relativity
demonstrate clearly the need for a non-perturbative treatment.
Furthermore besides the changes in electron energy levels and charge
distributions, spin-orbit coupling will cause the Russel-Saunders

Table 1. Valence one electron energies (ε) and mean
radius ($<r>$) in a.u. as given in reference 1

Potential (Core)		R	NR	R	NR
Kinematics (Valence)		R	NR	NR	R
Au 6s	ε	0.584	0.442	0.454	0.586
(Z=79)	$<r>$	3.06	3.70	3.65	3.02
Lu 5d	ε	0.376	0.487	0.352	0.520
(Z=71)	$<r>$	2.73	2.48	2.87	2.40

R stands for relativistic and NR for non-relativistic.
For the relativistic results of the 6d the average value
$\left[2(d_{3/2}) + 3(d_{5/2})\right] / 5$ is used.

coupling scheme to be less and less appropriate as the nuclear
charge increases. Also electron correlation will remain important
for ions essentially for low stages of ionization or for systems with
more than one or two valence electrons. If for few electron systems
it is possible to obtain accurate results by using a double series
expansion in Z^{-1} and $(\alpha Z)^2$, for many electron systems rather sophisticated methods are required. These methods will be described in the
first part of these lectures before discussing results. But as a
warm-up exercise it may be useful to summarize the properties of
the Dirac one-electron equation.

1. DIRAC EQUATION

In order to satisfy the requirements of the special theory of
relativity, the time and space variables should appear on an equal
footing in the wave equation. It would thus appear natural to start
from the relativistic relation between energy and momentum :

$$E = \left[c^2 p^2 + m^2 c^4\right]^{1/2} \tag{3}$$

and use the quantum mechanical substitutions :

$$E \rightarrow -i\partial/\partial t \qquad \vec{p} \rightarrow -i\nabla \tag{4}$$

to derive the relativistic analogue of the Schrödinger equation.
In doing so one obtains :

$$-i \frac{\partial \psi}{\partial t} = \left[-c^2 \nabla^2 + m^2 c^4 \right]^{1/2} \psi \tag{5}$$

where the square root operator is rather troublesome to handle. To avoid this difficulty, a better solution is first to square both sides of Eq. 3 and then replace E and p by the corresponding opera-tors to obtain :

$$\frac{\partial^2 \psi}{\partial t^2} = \left[c^2 \nabla^2 - m^2 c^4 \right] \psi \tag{6}$$

This equation, called the Klein-Gordon equation, is obviously invariant under a Lorentz transformation but was found at the beginning to have some undesirable feature connected with the fact that it is a second order equation in the time (its validity has been reestabli-shed as the equation describing spin-zero particles). To obtain a first order equation in the time and consequently in the space as well, Dirac tried to extract the square root of Eq. 3 by writing :

$$E = c\vec{\alpha}.\vec{p} + \alpha_4 mc^2 \tag{7}$$

and showed that it is indeed possible if α_x, α_y, α_z and α_4 are not numbers but 4x4 matrices. A convenient choice for these matrices is :

$$\vec{\alpha} = \begin{pmatrix} 0 & \vec{\sigma} \\ \vec{\sigma} & 0 \end{pmatrix} \qquad \alpha_4 = \begin{pmatrix} I & 0 \\ 0 & -I \end{pmatrix} \tag{8}$$

where the components of $\vec{\sigma}$ are the usual Pauli matrices and 0 and I the second order zero and unit matrices respectively. Subtracting the rest mass energy, the Dirac's equation for an electron moving in a central field V(r) is thus :

$$i \frac{\partial \psi}{\partial t} = H\psi = \left[c\vec{\alpha}.\vec{p} + \beta c^2 + V(r) \right] \psi, \qquad \beta = \alpha_4 - 1 \tag{9}$$

In the above expression and in what follows the fourth order unit matrix should be inserted in an obvious way when necessary. It can be shown that the eigensolutions of the Dirac equation are simul-taneous eigenfunctions of H, j^2, j_z and $\alpha_4 K$ (to be defined below) and can be written :

$$\psi = \frac{1}{r} \begin{bmatrix} P_{n,\kappa}(r) \; \chi_{\kappa,m}(\theta,\phi) \\ iQ_{n,\kappa}(r) \; \chi_{-\kappa,m}(\theta,\phi) \end{bmatrix} \tag{10}$$

where n is the principal quantum number and $\chi_{\kappa,m}$ are two component spinors :

$$\chi_{\kappa,m} = \sum_{\sigma=\pm 1/2} \langle \ell, m-\sigma, 1/2, \sigma | j, m \rangle Y_\ell^{m-\sigma} \phi^\sigma \tag{11}$$

simultaneous eigenfunctions of $\vec{\ell}^2$, $\vec{s}^2$, $\vec{j}^2$ and j_z. The operator K
mentioned above is defined by :

$$K = \vec{\ell}^2 + \vec{s}^2 - \vec{j}^2 - 1 = -(1 + \vec{\sigma}.\vec{\ell}), \quad \vec{s} = \frac{1}{2} \vec{\sigma} \tag{12}$$

so that :

$$K \chi_{\kappa,m} = \kappa \chi_{\kappa,m} \text{ with } \kappa = -(j+1/2)a \text{ if } j = \ell + a/2 \tag{13}$$

The values of the various quantum numbers and the labels most commonly found in the literature are listed in Table II.

From the expression of the Dirac orbitals (Eqs. 10 and 11) and
the properties of the spherical harmonics, we find that $\alpha_4 \pi$ is the
Dirac equivalent of the parity π . As this operator also commutes
with the Hamiltonian it can be used together with $\vec{j}^2$ and j_z to clas-
sify the eigenstates. Another equivalent possibility, as can be seen
from Eq. 13 and the values given in table II is to use K and j_z.

To obtain the radial equations fulfilled by P and Q, we invoke
the relation :

$$(\vec{\sigma}.\vec{p}) \left[\frac{F(r)}{r} \chi_{\kappa,m}\right] = \frac{i}{r} \left[\frac{dF}{dr} + \frac{\kappa F}{r}\right]\chi_{-\kappa,m} \tag{14}$$

to get

$$\frac{dP}{dr} + \frac{\kappa P}{r} = \left[2c + \frac{1}{c}\{\varepsilon + V(r)\}\right]Q$$
$$\frac{dQ}{dr} - \frac{\kappa Q}{r} = -\frac{1}{c}\left[\varepsilon + V(r)\right] P \tag{15}$$

Table II
Quantum numbers and orbital labels

κ	-1	$+1$	-2	$+2$	-3	$+3$	-4	....
j	1/2	1/2	3/2	3/2	5/2	5/2	7/2	....
ℓ	0	1	1	2	2	3	3	....
a	+1	−1	+1	−1	+1	−1	+1	....
Labels	$s_{1/2}$	$p_{1/2}$	$p_{3/2}$	$d_{3/2}$	$d_{5/2}$	$f_{5/2}$	$f_{7/2}$	....
	s	$\bar{p}$	p	$\bar{d}$	d	$\bar{f}$	f	....
	s	p*	p	d*	d	f*	f	....

The sign convention in these equations is a matter of convenience.
It depends on the way that orbital and spin angular momenta are

coupled (Eq. 11) to obtain the total angular momentum $(\vec{j} = \vec{l} + \vec{s})$
and on the position of the factor i (Eq. 10) introduced to have
both P and Q as real functions. If the convention is of no matter as
long as internal consistency is maintained in all formulas, the
various choices between different authors may be quite troublesome.

Elimination of Q between the two equations (15) gives :

$$\frac{d^2P}{dr^2} = \left[\frac{\kappa(\kappa+1)}{r^2} - 2(\varepsilon + V)\right]P + 0(\frac{1}{c^2}) \tag{16}$$

which because of the identity :

$$\kappa(\kappa + 1) = \ell(\ell + 1) \tag{17}$$

is just the Schrödinger equation at the formal limit $c \to \infty$. Thus
the so called large component P(r) tends to the non-relativistic
radial amplitude while to lowest order the small component is given
by :

$$c\ Q(r) = \frac{1}{2}\left[\frac{dP}{dr} + \frac{\kappa P}{r}\right] + 0(\frac{1}{c^2}) \tag{18}$$

In fact the small component may be locally the largest one. For
$r \to 0$ it can be shown that :

$$Q/P = c(\lambda + \kappa)/Z \text{ with } \lambda = \left[\kappa^2 - Z^2/c^2\right]^{1/2} \tag{19}$$

which for positive valves of κ is larger than one. This reflects
the fact that in the present case the angular momentum of the small
component is 1-1 when that of the large is 1 (the small component
of a $p_{1/2}$ orbital is of s character). On the other hand for $r \to \infty$

$$Q/P \cong \left[\varepsilon/(2c^2 - \varepsilon)\right]^{1/2} \cong Z/2nc \tag{20}$$

which is lower than unity but certainly not small for high   atomic
numbers. Using series expansions near the nucleus for the radial
components both of them behave like r^λ with λ given in Eq. 19.
Thus near the origin, the Dirac orbitals for $|\kappa| = 1$ diverge like
$r^{\lambda-1}$. This singularity is an artifact due to the unphysical pure
Coulomb potential, when the finite extent of the nuclear charge
distribution is taken into account the divergence is removed.

The energy levels of the bound states are given by the
Sommerfeld formula :

$$\begin{aligned}
\varepsilon_{n\kappa} &= c^2\{\left[1 - (\alpha Z)^2/N^2\right]^{1/2} - 1\} \\
&= \frac{Z^2}{N^2}\{\left[1 - (\alpha Z)^2/N^2\right]^{1/2} + 1\}^{-1}
\end{aligned} \tag{21}$$

where the apparent principal quantum number N is related to the principal quantum number n by :

$$N = \left[n^2 - 2(n - |\kappa|)(|\kappa| - \lambda) \right]^{1/2} \leq n \tag{22}$$

Evidently $\varepsilon_{NK} < -Z^2/2n^2$. Expanding (21) in powers of αZ we obtain :

$$\varepsilon_{n\kappa} = - \frac{Z^2}{2n^2} \left[1 + \frac{(\alpha Z)^2}{n} (\frac{1}{|\kappa|} - \frac{3}{4n}) + 0(\alpha Z)^4 \right] \tag{23}$$

where the first term is the energy eigenvalue of the Schrödinger equation and the second term is exactly the correction obtained in the Pauli approximation. Thus the difference between the familiar fine structure formula (note that $|\kappa| = j + 1/2$) and the exact Dirac expression is of order $\alpha^4 Z^6$. On the other hand radiative corrections which are not included in the Dirac theory are of order $\alpha^3 Z^4 \log\alpha$ and for small values of the nuclear charge are more important since $(\alpha Z)^2 < \alpha \log\alpha$.

The first equation of (21) defined the apparent quantum number N as a function of the energy given by :

$$N = \alpha Z \left[1 - (1 + \varepsilon/c^2)^2 \right]^{-1/2} \tag{24}$$

which is a complex number for $\varepsilon > 0$ or $\varepsilon < -2c^2$. Consequently the discrete spectrum covers the energy range $-2c^2 < \varepsilon < 0$. For $\varepsilon > 0$ one finds a continuous spectrum and the corresponding wave functions are oscillating at large distance from the origin. Up to now everything is in complete analogy with the non-relativistic case, but it appears that the Dirac equation exhibits a second continuous spectrum for $\varepsilon < -2c^2$ as illustrated in Fig. 1. From the beginning, the existence of these "negative" energy levels was the source of problems. As a parenthesis it is rather funny to notice that, in all relativistic theories, these negative levels are always present when remembering that one of the early motives for rejecting the Klein-Gordon equation was precisely to exclude thes levels. In the one-electron case, Dirac overcomes the difficulty by postulating that the vacuum state must be the one in which all these negative levels are filled according to the Pauli exclusion principle. Reinterpretation of holes in the negative continuum as positrons (and that before their experimental discovery) was later considered the most obvious justification for the validity of the Dirac theory.

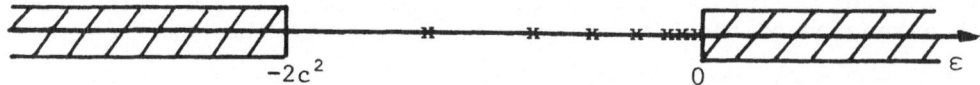

Fig. 1 Eigenvalue spectrum of the Coulomb Dirac Hamiltonian.

To conclude this section we summarize some interesting points about the hydrogenic relativistic solution :
1) The relativistic binding energies are larger than the non-relativistic one. This increase is particularly important for $j = 1/2$ levels.
2) The non-relativistic l degeneracy is partially removed since levels with the same l but different j have different energies. On the other hand levels with the same j but different l are degenerate . Radiative corrections remove this degeneracy and are responsible of the Lamb-shift.
3) For $|\kappa|$ = 1 the Coulomb wave functions have a singularity at origin.
4) As the nodes of the large and small components are interlaced, the radial electron density $\rho(r) = P^2 + Q^2$ never vanishes except obviously at the origin and infinity).
5) The radial electron distribution shrinks towards the nucleus.

2. TWO ELECTRON SYSTEMS

All the methods we shall describe in the next section are built upon a suitably defined set of one-electron wave functions which, as we want to introduce relativistic corrections in a non-perturbative way, must be eigenstates of some generalized form of the Dirac Coulomb operator discussed above. This extension to many-particle systems has recently been the subject of many discussions since high precision calculations which are in remarkable agreement with experiment have been critized for lacking theoretical justification. This rather disturbing situation has been widely discussed in recent papers[2-4] and we refer the interested reader to them for the extensive discussion of the tricky arguments needed to resolve this apparent contradiction. We shall adopt here a more pedestrian approach to show that the methods used in practical calculations are indeed free from fundamental disease.

2.1 The negative energy problem

As may be expected, the problem of the negative energy states we found in the one-electron case will not vanish when going to many particle systems. Brown and Ravenhall[5] seem to be the first ones, about twenty years after Dirac's equation, to have pointed out that no Hamiltonian including any kind of electron-electron interaction can really describe bound states unless something is done in the formalism to explicitly handle the negative energy levels. To illustrate what is the problem, let us consider the two-electron case with any kind of interaction g(1,2) between the electrons. The natural extension of the Dirac theory would lead to the Hamiltonian

$$H = H_D(1) + H_D(2) + g(1,2) \tag{25}$$

as in fact was suggested by Breit[6] very soon after Dirac's equation. The proof that the above Hamiltonian cannot describe bound states is very simple and proceeds as follows : assume first that we neglect the interaction ($g = 0$), then the zero order solution of (25) is the antisymmetrized product :

$$\phi_0 = A \ \phi_1(1)\phi_2(2) \tag{26}$$

where ϕ_1 and ϕ_2 are eigenstates of the Dirac Hamiltonian (Eq. 9) with eigenvalues ε_1 and ε_2 belonging to the discrete spectrum. If we now turn on the interaction, the two electron wave function can be expanded in terms of products like (26), i.e

$$\psi = \sum_n c_n \phi_n \text{ with } \sum_n |c_n|^2 = 1 \tag{27}$$

In this expansion we will always find terms, say $\emptyset_m$, such that one or both of the one electron energies $\varepsilon_{m1} = \varepsilon_2 - \Delta$ and $\varepsilon_{m2} = \varepsilon_2 + \Delta$ belong to the continuum and because of the existence of the negative continuum there will be an infinity of these terms. We have consequently to deal with the well known problem in atomic physics of discrete levels embedded in a sea of continuous ones, a situation which results in autoionization and which in the present case conflicts with the physical facts (the helium atom is stable!). At this stage it cannot be objected that the negative energy levels are filled since this is the way that Dirac solved the problem but our poor Hamiltonian (25) has never met Dirac and the trouble is obviously that nothing in the formalism prevents variational collapse into the negative energy states. This has often been used as a commandment to use the Breit operator in first order perturbation theory only [7] (its use in second ordre perturbation was rapidly found to give large and wrong contributions) even if it is quite clear that the detailed nature of the interaction is of no importance.

Maybe the most natural way to eliminate the unwanted contributions of the positron states is to introduce a projection operator in order to ensure that the electron-electron interaction $g(1,2)$ will connect only positive energy states among themselves. This method has been used by Brown and Ravenhall[5] followed by Sucher [2] and Mittleman[3] and proceeds as follows : equation (25) is replaced by :

$$H^+ = H_D(1) + H_D(2) + \Lambda_+ g(1,2)\Lambda_+ \tag{28}$$

where the positive energy state projection operator Λ_+ is given by :

$$\Lambda_+ = \Lambda_+(1)\Lambda_+(2)$$

with $\quad \Lambda_+(i) = \sum_n |\mu_n^+(i)><\mu_n^+(i)>$

$$\tag{29}$$

the u_n^+ being the positive energy eigenfunctions (belonging in general to both the discrete and continuum spectra) of some one-electron
Hamiltonian H_o. Then we have to solve

$$H^+\psi = E\,\psi \qquad \text{with } \Lambda_+\psi = \psi \tag{30}$$

Many choices are in principle possible for the Hamiltonian H_o used
to construct the projection operator : the free particle Dirac
Hamiltonian, the Coulomb one or, as many-body perturbation theory,
an operator chosen a posteriori in order to annihilate certain terms
(remember for example that the choice of the Hartree-Fock potential
results in a zero contribution of some diagrams in many-body perturbation theory [8]). This freedom in the choice of H_o has been
investigated by Mittleman [3] from a variational point of view. Requiring the total energy to be stationary with respect to the potential used to define H_o and restricting the total wave function
to a single Slater determinant he reaches the conclusion that the
best possible choice is the Hartree-Fock potential obtained by
solving the usual Dirac-Fock equations, i.e. the relativistic counterpart of the Hartree-Fock equations obtained by the usual variational principle applied to the Hamiltonian (25) without projection
operator. However this choice introduce new terms in the energy expression which reflect the fact that electron-positron pairs can be
created by the Hartree-Fock potential (stated in other terms is
a consequence of the fact that the Dirac theory is not a one-par-
ticle theory since if the charge of the system is conserved the
number of particles is not). In the absence of concrete numerical
calculations of the contribution of the pair terms it is hard to
say how significant they will be. Even if the argument of Mittleman
has not been extended yet to more complex wave functions like the
multiconfiguration ones, it is at least an indication why numerical
calculations give good results. Another approach [4] based on the
study of the self-adjointness of the Dirac operator and on the
properties of the quadratically integrable one-electron Dirac orbitals reaches the same conclusions concerning the theoretical fun-
dations of present relativistic calculations.

        To conclude this section we want to emphasize that the numeri-
cal methods we shall consider below never solve in practice
equations for the many-body problem but instead a set of one-electron equations in some kind of self-consistent process. These
equations do have positive energy solutions and since to properly
define the numerical orbitals it is necessary to specify the boundary conditions at both $r \to 0$ and $r \to \infty$, this amounts to reducing
the variational space to electron like solutions only. This is the
reason why numerical programs do not suffer from variational collapse contrary to what was found in the use of basis set expansions.
Also this variational collapse has not to be confused with the ability of Eq(25) to describe bound states even if the underlying problem in both cases is the existence of the negative energy levels
of the Dirac equation.

2.2 Electron-electron interaction

Up to now we have deliberately avoided explicitly specifying the form of the electron-electron interaction g(1,2), the correct form of which should be obtained from quantum electrodynamics (QED). For minimal coupling between the particle and photon fields, the lowest order contribution to the scattering matrix without change in the number of particles is given by :

$$\int j^{\mu}(x_1)\ A_{\mu}(x_1)d^4x_1 \tag{31}$$

where summation over Greek indices is implied and j and A are respectively the charge-current density and electromagnetic 4 vectors. This is just the interaction between an electron located at x_1 with the electromagnetic field created at this point by the other electron or stated in another way the process involving a photon emitted by one electron and absorbed by the other as illustrated in fig. 2. When the photon is absorbed by the same electron this gives rise to higher order corrections (self-energy) which we shall consider later. For practical purposes it is necessary to rewrite the interaction in terms of a matrix element between one-electron Dirac orbitals and an effective interaction in the form :

$$\int\int\Phi_A^+(1)\Phi_B^+(2)g(1,2)\Phi_C(1)\Phi_D(2)d^3x_1d^3x_2 \tag{32}$$

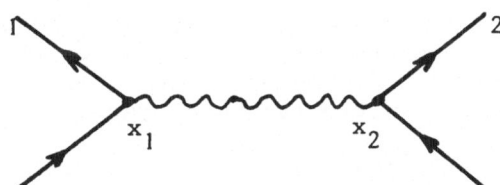

Fig. 2. Electron-electron interaction.

The explicit form of g(1,2) is given in the Lorentz gauge by the Møller [10] formula :

$$g^L(1,2)\ =\ \frac{1\ -\ \vec{\alpha}(1).\vec{\alpha}(2)}{R}\ e^{i\omega R} \tag{33}$$

while in the Coulomb gauge one obtains :

$$g^C(1,2)\ =\ \frac{1}{R}\ -\ \frac{\vec{\alpha}(1).\vec{\alpha}(2)}{R}\ e^{i\omega R}\ +\ \frac{(\vec{\alpha}(1).\vec{\nabla})(\vec{\alpha}(2).\vec{\nabla})}{\omega^2 R}(e^{i\omega R}-1) \tag{34}$$

where in both of the above expressions $R = |\vec{r}_1 - \vec{r}_2|$ is the inter-electronic distance and ω the frequency of the virtual exchanged photon which because of energy conservation is given by :

$$\omega = |\varepsilon_A - \varepsilon_C|/c = |\varepsilon_B - \varepsilon_D|/c \qquad (35)$$

the ε's being the electron energies. We expect obviously that QED gives results that are gauge independent even though it is well known that the electron-electron interaction potential has an expression which is gauge dependent. In fact almost all atomic calculations have been carried out in the Coulomb gauge the physical justification[4] being that the dominant interaction, the Coulomb one, is handled almost exactly. The degree to which the numerical values agree between the two gauges for approximate wave functions (remember the length and velocity form of the transition probability) has never been really tested. The two expressions (33) and (34) are identical only at the low frequency limit ($\omega = 0$) and give :

$$(1 - \alpha(\vec{1}).\alpha(\vec{2}))/R \qquad (36)$$

known as the Gaunt interaction. The next non vanishing term which corresponds to the retardation, in the interaction, is given in the two gauges by :

$$-\omega^2 R/2 \qquad (37)$$

and

$$\vec{\alpha}(1).\vec{\alpha}(2)/2R - (\vec{\alpha}(1).\vec{R})(\vec{\alpha}(2).\vec{R})/2R^3 \qquad (38)$$

respectively. The sum of (36) and (38) is just the well known Breit interaction[6]. Note that the expectation value of (37) and (38) for a direct integral (A = C, B = D) is zero since $\omega = 0$ as expected for the interaction between two static charges.

To conclude this section we would like to make two comments. First the operator defined by (33) or (34) is strictly restricted to direct or exchange integrals because of the energy conservation constraint (35). For the general case with $\omega_{AC} \neq \omega_{BD}$ Mittleman[11] using a series of unitary transformations to decouple the electron and radiation fields, suggested to use one half of the sum of two operators, the first with $\omega = \omega_{AC}$ and the second with $\omega = \omega_{BD}$. The second remark is to point out that expressions like (33) or (34) do not define operators in the ordinary sense since they implicitly depend on the wave functions due to the presence of the electron energies in the definition of the ω's.

2.3 Radiative corrections

Besides the corrections to the Coulomb interaction we have just discussed and which are of order $\alpha^2 Z^2$ compared to the instantaneous Coulomb repulsion we need now to consider the modifications to the

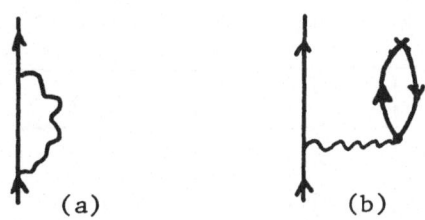

Fig. 3. Radiative corrections

Dirac theory introduced by the quantization of the electronmagnetic
field. These modifications are associated with the processes represented by the lowest order diagrams shown in Fig. 3. Diagram (a)
corresponds to the emission and reabsorption of a photon by the same
electron and depicts the interaction of the electron with its own
(virtual) radiation field. This contribution is known as the self-
energy correction and is, for electrons, the dominant contribution
to the Lamb-shift. The vacuum polarization diagram (b) refers to the
creation of a virtual electron-positron pair by the nuclear potential, followed by the annihilation of the pair giving a photon absorbed by the electron. We shall not attempt to describe the techniques of QED used to calculate these contributions but merely
summarize the known results. Up to now, few calculations have been
performed for many electron atoms. If we except two electron systems[12], the only other results are the screened self-energy
results for the K shell of few atoms [13-14]. Furthermore if the
calculation of the self-energy for two electron systems in the frame
work of the scattering formulation of QED seems straightforward it
will be quite time consuming and the extension to the N-electron
case may be intractable[12] due to the need to include diagrams
which involve also the electron-electron interaction. Consequently
in most of the cases we have to rely on the hydrogenic results of
Mohr[15] which are then corrected for electron screening. In most
of the published results the screening coefficients are obtained
by comparing the mean radius of a given orbital to that of its
hydrogenic counterpart. The reliability of this method can
be estimated for K electrons from the comparison given in Table 3
between Cheng and Johnson [14] results and those deduced from hydrogenic values with the screening corrections obtained from the
mean radius of the orbitals [16]. As both calculations used the
local Slater exchange approximation and finite nuclear charge distributions, the small discrepancies (less than 2 %) are very likely
due to the screening effect. On the other hand, these differences
may quite well be less important than the error due to the neglect
of more than one body terms.

     The vacuum polarization contribution (Fig. 3b) is easier to
obtain since it has been shown by Wichmann and Kroll[17] that the
main contribution is given by the expectation value of the Uehling
potential [18] which for a spherical nuclear charge distribution

Table 3. K-electron self-energy (in eV)

Z	Ref 13	Ref 15
70	123.9	121.3
80	202.0	198.7
90	318.0	314.5
100	492.3	489.5

$\rho(r)$ is given by :

$$V(r) = -\frac{2\alpha\lambdabar_e}{3r} \int_0^\infty u\, \rho(u)\left[K(\frac{2}{\lambdabar_e}|r-u|) - K(\frac{2}{\lambdabar_e}|r+u|)\right]du \qquad (39)$$

with $\lambdabar_e$ the electron Compton wavelength and the function K :

$$K(x) = \int_1^\infty (\frac{1}{t^3} + \frac{1}{2t^5})(t^2-1)^{1/2}\, e^{-xt}dt \qquad (40)$$

If higher order contributions are needed they can also be obtained as the expectation values of potentials as given by Källén and Sabry [19].

3. NUMERICAL VARIATIONAL METHODS

We turn now to a short description of some of the methods which are presently used to include both the relativistic and the dominant correlation effects for systems with in principle any number of electrons. We restrict ourselves to the methods which solve exactly the Dirac equation and consequently include to all orders the relativistic one body contributions. All of them start from the Breit-Dirac Hamiltonian

$$H_{BD} = \sum_i H_D(i) + \sum'_{i,j}\left[1/r_{ij} + B(i,j)\right] \qquad (41)$$

i.e. a sum of one-electron Dirac operators as given by Eq. 19 eventually corrected for finite nuclear charge distribution plus the sum over electron pairs of the instantaneous Coulomb repulsion corrected by the Breit interaction $B(i,j)$ which in the most elaborate case is taken as :

$$B(i,j) = \frac{\vec{\alpha}(1).\vec{\alpha}(2)}{R}\cos\omega R + \left[\vec{\alpha}(1).\vec{\nabla}\right]\left[\vec{\alpha}(2).\vec{\nabla}\right]\frac{\cos\omega R - 1}{\omega^2 R} \qquad (42)$$

to introduce the main contribution of the non zero frequency limit

($\omega \neq 0$) of the transverse interaction. In virtually all calculations the Breit interaction is excluded from the one-electron orbital determination process and included only as a first perturbation to the total energy. As we have discussed in the previous
chapter, ther are no reasons, except practical ones of not including
the Breit interaction in the variational process. The radiative
corrections may be introduced as suggested at the end of the
previous section.

     The total wave function which must be an eigenstate of the
total angular momentum J^2, its projection J_z and of the parity
(the only good quantum numbers in the present situation) is written
as the fully antisymmetric product :

$$|\nu, J, M, \pi> = A \, \phi_1(1)\phi_2(2) \ldots \Phi_N(N) \tag{43}$$

where the Φ's are four component central field Dirac spinors as
defined by Eq. 10 and ν stands for all other quantities (angular
momentum recoupling scheme, seniority numbers, etc...) necessary
to define unambiguously the configuration state function (CSF). The
requirement to use jj coupled states imposes that any reliable calculation for an open shell system must not be restricted to a single configuration since this would imply that Coulomb repulsion is
much smaller than spin-orbit interaction. It is thus necessary to
use, in the relativistic case, the many configuration machinery
even for what will be considered as the standard Hartree-Fock level
in non-relativistic calculations. To be more explicit, consider the
simple case of the configuration p^2 : because of spin-orbit interaction the p orbital is split into the p 1/2 and p 3/2 ones. In
pure j-j coupling the lowest level is the $p^2_{1/2}$ (J = 0) one, but
the configuration $p^2_{3/2}$ has also a J = 0 level which will be mixed
with the previous one by Coulomb interaction to give at the non
relativistic limit the $^3P_0$ ground level and the excited $^1S_0$ one.
In fact, due to both Coulomb repulsion and spin-orbit interaction,
the physical ground state is a mixing between the $^3P_0$ and $^1S_0$
levels. For a heavy atom like lead the mixture is about 65% $^3P_0$
with 35 % $^1S_0$, analogous mixing is found between the $^3P_2$ and
$^1D_2$ levels. It might seem that considering only the j-j configurations arising from a single LS one should be equivalent for low
Z atoms to the Pauli approximation but in fact some care has to be
taken since the situation is a little more subtle. In the relativistic case the radial functions of the $j = \ell + 1/2$ and $j = \ell - 1/2$
orbitals are allowed to vary independently while in the non-relativistic case they are generally restricted to be the same. Thus
to recover the non-relativistic limit the two relativistic radial
functions must become identical at the low Z limit a condition
which is fulfilled only if the level under consideration has a
single parent in the parentage expansion. Intermediate coupling
and correlation may thus be taken into account if the total wave

function is written as a linear combination of CSF, i.e. of the
form :

$$\psi(J,M,\pi) = \sum_n C_n |v_n, J, M, \pi\rangle \tag{44}$$

the amount of correlation included will obviously depend on the
more or less clever choice of the CSF's.

3.2 The Relativistic Parametric Potential Method

The first method we outline is the relativistic version of
the parametric potential method of Klapisch [20] which has been
used to study systematic trends along various isoelectronic sequences [21-24]. In this method approximate eigenstates of the Breit-
Dirac Hamiltonian H_{BD} (Eq. 41) are obtained by first introducing a
zero-order Hamiltonian H_0 of the central type :

$$H_{BD} = H_0 + H_1 \tag{45}$$

with $\quad H_0 = \sum_i c\, \vec{\alpha}_i \vec{p}_i + \beta_i c^2 - U(r)$

The single central potential $U(r)$ is used to generate all the one-
electron orbitals needed. Then H_1 is treated as a perturbation. The
key-point of the method is the determination of the potential $U(r)$
which is expressed as an analytic function depending on a set of
parameters. For an atom Z with N electrons the analytical form is
the following :

$$U(r) = - \sum_{n\ell} \left[2(2\ell + 1) f_{n\ell}(r) + Z - N + 1 \right]/r \tag{46}$$

where $f_{n\ell}$ is the potential due to the spherical charge distribution
$\rho_{n\ell}$ of the $(n\ell)$ shell. This distribution is approximated by a node-
less Slater function

$$\rho_{n\ell} = C_{n\ell}\, r^{2\ell+2}\, e^{-\theta_{n\ell} r} \tag{47}$$

where $C_{n\ell}$ is a normalization constant and $\theta_{n\ell}$ the parameter to be
determined. The potential $f_{n\ell}$ is thus given by :

$$f_{n\ell}(r) = e^{-\theta_{n\ell} r} \sum_{m=0}^{2\ell+1} (1 - \frac{m}{2\ell+2}) \frac{1}{m!} (\theta_{n\ell} r)^m \tag{48}$$

In practice it is enough to consider a single variable parameter
θ_n and use a semi-empirical relation due to Kapliseh to connect the
various $\theta_{n\ell}$ with the same principal quantum number n. The analytical expression (46) fulfils the conditions :

$$\begin{array}{ll} r\, U \to - Z & \text{for } r \to 0 \\ r\, U \to -(Z - N + 1) & \text{for } r \to \infty \end{array} \tag{49}$$

Various criteria can be used to determine the θ parameters [25].
When no experimental results are available the optimal potential is
obtained by minimization of the total energy of the ground state.
This energy is calculated to first order of perturbation theory
often excluding the Breit term from H_1 . The set of one-electron
orbitals being determined, the total wave function is expressed as
a linear combination of CSF's as given by Eq. 44 and the matrix of
the Hamiltonian diagonalized. The eigenvalues of the matrix are thus
the energies of the levels and the eigenvectors give the states in
intermediate coupling. To summarize, this method is essentially a
configuration interaction method using a numerical basis of one-electron functions determined in some optimum way. In this method a
given orbital is described by the same radial functions (large and
small components) for all the states and all states with the same
angular momentum and parity are obtained simultaneously by a single
diagonalization. Consequently orthogonality between wave functions
is obtained automatically and no overlap integrals are needed for
the calculation of off diagonal quantities like transition probabilities.

3.2 The multiconfiguration Dirac-Fock method

In this method (MCDF), use is made again of the expansion
(44) but now both the one electron orbitals and the mixing coefficients C_n are simultaneously optimized. In practice the two following problems are solved sequentially :
1) given a set of mixing coefficients C_n between the CSF's the total
   energy is minimized with respect to arbitrary variations in the
   large and small radial components of the Dirac spinors.
2) with the radial functions thus obtained the energy matrix is
   diagonalized and a new set of C_n coefficients determined. This
   two-step process is repeated until convergence is hopefully
   reached. Details of the methods used to solve the equations may
   be found in a recent paper by the author [26].

Besides the intuitive feeling that as more degrees of freedom
are allowed, the better should be the quality of the wave function;
the multiconfiguration method has the advantage of greatly reducing
the number of CSF's needed to reach a given accuracy. As shown by
Fischer [27] this results from the following observation : suppose
that we have a zero order wave function say of Hartree-Fock quality, let ij be a pair of electrons and write the state under consideration in the form ij $J_1M_1, \nu_2J_2M_2$ where the electrons ij are
coupled to form the J_1M_1 state, the remaining (N-2) electrons being
coupled to form the state denoted by $\nu_2J_2M_2$ and finally these two
states are coupled to give the correct value of the total angular
momentum JM. Then a systematic improvment of our zero order wave
function can be obtained by adding states of the type $\ell m\, J_1M_1,$
$\nu_2J_2M_2$ where now ℓm differs from ij by at least one electron. We
have thus to handle functions which consists of doubly sums such as :

$$\sum_{\ell m} C_{\ell m} |\ell m \ J_1 M_1, \nu_2 J_2 M_2 > \tag{50}$$

which, with even a rather small basis of radial functions, involve
a large number of Slater integrals in the energy expression essentially because many of the configurations differ by only one electron. Using a series of orthogonal transformations[27] it is possible
to reduce the number of configurations needed in the expansion and
at the same time lead to configurations which differ usually by two
electrons resulting in a great saving of the amount of computation
required. Examples (in LS coupling for simplicity) are :

$$\sum_{n \geq m} \sum_{n' \geq m'} C_{nn'} |n\ell n'\ell'> = \sum_{n \geq m} a_n \ \phi(n\ell(n + m' - m)\ell')$$

$$\sum_{n} \sum_{n' \geq n} C_{nn'} |n\ell n'\ell> = \sum_{n} a_n \ \phi(n\ell n\ell) \qquad \text{if } L + S \text{ even} \tag{51}$$

$$= \sum_{\substack{n \\ \Delta n = 2}} a_n \ \phi(n\ell(n+1)\ell) \ \text{if } L + S \text{ odd}$$

where the ϕ's are CSF's defined in terms of the transformed radial
functions, i.e. those to be determined in the MCDF process.

3.3 The Relativistic Random Phase Approximation

In the two methods outlined above one tries to optimize independently the wave functions of the various states under consideration. Another approach is to concentrate on the operator which
connects two states, like the initial and final ones in a photo-
ionization process or radioactive decay, rather than on the states
themselves. This is the basic idea of the so-called Random Phase
Approximation (RPA) and of its relativistic extension (RRPA).
Approximations are then introduced in the form of the operator and/
or in the wave functions restricted to belong to a given class. The
RPA method may be introduced from different points of view: the Green's
function method[25], the linearized equation of motion[29], the diagrammatic perturbation theory[30] and the time-dependent Hartree-Fock
(TDHF) theory[31]. In order to emphasize that the RPA as applied to a
finite number of particles in the atomic case is not to be considered
as a theory valid only for an electron gas as the name first introduced by Bohm and Pines[32] in that context may suggest, we will
adopt the TDHF point of view used by Johnson and Lin[33] in their
relativistic derivation. Exchange is also included so that this
method represents in fact a certain level of approximation the
level of which is justified or not by the physical system under
consideration. In its most common form the RPA relies on the assumption that the ground state of a N-electron system (generally a closed

shell atom or ion) can be adequately represented by a single Slater determinant whose one-electron orbitals satisfy, in the relativistic case, the Dirac-Fock (DF) equations :

$$\left[H_D + V_{DF}\right]\Phi_i = \varepsilon_i \Phi_i \tag{52}$$

with the DF potential given by :

$$V_{DF}\Phi_i = \sum_j \int \left[\Phi_j^+(r')\Phi_j(r')\Phi_i(r) - \Phi_j^+(r')\Phi_i(r')\Phi_j(r)\right] \frac{d^3r'}{|\vec{r} - \vec{r}'|} \tag{53}$$

If a time dependent external perturbation $V(t)$ is introduced, the Hamiltonian becomes :

$$H(t) = H + V(t)$$
$$\text{with } V(t) = v_+ e^{-i\omega t} + v_- e^{i\omega t} \tag{54}$$

Approximate solutions can be obtained with the help of the time dependent variational principle[34] :

$$\langle \delta\psi(t) | i \frac{\partial}{\partial t} - H(t) | \psi(t) \rangle = 0 \tag{55}$$

which states that the error in the wave function remains small at all time if it is at the initial time. The time dependent potential $V(t)$ induces a time dependent perturbation in each of the DF orbitals. Retaining only the first order corrections they become :

$$\Phi_i(r) \to \Phi_i(r) + u_i^+(r)e^{-i\omega t} + u_i^-(r)e^{+i\omega t} \tag{56}$$

and satisfy the set of TDHF equations :

$$\left[H_D + V_{DF} - \varepsilon_i \mp \omega\right]u_i^\pm = (v_\pm - V_\pm)\Phi_i + \sum_j \lambda_{ij}^\pm \Phi_j \tag{57}$$

where $V_\pm$ is analogous to the DF potential V_{DF} and the Lagrange multipliers $\lambda_{ij}^\pm$ are introduced to insure the orthogonality between the u's and Φ's orbitals. An extensive discussion of the RRPA equations is given in reference 22.

3.4 Comparison between the methods

At this point the non expert reader may be somewhat alarmed and starts to worry about the contributions included or omitted when a calculation is performed with one of the three methods we have just described. The answer to this is not simple and one has in fact to know in great detail what has really been done. In principle both the parametric and MCDF methods are able to provide the exact solution (except for the approximations in the choice of the Hamiltonian) but the necessity to cope with finite basis sets in any practical calculation implies a non trivial assessment of the convergence. In fact both methods are designed and used so that the dominant correlation effects arise from a limited number of well optimized states and if the practitioners of each method are smart enough one may expect quite equivalent results. In my opinion the parametrized potential method should be considered as a way to generate adapted single particle wave functions which insure a fast convergence in the configuration interaction (CI) description of the system while the MCDF method is designed to reduce the size of the CI. More mathematical justifications can certainly be found but at the end it is merely a question of physical intuition and know how.

On the other side, the RPA method may appear to suffer from more fundamental problems since it is well known that in terms of diagrammatic language only the bubble diagrams are taken into account and to all orders in perturbation theory while other types of diagrams are completely neglected. The bubble diagrams are those which at any vertex have only a single particle line and a single hole line so that at any given time many electrons can be excited out of their ground state but they never interact with each other. Thus the usual restriction of the RPA, as applied for calculating transition probabilities, to include correlation contributions from only two electrons two holes in the initial state and one electron one hole in the final state can be justified only by the physics of the system under consideration. To overcome this difficulty a recent extension has been proposed[35] to handle multiconfiguration reference states instead of single Slater determinants. Preliminary results for the Beryllium sequence[36] (some of them are reproduced in Table IV) clearly show the improvments which may be expected. Certainly one point in favor of the RPA is its ability to handle as efficiently transitions to continuum than the ones to discrete levels which is not true for neither of the two other methods. As a parenthesis we recall that RPA is gauge invariant so that transition probabilities calculated in the length and velocity forms are identical. But in practical calculations the full set of RPA equations (57) is almost never solved completly (only the active orbitals and their closest neighbours are expanded as given by (56) while the others are frozen at the Hartree-Fock level) so that the difference between the two values is a measure of the approximations introduced.

4. NUMERICAL RESULTS

Before considering many electron systems, it may be worthwhile to come back to the hydrogenic case in order to have a feeling for the order of magnitude of the various one-electron relativistic and QED corrections. Table V gives the lowest energy levels of high Z hydrogen-like atoms. For completeness the last column gives the finite nuclear size correction calculated by assuming a uniform proton charge distribution inside a sphere of radius $1.2 \, A^{\frac{1}{2}}$ Fm

Table IV

Be-Like Ions : $1s^2 2s^2 (^1S_0) \rightarrow 1s^2 2s 2p (^1P_1)$ transitions

		MCDF[1]	RRPA[2]	MCRRPA[2]	Experiment
Be	E (a) F (b)	0.222 1.28	0.177 1.38	0.201 1.42	0.194 1.38 ± 0.1
Ne^6	E F	1.04 0.399	0.885 0.375	1.01 0.404	0.939 0.336 ± 0.02
Fe^{22}	E F	3.49 0.155	3.15 0.150	3.47 0.157	3.43 –

(1) Multiconfiguration Dirac-Fock results from Ref. 37
(2) Relativistic RPA and multiconfiguration relativistic RPA
    results from Ref. 36
(a) E : transition energies in atomic units
(b) F : oscillator strength (dimensionless)
Sources of experimental results are given in Ref. 36.

(A is the atomic mass). Looking at the number two comments emerge :
first, if the Dirac correction is always the dominant one, none of
the other corrections can be neglected in high quality calculations
of heavy systems. Our second comment concerns the finite nuclear size
correction which for very high Z and j = 1/2 states is as important
as the sum of the QED terms. Consequently it is necessary in high
precision calculations to use actual nuclear parameters (for example
Fermi distributions) as obtained experimentally. The negligible value of that correction for energy levels with angular momentum
greater than 1/2 has to be related to the fact that only j = 1/2
wave functions display an s behaviour either in their large or small
component and are thus the only ones with finite density at origin.
On the other hand the Z dependence is much stronger than the Z^3

J. P. DESCLAUX

Table V

Hydrogenic energies (in eV)

		(a)	(b)	(c)	(d)	(e)	(f)
Z=60	1s	−48980.89	−2603.67	73.45	−11.70	−0.23	6.16
	2s	−12245.22	−817.89	11.42	−1.68	−0.03	0.92
	2p*	−12245.22	−817.89	0.30	−0.10	−	0.04
	2p	−12245.22	−150.34	1.21	−0.01	−	−
Z=80	1s	−87077.15	−9039.48	207.24	−45.52	−0.67	54.30
	2s	−21769.29	−2853.08	35.59	−7.33	−0.12	9.27
	2p*	−21769.29	−2853.08	3.25	−0.89	−0.01	0.75
	2p	−21769.29	−484.57	4.60	−0.06	−	−
Z=100	1s	−136058.04	−25557.02	515.55	−157.80	−5.43	451.45
	2s	−34014.51	−8126.92	100.43	−29.83	−1.06	93.01
	2p*	−34014.51	−8126.92	19.06	−6.58	−0.21	13.93
	2p	−34014.51	−1214.35	13.21	−0.21	−0.01	−

a : non-relativistic value
b : Dirac correction
c : self energy correction
d : vacuum polarization as given by the Uehling potential
e : higher order vacuum polarization corrections
f : nuclear size correction

Empty entries indicate that the correction is less than
0.01 eV.

behaviour expected from non relativistic density at origin and this
is due to the relativistic contraction of the orbitals.

4.1 Importance of the Breit correction

The contribution of the Breit operator (Eq. 42) is of order
α^2 compared to the Coulomb repulsion but increases approximately
proportional to $Z^{3.6}$[39] and consequently its influence has mainly
been studied for inner shell properties of heavy atoms[40-43].
Recently Chen et al.[44] found that the effect of the Breit interaction on the multiplet splittings of double-inner-shell-vacancy
states is quite significant for atomic numbers greater than 65. For
example in the case of the two hole configuration $(1s2p_{1/2})$, the
inclusion of the Breit interaction in addition to the Coulomb

repulsion not only increases greatly the splitting between the two
J = 0 and J = 1 levels (by a factor of 3 at Z = 90) but also inverts
the order of these two levels. The importance of the magnetic interaction was previously pointed out even for low atomic numbers[45] and
we shall discuss this point more in detail here. It has been observed[46] that the fine structure intervals in the $1s2p^2$ and $1s2s2p$
^4P states of lithium-like carbon, nitrogen and oxygen ions deviate
substantially from the Landé interval rule. This is understandable
since the Landé rule is based on the spin-orbit interaction only.
As for the quartet states the three electron spins are lined up and
other effects such as spin-spin and spin-other-orbit interactions
may become important. Table VI gives the results obtained for the
relative fine structure levels of O^{5+} with respect to the centre of
gravity (i.e. the 2J+1 statistically weighted sum) of the $1s2p^2$ ^4P
states at three levels of approximation. First of all, the Breit
interaction is not included. The fine structure is thus due to spin-
orbit interaction only and follows closely the 5 : 3 Landé interval
rule and is in disagreement with experiment. Secondly the Breit
interaction is introduced at the average configuration approximation
level, i.e. one considers only the statistically weighted contribution :

Table VI

$$E - E_{CG} \text{ (in cm}^{-1}\text{) relative fine structure levels}$$
of the $1s2p^2$ ^4P states in O^{5+}

	J = 5/2	J = 3/2	J = 1/2
Dirac-Fock (DF)	286	−189	−479
DF + Average Breit	270	−186	−439
DF + Exact Breit	167	− 69	−362
Experiment (Ref. 46)	177	− 81	−371

$$E^B_{Av} = \Sigma(2J + 1)E^B(J) / \left[\Sigma(2J + 1)\right] \tag{58}$$

which leaves the relative fine structure levels virtually unchanged.
This is understandable since in the average configuration approximation the contribution of the Breit term is almost the same for
all the levels except   that the effective occupation numbers of
the $p_{1/2}$ and $p_{3/2}$ orbitals are different between the various
levels. On the other hand for a light system the difference between
$p_{1/2}$ and $p_{3/2}$ radial functions is negligible and no important effect is expected from the transfer of electrons between the two
orbitals. Finally the Breit term calculated exactly introduces

dramatic changes to the fine structure and brings theoretical and
experimental results in close agreement. The need to calculate
exactly the Breit term is further illustrated by looking at the
fine structure parameters given in Table III. Assuming LS coupling
to be a good approximation for such a light system it is possible[46] to express the spin-orbit (SO) and spin-other-orbit (SOO)
operators as

$$H_{SO} + H_{SOO} = C_{SO} \vec{L} \cdot \vec{S} \tag{59}$$

and the spin-spin operator as :

$$H_{SS} = C_{SS}\left[3(\vec{L} \cdot \vec{S})^2 + \frac{3}{2}(\vec{L} \cdot \vec{S}) - L(L+1)S(S+1)\right] \tag{60}$$

which define the fine structure constants C_{SO} and C_{SS} related to
the fine structure intervals $E_{53}(= E_{5/2} - E_{3/2})$ and
$E_{31}(= E_{3/2} - E_{1/2})$ of the quartet states by :

$$C_{SO} = (9E_{53} + 5E_{31})/30 \; ; \; C_{SS} = (3E_{53} - 5E_{31})/90 \tag{61}$$

Table VII

Fine structure parameters (in cm^{-1}) for
the $1s2p^2 \; ^4P$ states of O^{5+}

	C_{SO}	C_{SS}
Average Breit	179.0	1.1
Exact Breit	119.5	−8.5
Experiment (Ref. 46)	124.8 ± 1.5	−8.0 ± 0.3

Note that the present values differ slightly from those given in
reference 45 since the retardation part of the Breit interaction
has now been calculated exactly instead of evaluating it in an
average configuration approximation. Nevertheless as the retardation contribution is almost an order of magnitude smaller than the
magnetic one its influence on fine structure splittings of light
Z systems is of minor importance.

We have just seen the prime importance of the Breit interaction to obtain accurate fine structure splittings and we may of
course wish to know how it compares with the contribution of QED
corrections. This is illustrated by the results obtained for carbon-
like iron[47] some of them being reproduced in Table VIII which

gives all the levels of the $1s^2 2s^2 2p^2$ configuration while for the configurations $1s^2 2s2p^3$ and $1s^2 2p^4$ only the lowest and highest levels are quoted.

Table VIII

Some energy levels of Fe XXI (in cm^{-1})

Config.	State	MCDF[a]	Breit	QED	Total - Exp.[b]
$2s^2 2p^2$	$^3P_0$	0	0	0	0
	$^3P_1$	76 789	−2834	22	67
	$^3P_2$	123 972	−5861	−15	796
	$^1D_2$	250 396	−6981	−121	−656
	$^1S_0$	377 432	−4564	−92	1256
$2s2p^3$	$^5S_2$	494 980	−3405	−4646	−101
	$^1P_1$	1 258 975	−4841	−5250	−1675
$2p^4$	$^3P_2$	1 656 115	−768	−8325	702
	$^1S_0$	2 057 194	−2365	−8116	−1087

(a) The MCDF results are corrected for intershell correlation as given in reference 48.
(b) Experimental results of Lawson and Peacock (Ref. 49).

Again we observe the dominant role of the Breit term in describing correctly the fine structure splittings within a given configuration while QED corrections are very important for transitions between configurations which differ in the number of s electrons. To achieve this quite good agreement between theoretical and experimental results, it was necessary to allow sufficiently for correlation contribution. On the other hand, such comparisons are essential to assess the quality of relativistic atomic theory, and we shall come back to that point when considering trends along isoelectronic sequences. Before leaving the fine structure splittings I would like to draw attention to the difficulty in extracting the true relativistic contribution to their values[50].

Consider the two $^2P_{1/2}$ and $^2P_{3/2}$ levels of the ground state configuration $1s^2 2s^2 2p^5$ of F-like ions. It would seem obvious that the fine structure splitting is given simply by the difference between the two relativistic energies $E(^2P_{1/2})$ and $E(^2P_{3/2})$. To understand why this is not true let us write the total energy as :

$$E(^2P_J) = u_J(Z) + \alpha\, v_J(Z,\alpha) \tag{62}$$

where v denotes formally all corrections of relativistic origin which involve at least one power of the fine structure constant α. The calculated splitting is thus given by :

$$\Delta E = \left[u_{1/2}(Z) - u_{3/2}(Z) \right] + \alpha \left[v_{1/2}(Z,\alpha) - v_{3/2}(Z,\alpha) \right] \tag{63}$$

$$= \Delta u + \alpha\, \Delta v$$

Because the fine structure splitting is purely of relativistic origin care should be taken that Δu is zero in order to avoid spurious contributions. Obviously inconsistent choices of configurations can lead to different non relativistic correlation energies but as already mentioned in the beginning of section III, there is another mechanism in relativistic calculations that introduce non zero Δu. Relativistically the $2p_{1/2}$ and $2p_{3/2}$ orbitals have different radial functions which are expected to converge to the same non-relativistic function in the formal limit $\alpha = 0$. However as pointed out previously[51,52] this is true only for configurations having a single parent core state. In the present case we have the $(2p^4)\,^3P$, ^1D and ^1S core states and the two radial functions will not converge to the same non-relativistic limit and consequently Δu will not vanish. These spurious contributions must be removed from theoretical values before comparisons are made with experiment. Results are given in Table IX and we can see that the corrected results agree better with experiment for ions of low ionicity.

Table IX

Fine structure splittings (in cm^{-1}) for F-like ions

Ion	Uncorrected	Corrected	Experiment
F	375.2	399.0	404.1
Ne$^+$	755.7	774.3	780.34
Mg^{3+}	2206	2221	2228
Si^{5+}	5071	5084	5100

If the spin-orbit splittings are of relativistic origin they are modified by electron correlation. When studying the B-like ions[37] we include all configurations in the same complex to account for the most important correlation contribution. In LSJ notation the multiconfiguration (MC) representation can be written :

$$(2s^2 2p)^2 P_{1/2} + (2p^3)^2 P_{1/2} \tag{64}$$

for the $^2P_{1/2}$ levels and :

$$(2s^2 2p)^2 P_{3/2} + (2p^3)^2 P_{3/2}, \; ^4S_{3/2}, \; ^2D_{3/2} \tag{65}$$

for the $^2P_{3/2}$ one. Single configuration (SC) calculations include only the first term. MC calculations should be better than SC ones because correlation effects are taken into account. In fact our first MC results[37] show a large discrepancy with experiment and the reason is the same as for F-like ions. If the $2s^2 2p$ configuration has the single core parent $2s^2$, the addition of the $2p^3$ configuration which has three core parents $(2p^2)^3P$, ^1S and ^1S results in a dramatic large non zero value of Δu. After subtraction of this spurious contribution the corrected MC results[50] provided a slightly better agreement with experiment (see Table X) as expected. Because the SC's have a unique parent no correction is needed.

Table X

Fine structure splittings (in cm^{-1}) for B-like ions

Ion	SC	MC		
		Uncorrected	Corrected	Experiment
B	15.7	435.3	15.7	16
C^+	64.4	271.5	62.7	63.42
Ne^{5+}	1346	1472	1298	1310
Si^{9+}	7194	7183	6968	6990
Ar^{13+}	23286	22856	22612	22655.9

4.2 He-like ions

If it is well established that Dirac's theory associated with QED predicts very accurately the energy levels of one-electron systems, it is not yet possible to put forward a similar statement for the many electron case. For this reason a great deal of experimental and theoretical work is devoted to the two-electron systems which can provide crucial tests for relativistic atomic theory. Recently precise measurements of the $(1s^2)^1S_0 - (1s2p)^1P_1$, $^3P_1$, $^3P_2$ transitions energies have been obtained for argon[52] and iron[53] He-like ions. These transitions energies have been calculated by various authors at different levels of approximations out of which it is rather difficult to extract the various relativistic effects.

Either the non relativistic contribution was calculated very accurately
ly and then relativistic hydrogenic corrections were added without
taking into account the effect of electron screening (or only in a
very crude way) or no systematic study of the correlation effects
was carried out. In these circumstances we shall use the MCDF method
to discuss the various contributions to the transition energies.

   Making use of the CSF's expansion given in Eq. 51, the ground
state is well described by the expansion

$$1s^2 + 2s^2 + 2p^2 + 3s^2 + 3p^2 + 3d^2 \tag{66}$$

while for the 1s2p levels we use the multiconfiguration expansion :

$$1s2p + 2s3p + 2p3d + 3s4p \tag{67}$$

Both of these wave functions are good enough to reproduce more than
90 % of the correlation energy of each of the levels considered. In
Table XI we list the various contributions to the transition energies

Table XI

$(1s^2)^1S-(1s2p)^1P$, ^3P transition energies (in eV) of

He-like argon and iron

		$^1P_1 \rightarrow {}^1S_0$	$^3P_1 \rightarrow {}^1S_0$	$^3P_2 \rightarrow {}^1S_0$
Ar^{16+}	Hartree-Fock	3126.39	3111.78	3111.78
	Dirac correction	15.19	13.43	16.35
	Breit interaction	-2.04	-1.79	-1.97
	QED corrections	-0.99	-1.02	-1.01
	Correlation	1.18	1.18	1.17
	Total	3139.73	3123.58	3126.32
	Experiment[53]	3139.6 (±.3)	3123.6 (±.3)	3126.4 (±.4)
Fe^{24+}	Hartree-Fock	6638.81	6616.77	6616.77
	Dirac correction	70.22	59.00	74.06
	Breit interaction	-6.10	-5.70	-6.08
	QED corrections	-3.52	-3.58	-3.55
	Correlation	1.29	1.30	1.28
	Total	6700.70	6667.79	6682.48
	Experiment[54]	6700.9 (±.4)	6667.8 (±.4)	6682.7 (±.4)

and compare the total value with experiment. The net accuracy of
the theoretical results may be estimated to be .2 or .3 eV[54] and
the agreement with experiment is more than satisfactory. One point
worthwhile noticing is again the influence of the Breit term not only
because of its quite significant value but also because of its
contribution to the correlation energies given in Table XII. For
the ground state configuration $(1s^2)$ of Fe^{24+} the contribution of

Table XII

Correlation energies of He-like ions (in eV)

	$(1s^2)$	1s2p		
	$^1S_0$	$^1P_1$	$^3P_1$	$^3P_2$
Ar^{16+} Coulomb only	1.164	0.085	0.085	0.101
Coulomb + Breit	1.268	0.087	0.085	0.102
Fe^{24+} Coulomb only	1.165	0.089	0.090	0.103
Coulomb + Breit	1.389	0.096	0.093	0.105

the Breit interaction amounts to almost 20 % of the Coulomb one
and this is, to our knowledge, the first time that the importance
of the Breit term in accurate calculations of correlation energies
is clearly pointed out. As this contribution increases rapidly with
Z, further experiments on heavier systems will be of great interest
to check the present accuracy to which relativistic correlation
energy can be calculated. The MCDF method has also been shown to
be quite adequate for estimating relativity-correlation corrections
to the 1s2s-1s2p energies of the helium isoelectronic sequence[55].

4.3 Isoelectronic sequences

Up to now we have considered the influence of relativistic
corrections to the energy levels of some specific systems. In this
last section we shall try to give a more general overview by studying
trends along isoelectronic sequences. The first consequence of introducing explicitly the spin-orbit interaction is obviously the breakdown of the Russel-Saunders coupling as illustrated in Table XIII
for the $(1s2p)^1P_1$ level of the He sequence. Writing the J = 1 wave
functions as

$$a|1s2p_{1/2}, J = 1> + \left[1 - a^2\right]^{1/2}|1s2p_{3/2}, J = 1> \qquad (68)$$

Table XIII

Configuration mixing coefficients of the

$(1s2p)^1P_1$ level for the He sequence

Z	Coefficient[a]	Mixing with $^3P_1$ (in %)
2	-0.5771	0.
7	-0.5717	0.
10	-0.5603	0.02
20	-0.4460	1.2
30	-0.2564	6.3
40	-0.1289	11.6
50	-0.0068	17.9
60	-0.0038	18.1
70	-0.0023	18.2
80	-0.0014	18.3
90	-0.0009	18.3
100	-0.0006	18.3

(a) The LS coupling limit is $-1/\sqrt{3} = -0.57735$.

we identify the $^1P_1$ level with the upper one. The coefficient a is listed in Table XIII together with the mixing between $^1P_1$ and $^3P_1$ levels. As can be seen the transition between LS and jj couplings is rather sharp and occurs between Z = 20 and Z = 50. For such a configuration where the maximum mixing between triplet and singlet levels is less than 20 % we will never observe strong spin-forbidden transitions which will remain weak compared to allowed ones.

Beside this change in the character of the wave functions, the relativistic behaviour of the transition probabilities, when plotted as a function of Z, is completly different from the non-relativistic case. This has been observed for many sequences and discussed in detail for the first time in the case of the magnesium sequence[56]. Let us consider the resonance transition $(1s^2 2s^2)^1S_0 \rightarrow (1s^2 2s2p)^1P_1$ of the Be-like ions as a typical example of the modifications induced by relativistic effects. As shown by Layzer and Bahcall[57] the relativistic energies along an isoelectronic sequence may be expressed in terms of the double series expansion :

$$E = Z^2 \sum_{n=0} \sum_{m=0} E_{nm} (\alpha^2 Z^2)^n Z^{-m} \qquad (69)$$

where the terms n = 0, m ≠ 0 correspond to the non relativistic energy. Those with n ≠ 0, m = 0 give the series expansion of the Sommerfeld one-electron formula (Eq. 21) and terms with n ≠ 0, m ≠ 0 take into account the relativistic electron-electron interaction. For the transition considered here ($\Delta n = 0$, $\Delta j = 1$) the Z dependence of the excitation energy is given by :

$$\Delta E = Z \, \Delta E_{01} + \alpha^2 Z^4 \, \Delta E_{10} + \ldots \tag{70}$$

where the first term describes the non-relativistic electrostatic interaction among electrons with the same principal quantum number and the second term is the j dependence of the relativistic Dirac correction. In the length form, the oscillator strengths (f values) are proportional to the product of the excitation energy and of the square of the dipole transition matrix element :

$$\langle n\ell j | r | n'\ell' j' \rangle = A/Z + \ldots \tag{71}$$

where the Z dependence is for hydrogenic orbitals. Combining (70) and (71) we see that the f values are expected to vary like :

$$\text{Low } Z \qquad f \propto \Delta E_{01}/Z$$
$$\text{High } Z \qquad f \propto \Delta E_{10} \, Z^2 \tag{72}$$

Sample results for the beryllium sequence given in Table XIV show that this is indeed the case. While the non relativistic f values give a monotonically decreasing function of Z, the relativistic ones show a minimum around Z = 35 and then begin to increase. It can also be observed that the line strength decreases

Table XIV

Transition energies (E) and line strength (S) for the f values of the beryllium sequence

Z	E (eV)		$Z^2 S$		f	
	R	NR	R	NR	R	NR
10	28.31	28.10	57.44	57.32	0.398	0.395
42	251.9	136.0	40.16	41.90	0.140	0.079
74	175.8	243.0	33.80	40.33	0.266	0.044

N and NR stand for relativistic and non relativistic results respectively.

faster than Z^{-2} due to the relativistic contraction of the wave
functions. The preceeding analysis can be carried out for other
types of transitions to predict the low Z and high Z behaviour[23].
For example if $\Delta_j = 0$ the increase will be much slower since E_{10} is
ℓ independent and consequently the high Z behaviour is proportional
to $\alpha^2 Z \, \angle E_{11}$ only.

If the preceding analysis in terms of the double series expansion is quite helpful in understanding asymptotic behaviour, it
does not tell anything about the intermediate region and in particular where level crossings or anticrossings will hapen. These
crossings are a consequence of the transition from LS to jj coupling
scheme which introduces a reordering of the levels between the low
Z and high Z regions as discussed by Cheng and Johnson[58] in the
case of the Mg sequence. As an example consider the first four levels of even parity and with total angular momentum J = 2 which
arise from the $3p^2$ and 3s3d configurations. At low Z when spin-
orbit interaction is small the ordering is the following :
$(3p^2)^1D_2$, $(3p^2)^3P_2$, $(3s3d)^3D_2$ and $(3s3d)^1D_2$. When spin-orbit interaction is dominant the ordering of the (j,j) levels follows the
sequence : (1/2,1/2), (1/2,3/2), (1/2,5/2), (3/2,3/2)... Thus since
in the jj coupling limit the $(3s3d_{3/2})$ level is lower than the
$(3p^2_{3/2})$ one, these two levels have to cross at some value of Z.
Even if true crossings are forbidden the two levels will be very
close to each other in energy and strong interaction between them
will take place over a more or less wide range of Z values. Such
strong interaction will result either in a decrease of the transition probability due to a destructive interference between the
transition channels or on the contrary to the enhancement of the
transition because of coherent mixing. For the Mg sequence it was
found that such a situation arises around Z = 45 and that the mixing
of the $(3p^2)$ and (3s3d) configurations in the $^3P_2$ level strongly
supresses, in the high Z region, the transitions to the $^3P_2$ and $^1P_1$
levels of the (3s3p) configurations while the transition to the $^3P_1$
level was enhanced. Analogous effects where found in the sequences
of Ne-like[59], B-like[23] and Al-like[24] ions and result in important irregularities in the trend of the oscillator strengths.

The few examples given above were used to illustrate the changes
induced by relativistic corrections along the isoelectronic sequences. We shall conclude by taking the opposite point of view and show
how comparison between experiment and theory is of great interest
to check the present status of the relativistic atomic theory. In
Table XV we make a comparison of the transition energies of
$(1s^2 2s^2)^1S_0 - (1s^2 2s2p)^1P_1$ in the Be isoelectronic sequence. To analyze these results we again invoke the Layzer and Bahcall double
series expansion (Eq. 69). In our previous study of these transitions[37] all the configurations within the n = 2 complex were
included so that the E_{00} and E_{01} nonrelativistic coefficients are

Table XV

Be sequence $(2s^2)^1S_0-(2s2p)^1P_1$ transition energies (in cm^{-1})

Z	Theory[a]	Experiment[b]	Difference (%)
6	99 411	102 351	- 2.87
8	156 784	158 798	- 1.27
10	213 382	214 952	- 0.73
12	270 353	271 694	- 0.49
14	328 437	329 679	- 0.38
16	388 357	389 596	- 0.32
18	450 936	452 227	- 0.29
20	517 162	518 516	- 0.26
22	588 080	589 658	- 0.27
24	665 192	667 230	- 0.31
26	750 084	752 760	- 0.36
28	844 710	848 100	- 0.40

(a) Ref. 48.
(b) Z = 6-22 Ref. 60 and Z = 24-28 Ref. 49.

calculated exactly. Using the non-relativistic multiconfiguration method together with the many-body calculations of Ivanova and Safnonova[61] Cheng et al.[48] were able to estimate the next contribution (i.e. the E_{02} coefficient) to the correlation energy. If all relativistic corrections would have been taken into account exactly one should see a monotonous decrease in the difference between theoretical and experimental results since the next order correlation contribution behaves as E_{03}/Z. In fact the results of Table XV show that the agreement steadily improves with increasing Z up to Z $\cong$ 20 and then starts to get worse. The origin of the increasing discrepancy at high Z is not obvious. First the shift from LS to jj coupling introduces an effective Z dependance of the E_{02} coefficient not taken into account. But as the dominant changes are included in the E_{n0} terms it is unlikely that this effect will explain the whole discrepancy. More likely the fact that the Breit interaction was included only as a perturbation in the low frequency limit (ω = 0 in expression 42) and that QED corrections are restricted to one-electron order of magnitude estimates only will be the source of uncertainties in the theoretical values. Clearly further refinements in the theory as well as extension of the experimental results to higher Z are desirable since the Z dependance of the discrepancy will be of great help in making more reliable tests of higher order relativistic and QED effects.

ACKNOWLEDGEMENTS

    I would like to thank D. Vernhet and M. Tavernier for stimula-
ting discussions which influenced the written version of these
lectures. I am grateful to all the organizers of the institute for
their warm hospitality in Cargèse.

REFERENCES

(1) S.J. Rose, I.P. Grant and N.C. Pyper, J. Phys. B 11, 1171
     (1978).

(2) J. Sucher, Phys. Rev. A 22, 348 (1980).

(3) M.H. Mittleman, Phys. Rev. A 24, 1167 (1981).

(4) I.P. Grant in "Relativistic Effects in Atoms, Molecules and
     Solids" (ed. G.L. Malli), Plenum Press, New-York (1982) (to
     appear).

(5) G.E. Brown and D.G. Ravenhall, Proc. Roy. Soc. London A 208,
     552 (1951).

(6) G.E. Breit, Phys. Rev. 34, 553 (1929).

(7) H.A. Bethe and E.S. Salpeter, "Quantum Theory of One- and
     Two-electron Atoms", Springer-Verlag, Berlin (1957).

(8) H.P. Kelly, Phys. Scripta 21, 448 (1980).

(9) V.B. Berestetskii, E.M. Lifshitz and L.B. Pitaevskii, "Relati-
     vistic Quantum Theory" Pergamon Press, Oxford (1971 and 1974).

(10) C. Møller, Z. phys. 70, 786 (1931) and Ann. Phys. 14, 531 (1932).

(11) M.H. Mittleman, Phys. Rev. A 6, 2395 (1972).

(12) P.J. Mohr, "Proceedings of the Workshop on Foundations of
     the Relativistic Theory of Atomic Structure", Argonne National
     Laboratory report ANL-80-126 (1981) and references therein.

(13) A.M. Desiderio and N.R. Johnson, Phys. Rev. A 3, 1267 (1971).

(14) K.T. Cheng and W.R. Johnson, Phys. Rev. A 14, 1943 (1976).

(15) P.J. Mohr, Ann. Phys. (New York) 88, 52 (1974) ; Phys. Rev.
     Lett. 34, 1050 (1975).

(16) M.H. Chen, B. Crasemann, M. Aoyagi, K.N. Huang and H. Mark, At. Data Nucl. Data Tables 26, 561 (1981).

(17) E.H. Wichmann and N.M. Kroll, Phys. Rev. 101, 843 (1956).

(18) E.A. Uehling, Phys. Rev. 48, 55 (1935).

(19) G. Källen and A. Sabry, K. Dan. Vidensk. Selsk. Mat. Fys. Medd. 29, 17 (1955) see also J. Blomqvist Nucl. Phys. B 48, 95 (1972).

(20) M. Klapisch, Comput. Phys. Commun. 2, 239 (1971).

(21) M. Aymar and E. Luc-Koenig, Phys. Rev. A 15, 821 (1977).

(22) A Farrag, E. Luc-Koenig and J. Sinzelle, At. Data Nucl. Data Tables 24, 227 (1979).

(23) A. Farrag, E. Luc-Koenig and J. Sinzelle, J. Phys. B 13, 3939 (1980).

(24) A. Farrag, E. Luc-Koenig and J. Sinzelle, J. Phys. B 14, 3325 (1981).

(25) E. Koenig, Physica 62, 393 (1972).

(26) J.P. Desclaux in "Relativistic Effects in Atoms, Molecules and Solids" (ed. G.L. Malli) Plenum Press, New-York (1982) (to appear).

(27) C.F. Fischer "The Hartree-Fock Method for Atoms", Wiley, New-York (1977).

(28) G. Scanak, H.S. Taylor and R. Yaris, Adv. At. Mol. Phys. 7, 287 (1971).

(29) P.L. Altick and A.E. Glassgold, Phys. Rev. 133, A632 (1964).

(30) G. Wendin, J. Phys. B 5, 110 (1972).

(31) A. Dalgarno and G.A. Victor, Proc. Roy. Soc. A291, 291 (1966).

(32) D. Bohn and D. Pines, Phys. Rev. 82, 625 (1951).

(33) W.R. Johnson and C.D. Lin, Phys. Rev. 82, 625 (1951).

(34) P.W. Langhoff, S.T. Epstein and M. Karplus, Rev. Mod. Phys. 44, 602 (1972).

(35) K.N. Huang and W.R. Johnson, Phys. Rev. A25, 634 (1982).

(36) W.R. Johnson and K.N. Huang, Phys. Rev. Lett. 48, 315 (1982).

(37) K.T. Cheng, Y.K. Kim and J.P. Desclaux, At. Data Nucl. Data Tables 24, 111 (1979).

(38) P.J. Mohr, Ann. Phys. (New-York) 88, 26 (1974) ; Phys. Rev. Lett. 34, 1050 (1975) and to be published.

(39) J.B. Mann and W.R. Johnson, Phys. Rev. A 4, 41 (1971).

(40) J.P. Desclaux, Ch. Briançon, J.P. Thibaud and R.J. Walen, Phys. Rev. Lett. 32, 447 (1974).

(41) I.P. Grant and B.J. Mc Kenzie, J. Phys. B 13, 2671 (1980).

(42) N. Beatham, I.P. Grant, B.J. Mc Kenzie and S.J. Rose, Physica Scripta 21, 423 (1980).

(43) J.P. Desclaux, "Proceedings of the Workshop on the Foundations of the Relativistic Theory of Atomic Stucture" Argonne National Laboratory Report ANL 80-126 (1981).

(44) M.H. Chen, B. Crasemann and H. Mark, Phys. Rev. A 25, 391 (1982).

(45) K.T. Cheng, J.P. Desclaux and Y.K. Kim, J. Phys. B 11, L359 (1978).

(46) A.E. Livingston and H.G. Berry, Phys. Rev. A 17, 1966 (1978).

(47) J.P. Desclaux, K.T. Cheng and Y.K. Kim, J. Phys. B 12, 3819 (1979).

(48) K.T. Cheng, C.F. Fischer and Y.K. Kim, J. Phys. B 15, 181 (1982).

(49) K.D. Lawson and N.J. Peacock, J. Phys. B 13, 3313 (1980).

(50) K.N. Huang, Y.K. Kim, K.T. Cheng and J.P. Desclaux, Phys. Rev. Lett. 48, 1245 (1982).

(51) C.P. Wood and N.C. Pyper, Mol. Phys. 41, 149 (1980).

(52) J. Bauche and M. Klapisch, J. Phys. B 5, 29 (1972).

(53) J.P. Mossé, Thesis, Paris (1982) and to be published.

(54) J.P. Briand et al., to be published.

(55) J. Hata and I.P. Grant, J. Phys. B (to be published).

(56) M. Aymar and E. Luc-Koenig, Phys. Rev. A 15, 821 (1977).

(57) D. Layzer and J. Bahcall, Ann. Phys. (New-York) 17, 177 (1962).

(58) K.T. Cheng and W.R. Johnson, Phys. Rev. A 16, 263 (1977).

(59) P. Shorer, Phys. Rev. A 20, 642 (1979).

(60) B. Edlen, Phys. Ser. 20, 129 (1979) and 22, 593 (1981).

(61) E.P. Ivanova and U.I. Safronova, J. Phys. B 8, 1591 (1975).

RELATIVISTIC AND QED EFFECTS IN HIGHLY IONIZED SYSTEMS

G. W. F. Drake*

Research and Engineering Staff
Ford Motor Company
Dearborn, Michigan 48121

I. INTRODUCTION

Research in atomic physics has traditionally focussed on the properties of neutral atoms, or atomic ions from which at most a few electrons have been removed. However, over the past ten or fifteen years, interest in highly ionized atomic systems has risen dramatically. For example, it is now possible, with the help of plasma sources or high energy accelerators such as the Berkeley Super-Hilac, to study krypton ions with all but one or two of the normal thirty-six electrons stripped away. Experiments with still heavier systems are being planned.

The growth and continuing vitality of this area results from the combined influences of experimental techniques, theoretical developments, and applications to other fields such as plasma diagnostics. The data emerging from studies of highly ionized systems have presented theorists with challenging questions which cannot be answered by simple extrapolations from low energy experiments. In particular, the orbital velocities of the electrons are typically relativistic. The Bethe-Salpeter equation provides a theoretical framework for simultaneously treating relativistic effects and electron-electron interactions, but no exact solutions to this equation are known. All practical calculations involve successive approximations, starting with the Schrödinger equation in zero-order. Since higher iterations became exceedingly complex, only first-order corrections (i.e.

*Permanent address: Department of Physics, University of Windsor, Windsor, Ontario N9B 3P4, Canada.

the Breit interaction) have been obtained for systems other than
neutral helium. In many cases, the measurements are substantially more accurate than what has been achieved by theoretical
calculation. Since two-electron ions provide the fundamental
testing ground for understanding many-electron systems, it is
particularly important that the residual discrepancies resulting
from relativistic and QED affects be resolved.

The theory of QED effects in one-electron ions forms a topic
of especially fundamental interest. Highly ionized one-electron
systems provide a particularly important testing ground for the
computational techniques of QED in the presence of strong external fields (i.e. the Coulomb field of the nucleus). Differences
between theory and experiment, and between different theoretical
calculations, become progressively larger with increasing Z.
Accurate experiments in the high Z region are therefore particularly significant.

The aim of these lectures is to provide an overview of the
theoretical techniques for incorporating relativistic and QED
effects in few electron systems, and to summarize the comparison
with experiment. The energies of one-electron systems are
discussed in Section II, and the comparison with light muonic
systems is outlined in Section III. The extension to two- and
three-electron systems is discussed in Sections IV and V respectively. Finally, a method for including relativistic effects in
the theory of radiative transitions is presented in Section VI.
Relativistic Hartree-Fock calculations are not discussed in
detail since they are covered in the lectures by Professor J. P.
Desclaux. In addition, Professor Soff discusses applications to
super heavy atomic systems with $Z > 137$. The experimental techniques relating to the present work are described by Professor R.
Marrus. Further details can be found in a review article by
Drake (1982) and earlier references therein.

II.   ONE-ELECTRON SYSTEMS

The study of QED effects in one-electron systems has now
reached a high state of refinement, both theoretically and
experimentally. Some of the material in this section is contained in the review articles of Lautrup et al. (1972), Kugel and
Murnick (1977) and Brodsky and Mohr (1978).

II.A.   Lamb Shift Theory

The development of modern QED began in 1947 with the discovery of the Lamb shift in atomic hydrogen (Lamb and Retherford,
1947). According to one-electron Dirac theory, the $2s_{1/2}$ and
$2p_{1/2}$ states should be exactly degenerate. However, as shown in

Fig. 1, QED corrections raise the $2s_{1/2}$ state relative to the $2p_{1/2}$ state by about one tenth of the $2p_{1/2} - 2p_{3/2}$ fine structure splitting. The direct observation of the $2s_{1/2}-2p_{1/2}$ transition frequency (Lamb shift) was rapidly followed by Bethe's (1947) estimate of the electron self-energy, with a result in rough agreement with experiment.

The physical origin of the Lamb shift can be understood as follows. In lowest order, the two contributions are the electron self-energy and vacuum polarization, as illustrated by the Feynman diagrams in Fig. 2. The electron self-energy, which arises from the emission and reabsorption of virtual photons, can be treated qualitatively as the interaction of the electron with the zero point oscillations of the electromagnetic field (Welton, 1948). This tends to smear the electron charge over a mean square radius of (Bjorken and Drell, 1964)

$$\langle (\delta r)^2 \rangle = (2\alpha^3 a^2/\pi)\ln(Z\alpha)^{-1}$$
$$= (5.838 \times 10^{-12} \text{ cm})^2 \text{ for } Z = 1 \ . \tag{1}$$

Here $a = \hbar^2/me^2$ is the Bohr radius, $\alpha = e^2/\hbar c$ is the fine structure constant and Z is the nuclear charge. The corresponding correction to the interaction energy with the Coulomb field of the nucleus is

$$\langle \delta V \rangle = \langle V(r + \delta r) - V(r) \rangle \approx \frac{1}{6} (\delta r)^2 \langle \nabla^2 V \rangle \ . \tag{2}$$

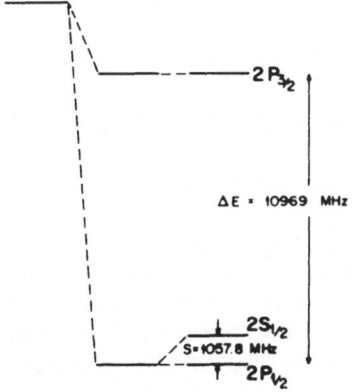

$2P_{3/2}$

$\Delta E = 10969$ MHz

$2S_{1/2}$
S = 1057.8 MHz
$2P_{1/2}$

Fig. 1.   Energy level diagram for the n=2 states of hydrogen showing the progressive splittings produced by first, relativistic corrections in one-electron Dirac theory, and then by QED corrections. The diagram is drawn roughly to scale.

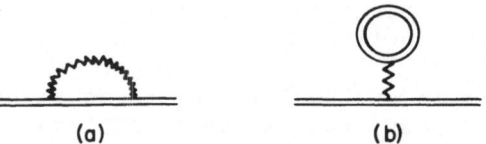

Fig. 2.   Feynman diagrams for (a) the electron self-energy and
           (b) the vacuum polarization. Double solid lines repre-
           sent a bound electron in the Coulomb field of the
           nucleus.

Using

$$\nabla^2 V = 4\pi Z e^2 \delta^3(\vec{r}) \tag{3}$$

$$|\psi_{n,\ell}(0)|^2 = \frac{Z^3 \delta_{\ell,0}}{\pi n^3 a^3} \tag{4}$$

for the electron probability density at the nucleus, the energy
shift for s-states is

$$\Delta E_n = 8Z^4 \alpha^3/(3\pi n^3)\ell n(Z\alpha)^{-1} R_\infty \tag{5}$$

where $R_\infty = e^2/2a$ is the Rydberg unit of energy for infinite
nuclear mass. It is usual to express Lamb shifts in Rydberg frequency units defined by $Ry = R_\infty/\hbar$. Equation (5) for n = 2 and Z
= 1 then gives a frequency shift of ~1000 Mc/sec.

The vacuum polarization correction can be thought of as
arising from a polarization of the virtual electron-positron
pairs surrounding a bare charge. The physically measured charge
is therefore smaller than the bare charge since it is screened by
the polarized vacuum. The screening distance is of the order of
the Compton wavelength $\hbar = \alpha a$. The resulting correction to
Coulomb's law was first investigated by Uehling (1935). To
lowest order in α, the interaction energy between an electron and
a point nucleus of charge Z is (Akhiezer and Berestetskii, 1965;
Schwinger, 1949)

$$V(r) = -\frac{Ze^2}{r} \{1 + \frac{2\alpha}{3\pi} \int_1^\infty e^{-2r\zeta/\lambda} (1 + \frac{1}{2\zeta^2})\frac{(\zeta^2-1)^{1/2}}{\zeta^2} d\zeta \}$$

$$= V_0(r) + \Delta V_{VP}(r) \tag{6}$$

and the first order perturbation correction to the $2s_{1/2}$–$2p_{1/2}$ energy splitting is

$$\Delta E_{VP} = \int d^3\vec{r} \, \Delta V_{VP}(r)[\,|\psi(2s_{1/2})|^2 - |\psi(2p_{1/2})^2|\,] \, . \tag{7}$$

The Lamb shift has been measured in the muonic system μ^- – He^{++} as well as in ordinary hydrogenic ions. If μ is the reduced mass, $m_\mu m_N/(m_\mu + m_N)$ of the muon-nucleon system and m_e is the electron mass, then the vacuum polarization term scales roughly as $\mu^3 c^2/m^2_e$, while the self energy scales only as μc^2. Thus, vacuum polarization dominates the Lamb shift of muonic systems, while self energy terms dominate the Lamb shift of electronic systems. For a particle of reduced mass μ and Bohr radius $a_\mu = \hbar^2/\mu e^2$, the nonrelativistic wave functions in (7) are

$$\psi(2s) = (\frac{Z}{a_\mu})^{3/2} \frac{1}{2\sqrt{2}} (2-\rho)e^{-\rho/2} \tag{8}$$

and

$$\psi(2p) = (\frac{Z}{a_\mu})^{3/2} \frac{1}{2\sqrt{2}} \rho e^{-\rho/2} \tag{9}$$

where $\rho = Zr/a_\mu$. Substituting (8) and (9) into (7), integrating over r and changing the remaining variable of integration in (6) from ζ to $z = 1/\zeta$ yields

$$\Delta E_{VP} = -\frac{\alpha Z^4 R_\mu}{15\pi} (\frac{\lambdabar}{a_\mu})^2 I(\beta) \tag{10}$$

where

$$I(\beta) = \frac{5}{2} \int_0^1 \frac{(1 + z^2/2)(1-z^2)^{1/2}}{(1 + \beta z)^4} z\,dz \tag{11}$$

$R_\mu = e^2/2a_\mu$ is the Rydberg for a particle of reduced mass μ and $\beta = Z\lambdabar/(2a_\mu)$ measures the extent to which the orbital radii lie within the vacuum polarization region. $I(\beta)$ can be expanded in powers of β as

$$I(\beta) = \sum_{n=0}^{\infty} T_n \beta_n \tag{12}$$

with $T_0 = 1$, $T_1 = -25\pi/32$, and the remaining T_n are given by the recurrence relation

$$T_n = \frac{(n+3)(n+4)}{(n-1)(n+5)} T_{n-2} . \tag{13}$$

For an electron, $\beta = \alpha Z/2$, $\lambdabar/a_\mu = \alpha$ (neglecting reduced mass corrections) and $I(\beta) \simeq 1$. Then (10) reduces to

$$\Delta E_{VP} = -\frac{\alpha^3 Z^4}{15\pi} R_\mu \tag{14}$$
$$\simeq -27 \text{ MHz for } Z = 1.$$

However, for a muon ($m_\mu/m_e = 206.769$), $\lambdabar/a_\mu = \alpha m_\mu/m_e$ and $\beta = Z\alpha m_\mu/(2m_e) \sim 1$. In this case, the higher terms in (12) make an important contribution. The lowest order vacuum polarization correction is given in Table 1 for several muonic systems. For $\beta < 1$ the power series (12) is convergent, but for the heavier ions, $\beta > 1$ and the integral (11) must be calculated numerically. An accurate calculation requires also relativistic self-energy and finite nuclear size corrections (see Section III). The point of Table 1 is that for increasing Z, the small values of $I(\beta)$ cause ΔE_{VP} to increase much more slowly than the Z^4 dependence indicated by Eq. (10). In fact, for $Z > 4$, the upward correction due to finite nuclear size exceeds the downward vacuum polarization correction, causing the overall Lamb shift to reverse sign. The results in Table 1 for the muonic systems up to $\mu^- - {}^4He_2$ are in good agreement with calculations by DiGiacomo (1969, 1970), Campani (1970) and Borie (1975).

II.B   Precise Lamb Shift Calculations

The traditional method of calculating the dominant self-energy and vacuum polarization corrections has been to evaluate successively higher order terms in powers of $Z\alpha$. This approach is discussed in detail by Erickson and Yennie (1965) and reviewed by Taylor et al. (1969). Since then, a number of important advances have been made. We start by summarizing the low order terms which are known exactly.

The expansion of the radiative corrections in powers of $Z\alpha$ for a bound atomic electron has the form

$$\Delta E = \frac{8mc^2\alpha(Z\alpha)^4}{6\pi n^3} \{A_{40} + A_{41} \ln(Z\alpha)^{-2} + A_{50}(Z\alpha)$$

$$+ (Z\alpha)^2 [A_{62}\ln^2(Z\alpha)^{-2} + A_{61}\ln(Z\alpha)^{-2} + G(Z\alpha)]$$

$$+ (\alpha/\pi)[B_{40} + O(Z\alpha)] + O(\alpha^2/\pi^2)\} + \Delta E_M + \Delta E_R \qquad (15)$$

Table 1. Lowest order vacuum-polarization corrections for several muonic systems.

System	μ/m_e	β	$I(\beta)$	ΔE_{VP}
$e^--{}^1H_1$	0.999456	3.64669×10^{-3}	0.991106	-26.8435 MHz
$\mu^--{}^1H_1$	185.841	0.678076	0.284695	- 0.20502 eV
$\mu^--{}^2D_1$	195.739	0.714189	0.270532	- 0.22763
$\mu^--{}^3He$	199.271	1.45415	0.115589	- 1.64189
$\mu^--{}^4He$	201.068	1.46726	0.114154	- 1.66577
$\mu^--{}^6Li_3$	202.940	2.22139	0.0614161	- 4.66497
$\mu^--{}^7Li$	203.478	2.22727	0.0611584	- 4.68242
$\mu^--{}^1Be_4$	204.198	2.98020	0.0378325	- 9.25583
$\mu^--{}^{11}B_5$	204.659	3.73367	0.0255789	-15.3756
$\mu^--{}^{12}C_6$	204.832	4.48418	0.0184046	-22.9986
$\mu^--{}^{14}N_7$	205.107	5.23856	0.0138300	-32.1464
$\mu^--{}^{16}O_8$	205.312	5.99294	0.0107579	-42.7867
$\mu^--{}^{19}F_9$	205.541	6.74960	0.0085935	-54.9310
$\mu^--{}^{20}Ne_{10}$	205.602	7.50176	0.0070260	-68.5124

where

$$G(Z\alpha) = A_{60} + O(Z\alpha) .$$ (16)

Except for A_{60} in $G(Z\alpha)$, all of the coefficients shown in (15) have been calculated exactly (Lautrup et al. 1972, Brodsky and Mohr, 1978, Drake, 1982). The terms ΔE_M and ΔE_R in (15) are finite nuclear mass and nuclear size corrections discussed by Erickson and Yennie (1965) and Brodsky and Mohr (1978).

The principal uncertainty is the contribution from higher order binding energy corrections, represented by $G(Z\alpha)$ in (16). The difficulty is that in an exact relativistic calculation, the exact Dirac Coulomb Green's function must be used. Although a form involving an infinite sum over partial waves is known (Wichmann and Kroll, 1956, Zon et al. 1972, Zapryagaev and Manakov, 1976), no closed form expression exists. The problem is severe because the uncalculated terms in the expansion (15) become of the order of magnitude of the calculated terms for Z > 10, and the expansion seems to diverge for $\alpha Z \geq 1$. Erickson (1971) and Mohr (1976) have done independent calculations of the self energy part of $G(Z\alpha)$, by performing a numerical summation over the partial wave expansion of the exact Dirac Coulomb Green's function and subtracting off the known lower order contributions, thereby avoiding an expansion in powers of $Z\alpha$. The results obtained by Mohr (1976) for Z = 10, 20, ..., 50 are shown in Table 2. Numerical difficulties prevented explicit calculations for Z < 10, but Mohr estimated the small Z behavior by fitting the results for Z = 10, 20, 30 to a series of the form

Table 2. Values of $G^{SE}(Z\alpha)$ and $G^U(Z\alpha)$ Tabulated by Mohr (1976)

Z	$G^{SE}(Z\alpha)$	$G^U(Z\alpha)$
1		−0.5587
10	−20.13(34)	−0.5015
20	−17.674(28)	−0.4711
30	−15.776(11)	−0.4576
40	−14.1376(62)	
50	−12.6650(28)	

$$G^{SE}(Z\alpha) = A_{60}^{SE} + b(Z\alpha)\ell n(Z\alpha)^{-2} + c(Z\alpha) \tag{17}$$

in analogy with the corresponding high order terms in the vacuum polarization. The results of the fitting procedure are

$$a = -24.064 \pm 1.2 \qquad b = 7.3071 \qquad c = 15.6609$$

for the $2s_{1/2}-2p_{1/2}$ energy difference. The result $G^{SE}(0) = -24.064 \pm 1.2$ is within the error limits of the earlier estimate -19.08 ± 5 by Erickson and Yennie (1965), but disagrees with Erickson's (1977) value -17.246 ± 0.5. The difference corresponds to a Lamb shift difference of 0.049 MHz in hydrogen, which is much greater than the accuracy of recent experiments. A recent recalculation by Sapirstein (1981), which treats as perturbations certain small terms in the equation for the Dirac Coulomb Green's function, yields $G^{SE}(0) = -24.9 \pm 0.9$ in agreement with Mohr's value. Although Sapirstein's (1981) calculation is for the $1s_{1/2}$ ground state, the n dependence is expected to be small. His result therefore indicates that Mohr's values for $G^{SE}(Z\alpha)$ are probably substantially correct.

Mohr (1976) has similarly calculated the vacuum polarization part $G^{VP}(Z\alpha)$ of G by writing it in the form

$$G^{VP}(Z\alpha) = G^{U}(Z\alpha) + G^{WK}(Z\alpha) \tag{18}$$

where $G^{U}(Z\alpha)$ is the contribution from the integral over the Uehling potential as shown in Eq. (7) after subtracting the known lower order terms contained in (15), and $G^{WK}(Z\alpha)$ is the contribution of third order in the external potential calculated by Wichmann and Kroll (1956). The latter is given by

$$G^{WK}(Z\alpha) = \frac{19}{60} - \frac{\pi^2}{36} + (\frac{3}{64} - \frac{31\pi^2}{3840})\pi(Z\alpha) + \ldots$$

$$= 0.04251 - 0.10305(Z\alpha) + \ldots \tag{19}$$

and Mohr's numerical values for $G^{U}(Z\alpha)$ at Z = 10, 20 and 30 are shown in Table 2. Fitting a functional form analogous to Eq. (17) yields

$$G^{U}(Z\alpha) \simeq -0.557 \pm 0.003 + 0.221(Z\alpha)\ell n^{-2}(Z\alpha)$$

$$-0.128(Z\alpha) . \tag{20}$$

The sum of (17), (19) and (20) is

$$G(Z\alpha) \simeq -24.598 \pm 1.2 + 7.528(Z\alpha)\ell n(Z\alpha)^{-2}$$

$$+ 15.429(Z\alpha) \pm 0.1(Z\alpha) \tag{21}$$

for the $2s_{1/2} - 2p_{1/2}$ energy splitting. The last term in (21) is an approximation for the uncertainty due to higher order omitted terms.

A further recent contribution to the theory of Lamb shifts is an argument by Borie (1981) that finite nuclear size effects should be included directly in the calculation of the low order level shift of $ns_{1/2}$ states. Referring back to the arguments leading to Eq. (5), this means that $\langle \nabla^2 V \rangle$ should be replaced by $\langle 4\pi e\rho(\vec{r}) \rangle$, where $\rho(\vec{r})$ is the nuclear charge density. For a point nucleus, $\rho(\vec{r}) = Ze\delta^3(\vec{r})$, and the usual results are recovered. Using a calculation analogous to that of Zemach (1956) for the size correction to hyperfine structure, and of Friar (1979) for the size correction in muonic atoms, Borie finds that in lowest order

$$\langle \rho(r) \rangle_{ns} = Ze |\psi_{ns}^{(0)}|^2 (1 - \frac{2\alpha2}{\pi} \langle r \rangle_{(2)}) \tag{22}$$

where $\langle r \rangle_{(2)} = \int r\rho(|\vec{r} - \vec{u}|)\rho(u)d^3ud^3r.$ \tag{23}

Assuming an exponential form for the charge distribution, then

$$\langle r \rangle_{(2)} = \frac{35R}{16\sqrt{3}} \tag{24}$$

where R is the nuclear radius. Thus, the s-state Lamb shift terms $A_{40} + A_{41}\ell n(Z\alpha)^{-2}$ in (15) should be multiplied by the factor $[1 - 35\alpha ZR/(16\sqrt{3}\pi)]$ to give an additional additive correction of

$$\Delta E_B = \frac{8mc^2\alpha(Z\alpha)^4}{6\pi n^3} (\frac{-35\alpha ZR}{8\sqrt{3}\pi})[A_{40} + A_{41} \ell n(Z\alpha)^{-2}]\delta_{\ell,0} \tag{25}$$

For hydrogen and deuterium, the additional shifts are -0.042 MHz and -0.104 MHz, respectively, assuming $R_H = 0.86$ fm and $R_D = 2.10$ fm. The validity of the Borie correction has been questioned by Lepage et al. (1981) on the grounds that it may be cancelled by higher order and relativistic effects.

In summary, the $2s_{1/2} - 2p_{1/2}$ Lamb shift for hydrogenic ions is obtained from (15) by calculating $S = \Delta E(2s_{1/2}) - \Delta E(2p_{1/2})$ with the result

$$S = \frac{mc^2\alpha(Z\alpha)^4}{6\pi} \left\{ \ln(Z\alpha)^{-2} + \ln\left(\frac{\epsilon_{2,1}}{\epsilon_{2,0}}\right) + \frac{91}{120} \right.$$

$$+ \; 0.32208 \;\alpha/\pi \;+\; 2.29622\pi\alpha Z$$

$$+ \; (Z\alpha)^2 \left[-\frac{3}{4}\ln^2(Z\alpha)^{-2} + 3.91842\ln(Z\alpha)^{-2} + G(Z\alpha)\right]\bigg\}$$

$$+ \; S_M + S_R + S_B \tag{26}$$

where

$$S_M = \frac{mc^2\alpha(Z\alpha)^4}{6\pi}\left\{\left(-\frac{3m}{M}\right)\left[\ln(Z\alpha)^{-2} + \ln\left(\frac{\epsilon_{2,1}}{\epsilon_{2,0}}\right) + \frac{23}{60}\right]\right.$$

$$+ \; \left(\frac{Zm}{M}\right)\left[\frac{1}{4}\ln(Z\alpha)^{-2} + 2\ln\left(\frac{\epsilon_{2,1}}{\epsilon_{2,0}}\right) + \frac{97}{12}\right]\bigg\} \tag{27}$$

$$S_R = [1 + 1.70(Z\alpha)^2]\,\frac{(Z\alpha)^2}{12}\,mc^2\left(\frac{Z\alpha R}{\lambda}\right)^{2s} \tag{28}$$

with $\quad s = [1 - (Z\alpha)^2]^{1/2}$,

$$S_B = -\frac{mc^2\alpha(Z\alpha)^4}{6\pi}\,\frac{35 Z\alpha R}{8\sqrt{3}\lambda}\left[\ln(Z\alpha)^{-2} + \ln\frac{Z^2 R_\infty}{\epsilon_{2,0}} + \frac{19}{30}\right] \tag{29}$$

and $G(Z\alpha)$ is given by Eq. (21). The quantity $\epsilon_{n,\ell}$ is the Bethe mean excitation energy calculated to high accuracy by Klarsfeld and Maquet (1973). Values for several low lying hydrogenic states are given in Table 3.

In addition to uncertainties arising from the value of $G(Z\alpha)$, an additional error is introduced by uncertainties in the value of R appearing in the finite nuclear radius correction S_R. Mohr (1976) has estimated that the error ΔS_R in S_R arising from an error ΔR in R is

$$\frac{\Delta S_R}{S_R} \simeq 0.7(Z\alpha)^2 + \frac{2\Delta R}{R} \;. \tag{30}$$

It is significant that ΔS_R tends to become a decreasingly small fraction of the total uncertainty in the Lamb shift as Z increases. Experiments at high Z are therefore better able to isolate QED effects from nuclear size uncertainties.

Table 3. Values of the Bethe Logarithm Calculated by Klarsfeld
         and Maquet (1973).

State	$\ln(\epsilon_{n\ell}/Z^2 R_\infty)$
1s	2.9841285558
2s	2.8117698931
2p	−0.0300167086
3s	2.7676636125
3p	−0.0381902294
3d	−0.0052321481
4s	2.7498118405
4p	−0.0419548946
4d	−0.0067409
4f	−0.0017337

The calculated values of the Lamb shift are given in Table 4
for Z from 1 to 30. The tabulated numbers are essentially the
same as those given by Mohr (1976), with the following excep-
tions. For hydrogen, the revised proton charge radius of R =
0.862±0.012 fm (Simon et al. 1980) has been used in place of the
value 0.81±0.02 fm used by Mohr. This change increases S by
0.016 MHz. Also for He$^+$, the revised nuclear radius R =
1.674±0.012 fm (Sick et al. 1976; Borie and Rinker, 1978) has
been used in place of 1.644 fm. The other nuclear radii are
representative values derived from electron scattering data
(Elton et al. 1967 and de Jager et al. 1974), and from muonic
atom transition energies (Engfer et al. 1974). The additional
Borie correction term S_B is not included, but is tabulated
separately since its validity is still open to question. Its
effect (relative to the experimental precision) is particularly
great in the case of deuterium. The values of the fundamental
constants used are α^{-1} = 137.03604, $\lambdabar$ = 386.159 fm and $mc^2\alpha^5/6\pi$
= 135.643665 MHz.

II.C. Comparison with Experiment

The uncertainties in the calculated Lamb shifts shown in
Table 4 arise principally from two sources: (i) higher
order binding energy corrections, represented by G(Zα),
which increase in proportion to Z^6, and (ii) nuclear radius
corrections, which increase in proportion to Z^4. Since the
Lamb shift itself increases roughly in proportion to Z^4, the

Table 4. Values of the $2S_{1/2} - 2P1/2$ Lamb shift S excluding the S_O contribution, and of S_B.

Z	A	R(fm)	S[a]	$-S_B$
1	1	0.862(12)	1057.883(13	0.032 MHz
1	2	2.10(2)	1059.241(27)	0.104
2	4	1.674(5)	14043.36(55)	2.18
3	6	2.56(5)	62737.5(6.6)	22.0
4	9	2.52(2)	179.791(25)	0.081 GHz
5	11	2.4(1)	404.57(10)	0.22
6	12	2.45(1)	781.99(21)	0.50
7	14	2.54(2)	1361.37(47	1.04
8	16	2.72(3)	2196.21(92)	2.02
9	19	2.90(2)	3343.1(1.6)	3.6
10	20	3.02(4)	4861.1(2.7)	6.0
11	23	2.94(4)	6809.0(4.0)	8.8
12	24	3.01(3)	9256.0(5.8)	13.1
13	27	3.03(3)	12264.7(8.0)	18.4
14	28	3.09(2)	15907.(11)	26.
15	31	3.19(2)	20254.(13)	35.
16	32	3.24(2)	25373.(17)	47.
17	35	3.34(3)	31347.(20)	61.
18	40	3.45(5)	38250.(25)	79.
19	39	3.41(3)	46133.(29)	97.
20	40	3.48(3)	55116.(37)	120.
21	45	3.54(8)	65259.(55)	147.
22	48	3.60(1)	76651.(56)	178.
23	51	3.60(5)	89345.(78)	209.
24	52	3.66(5)	103.482(98)	0.246 THz
25	55	3.72(7)	119.12(13)	0.29
26	56	3.73(6)	136.32(15)	0.33
27	59	3.80(5)	155.25(18)	0.38
28	58	3.78(3)	175.85(21)	0.42
29	63	3.93(3)	198.54(26)	0.48
30	64	3.95(4)	223.03(32)	0.53

a

Calculated with Mohr's (1976) values for $G^{SE}(Z\alpha)$ given by Eq. (32). To obtain the corresponding Lamb shifts with Erickson's (1977) values, add $Z^6[0.04925-0.05278(Z\alpha)\ln(Z\alpha)^{-2} + 0.02961(Z\alpha)]$MHz.

experimental precision required for a significant test of the
$G(Z\alpha)$ term is roughly $\pm$ 10 Z^2 ppm. Thus, less precision is
required at high Z than low Z for the same theoretical significance. Also, uncertainties arising from the nuclear radius
correction, which is predominantly a non-QED effect, become relatively less important with increasing Z. It is therefore of
importance to perform both high precision measurements at low Z
and lower precision measurements at high Z. The overall comparison between theory and experiment is summarized in Table 5
and Fig. 3. In Fig. 3, $(S-\bar{S})/Z^6$ is plotted for the various hydrogenic ions, where S is a Lamb shift measurement and $\bar{S}$ is the Lamb
shift calculated as in Table 4, except that $G_{Mohr}(Z)$ given by Eq.
(17) is replaced by the average value

$$\bar{G}^{SE}(Z\alpha) = \frac{1}{2} [G^{SE}_{Mohr}(Z\alpha) + G^{SE}_{Erickson} (Z\alpha)]. \qquad (31)$$

The lower solid bars are then Mohr's values for the Lamb shift as
tabulated in Table 4, and the upper solid bars are Erickson's
values as calculated by adding the expression at the bottom of
Table 4. These are not quite the same as those tabulated
directly by Erickson (1977) since he uses slightly different
values for the nuclear radii. The same remarks apply to the
entries in Table 5. The dashed lines in Fig. 3 are the
corresponding theoretical values when the Borie correction S_B is
added.

The most significant points emerging from the comparison between theory and experiment shown in Fig. 3 are as follows.
First, the most severe tests of theory are the high precision
measurement for H by Lundeen and Pipkin (1981), and the high Z
measurements for Cl^{16+} (Wood et al. 1982) and Ar^{17+} (Gould and
Marrus, 1978). The H measurement agrees well with Mohr's theoretical value if the Borie correction for finite nuclear size is
included, but lies below theory if the Borie correction is
omitted. The validity of the Borie term could most easily be
tested in deuterium, but no experimental data of comparable
accuracy are available. At high Z, where the Borie correction is
negligible, the experimental data are in good agreement with
Mohr's theoretical values.

III. LIGHT MUONIC SYSTEMS

Interest in light muonic Lamb shifts arises primarily from a
measurement in the exotic system $\mu^- - {}^4He^{2+}$ by Bertin et al.
(1975). Their experiment is basically of the laser resonance
type, using a ruby-pumped infrared dye laser to stimulate the
$2s_{1/2}$–$2p_{3/2}$ transition at $\lambda \simeq$ 8120 Å.

Table 5. Comparison of Theoretical and Experimental Lamb Shifts in Hydrogenic Ions

Ion	Reference	Technique[a]	Value	Mohr Theory	Erickson Theory
H	Lundeen & Pipkin (1981)	SOF	1057.845(9)	1057.883(13)	1057.929(13) MHz
	Newton et al. (1979)	rf	1057.862(20)	1057.840(13)	1057.886(13)
	Robiscoe et al. (1970)	rf	1057.90(6)		
	Triebwaser et al.(1953)	rf	1057.77(6)		
D	Cosens (1968)	rf	1059.24(6)	1059.241(27)	1059.287(27)
	Triebwasser et al. (1953)	rf	1059.00(6)	1059.137(27)	1059.183(27)
	vanWijngaarden et al.(1978)	A	1059.36(16)		
^{4}He$^+$	Narasimham et al. (1971)	rf	14046.2(1.2)	14042.36(55)	14034.12(55)
	Lipworth et al. (1957)	rf	14040.2(1.8)	14040.18(55)	14042.94(55)
	Drake et al. (1979)	A	14040.2(2.9)		
^{6}Li^{2+}	Dietrich et al. (1976)	rf	62790.(70)	62737.5(6.6)	62767.4(6.6)
^{12}C^{5+}	Leventhal (1975)	rf	62765.(21)	62715.5(6.6)	62745.4(6.6)
	Kugel et al. (1972)	QR	780.1(8.0)	781.99(21)	783.67(21)GHz
				781.49(21)	783.17(21)
^{16}O^{7+}	Curnutte et al. (1981)	A	2192.(15)	2196.21(92)	2204.98(92)
	Lawrence et al.(1972)	QR	2215.6(7.5)	2194.19(92)	2202.96(92)
	Leventhal et al.(1972)	QR	2202.7(11.)		
^{19}F^{8+}	Kugel et al. (1975)	LR	3339.(35)	3343.1(1.6)	3360.3(1.6)
	Murnick et al. (1972)	QR	3405.(75)	3339.5(1.6)	3356.7(1.6)
^{35}Cl^{16+}	Wood et al. (1982)	LR	31190.(220)	31347.(20)	31965.(20)
				31286.(20)	31904.(20)
^{40}Ar^{17+}	Gould et al. (1978)	QR	38100.(600)	38250.(25)	39100.(25)
			38171.(25)	39021.(25)	

a SOF, separated oscillatory fields; rf, r.f. resonance; A, anisotropy; QR, quench rate; LR, laser resonance. b The second entry of each pair includes the Borie correction term S_B.

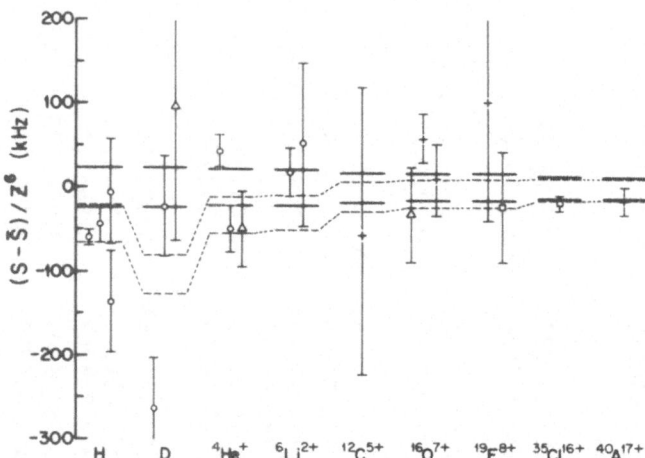

Fig. 3 Comparison between scaled theoretical and experimental
one-electron Lamb shifts, expressed as deviations form
the theoretical mean. The upper horizontal line for
each ion is Erickson's (1977) theory, the lower horizontal line is Mohr's (1976) theory and S is the average of
the two. The experimental data (see Table V) are
labelled by method of measurement according to o –
microwave resonance; Δ – anisotropy measurement; + –
quench rate measurement;   □ – laser resonance. The
dashed lines are the Borie (1981) corrections to the
theoretical values.

As discussed in Section II.A, the Lamb shift of light muonic
atoms is dominated by the vacuum polarization term, and approximate numerical values (using nonrelativistic wave functions) are
given in Table 1. The other important terms to be considered are
the $2p_{3/2} - 2p_{1/2}$ fine structure splitting

$$\Delta E_{FS} = \frac{(\alpha Z)^4 \mu c^2}{32} [1 + \frac{5\alpha^2 Z^2}{8}]$$

$$= 0.045283 (\frac{\mu}{m_e})[1 + \frac{5\alpha^2 Z^2}{8}] \text{ meV} \tag{32}$$

the finite nuclear size correction to the $2s_{1/2}$ state

$$\Delta E_R = \frac{e^2 Z^4}{12a_\mu^3} \langle r^2 \rangle$$

$$= 0.80979 \times 10^{-6} Z^4 (\frac{\mu}{m_e})^3 \langle r^2 \rangle \text{ meV/fm}^2 \tag{33}$$

and the muon self-energy (including anomalous magnetic terms, but excluding the vacuum polarization terms corresponding to A^{VP} and A^{VP} which are included in the numerical integration results in Table 1)

$$\Delta E_{SE} = - \frac{\alpha(Z\alpha)^4}{6\pi} m_\mu c^2 \{ (1 + \frac{m_\mu}{m_N})^{-3} [\frac{19}{30} + \ell n(Z\alpha)^{-2}$$

$$- \ell n \frac{\varepsilon_{20}}{\varepsilon_{21}}] - (1 + \frac{m_\mu}{m_N})^{-2} \frac{c_{2,j}}{8} + 3\pi\alpha Z (\frac{139}{128} - \frac{1}{2} \ell n2) \} \qquad (34)$$

where

$$\frac{\alpha(Z\alpha)^4 m_\mu c^2}{6\pi} = 0.11599 \text{ meV} \quad \text{and}$$

$\Delta E_{SE} = S^{SE}(2p_j) - S^{SE}(2s_{1/2})$. The expression for ΔE_{SE} assumes a point nucleus. It can be corrected for finite nuclear size by replacing the wave function at the origin $| \psi(0) |^2$ by the expectation value of the nuclear charge density $\langle \rho_N(r) \rangle$, as done for example by Rinker (1976) and Borie and Rinker (1978). The corresponding correction for electronic systems leads to the S_B term discussed in Section II.B Eq. (25). Also, the Bethe logarithms with finite nuclear size included have been calculated by Klarsfeld (1977) for heavy nuclei.

The largest effect not included above is the Källen–Sabry correction of order $\alpha^2(Z\alpha)^4$ to the vacuum polarization part. The leading term gives the B_{40}^{VP} contribution to the electronic Lamb shift, but for muonic systems a numerical integration must be done similar to the one for the Uehling term (Barbieri et al. 1973; DiGiacomo, 1969). There is also a relativistic recoil term analogous to the second term of (27), but with m_e replaced by m_μ, together with a finite radius recoil correction to the $2s_{1/2}$ state given by (Rinker, 1976; Friar and Negele, 1973)

$$\Delta B = \frac{9}{70} \frac{\mu}{m_N}(Z\alpha)^4(\mu c^2)[\beta + O(\beta \ell n \beta)] \qquad (35)$$

with $\beta = \alpha Z \mu R_0 / m_e \lambdabar$ for a uniform nucleus of radius R_0. These terms plus other small corrections are discussed by Borie and Rinker (1978). Their results for $\mu^- - {}^4He^{2+}$ are shown in Table 6. These calculations are particularly significant for electronic Lamb shift measurements in ${}^4He^+$ because they provide an independent check on the value $\langle r^2 \rangle^{1/2} = 1.674 \pm 0.012$ fm for the nuclear radius obtained from electron scattering data (Sick et al. 1976, Rinker, 1976). The 3p-3d transition frequencies (Borie and Rinker, 1980) and hyperfine structure in ${}^3He$ muonic states (Borie, 1976) have also been calculated.

Table 6. Contributions to the 2s-2p splittings in the $(\mu^4\text{He})^+$ system, in meV.   r is the rms charge radius of ^4He in fm. All formulas are evaluated using the value r^2 = 2.802 fm^2. (From Borie & Rinker, 1978).

Contributions	$S_1 = 2p_{3/2}-2s_{1/2}$	$S_2 = 2p_{1/2} -2s_{1/2}$
Electronic VP		
Uehling $\alpha Z\alpha$ first iteration		
	$1666.57-0.76r^2=1664.44$	$1666.30-0.76r^2=1664.17$
higher iterations	1.70	1.70
Källen-Sabry $\alpha^2 Z\alpha$	11.55	11.55
$\alpha(Z\alpha)^n$, $n \geqslant 3$	−0.02	−0.02
$\alpha^2(Z\alpha)^2$	0.02	0.02
muon VP $\alpha Z\alpha$	0.33	0.33
$\mu - e$ VP $\alpha^2 Z\alpha$	0.02	0.02
hadron VP	0.05	0.05
Total	$1680.22-0.76r^2=1678.09$	$1679.95-0.76r^2=1677.82$
Vertex corrections		
$\alpha Z\alpha$	$-10.90+0.23r=-10.52$	$-11.23+0.23r=-10.85$
$\alpha(Z\alpha)^n$, $n \geqslant 2$	−0.16	−0.16
$\alpha^2 Z\alpha$	−0.03	−0.03
Total	$-11.09+0.23r=-10.71$	$-11.42+0.23r=-11.04$
Recoil Breit	$+0.17r= 0.28$	$0.17r= 0.28$
two-photon	−0.44	−0.44
Total	$-0.44+0.17r= -0.16$	$0.44+0.17r= -0.16$
Point Coulomb (fine structure)	145.70	0.00
Total QED	$1814.39+0.40r-0.76r^2$ $=1812.92$	$1668.09+0.40r-0.76r^2$ $=1666.62$
Nuclear polarization	3.1±0.6	3.1±0.6
Finite size	$-105.46r^2+1.40r^3$ $=-288.9\pm4.1$	$-105.46r^2+1.40r^3$ $=-288.9\pm4.1$
Total	$1817.5+0.4r-106.2r^2+1.4r^3$ $=1527.1\pm4.2$	$1671.2+0.4r-106.2r^2$ $+1.4r^3=1380.8\pm4.2$

Since Lamb shift measurements in other light muonic atoms may become possible, estimates obtained from (32), (33), and (34), together with the nuclear radii in Table 4, are given in Table 7. The higher order terms discussed above could also be included, but the uncertainties are dominated by the nuclear radius correction. The important conclusion is that $\Delta E(2p_{3/2} - 2s_{1/2})$ for ^{6}Li ($\lambda \simeq 6900$ Å), and $\Delta E(2p_{1/2} - 2s_{1/2})$ for ^{9}Be ($\lambda \simeq 6000$ Å) lie in the visible region of the spectrum accessible to tunable dye lasers.

The calculation of transition energies in heavy muonic systems is reviewed by Brodsky and Mohr (1978), and detailed tabulations have been given by Engfer et al. (1974) and Rinker et al. (1977). Self energy corrections for $1s_{1/2}$ levels have been calculated to all orders in the external field by Cheng et al. (1978). The effects of electron screening (von Edidy et al. 1978) and the Breit interaction (Rashid and Fricke, 1980) on the fine structure of excited muonic states have also recently been discussed.

Table 7. Estimates of energy differences in light muonic atoms (meV)

Ion	ΔE_{VP}	$\Delta_{VP}^{(2)a}$	ΔE_{SE}	ΔE_R	ΔE_{FS}	Total
			$E(2p_{1/2}) - E(2s_{1/2})$			
$^1_1$H	0.2050	0.0017	−0.006	−0.0038	0	0.2023(1)
$^4_2$He	1.666	0.011	−0.011	−0.289	0	1.377(4)
$^6_3$Li	4.665	0.032	−0.051	−3.593	0	1.053(70)
$^9_4$Be	9.256	0.064	−0.148	−11.21	0	−2.037(89)
$^{11}_5$B	15.376	0.11	−0.336	−24.99	0	−9.8(1.0)
$^{12}_6$C	22.999	0.16	−0.653	−54.14	0	−31.63(22)
			$E(2p_{3/2}) - E(2s_{1/2})$			
$^1_1$H	0.2050	0.0017	−0.006	−0.0038	0.0085	0.2108(1)
$^4_2$He	1.666	0.011	−0.011	−0.289	0.146	1.523(4)
$^6_3$Li	4.665	0.032	−0.049	−3.593	0.795	1.800(70)
$^9_4$Be	9.256	0.064	−0.143	−11.21	2.368	0.336(89)
$^{11}_5$B	15.376	0.11	−0.323	−24.99	5.797	−4.0(1.0)
$^{12}_6$C	22.999	0.16	−0.626	−54.14	12.04	−19.53(22)

a
The Källen–Sabrey term. The values for ^{6}Li to ^{12}C are estimates.

IV. TWO—ELECTRON SYSTEMS

High precision calculations for two—electron atoms and ions
are complicated by the necessity of simultaneously taking into
account relativistic, QED and electron correlation effects.
There is no unique way of specifying an exact relativistic two
electron Hamiltonian analogous to the Dirac Hamiltonian without
at the same time including QED effects to all orders. The
rigorous starting point is generally accepted to be the fully
covariant Bethe—Salpeter equation derived from the Feynman form
of QED by Salpeter and Bethe (1951), and from quantum field
theory by Gell—Mann and Low (1951). Exact solutions to this
equation have not been obtained. All practical calculations are
based (explicitly or implicitly) on expansions of the
Bethe—Salpeter equation in powers of α, αZ and Z^{-1}. Early work
of this type for two—electron atoms was done by Sucher (1958) and
Araki (1957). Unlike the Lamb shift in one—electron systems,
there are no QED effects which manifest themselves directly in
lowest order. It is first necessary to subtract the nonrelativistic two—electron energies and relativistic corrections before
specifically QED effects such as the Lamb shift are revealed.

Distinctly different types of approximation are useful in
the low Z and high Z regions. For low Z, the best strategy is to
obtain highly accurate solutions to the nonrelativistic
Schrodinger equation, and then include relativistic effects by
perturbation theory. For high Z, better results are obtained by
starting with exact solutions to the one—electron Dirac equation
and then including the electron—electron interaction and higher
order radiative corrections by perturbation theory.

Because of recent developments in beam foil spectroscopy, and
other highly stripped ion sources such as tokamaks and fusion
plasmas, high precision energy level measurements in heavy two—
electron ions are becoming available for comparison with theory.
The two types of approximation in the low Z and high Z ranges are
discussed separately in the following sections, together with
comparisons with experiment.

IV.A. Calculations and Results for He and Li+

Fine Structure Splittings. Until recently, most calculations for low Z have been based on an approximately covariant
generalization of the one—electron Dirac equation proposed by
Breit (1929). The non-relativistic energy E_{NR} is of order $\alpha^2 mc^2$
and the fine structure splittings are of order $\alpha^4 mc^2$. The terms
to this order are given correctly by the eigenvalues of the
16—component spinor equation

$$H_B \psi = E_B \psi \tag{36}$$

where $H_B = H_D(1) + H_D(2) + V_{12} + B$, $\tag{37}$

$$H_D(i) = c\vec{\alpha}_i \cdot \vec{p}_i + \beta mc^2 - Ze^2/r_i \quad , \tag{38}$$

$$V_{12} = e^2/r_{12} \quad , \tag{39}$$

$\vec{\alpha}$ and β are 4 x 4 Dirac matrices (Bjorken and Drell, 1964), and B is the Breit operator repesenting the retarded Coulomb and magnetic interactions between the two electrons. For diagonal matrix elements in lowest order, B reduces to (see, e.g. Akhiezer and Berestetskii, 1965)

$$B = \frac{-e^2}{2r_{12}} (\vec{\alpha}_1 \cdot \vec{\alpha}_2 + \vec{\alpha}_1 \cdot \hat{r}_{12} \vec{\alpha}_2 \cdot \hat{r}_{12}) \tag{40}$$

where $\vec{r}_{12} = | \vec{r}_1 - \vec{r}_2 |$.

Mittleman (1971, 1972) has derived a state dependent operator which is valid for off-diagonal matrix elements and which also includes higher order Coulomb retardation corrections. Defining two particle matrix elements of an operator $\beta_{1,2}$ by

$$\beta_{n\ell,n'\ell'} = \iint d\vec{r}_1 d\vec{r}_2 \psi_n^*(\vec{r}_1) \psi_\ell^*(\vec{r}_2) \beta_{1,2} \psi_{n'}(\vec{r}_1) \psi_{\ell'}(\vec{r}_2) \tag{41}$$

then Mittleman's operator for the matrix elements $(V_{12} + B)_{n\ell,n'\ell'}$ can be written in the form (Drake, 1979)

$$V_{12} + B = \frac{e^2}{r_{12}} [F(r_{12}) - \vec{\alpha}_1 \cdot \vec{\alpha}_2 \, G(r_{12})] \tag{42}$$

where

$$F(r_{12}) = 1 + \frac{E_{\ell,\ell'}}{2E_{n,n'}} (1 - \cos\Omega_{n,n'})$$

$$+ \frac{E_{n,n'}}{2E_{\ell,\ell'}} (1 - \cos\Omega_{\ell,\ell'}) \tag{43}$$

$$G(r_{12}) = \frac{1}{2} (\cos\Omega_{n,n'} + \cos\Omega_{\ell,\ell'}) \tag{44}$$

$\Omega_{n,n'} = E_{n,n'} r_{12}/\hbar c$ and $E_{n,n'} = E_n - E_{n'}$. For diagonal matrix elements, (42) reduces to the more familiar form

$$V_{12} + B = \frac{e^2}{r_{12}} (1 - \vec{\alpha}_1 \cdot \vec{\alpha}_2) \cos\Omega_{n,n'} \tag{45}$$

used in Dirac–Hartree–Fock calculations (Bethe and Salpeter, 1957; Mann and Johnson, 1971). If the $\cos\Omega_{n,n'}$ part of (45) is expanded into a power series and the term $(eE_{n,n'}/hc)^2 r_{12}$ replaced by the double commutator $- (e/\hbar c)^2 [H_D(1),[H_D(2),r_{12}]]$, then (45) reduces to the state-independent form (40). Further reducing (37) to the two component Pauli form results in

$$H_P = H^0 + B_P \tag{46}$$

where H_0 is the nonrelativistic Schrödinger Hamiltonian

$$H^0 = \frac{-\hbar^2}{2m} (\nabla_1^2 + \nabla_2^2) - \frac{Ze^2}{r_1} - \frac{Ze^2}{r_2} + \frac{e^2}{r_{12}} \tag{47}$$

and B_P is a sum of five operators of $O(\alpha^4 mc^2)$ representing the well known spin-orbit, spin-spin, etc. terms (see, e.g. Bethe and Salpeter, 1957, Eq. 39.14). High precision calculations of the eigenvalues of H_0 and the matrix elements of B_P have been done by Accad et al. (1971) and Schiff et al. (1973) for the 1sns ^{1}S, 1sns ^{3}S, 1snp ^{1}P and 1snp ^{3}P states with n up to 5 and Z up to 10. The most accurate calculations for the 1s3d ^{1}D and ^{3}D states are those of Blanchard and Drake (1973), Drake (1981) and Sims et al. (1982). Results for higher D and F states are given by Chang and Poe (1976), and Brown and Cortez (1971). These corrections, together with mass polarization effects, must be subtracted from experimental transition frequency measurements before higher order QED effects can be observed.

The experimental measurements in helium most sensitive to higher order effects are the fine structure splittings $J = 0 \to 1$ and $J = 1 \to 2$ in the 1s2p ^3P_J level (Lewis et al. 1970, Kpanou et al. 1971) (see Fig. 4). The experimental precision of about 1 ppm is sufficient to determine the fine structure constant to the same accuracy if the theoretical splitting can be calculated. Since corrections of $O(\alpha^6 mc^2)$ to B_P can be expected to contribute about 100 ppm, they must be included in the calculation. Schwartz (1964) initiated a program to carry the theoretical calculations to the 1 ppm level of accuracy.

The necessary corrections of $O(\alpha^6 mc^2)$ cannot be obtained directly from (36) because the equation becomes ill-behaved if relativistic effects are treated as other than a first order perturbation. As pointed out by Brown and Ravenhall (1951),

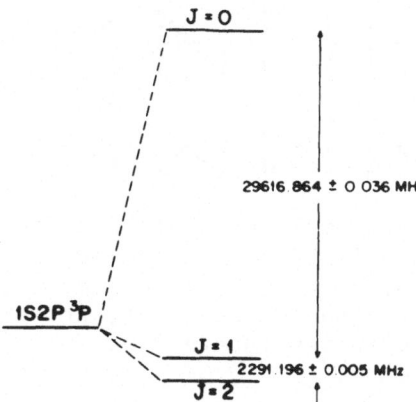

Fig. 4.   Energy level diagram showing the fine structure
           splitting of the 1s2p ^{3}P state of neutral helium.
           $\vec{J} = \vec{L} + \vec{S}$ is the total angular momentum.

Salpeter (1952), and recently discussed by Sucher (1981), the
Breit equation is inconsistent with Dirac hole theory in that it
allows transitions into negative energy states, which in hole
theory are filled.   To avoid these difficulties, Douglas (1971),
and Douglas and Kroll (1974) carried through a systematic reduction of the Bethe-Salpeter equation to obtain all the spin-
dependent terms which contribute to the 2 ^{3}P$_J$ fine structure up
to order $\alpha^6 mc^2$.   The results can be expressed symbolically in the
form

$$E_J = E^0 + \alpha^2 \langle B_P \rangle_J + \alpha^4 \langle B_P \frac{1}{E^0 - H^0} B_P \rangle_J$$

$$+ \alpha^4 \langle H_6 \rangle_J \tag{48}$$

where E^0 is the nonrelativistic energy, the second term is the
usual Breit interaction, the third term represents the Breit
interaction taken to second order and the fourth term containing
H_6 is the sum of higher order QED corrections to B_P obtained by
Douglas and Kroll.   The terms included are represented by the
Feynman diagrams in Fig. 5.   (Their result is not completely
general in that only the spin-dependent terms are retained.)   In
addition, there are anomalous magnetic moment corrections of
leading order $\alpha^5 mc^2$ (Araki et al. 1959; Stone, 1961, 1963;
Schwartz, 1964; Daley, 1972) and small relativistic recoil
corrections of order $(m/M)\alpha^4 mc^2$ (Stone, 1961, 1963; Daley et al.
1972).   If the spin dependent part of B_P is written in the form

Fig. 5.   Feynman diagrams contributing to the $\alpha^6 mc^2$ fine
structure splitting of two-electron systems. The
dashed line represents a Coulomb interaction between the two bound electrons, and the wavy line a
transverse photon interaction.

$$B_P = H_{Z,so} + H_{e,so} + H_{e,ss} \tag{49}$$

where $H_{Z,so}$ is the spin-orbit interaction with the nucleus,
$H_{e,so}$ is the spin-orbit interaction between the electrons and
$H_{e,ss}$ is the spin-spin interaction between the electrons, then
the anomalous magnetic moment terms are

$$H_5 = (\alpha/2\pi)(2H_{Z,so} + \tfrac{4}{3} H_{e,so} + 2H_{e,ss}) \tag{50}$$

and the spin-dependent relativistic recoil terms are (Stone,
1961; Daley et al. 1972)

$$H_N = - \frac{m}{M_{He}} [H_{Z,so} + 3H_{e,so} + 3H_{e,ss}$$

$$- \frac{Z\alpha^2}{r_1^3} \vec{\alpha}_1 \cdot (\vec{r}_1 \times \vec{p}_2)] . \tag{51}$$

The next higher anomalous magnetic moment terms are included in
H_6. H_6 contains a total of 16 terms, which are written out in
full by Daley et al.(1972).

The terms in (48) were calculated to high accuracy with
correlated variational wave functions (containing up to 455
terms) in Hylleraas coordinates. The second order perturbation
term was calculated variationally by solving an inhomogeneous
perturbation equation according to the method of Dalgarno and
Lewis (1956). The results of these extremely lengthy calculations are summarized in Table 8 (Lewis and Serafino, 1978). If

Table 8. Theoretical contributions to the fine-structure of 2^3P Helium in MHz. (From Lewis and Serafino, 1978).[a]

Term	ν_{01}	ν_{12}
$\alpha^4 mc^2$	29564.577(6)	2317.203(2)
$\alpha^5 mc^2$	54.708	-22.548
$(m/M)\alpha^4 mc^2$	-10.707	1.952
Second order	11.657(42)	-6.866(81)
$\alpha^6 mc^2$	-3.331(4)	1.542(7)
ν_{theory}	29616.904(43)	2291.283(81)
ν_{expt}	29616.864(36)[b]	2291.196(5)[c]
$\nu_{theory} - \nu_{expt}$	0.040	0.087
	=1.35 ppm	=37. ppm

[a] The values of α^{-1}, c, R_∞ and m/M used are 137.035987(29), 2.99792458(12) x 10^{10} cm/sec, 109737.3143 cm^{-1} and 1.370934 x 10^{-4}, respectively.

[b] Kponou et al. (1971).

[c] Lewis et al. (1970).

the theoretical value for ν_{01} is taken as correct, then the derived value for α^{-1} is 137.03608(13) in accord with other measurements as discussed by Lewis and Serafino (1978).

The fine structure splittings of the helium 1s3p ^3P states are discussed by Tam (1979). He recommends the values $\nu_{02} = 8772.55 \pm 0.04$ MHz (Kramer and Pipkin, 1978) and $\nu_{12} = 658.67 \pm 0.09$ MHz. The best theoretical values available for comparison are the calculations of Schiff et al. (1973) which include a correction for mixing between 1s3p $^3P_1$ and 1s3p $^1P_1$, but not for mixing with other intermediate states introduced by the third term of (48) or for the H_6 terms. The theoretical values are 8772.5 MHz and 658.04 MHz, respectively. The poor agreement for ν_{12} relative to ν_{02} indicates that the J = 1 state may be substantially shifted by further intermediate mixing terms.

In Li$^+$, the 2^3P_J fine structure splittings have recently been measured to high precision by Bayer et al. (1979) and Holt et al. (1980). The fine structure splittings of the 2^3P_J manifold are obtained by first extracting the effects of hyperfine structure (Jette et al. 1974; Aashamar and Hambro, 1977), and then taking differences of the transition frequencies to the common 2^3S_1 state. The results for a hypothetical spin zero nucleus are compared with each other and with theory in Table 9. The experimental results for ν_{02} do not lie within the quoted error limits of each other, although the ν_{01} and ν_{21} intervals agree reasonably well.

Table 9. Comparison of theory and experiment for the fine structure splitting of the 1s2p $^3P_{0,1,2}$ state in Li$^+$(in cm^{-1}).

	$\Delta \nu_{02}$	$\Delta \nu_{01}$	$\Delta \nu_{21}$
Theory[a]	3.1046	5.1944	2.0898
ΔE_{rel}[b]	0.0131	0.0104	-0.0027
Total	3.1177	5.2048	2.0871
Experiment	3.1051(12)[c]	5.1948(12)[c]	2.0897(12)[c]
	3.1028(2)[d]	5.1934(8)[d]	2.0906(8)[d]

[a] Schiff et al. (1973)
[b] Relativistic corrections of order $\alpha^4 Z^6$ and $\alpha^4 Z^5$.
[c] Holt et al. (1980)
[d] Bayer et al. (1979)

The calculation of the $\alpha^6 mc^2$ terms in Eq. (48) has not yet
been extended to Li$^+$. The theoretical results of Schiff et al.
(1973) shown in Table 9 include the $\alpha^2 \langle B_P \rangle_J$ term (of order
$\alpha^4 mc^2$), together with a 2 $^1P_1$ - 2 $^3P_1$ mixing correction and the
lowest order anomalous magnetic moment terms. These values are
in reasonably good agreement with experiment. However, addition
of the term ΔE_{rel} containing the leading corrections of order
$\alpha^6 Z^6 mc^2$ and $\alpha^6 Z^6 mc^2$ in (48) puts the totals into strong
disagreement with experiment. The source of the discrepancy
appears to come from higher order terms in Z^{-1}. For example,
ΔE_{rel} for the 0 → 2 transition is approximately

$$\Delta R_{rel}(0 \to 2) = -0.01953\alpha^6 Z^6 mc^2 (1 - 7.419Z^{-1}) .$$

The large contribution from the Z^{-1} term will presumably be
reduced by higher order terms in the series. The accuracy for
low Z could be improved by introducing a screening approximation.

Lamb Shifts. A number of transitions in helium have been
measured to sufficient accuracy to be sensitive to the Lamb shift
(Accad et al. 1971). The leading terms in the Lamb shift of two-
electron systems have been discussed by Araki (1957), Ermolaev
(1973, 1975) and Aashamar et al. (1976), among others. The
results can be expressed in the form

$$E_{L,2} = E'_{L,2}(nLS) + E''_{L,2}(nLS) + E'''_{L,2}(nLSJ) \tag{52}$$

where $E'_{L,2}$ is the proper Lamb and vacuum polarization correction given by

$$E'_{L,2}(nLS) = \frac{8\alpha^5 Zmc^2}{6} <\delta^3(\vec{r}_1) + \delta^3(\vec{r}_2)> \left[\ln(Z\alpha)^{-2}\right.$$

$$\left. + \frac{19}{30} + \ln\frac{Z^2 R_\infty}{\varepsilon_{nLS}} + 3\pi\alpha Z \left(\frac{427}{384} - \frac{1}{2}\ln 2\right)\right] \quad (53)$$

$E''_{L,2}(nLS)$ is an electron-electron interaction correction given by

$$E''_{L,2}(nLS) = \frac{\alpha^5 mc^2}{2}[<\delta^3(\vec{r}_{12})> (\frac{28}{3}\ln\alpha + \frac{178}{15} - \frac{40}{3}\vec{s}_1\cdot\vec{s}_2)$$

$$- \frac{28}{3}Q] \quad (54)$$

where Q is the principal part of $<r^{-3}_{12}>$, and $E'''_{L,2}(nLSJ)$ is the anomalous magnetic moment correction given by the expectation values of the operator H_5 defined by Eq. (50). Specifically,

$$H_5 = \frac{\alpha e \mu_B}{\pi mc} [\frac{Z}{r_1^3}(\vec{r}_1 \times \vec{p}_1)$$

$$+ \frac{Z}{r_2^3}(\vec{r}_2 \times \vec{p}_2) - \frac{2}{r_{12}^3}\vec{r}_{12} \times (\vec{p}_1 - \vec{p}_2)]\cdot(\vec{s}_1 + \vec{s}_2)$$

$$+ \frac{4\alpha\mu_B^2}{\pi r_{12}^3} [\vec{s}_1\cdot\vec{s}_2 - \frac{3}{r_{12}^2}(\vec{r}_{12}\cdot\vec{s}_1)(\vec{r}_{12}\cdot\vec{s}_2)] \quad (55)$$

with $\mu_B = e\hbar/2mc$. If $<\delta^3(\vec{r}_1)>$ in (53) is replaced by the hydrogenic value Z^3/π (in atomic units), then (53) reduces to the corresponding expression for the one-electron Lamb shift. The $<\delta^3(\vec{r}_{12})>$ terms vanish for triplet states. The greatest uncertainty in evaluating (52) arises from the value of the two-electron Bethe logarithm $\ln(\varepsilon_{nLS}/R_\infty)$. It has been accurately calculated by Schwartz (1961) for the ground state of helium with the result

$$\ln(\varepsilon_{100}/R_\infty) = 4.370 \pm 0.004 .$$

The resulting ground state Lamb shift of about 1.35 cm^{-1} combined with the accurate calculations of Pekeris (1959) brings the theoretical ionization potential for helium

$$J_{theo} = 198310.685 \pm 0.005 \text{ cm}^{-1}$$

into agreement with the experimental value

$$J_{exp} = 198310.82 \pm 0.15 \text{ cm}^{-1}$$

(Herzberg, 1958). More recent calculations by Aashamar and Austvik (1976) have confirmed Schwartz's value and extended the calculations to higher values of Z. Their values are given in Table 10, along with 1/Z expansion values discussed further in the next section.

The only other results of comparable accuracy are from a calculation by Suh and Zaidi (1965) for the 1s2s ^{1}S and 1s2s ^{3}S states of He. They used an elaboration of an oscillator strength sum method proposed by Pekeris (1959) to obtain

$$\ln(\varepsilon_{200}/R_\infty) = 4.345 \pm 0.020 \ (2\ ^1\text{S})$$

$$\ln(\varepsilon_{210}/R_\infty) = 4.380 \pm 0.020 \ (2\ ^3\text{S}) \ .$$

For Li$^+$, Dalgarno (quoted by Accad et al. 1971) estimated that the Lamb shift corrections to the ionization potentials of the 2 ^{1}S and 2 ^{3}S states are $I_L = -0.69 \text{ cm}^{-1}$ and -1.14 cm^{-1}, respectively. (These values differ in sign from the corresponding energy level shifts.) Berry and Bacis (1973) used instead a Z^4 scaling of Suh and Zaidi's (1965) calculation for the 2 ^{3}S$_1$ state of He to obtain $I_L = 0.99 \text{ cm}^{-1}$ for Li$^+$. They then derived from the experimental 2 ^{3}S$_1$ - 2 ^{3}P$_1$ transition frequency, the value $0.28 \pm 0.05 \text{ cm}^{-1}$ for the residual Lamb shift of the 2 ^{3}P state. Although this value is unexpectedly large,

Table 10. Values of $\ln(\varepsilon_{100}/Ry)$ for the ground state of the helium isoelectronic sequence. (Aashamar and Austvik, 1976).

Z	$\ln(\varepsilon_{100}/Ry)$	$\ln[19.77x (Z-0.0063)^2]$	$Z^2\Delta$
2	4.37 ± 0.01	4.364	0.02
3	5.21 ± 0.01	5.177	0.29
4	5.777 ± 0.003	5.754	0.37
5	6.214 ± 0.002	6.201	0.33
6	6.565 ± 0.002	6.566	-0.02
7	6.864 ± 0.002	6.874	-0.50
8	7.115 ± 0.002	7.141	-1.70
9	7.334 ± 0.002	7.377	-3.50
10	7.525 ± 0.002	7.588	-6.31

Ermolaev (1975) showed that it can be explained by an anomalously low electron density at the nucleus in the $2\,^3P$ state. The values calculated by Ermolaev for Li^+ are

$$I_L(2\,^3S_1) = 1.025 \pm 0.055 \ cm^{-1}$$

$$I_L(2\,^3P_1) = 0.291 \pm 0.041 \ cm^{-1} \ .$$

The predicted Lamb shift contribution to the $2\,^3S_1 - 2\,^3P_1$ transition frequency of $1.316 \pm 0.069\ cm^{-1}$ agrees with the value $1.2543 \pm 0.0016\ cm^{-1}$ derived from the measurements of Holt et al. (1980). Experimental and theoretical values for higher Z ions are discussed in the following section.

IV.B. High Z Extrapolations

Theory. For higher values of Z, it becomes advantageous to perform a double perturbation expansion in powers of Z^{-1} as well as Z. Early calculations of this type were done by Dalgarno and Stewart (1960), Layzer and Bahcall (1962) and Doyle (1969). Extensive numerical results have been obtained by Labsovskii (1970), Klimchitskaya et al. (1971), Ermalaev and Jones (1974), Ivanov et al. (1975) and Vainshtein and Safranova (1978). A useful review of this work and the experimental data for helium-like ions in the range Z = 11-18 has recently been published by Martin (1981). The discussion in this section is based on a formulation by Drake (1979).

The origin of the Z^{-1} expansion can be seen as follows. If the units of distance and energy are rescaled so that $\varepsilon = E/Z^2$, $\rho = Zr$, then the nonrelativistic Schrödinger equation becomes (see, e.g. Dalgarno and Stewart, 1957)

$$H^0 = \varepsilon^0 \phi \tag{56}$$

with $\quad H^0 = H_0^0 + Z^{-1}\,H_1^0 \tag{57}$

$$H_0^0 = -\frac{\hbar^2}{2m}(\nabla_{\rho_1}^2 + \nabla_{\rho_2}^2) - \frac{e^2}{\rho_1} - \frac{e^2}{\rho_2} \tag{58}$$

and $\quad H_1^0 = \dfrac{e^2}{\rho_{12}} \quad . \tag{59}$

Thus, Z has been scaled out of the zero-order Hamiltonian $H^0{}_0$. The term $Z^{-1}H^0_1$ can now be treated as a perturbation with Z^{-1} playing the role of a perturbation parameter. E^0 and ϕ then have the perturbation expansions

$$E^0 = Z^2 [E^0_0 + Z^{-1}E^0_1 + Z^{-2} E^0_2 + \ldots \;] \tag{60}$$

$$\phi = \phi_0 + Z^{-1} \phi_1 + Z^{-2} \phi_2 + \ldots \tag{61}$$

such that $(H^0_0 - E^0_0) \phi_0 = 0$ (62)

$$(H^0_0 - E^0_0)\phi_1 + H^0_1\phi_0 = E^0_1\phi_0 \tag{63}$$

etc.

Accurate variational solutions to the above perturbation equations have been used to calculate a wide range of atomic properties in the form of a Z^{-1} expansion series (see, for example, Sanders and Scherr, 1969; Dalgarno and Drake, 1969; Doyle et al. 1972; Aashamar et al. 1970). Using the ϕ expansion to evaluate the matrix element $\langle B_p \rangle$ in (48) results in an expansion of the form

$$\alpha^2 \langle B_p \rangle = \alpha^2 Z^4 [E^2_0 + Z^{-1}E^2_1 + Z^{-2}E^2_2 + \ldots] \tag{64}$$

and the α^4 terms in (4.13) can be expanded

$$\alpha^4 \langle B_P \frac{1}{H^0 - E^0} \; B_P \rangle + \alpha^4 \langle H_6 \rangle$$

$$= \alpha^4 Z^6 [E^4_0 + Z^{-1}E^4_1 + Z^{-2}E^4_2 + \ldots] \tag{65}$$

and similarly for higher order terms. Thus, the total energy can be written in the form

$$E = Z^2 E^0_0 + \alpha^2 Z^4 E^2_0 + \alpha^4 Z^6 E^4_0 + \ldots$$

$$+ Z \; E^0_1 + \alpha^2 Z^3 E^2_1 + \alpha^4 Z^5 E^4_1 + \ldots$$

$$+ Z^0 E^0_2 + \alpha^2 Z^2 E^2_2 + \ldots$$

$$+ Z^{-1} E^0_3 + \alpha^2 Z^1 E^1_3 + \ldots \tag{66}$$

excluding Lamb shift and nuclear motion type corrections. These can be added in separately at the end. For convenience, the anomalous magnetic moment correction (55) is included in B_p. The first column sum is the exact nonrelativisitic energy E^0, the second column sum is $\alpha^2 \langle B_P \rangle$, the first row sum is the sum of the exact single particle Dirac energies for the two electrons, and the second row sum corresponds to the matrix element $\langle e^2/r_{12} + B\rangle$ of the operator (42), using products of hydrogenic Dirac spinors as wave functions. Both the rows and the columns are summed to infinity, and off-diagonal mixing effects are included, by diagonalizing the matrix

$$\underline{H} = (\underline{H}^0 + \underline{B}_P)_{LS} + \underline{R}(\underline{H}_D + \underline{V}_{12} + \underline{B})_{jj} \underline{R}^{-1} - \underline{\Delta} \tag{67}$$

in the basis set of states which are degenerate in zero order for a given total angular momentum $\vec{J} = \vec{L} + \vec{S}$ (or $\vec{J} = j_1 + j_2$ in jj-coupling) and parity. $\underline{R}$ is the jj → LS recoupling transformation, and $\underline{\Delta}$ is the double counting correction for the four terms in the upper left hand corner of (66) which are otherwise counted twice. For consistency, we define

$$H_D = H_D(1) + H_D(2) - 2mc^2 . \tag{68}$$

Consider as a particular example the zero-order degenerate states $1s2p\ ^3P_1$ and $1s2p\ ^1P_1$, which have the same parity and total angular momentum $J = 1$. Then each of the terms in (67) is a 2 x 2 matrix in the basis set of states $|\ 2^3P_1\rangle$ and $|2^1P_1\rangle$. The subscript LS on the first term means that the matrix elements are to be evaluated with the best available LS-coupled variational wave functions. For $Z \leqslant 10$, the results tabulated by Accad and Pekeris (1971) can be used. The subscript jj on the second term means that the matrix elements are to be calculated exactly, using hydrogenic products of jj-coupled Dirac spinors for wave functions; i.e. $|\ 1s_{1/2}\ 2p_{1/2},\ 1\rangle$ and $|\ 1s_{1/2}\ 2p_{3/2},\ 1\rangle$. For this example, the jj → LS recoupling transformation is

$$\begin{pmatrix} |\ 1s2p\ ^3P_1\rangle \\ |\ 1s2p\ ^1P_1\rangle \end{pmatrix} = \underline{R} \begin{pmatrix} |\ 1s_{1/2}2p_{1/2},\ 1\rangle \\ |\ 2s_{1/2}2p_{3/2},\ 1\rangle \end{pmatrix} \tag{69}$$

with $R = \dfrac{1}{3} \begin{pmatrix} \surd 2 & -1 \\ 1 & \surd 2 \end{pmatrix} .$ \hfill (70)

For the states 3^3D_2 and 3^1D_2, it is

$$\begin{pmatrix} |\ 1s3d\ ^3D_2\rangle \\ |\ 1s3d\ ^1D_2\rangle \end{pmatrix} = \underline{R} \begin{pmatrix} |\ 1s_{1/2}\ 3d_{3/2},\ 1\rangle \\ |\ 1s_{1/2}\ 3d_{5/2},\ 1\rangle \end{pmatrix} \tag{71}$$

with $\underline{R} = \dfrac{1}{\surd 5} \begin{pmatrix} \surd 3 & -\surd 2 \\ \surd 2 & \surd 3 \end{pmatrix} .$ \hfill (72)

For states which are non-degenerate in zero order, such as $1s^2$ $^1S_0$, $1s2p$ $^3P_0$, $1s2p$ $^3P_2$ ..., Eq. (67) contains only 1 x 1 matrices, and R = 1.

The matrix $\underline{\Delta}$ in (67) subtracts those contributions which are counted twice in the first two terms. Specifically, these are the contributions to the matrix elements of orders Z^2, Z, α^2Z^4 and α^2Z^3 displayed in Table 11 calculated in LS coupling. The entries up to $1s2p$ $^3P_2$ are equivalent to those evaluated by Doyle (1969) in jj coupling. As examples,

$$\Delta(1s^2\ ^1S_0) = -Z^2 + 0.625Z - 1/4\alpha^2Z^4$$

$$+ 0.4801396\alpha^2Z^3 \tag{73}$$

and $\underline{\Delta}$ for the $1s2p$ $^3P_1$ and $^1P_1$ states is a 2 x 2 matrix with elements

Table 11. Leading Terms in the Z^{-1} and $(\alpha Z)^2$ Expansions of the Hamiltonian Matrix in LS Coupling (a.u.).

State	Z^2	Z	α^2Z^4	α^2Z^3
$1s^2$ $^1S_0$	-1	0.625000000	$-1/4$	0.4801396
$1s2s$ $^1S_0$	$-5/8$	0.231824417	$-21/128$	0.1694781
$1s2s$ $^3S_1$	$-5/8$	0.187928669	$-21/128$	0.0769352
$1s2p$ $^3P_0$	$-5/8$	0.225727785	$-21/128$	0.2197682
$1s2p$ $^3P_1$	$-5/8$	0.225727785	$-59/384$	0.1304287
$1s2p$ $^1P_1$	$-5/8$	0.259868922	$-55/384$	0.0554030
$2\,^3P_1-2\,^1P_1$	0	0	$-\sqrt{2}/96$	0.0288508
$1s2p$ $^1P_2$	$-5/8$	0.225727785	$-17/128$	0.0406387
$1s3s$ $^1S_0$	$-5/9$	0.105255127	$-5/36$	0.0581620
$1s3s$ $^3S_1$	$-5/9$	0.093719482	$-5/36$	0.0331906
$1s3p$ $^3P_0$	$-5/9$	0.104293823	$-5/36$	0.0715549
$1s3p$ $^1P_1$	$-5/9$	0.104293823	$-11/81$	0.0460209
$1s3p$ $^1P_1$	$-5/9$	0.113357543	$-43/324$	0.0269123
$3\,^3P_1-3\,^1P_1$	0	0	$-\sqrt{2}/324$	0.0088266
$1s3p$ $^3P_2$	$-5/9$	0.104293823	$-7/54$	0.0189765
$1s3d$ $^3D_1$	$-5/9$	0.110775757	$-7/54$	0.0238621
$1s3d$ $^3D_2$	$-5/9$	0.110775757	$-52/405$	0.0140516
$1s3d$ $^1D_2$	$-5/9$	0.111270142	$-23/180$	0.0108856
$3\,^3D_2-3\,^1D_2$	0	0	$-\sqrt{6}/1620$	0.0030447
$1s3d$ $^3D_3$	$-5/9$	0.110775757	$-41/324$	0.0044280
$3\,^3S_1-3\,^3D_1$	0	0	0	-0.00090986

$$\Delta_{1,1} = -\frac{5}{8} Z^2 + 0.225727785Z - \frac{59}{384} \alpha^2 Z^4$$

$$+ 0.1303187 \alpha^2 Z^3 \qquad\qquad (74)$$

$$\Delta_{2,2} = -\frac{5}{8} Z^2 + 0.259868922Z - \frac{55}{384} \alpha^2 Z^4$$

$$+ 0.05540 \alpha^2 Z^3 \qquad\qquad (75)$$

$$\Delta_{1,2} = \Delta_{2,1} = -\frac{\sqrt{2}}{96} \alpha^2 Z^4 + 0.0288508 \, \alpha^2 Z^3 . \qquad (76)$$

In the latter example, the states $1s2p \; ^3P_1$ and $^1P_1$ are progressively mixed by the spin-dependent interactions in H as Z increases until ultimately the jj coupling limit is reached at high Z. The eigenvectors of $\underline{H}$ (labeled by the eigenvalues E_λ, $\lambda = 1,2$) can be written

$$\begin{pmatrix} |E_1\rangle \\[1em] |E_2\rangle \end{pmatrix} = \underline{T} \begin{pmatrix} |\, 1s2p \; ^3P_1\rangle \\[1em] |\, 1s2p \; ^1P_1\rangle \end{pmatrix} \qquad\qquad (77)$$

with

$$\underline{T} = \begin{pmatrix} \cos\theta & \sin\theta \\[0.6em] -\sin\theta & \cos\theta \end{pmatrix} \qquad\qquad (78)$$

and θ is the singlet-triplet mixing angle. In the limit of low Z, $\underline{T} \to \underline{1}$ (LS coupling) since the second and third terms of (67) nearly cancel, and $\underline{H}_{NR} \gg \underline{B}_P$ in the first term. In the limit of large Z, $\underline{T} \to \underline{R}^{-1}$ (jj coupling) since the first and third terms of (67) nearly cancel, and $\underline{H}_D \gg \underline{V}_{12} + \underline{B}$ in the second term. As Z varies, $\underline{T}$ generates a continuous range of transformations which tend to the correct limit in both extremes.

Results using the above techniques have been published for the $1s^2 \; ^1S_0 - 1s2p \; ^3P_1$ and $1s^2 \; ^1S_0 - 1s2p \; ^1P_1$ transitions for ions up to Z = 100 by Drake (1979), and for the $1s2s \; ^3S_1 - 1s2p \; ^3P_0$ and $1s2s \; ^3S_1 - 1s2p \; ^3P_2$ transitions for ions up to Z = 50 by DeSerio et al. (1981). The calculations have been extended to the $1s3d \; ^1D$ and ^3D states by Drake (1981). Results for the fine structure splittings of the $1s3d \; ^3D_{1,2,3}$ states are shown in Fig. 6. At Z = 10, $\sin\theta = 0.209$, indicating that strong singlet-triplet mixing is present in the $^3D_2$ configuration. Values up to Z = 25 are given in the above reference.

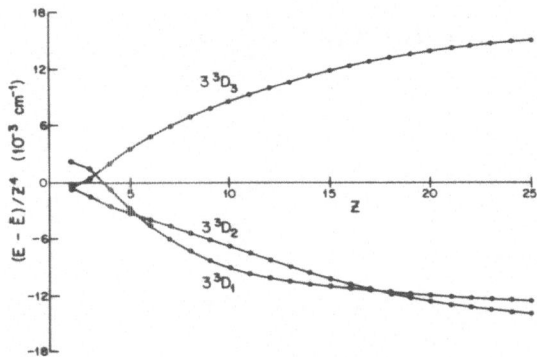

Fig. 6   Theoretical fine structure splittings of the 1s3d
          ^{3}D state of helium-like ions. $\bar{E}$ is the center of
          gravity of the configuration and Z is the nuclear
          charge.

Although a unique separation of relativistic and QED terms is
not well defined, energy differences as calculated above can be
taken as representing the relativistic but non-QED contribution
to the transition frequency. After adding the mass polarization
correction, given by the expectation value of the operator

$$H_M = \frac{1}{M} \vec{p}_1 \cdot \vec{p}_2 \tag{79}$$

and subtracting from the experimental transition frequency, what
remains is the Lamb shift of relative order $\alpha^3 Z^4 \ell n(\alpha Z)$, together
with terms not included in the expansion (66), the leading one
being $\alpha^4 Z^4 E'_2$. One can therefore expect about 1% accuracy in the
Lamb shift at low to moderate Z if the latter correction is not
taken into account.

Since accurate two-electron Bethe logarithms for excited
states are not available, one is forced to make estimates. The
simplest approximation is to use the sum of the individual one-
electron terms discussed in Section II.B. (Note that if anomalous magnetic moment corrections of relative order α/π are
included in B_p, then they should be excluded from the Lamb shift.
This means for example, omitting the term $E'''_{L,2}$ from Eq. (52).
Since $\langle \delta^3(r) \rangle = (Z^3/\pi n^3) \delta_{\ell,0}$, for an $n\ell$ configuration, DeSerio
et al. (1981) suggest defining the two-electron Bethe logarithm
to be

$$\lim_{Z \to \infty} \ell n \frac{\varepsilon(sn\ell)}{R_\infty} = \ell n \frac{\varepsilon(1s)}{R_\infty} + \frac{1}{n^3} \ell n \frac{\varepsilon(n\ell)}{R_\infty} \tag{80}$$

for a 1snℓ configuration so that the hydrogenic Lamb shift difference is recovered in the limit of large Z. Using the data in
Table 3, this leads to

$$\epsilon(1s2s) = 28.095 \ Z^2 \ R_\infty \tag{81}$$

$$\epsilon(1s2p) = 19.695 \ Z^2 \ R_\infty . \tag{82}$$

However, for Z = 2, this gives $\ell n \epsilon(1s2s)/R_\infty$ = 4.72 in poor
agreement with the value 4.38 ± 0.02 calculated by Suh and Zaidi
(1966). An alternative approach involving hydrogenic wave functions with adjustable effective nuclear charges is discussed by
Ermolaev (1975), but there is clearly a need for more accurate
calculations at higher Z.

One method to obtain accurate estimates of the Bethe
logarithm for high Z ions is to apply 1/Z expansion techniques.
The formal expression for the Bethe logarithm for a state ψ_0 is

$$\ell n \ \epsilon = A/B \tag{83}$$

where $$A = \sum_n (E_n - E_0)^3 \ | \langle \psi_0 | \sum_i \vec{r}_i | \psi_n \rangle |^2 \ \ell n \ | E_n - E_0 | \tag{84}$$

and $$B = \sum_n (E_n - E_0)^3 \ | \langle \psi_0 | \sum_i \vec{r}_i | \psi_n \rangle |^2 . \tag{85}$$

For computational purposes, we first make use of the com-
mutation relations

$$(E_n - E_0)^2 \langle \psi_0 | \sum_i \vec{r}_i | \psi_n \rangle = \langle \psi_0 | [H,[H, \sum_i \vec{r}_i]] | \psi_n \rangle$$

$$= Z \langle \psi_0 | \sum_i \frac{\vec{r}_i}{r_i^3} | \psi_n \rangle \tag{86}$$

to rewrite (84) and (85) in the form

$$A = Z^2 \sum_n | \langle \psi_0 | T | \psi_n \rangle |^2 \ \frac{\ell n \ | E_n - E_0 |}{E_n - E_0} \tag{87}$$

$$B = Z^2 \sum_n \frac{| \langle \psi_0 | T | \psi_n \rangle |^2}{E_n - E_0} \tag{88}$$

with $T = \sum_i \vec{r}_i/r_i^3$. Then the expansions

$$\psi_n = \psi_n^0 + Z^{-1} \psi_n^1 + \dots \tag{89}$$

$$E_n = Z^2 (\varepsilon_n^0 + Z^{-1} \varepsilon_n^1 + \dots) \tag{90}$$

are inserted to obtain

$$A = Z^4 [A_0 + 2B_0 \ln Z + A_1 Z^{-1} + \dots \tag{91}$$

$$B = Z^4 [B_0 + B_1 Z^{-1} + \dots] \tag{92}$$

where $A_0 = \sum_n \dfrac{| \langle \psi_0^0 | T | \psi_n^0 \rangle |^2}{\varepsilon_n^0 - \varepsilon_0^0} \ln | \varepsilon_n^0 - \varepsilon_0^0 | \tag{93}$

$$B_0 = \sum_n \dfrac{| \langle \psi_0^0 | T | \psi_n^0 \rangle |^2}{\varepsilon_n^0 - \varepsilon_0^0} \tag{94}$$

$$A_1 = 2\sum_n \dfrac{\langle \psi_0^0 | T | \psi_n^1 \rangle \langle \psi_n^0 | T | \psi_0^0 \rangle}{\varepsilon_n^0 \, \varepsilon_0^0} \ln | \varepsilon_n^0 - \varepsilon_0^0 |$$

$$+ 2\sum_n \dfrac{\langle \psi_0^0 | T | \psi_2^1 \rangle \langle \psi_0^0 | T | \psi_0^0 \rangle}{\varepsilon_n^0 - \varepsilon_0^0} \ln | \varepsilon_n^0 - \varepsilon_0^0 |$$

$$+ \sum_n | \langle \psi_0^0 | T | \psi_n^0 \rangle |^2 \dfrac{(\varepsilon_n^1 - \varepsilon_0^1)(1 - \ln | \varepsilon_n^0 - \varepsilon_0^0 |)}{(\varepsilon_n^0 - \varepsilon_0^0)^2} \tag{95}$$

$$B_1 = 2 \sum_n \dfrac{\langle \psi_0^0 | T | \psi_n^1 \rangle \langle \psi_n^0 | T | \psi_0^0 \rangle}{\varepsilon_n^0 - \varepsilon_0^0}$$

$$+ 2 \sum_n \frac{\langle \psi_0^0 | T | \psi_n^0 \rangle \langle \psi_n^0 | T | \psi_0^1 \rangle}{\varepsilon_n^0 - \varepsilon_0^0}$$

$$- \sum_n | \langle \psi_0^0 | \psi_n^0 \rangle |^2 \frac{(\varepsilon_n^1 - \varepsilon_0^0)}{\varepsilon_n^0 - \varepsilon_0^0} \tag{96}$$

$$\varepsilon_n^1 = \langle \psi_n^0 | \frac{1}{r_{12}} | \psi_n^0 \rangle$$

and
$$| \psi_n^1 \rangle = \sum_{k \neq n} \frac{| \psi_n^0 \rangle \langle \psi_n^0 | \frac{1}{r_{12}} | \psi_k^0 \rangle}{\varepsilon_n^0 - \varepsilon_k^0} .$$

The last two equations above are the standard results for the first order perturbation corrections to the two-electron energies and wave functions. The infinite sums over n and k include all singly and doubly excited states. However, since T is a one-electron operator, only the singly excited states contribute.

The direct evaluation of A_1 and B_1 is difficult because large contributions come from highly excited states lying in the continuum. An alternative method recently used by Goldman and Drake (1982) involves replacing the actual spectrum of hydrogenic eigenstates (bound states plus the continuum) by an entirely discrete set of Sturmian-like pseudostates. The pseudostates are generated by diagonalizing the one-electron Hamiltonian in a basis set of functions of the form

$$\phi_i(\alpha_j) = r^i e^{-\alpha_j r} Y_\ell^m(\theta, \phi) . \tag{97}$$

The the pseudostates $\tilde{\psi}_n^0$ are

$$\tilde{\psi}_n^0 = \sum_{i=0}^N \sum_{j=1}^M a_{i,n}^{(j)} \phi_i(\alpha_j) \tag{98}$$

with the linear variational coefficients $a_{i,n}^{(j)}$ determined by the conditions

$$\langle \tilde{\psi}_m^0 \mid \tilde{\psi}_n^0 \rangle = \delta_{m,n} \tag{99}$$

$$\langle \tilde{\psi}_m^0 \mid H_0 \mid \tilde{\psi}_n^0 \rangle = \tilde{\varepsilon}_n \, \delta_{m,n} \tag{100}$$

The ψ_n^0 and ε_n^0 are used in place of the ψ_n^0 and ε_n^0 in evaluating A, and B.

Ideally, the answer should become independent of the size of the basis set and the non-linear parameters α_j as the number of terms is progressively increased. However, we find no region of stability with respect to variations in α_1 if there is only a single non-linear parameter (ie. M=1 in Eq. 98). However with M>2, well defined regions of stability exist with respect to variations in the α_j.

Our preliminary results for the ground state of helium-like ions obtained from basic sets containing up to 15 terms are (in a.u.) $A_0 = 9.164$, $B_0 = 4$, $A_1 = -6.169\pm0.001$, $B_1 = -2.671$. Using (91) and (92) in (83), the expansion of the Bethe logarithm is

$$\ln \left(\frac{\varepsilon}{Ry} \right) = \frac{A_0}{B_0} + 2 \ln Z + \ln 2 + \frac{(A_1 B_0 - A_0 B_1)}{B_0^2} Z^{-1} + O(Z^{-2})$$

$$\simeq \ln \left[19.77 Z^2 \left(1 - \frac{0.0127\pm.0002}{Z} \right) \right]$$

$$\simeq \ln \left[19.77 (Z - 0.0063 \pm 0.0001)^2 \right] . \tag{101}$$

The error represents the estimated degree of convergence of the calculation. The only significant source of error is the value of A_1.

The results from (101) are compared with the calculations of Aashamar and Austik (1976) in Table 10. In the last column, Δ is the difference between the two values. Thus, $Z^2 \Delta$ should tend to a constant related to the coefficient of Z^{-2} in (101). Since $Z^2 \Delta$ does not appear to be tending to a constant, it may be that the Aashamar and Austvik values are inaccurate at high Z.

Comparison with Experiment. Transitions of the type 1s2s $^3S_1$ – 1s2p ^3P_J (J= 0,1,2) are particularly sensitive to the Lamb shift. Since the Lamb shift increases approximately in proportion to Z^4 while the nonrelativistic energy difference increases only as Z, the ratio is given approximately by

$$\Delta E_L / \Delta E_{NR} \simeq 1.4 \times 10^{-6} Z^3 .$$

Thus, for Fe^{24+}, the Lamb shift contributes about 2% of the total energy difference.

Several measurements have recently been made of the wavelengths of the $1s2s \ ^3S - 1s2p \ ^3P$ transitions in the range $Z = 8$ to 26 (Davis and Marrus, 1977; Berry et al. 1978, 1980; O'Brien et al. 1979; Armour et al. 1979; DeSerio et al. 1981; Buchet et al. 1981; Stamp et al. 1981). The experimental transition frequencies are summarized in Table 12. Both DeSerio et al. (1981) and Stamp et al. (1981) have made detailed comparisons with theory, including relativistic and one-electron Lamb shift corrections. They both find that there are systematic discrepancies which increase in proportion to Z^3 along the isoelectronic sequence. This comes predominantly from higher order terms in the expansion

$$\langle \delta^3 (\vec{r}_1) + \delta^3(\vec{r}_2) \rangle = \frac{2Z^3}{\pi} (\frac{1}{2} + \frac{\dot{\delta}_{\ell,0}}{2n^3} + d_1 Z^{-1}$$

$$+ d_2 Z^{-2} + ...) \qquad (102)$$

multiplying the other terms in (53). Values of the first few d_i, obtained from variational perturbation theory with 50 term correlated variational basis sets, (see, e.g. Drake and Dalgarno, 1970) are listed for several low-lying states in Table 13. DeSerio et al. have estimated the two electron corrections, using (81) and (82) to approximate the two-electron Lamb shifts, and find reasonably good agreement as shown by the dashed curve in Fig. 7.

It is also of interest to compare theory and experiment for the fine structure splittings of the $1s2p \ ^3P_0$, $^3P_1$ and $^3P_2$ states in helim-like ions. These are particularly sensitive to higher order one- and two-electron relativistic corrections. In addition, the $2 \ ^3P_1$ state becomes strongly perturbed by the $n \ ^1P_1$ states. Myers et al. (1981) have recently exploited the near coincidence of the CO_2 laser frequency with the $2 \ ^3P_1 - 2 \ ^3P_2$ M1 transition of $^{19}F^{8+}$ to measure directly the $J = 1 \to 2$ transition frequency. After subtracting hyperfine structure effects, they obtained the result $\nu_{21} = 957.88 \pm 0.03 \ cm^{-1}$. This direct measurement provides a severe test of theory as discussed below. In addition, several other less accurate fine structure splittings can be obtained by subtracting the transition frequencies to the $2 \ ^3S_1$ state shown in Table 12.

Table 12. Experimental values for the $1s2s\ ^3S_1$ – $1s2p\ ^3P_J$ transition frequencies (cm^{-1}).

Ion	$^3P_2$	$^3P_1$	$^3P_0$	
He I	9230.795	9230.871	9231.859	Meggers (1935)
Li II	18228.198(1)	18226.108(1)	18231.303	Holt et al. (1980)
Be III	26867.9(.2)	26853.1(.2)	–	Lofstrand (1972)
	26871.5(.7)	26856.3(.7)	26867.4(.7)	Eidelsberg (1972)
B IV	35429.5	35377.	35393.2	Edlen (1934)
C V	44021.6(1)	43886.1(.1)	43899.0(.1)	Edlen et al. (1970)
N VI	52719.5(.6)	52429.0(.6)	52413.9(1.4)	Baker (1973)
O VII	61588.3(1.5)	61036.6(1.1)	60978.2(1.5)	Stamp et al. (1981)
F VIII	70700.4(3.0)	69743.8(3.0)	69586.0(4.0)	Engelhardt et al.(1971)
Ne IX	80120.5(1.3)	78566.3(1.3)	78266.9(2.5)	Engelhardt et al.(1971)
Al XII	111110.(100)	–	104930.(100)	Denne et al.(1980)
Si XIII	122746.(3)	–	113815.(4)	DeSerio et al.(1981)
S XV	148493.(5)	–	132198.(10)	DeSerio et al.(1981)
Cl XVI	162923.(6)	–	141643.(40)	DeSerio et al.(1981)
Ar XVII	178500.(300)	–	151350.(250)	Davis et al.(1977)
Fe XXIV	368960.(125)	–	232558.(550)	Buchet et al.(1980)

The comparison of the above experimental data with theory is summarized in Table 14. The theoretical values were calculated as described in Section IV.B and Drake (1981). They therefore contain the following higher order relativistic corrections

$\alpha^6 z^6 mc^2 (1 + \alpha^2 z^2 + ...)$ from one-electron Dirac theory

and $\alpha^6 z^5 mc^2 (1 + \alpha^2 z^2 + ...)$ from Breit interaction terms not contained in the calculations of Accad et al. (1973). For example, these extra terms increase the value for ν_{21} of Ne^{8+} from 955.26 cm^{-1} to 957.51 cm^{-1}. Although the latter value is closer to the high precision measurement 957.88 cm^{-1} of Myers

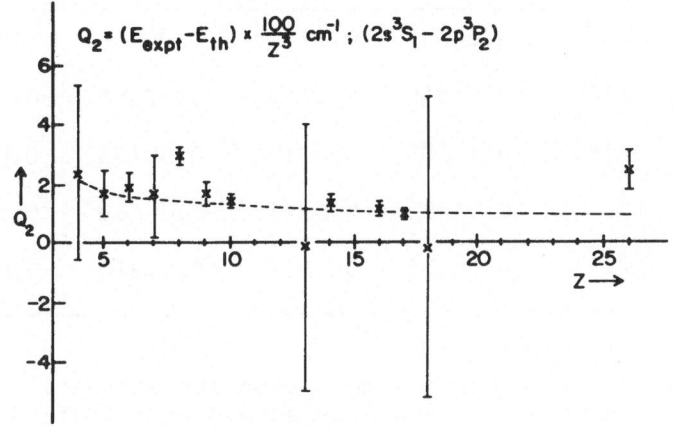

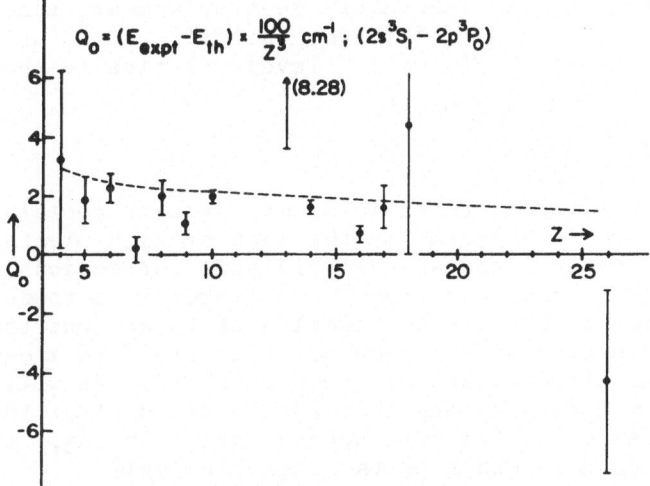

Fig. 7. Comparison of theory and experiment for the QED corrections to the $2s^3S_1 - 2p^3P_2$ and $2s^3S_1 - 2p^3P_0$ transitions of helium-like ions. E_{th} includes only one-electron QED terms, and the dashed curve represents an estimate of the two-electron corrections. (From DeSerio et al., 1981)

Table 13. Expansion coefficients for the calculation of
$\langle \delta^3(\vec{r}_1) + \delta^3(\vec{r}_2) \rangle$ (see Eq. 102).

State	d_0	d_1	d_2	d_3	d_4
$1s^2\ ^1S$	1	-0.66841	0.17805	0.00354	-0.00012
$1s2s\ ^1S$	9/16	-0.14546	0.11145	-0.02398	-0.01712
$1s2s\ ^3S$	9/16	-0.10583	0.03124	0.00733	0.00154
$1s2p\ ^1P$	1/2	0.01083	-0.02422	-0.00469	0.02370
$1s2p\ ^3P$	1/2	-0.04304	0.06070	0.01310	-0.00352

et al. (1981), the difference is more than ten standard
deviations. The discrepancy could be accounted for if the term
$\alpha^4 Z^4 E_2$ in (66), which was not included in the calculation,
contributed approximately $0.2\alpha^4 Z^4$Ry. The less accurate results
for the other ions are in reasonably good agreement, except for
N^{5+}, F^{7+} and S^{14+}. For all of these, the differences arise primarily from the location of the J=0 level relative to the other
two.

V. FEW ELECTRON SYSTEMS

The analysis of QED effects in atomic systems containing more
than two electrons is obscured by the lack of high precision
nonrelativistic eigenvalues and relativistic corrections
available for two-electron systems. It therefore becomes more
difficult to make a reliable subtraction of these contributions
from the observed transition frequencies in order to reveal the
specifically QED effects such as the Lamb shift. As will become
evident below, the conclusions that can be drawn about the Lamb
shift from the experimental data depend rather strongly on the
approximations used in other parts of the analysis.

Consider the $1s^2 2s\ ^2S_{1/2}$ - $1s^2 2p\ ^2P_{1/2}$ and $^2P_{3/2}$ transitions
of the Li isoelectronic sequence as an extensively studied
example. Large scale Hylleraas-type (HT) variational calculations, combined with configuration-interaction (CI) terms, have
been done for the nonrelativistic energies of the ^2S and ^2P states of neutral Li (Larsson, 1968; Sims and Hagstrom, 1975).
Unfortunately, the most accurate 150 term result for the ^2S state
is uncertain by ~10 cm^{-1}, which is large compared to even the
unscreened hydrogenic $2s_{1/2}$ - $2p_{1/2}$ Lamb shift of 2.1 cm^{-1}. Less

Table 14. Summary of Theroretical and Experimental Data for the Fine Structure Intervals of the $1s2p\ ^3P_{0,1,2}$ States (cm^{-1})

Ion	Transition	Theory	Experiment[a]	Difference
Be^{++}	2-1	14.89	14.8(3)	0.09(30)
B^{3+}	2-0	36.35	36.3(8)	0.05(80)
	1-0	-16.29	-16.0(8)	0.29(80)
C^{4+}	2-0	123.04	122.6(1.4)	0.4(1.4)
	1-0	-12.72	-12.9(1.4)	-0.2(1.4)
N^{5+}	2-0	299.18	305.6(1.4)	-6.4(1.4)
	1-0	8.26	15.1(1.4)	-6.8(1.4)
	2-1	290.92	290.5	0.4(0.6)
O^{6+}	2-0	609.48	610.1	-0.6(1.5)
	1-0	58.07	58.4(1.5)	-0.3(1.5)
F^{7+}	2-0	1107.56	1114.4(3.0)	-6.8(3.0)
	1-0	150.05	157.8(3.0)	-7.8(3.0)
	2-1	957.51	957.88(3)[b]	-0.37(3)
Ne^{8+}	2-0	1856.05	1853.6(2.4)	2.4(2.4)
	1-0	298.94	299.4(2.4)	-0.5(2.4)
Al^{11+}	2-0	6366.00	6180.(150.)	186.(150)
Si^{12+}	2-0	8924.38	8931.(5.)	-6.6(5.0)
S^{14+}	2-0	16261.5	16295.(11.)	-33.(11.)
Cl^{15+}	2-0	21286.2	21280.(40.)	6.(40.)
Ar^{16+}	2-0	27395.5	27150.(400.)	245.(400.)

[a] from data tabulated in Table 12.
[b] high precision measurement by Myers et al. (1981).

accurate variational calculations have been extended up to $Z = 8$ (Perkins, 1975). Even here, the ~200 cm^{-1} uncertainty in the calculation is large compared to the hydrogenic Lamb shift of 73.2 cm^{-1}. In the absence of more accurate calculations for low Z, the only hope is to investigate the region $Z > 12$, where QED effects may rise above the uncertainty in the high Z approximation methods described below. A large body of experimental data is available in this region, either from laboratory plasma observations (Fawcett, 1970), or from solar observations (Widing and Purcell, 1976; Behring et al. 1976, Sandlin et al. 1976; Dere, 1978). Edlén (1979) has obtained an accurate semi-empirical fit to the observational data. His results, which are listed in Table 15, provide a convenient comparison with the a priori calculations summarized in Table 16 and discussed in more detail in the following two sections.

Table 15. Comparison of calculated values of the $1s^2 2p$
$^2P_{1/2} - 1s^2 2s \ ^2S_{1/2}$ transition frequency (first three
columns) with a semi-empirical fit to the experimental
data (Edlén, 1979). E_L is the Lamb shift contribution
(1000 cm^{-1}).

Z	Cheng et al. (1979)	Shestakov (1979)	Berry et al. (1980)	Edlén (1979)[a]	$E_L(Z)$[b]	$E_L(Z-s)$[c]
10	129.06	127.978	128.068	128.152	-0.162	-0.085
12	161.00	159.831	159.953	160.012	-0.309	-0.180
14	193.09	191.870	192.031	192.063	-0.531	-0.334
16	225.48	224.170	224.373	224.377	-0.846	-0.561
18	258.1	256.783	257.030	257.020	-1.276	-0.880
20	291.2	289.786	290.068	290.046	-1.838	-1.304
22	324.8	323.209	323.539	323.516	-2.556	-1.850
24	358.8	357.124	357.494	357.486	-3.451	-2.529
26	393.4	391.582	391.988	392.019	-4.546	-3.354
28	428.4	426.635	427.076	427.182	-5.864	-4.329

[a] The entries in this column represent the experimental data with
an uncertainty of about $\pm$ 0.040 x 10^3 cm^{-1}.
[b] Unscreened hydrogenic Lamb shift used by Berry et al. (1980)
and Shestakov (1979).
[c] Screened hydrogenic Lamb shifts with s = 1.60 used by Edlén
(1979).

V.A. Relativistic Hartree-Fock Calculations

The generalization of the usual Hartree-Fock method to
include the single-particle Dirac Hamiltonian together with
electron-electron interaction terms is discussed in detail by
Grant (1970). The resulting Dirac-Fock (DF) equations have been
programmed by Desclaux (1975) and employed by Cheng et al.
(1979) in an extensive survey of transition wavelengths and rates
in the isoelectronic sequences of Li-like through F-like ions up
to Z = 92. They estimated the vacuum polarization term by calculating the expectation value of the Uehling potential in (6), and
the self-energy term by using the screened hydrogenic value for
each orbital (Desclaux et al. 1979). The screened nuclear change
Z-s was determined by requiring that the hydrogenic orbital
radius $\langle r \rangle$ have the same value as for the DF orbital. In addition, the Breit interaction term (40) was included as a first
order perturbation.

Table 16. Summary of computational methods for the $1s^2 2p$ $^2P_{1/2}$ - $1s^2 2s$ $^2S_{1/2}$ transition frequencies of high-Z Li-like ions.

Author	Method	Electron correlation correction	Relativistic correction	Lamb shift
Cheng et al. (1979)	Dirac-Fock	none	$\langle \psi_{DF} \mid B \mid \psi_{DF} \rangle$	screened hydrogenic
Shestakov (1979)	Dirac-Fock	ΔE_2^0	$\langle \psi_{DF} \mid B \mid \psi_{DF} \rangle$	unscreened hydrogenic
Berry et al. (1980)	Z^{-1} expansion	ΔE_2^0	screened Dirac total energies	unscreened hydrogenic
Edlén (1979)	Z^{-1} expansion with empirical corrections	ΔE_2^0	screened Dirac outer electron energies	screened hydrogenic

The results of these calculations shown in Table 15 are systematically larger than Edlén's semi-empirical fit to the experimental data and the other two calculations for two basic reasons. The first comes from the nonrelativistic correlation energy. The total nonrelativistic energy extracted from (66) is

$$E^0 = Z^2 E_0^0 + Z E_1^0 + E_2^0 + Z^{-1} E_3^0 + \ldots \qquad (103)$$

where

$$E_0^0 = -\frac{1}{2} \sum_i \frac{1}{n_i^2} \qquad (104)$$

$$E_1^0 = \langle \phi_0 \mid H_1^0 \mid \phi_0 \rangle \qquad (105)$$

$$E_2^0 = \langle \phi_1 \mid H_1^0 - E_1^0 \mid \phi_0 \rangle \qquad (106)$$

etc.

as obtained from (56) to (63). Hartree-Fock calculations contain the exact E_0^0 and E_1^0, but they do not include the electron correlation contributions to E^0_2, E^0_3, ... However, E^0_2 and E^0_3 can be expressed as sums of one- and two-particle matrix elements which have been calculated to high accuracy (Horak et al. 1969; Ivanova and Safranova, 1975 and earlier references therein). The exact results are (in atomic units)

$$E^0(2^2S) = -9Z^2/8 + 1.022805Z - 0.4081652$$

$$- 0.0230/Z + \ldots$$

$$E^0(2^2P) = -9Z^2/8 + 1.093526Z - 0.5285756 + \ldots$$

while the Hartree-Fock expansions are

$$E^0_{HF}(2^2S) = - 9Z^2/8 + 1.022805Z - 0.354549$$

$$- 0.04135/Z + \ldots$$

$$E^0_{HF}(2^2P) = -9Z^2/8 + 1.093526Z - 0.469462$$

$$- 0.10758/Z + \ldots$$

The corresponding expressions for the transition energies are

$$\Delta E^0(^2P - {}^2S) \quad = 0.070721Z - 0.1204104 + O(Z^{-1}) \qquad (107)$$

$$\Delta E^0_{HF}(^2P - {}^2S) = 0.070721Z - 0.114913 - 0.06623/Z$$
$$\qquad (108)$$

The difference of $-0.005497 + O(Z^{-1})$ is the error in the Hartree-Fock transition energy due to nonrelativistic electron correlation effects. For sufficiently large Z the error, therefore, tends to the constant value 0.005497 a.u. = 1207 cm^{-1}. Subtracting this amount from the results of Cheng et al. (1979) in Table 15 brings their values into better agreement for low Z, but discrepancies remain at high Z.

The second possible source of error comes from the screening approximation used to calculate the Lamb shift. Their use of the value of $\langle r \rangle$ as a criterion for choosing s may lead to an over-estimate of s and therefore an under-estimate of $|E_L|$. A larger Lamb shift would further decrease their transition frequencies at high Z as required.

The DF calculations of Shestakov (1979) include the above correlation correction. However, he uses unscreened hydrogenic

Lamb shifts for the QED correction which, as discussed further in
the next section, is probably an over-estimate. His results
therefore came out consistently lower than Edlén's. Screened
Lamb shifts with s ≈ 0.7 would bring his reults into close
agreement in the range 18 < Z < 28.

V.B. Z^{-1} Expansion Calculations

Calculations based entirely upon Z^{-1} expansion techniques
have been studied for many years (Layzer and Bahcall, 1962;
Dalgarno and Stewart, 1960; Doyle, 1969; McKibbon and Stewart,
1969; Snyder, 1971, 1974). The expansion of the nonrelativistic
energy E^0 has already been discussed in Section V.A. The above
work also makes use of the corresponding expansion

$$E^2 = \alpha^2 Z^4 [E_0^2 + E_1^2 Z^{-1} + E_2^2 Z^{-2} + \dots] \tag{109}$$

for the leading Breit interaction correction given by the second
column of terms in (66). For an N electron atom containing q_t
electrons of type t, E_0^2 is trivially calculated from the
Sommerfeld formula

$$E_0^2 = \sum_t q_t E_0^2 (t) \tag{110}$$

with $E_0^2(t) = - \dfrac{1}{2n_t^4} \left(\dfrac{n_t}{j_t + 1/2} - \dfrac{3}{4}\right) . \tag{111}$

E_1^2 can also be calculated exactly for an N electron atom by
taking linear combinations of two-electron values (Doyle, 1969).
Since the Breit interaction Bp also connects states of the same
total angular momentum and parity within the basis set of hydrogenic states which are degenerate in zero order, a diagonalization step as in Eq. (67) may be necessary (Layzer and Bahcall,
1962). No direct calculations of E_2^2 or higher order terms have
been attempted for systems containing more than two electrons.

The truncation of Eq. (109) after the first two terms is not
sufficiently accurate for most applications. Again consider as
an example the $1s^2 2p\ ^2P - 1s^2 2s\ ^2S$ transitions of Li-like ions.
Doyle (1969) obtained a substantial improvement in the $^2P_{1/2} -$
$^2P_{3/2}$ fine structure splitting by writing the relativisitic
energy difference in the form

$$\Delta E^2 = \alpha^2 \Delta E_0^2 (Z - s)^4 \tag{112}$$

with $s = -\dfrac{\Delta E_1^2}{4 \Delta E_0^2}$.

The expansion of (112) then correctly reproduces the leading two terms of (109), together with an allowance for higher order terms. The expansion coefficients listed in Table 17 yield the value s = 1.7449.

Snyder (1971, 1974) has suggested a more elaborate procedure in which the relativistic energy of each jj-coupled orbital is screened separately. He writes

$$E^2 = \alpha^2 \sum_t q_t \, E_0^2(t) \, (Z - \sigma_t)^4 \tag{113}$$

where the orbital screening parameters σ_t are determined from a two-electron screening matrix via

$$\sigma_t = \sum_{t'} (q_t - \delta_{tt'}) \sigma(t \mid t') \ . \tag{114}$$

The $\sigma(t \mid t')$ are determined by solving the pair of equations

$$q_t E_0^2(t) \sigma(t \mid t') \ q_{t'} E_0^2 \ (t') \sigma(t' \mid t) = -\tfrac{1}{4} E_1^2(t,t') \tag{115}$$

$$q_t E_0^2(t) [\sigma(t \mid t')]^2 + q_{t'} E_0^2(t')[\sigma(t' \mid t)]^2 = \tfrac{1}{6} E_2^2(t,t')) \tag{116}$$

Table 17. Relativistic energy expansion coefficients and Snyder screening parameters for Li-like ions (a.u.)

State	E_2^0	E_2^1	σ_{1s}	$\sigma_{2s \text{ or } 2p}$
$1s^2 2s_{1/2} \ ^2S_{1/2}$	$-37/128$	0.68029	0.47366	1.3224
$1s^2 2p_{1/2} \ ^2P_{1/2}$	$-37/128$	0.78896	0.50858	1.7944
$1s^2 2p_{3/2} \ ^2P_{3/2}$	$-33/128$	0.57084	0.49748	2.3476
$1s^2 \ ^1S_0$	$-1/4$	0.48014	–	–

where $E^2{}_1(t,t')$ and $E^2{}_2(t,t')$ are the first and second order two-electron expansion coefficients. Snyder extracted the $E^2{}_2(t,t')$ from the variational calculations of Accad et al. (1971) to obtain the screening parameters listed in Table 17. Although this method incorporates the two-electron $E^2{}_2$'s, there is no guarantee that the N-electron $E^2{}_2$ obtained by expanding (113) and collecting coefficients of Z^2 will be even close to the right value. However, his calculated fine structure splittings reproduce the experimental data to better than 1% for $Z \gtrsim 7$.

Berry et al. (1980) have used Snyder's screening parameters to estimate the total relativistic energy of Li-like systems by writing

$$E^{rel} = \sum_t [q_t \, E_t^{Dirac}(Z - \sigma_t) + \frac{(Z-\sigma_t)^2}{2n_t}] \tag{117}$$

where the first term contains the exact Dirac energy for nuclear charge $Z - \sigma_t$. Thus all higher order one-electron relativistic corrections are also included. The second term subtracts out the leading nonrelativistic energy contribution. Using (107) for the nonrelativisitic energy (augmented by the Hartree-Fock term $-0.06623/Z$) and adding unscreened hydrogenic Lamb shifts, they obtained the results for the $2^2P_{1/2} - 2^2S_{1/2}$ transition frequency shown in Table 15. The values agree with Edlén's experimental fit to within the experimental accuracy for $Z \gtrsim 14$. Berry et al. therefore concluded that the QED corrections are close to the unscreened hydrogenic values.

However, this conclusion does not agree with the parameters obtained from Edlén's (1979) semi-empirical fit to the experimental data. His expression for the $2^2P_{1/2} - 2^2S_{1/2}$ transition frequency can be written in the form

$$\Delta E(^2P_{1/2} - {}^2S_{1/2}) = \Delta E_{NR} - \Delta_r(^2S) + \Delta_r(^2P_c)$$

$$- \frac{2}{3} \Delta_f(^2P) - \Delta_L(^2S) + \Delta_L(^2P_c) \tag{118}$$

where ΔE_{NR} is the nonrelativistic energy containing (107), together with higher order adjustable parameters, the Δ_r's are relativistic corrections, Δ_f is the fine structure splitting, the Δ_L's are the Lamb shifts and ^2P_c refers to the centre of gravity of the ^2P configuration defined by

$$E(^2P_c) = \frac{1}{3} E(^2P_{1/2}) + \frac{2}{3} E(^2P_{3/2}) \ . \tag{119}$$

Edlén chose to calculate screening parameters separately for each term in (118) from the equation

$$s = - \delta E_1^2 / (4 \delta E_0^2) \tag{120}$$

where δE^2_n ($n = 0,1$) is the change in E^2_n when the outer electron is added to the $1s^2\ {}^1S_0$ core. Using the data in Table 15, this yields (in atomic units)

$$\Delta_r({}^2S) - - \frac{5\alpha^2}{128} (Z - 1.2808)^4 + 0(\alpha^4 z^6) \tag{121}$$

$$\Delta_r(2\ P_c) = - \frac{7\alpha^2}{384} (Z - 2.2410)^4 + 0(\alpha^4 z^6) \tag{122}$$

$$\Delta_f({}^2P) = \frac{\alpha^2}{32} (Z - 1.7449)^4 + 0(\alpha^4 z^6) \tag{123}$$

for the leading terms in powers of α^2. Edlén also included the hydrogenic terms of order $\alpha^4(Z - s)^6$ and $\alpha^6(Z - s)^8$ with the same screening parameters as in (121) to (123). The screening parameter in (123) is slightly different from his empirically determined value

$$s = 1.7415 + 0.633(Z - 0.80)$$

but this does not substantially alter the discussion to follow.

The dominant QED correction comes from $\Delta^L({}^2S)$. Edlén used the hydrogenic expression (15) with an adjustable nuclear charge $Z - s_L$ to obtain $s_L = 1.60$ for the ${}^2S$ state and $s_L = 2$ for the 2p state. The resulting total Lamb shifts shown in Table 15 are quite different from the unscreened values used by Berry et al. (1980), even though the final transition frequencies are nearly the same. The additional empirical corrections introduced by Edlén into ΔE_{NR} and $\Delta_f({}^2P)$ are too small at high Z to account for the discrepancy.

The source of the discrepancy lies in the values of the coefficients E^2_2, E^2_3, ... in (109) implied by the screening approximations. Expanding (121) to (123) yields

$$\Delta E^2_{Edlén} = - \Delta_r({}^2S) + \Delta_r({}^2P_c) - \frac{2}{3} \Delta_f({}^2P)$$

$$= \alpha^2(0z^4 + 0.10869z^3 - 0.54539z^2 + 0.93507z$$

$$- 0.54777) \tag{124}$$

Similarly expanding (117) yields the terms of order α^2

$$\Delta E^2_{Berry} = -\frac{\alpha^2}{4}(Z-0.50858)^4 - \frac{5\alpha^2}{128}(Z-1.7944)^4$$

$$+\frac{\alpha^2}{4}(Z-0.47366)^4 - \frac{5\alpha^2}{128}(Z-1.3224)^4$$

$$= \alpha^2(0Z^4 + 0.10869Z^3 - 0.39625Z^2 + 0.56672Z$$

$$- 0.28967) \tag{125}$$

Thus, the two screening approximations agree up to order α^2Z^3 as expected, but the higher order coefficients are different. The difference is

$$\Delta E^2_{Berry} - \Delta E^2_{Edlén} = (0.14944Z^2 - 0.36867Z + 0.25809)$$

$$\times 11.687 \text{ cm}^{-1} . \tag{126}$$

The above values for ΔE^2 and their differences are tabulated in Table 18. For high Z, where the Lamb shift is large enough to be noticeable, the differences largely compensate for the different Lamb shifts used in the two calculations. Each has allowed the error in ΔE^2 to be absorbed by the Lamb shift, resulting in very

Table 18. Comparison of lowest order relativistic and Lamb shift contributions to the $1s^22p\ ^2P_{1/2}-1s^22s\ ^2S_{1/2}$ transition frequency calculated with the screening approximations used by Berry et al., (1980) and by Edlén (1979). (1000 cm^{-1}).

Z	ΔE^2_{Berry}	$\Delta E^2_{Edlén}$	difference	$E_L(Z)-E_L(Z-s)$[a]
10	0.870	0.736	0.134	−0.077
12	1.604	1.402	0.202	−0.129
14	2.667	2.383	0.284	−0.197
16	4.120	3.740	0.380	−0.285
18	6.023	5.534	0.489	−0.396
20	8.439	7.825	0.614	−0.534
22	11.426	10.675	0.751	−0.706
24	15.047	14.145	0.902	−0.922
26	19.363	18.296	1.067	−1.192
28	24.434	23.189	1.245	−1.535

[a] $s = 1.60$ in the calculation of Edlén (1979).

different values for s_L. It is therefore not possible to draw
firm conclusions about the true behavior of the Lamb shift from
experimental data without knowing the value of ΔE^2_2 to an
accuracy of a few percent.

VI. RELATIVISTIC EFFECTS ON RADIATIVE TRANSITIONS

For one-electron atoms, relativistic effects are fully taken
into account by making the replacement

$$\frac{e}{mc} \langle \phi_f^{NR} \mid \vec{p} \cdot \vec{A} \mid \phi_i^{NR} \rangle \rightarrow e \langle \phi_f \mid \vec{\alpha} \cdot \vec{A} \mid \phi_i \rangle$$

where $\vec{A}$ is the vector potential for the emitted photon, and the
wave functions on the right are 4-component solutions to the
Dirac equation.

For many electron systems, the methods described in Section
IV.B.1 for the calculation of eigenvalues can be extended to the
calculation of radiative transition rates. Consider as an
example the transitions $1s2p$ ^3P $-$ $1s^2$ $^1S_0$ and $1s2p$ $^1P_1$ $-$ $1s^2$
$^1S_0$. For low Z, the former transition is spin-forbidden and proceeds only through the spin-orbit mixing of 2^3P_1 with 2^1P_1 while
the latter transition is fully allowed. However, for high Z, the
spin-orbit coupling becomes strong and jj-coupling provides a
better description. In this limit, both transitions are allowed.
It is therefore advantageous to consider both transitions
together over the range of intermediate coupling cases as Z
varies. In analogy with Eq. (67) for the eigenvalues, the column
vector of transition matrix elements from the two initial states
to the common ground state is

$$\vec{M} = \underline{T} \; [\; (\vec{M}_P)_{LS} + \underline{R} \; (\vec{M})_{jj} - \vec{\Lambda} \;] \tag{127}$$

where

$$(\vec{M}_P)_{LS} = \begin{pmatrix} \langle \; 1s2p \; ^3P_1 \mid \vec{M}_{ts} \mid 1s^2 \; ^1S0 \; \rangle \\[2ex] \langle \; 1s2p \; ^1P_1 \mid \vec{M}_{ss} \mid 1s^2 \; ^1S_0 \; \rangle \end{pmatrix} \tag{128}$$

$$(\vec{M})_{jj} = e \begin{pmatrix} \langle 1s_{1/2} \; 2p_{1/2}, 1 \mid \sum_i \vec{\alpha}_i \cdot \vec{A}_i^* \mid 1s^2_{1/2}, \; 0 \rangle \\[2ex] \langle 1s_{1/2} 2p_{3/2}, \; 1 \mid \sum_i \vec{\alpha}_i \cdot \vec{A}_i^* \mid 1s^2_{1/2} \; , \; 0 \rangle \end{pmatrix} \tag{129}$$

and the 2x2 matrices $\underline{R}$ and $\underline{T}$ are as defined in (67) and (77). $\bar{\Lambda}$ is the double counting correction term analogous to $\underline{\Delta}$ in (67). It subtracts the low order contributions which are counted twice in the first two terms of (127). The Pauli transition operators M_{ts} and M_{ss} for triplet-singlet (ts) and singlet-singlet (ss) transitions, respectively, are in lowest non-vanishing order (Drake, 1979)

$$M_{ss} = e \sum_i \vec{p}_i \cdot \vec{A}_i^* / mc \tag{130}$$

$$M_{ts} = \sum_{n>2}^{\infty} B_P \frac{\mid n\,{}^1P_1 \rangle \langle n^1P_1 \mid M_{ss}}{E(2^3P_1) - E(n^1P_1)}$$

$$+ \sum_{\text{all } n} M_{tt} \frac{\mid n^3P_0^e \rangle \langle n^3P_0 \mid B_P}{E(1^1S_0) - E(n^3P_0^e)}$$

$$- \frac{e^2}{4m^2c^3} \sum_i \vec{\sigma}_i \cdot \nabla_i V \times \vec{A}_i^*$$

$$- \frac{e^3}{4m^2c^3} \sum_{i \neq j} \vec{\sigma}_i \vec{r}_{ij} \times \vec{A}_i^* / r_{ij}^3 \tag{131}$$

and $M_{tt} = M_{ss}$. Here $V_i = -Ze/r_i$, $\vec{r}_{ij} = \vec{r}_i - \vec{r}_j$ and $\vec{\sigma}$ is the Pauli spin operator. The origin of the terms in (131) is as follows. The first term represents the first-order spin-orbit mixing of the 1s2p $^3P_1$ initial state with all 1snp $^1P_1$ states having n>2. The n=2 term is excluded since the $2^3P_1 - 2^1P_1$ mixing is included to all orders when (67) is diagonalized to obtain the $\underline{T}$ matrix. The second term represents the first order mixing by B_P of the $1s^2$ $^1S_0$ final state with all npn'p $^3P^e_0$ states. The last two terms are spin-dependent corrections to the velocity form of the transition operator (Drake, 1976).

The double counting term $\bar{\Lambda}$ can be determined as follows. The Z^{-1} expansion of $(M_P)_{LS}$ is

$$(M_P)_{LS} = \alpha Z \begin{pmatrix} \alpha^2 Z^2 \sum_{n=0}^{\infty} M_{ts}^{(n)} Z^{-n} \\ \sum_{n=0}^{\infty} M_{ss}^{(n)} Z^{-n} \end{pmatrix} \tag{132}$$

Since the n=0 terms are also contained in $\underline{R}$ $(M)_{jj}$, it follows
that

$$
\overline{\Lambda} = \alpha Z \begin{pmatrix} \alpha^2 Z^2 \ M_{ts}^{(0)} \\ \\ M_{ss}^{(0)} \end{pmatrix} . \tag{133}
$$

As in Section IV.B.1, the most accurate non-relativistic matrix
elements available are used to calculate $(M_p)_{LS}$, while $(M)_{jj}$ is
calculated with jj coupled products of Dirac spinors. In the
limit of low Z, $\underline{T} \to \underline{1}$ and $M \to (M_p)_{LS}$. In the limit of high Z, $\underline{T}$
$\to \underline{R}^{-1}$ and $M \to (\overline{M})_{jj}$.

In general, the photon vector potential is

$$
\vec{A} = \eta \hat{e} e^{i\vec{k}\cdot\vec{r}} \tag{134}
$$

where η is a normalization constant, $\hat{e}$ is the unit polarization
vector, $\vec{k}$ is the propagation vector $(|\vec{k}| = \omega/c)$ and ω is the fre-
quency. Since the components of $(M_p)_{LS}$ are to be calculated only
in lowest nonvanishing order, we use the long wavelength approximation

$$
\vec{A} \simeq \eta \hat{e} \tag{135}
$$

for electric dipole transitions. The matrix elements of M_{ss} can
then be expanded (in a.u.)

$$
\langle 1s2p \ ^1P_1 | M_{ss} | 1s^2 \ ^1S_0 \rangle = i\eta\alpha Z \sum_{n=0}^{\infty} M_{ss}^{(n)} Z^{-n} . \tag{136}
$$

The expansion coefficients $M^{(n)}{}_{ss}$ calculated up to ninth order by
the direct variation-perturbation method described by Drake and
Dalgarno (1970) and Doyle et al. (1972) are listed in by Drake
(1979). Similar results have been obtained by Scherr and Sanders
(1968). The sums yield oscillator strengths which agree to four
figures or better with the explicit calculations of Schiff et al.
(1971).

The calculation of M_{ts} is more complicated because of the
infinite summations over intermediate states in the first two
terms of (131). The infinite summations can be performed by
replacing them by finite summations over variationally determined
sets of discrete basis sets as described, for example, by Drake
and Dalgarno (1969). For Z>6, the results are well represented
by (Drake, 1979)

$$\langle 1s2p\ ^3P_1|\ M_{ts}\ |\ 1s^2\ ^1S_0\rangle$$

$$= i\eta\alpha^3 z^3(0.040217 + 0.0318/Z - 0.076/Z^2)$$

$$+ i\eta\alpha^3 z^2(0.01772 - 0.01460/Z)$$

$$+ i\eta\alpha^3 z^2(-0.0448 + 0.0567/Z - 0.007/Z^2) \tag{137}$$

The first term above comes from the sum of the first and third terms in (131), while the second and third terms above correspond to the second and fourth terms of (131) respectively. It follows from (133) and the above values that the double counting correction is

$$\tilde{\Lambda} = i\eta\alpha Z \begin{pmatrix} 0.040217\alpha^2 z^2 \\ \\ 0.395062 \end{pmatrix}. \tag{138}$$

The remaining quantities required for the evaluation of $\tilde{M}$ (Eq. 127) are the $\underline{T}$ and $\underline{R}$ matrices given by (72) and (78), and the $(M)_{jj}$ matrix elements given by (129). The integrals over hydrogenic products of Dirac spinors can easily be calculated analytically. The reduction to radial integrals is discussed by Grant (1974) and Drake (1979). The resulting $1s2p\ ^3P_1 - 1s^2\ ^1S_0$ transition rates are compared with other calculations in Table 19. The principal points are as follows. First, Laughlin's (1978) calculation, which treats B_P as a first order perturbation, is quite accurate in the range $3 < Z < 10$, but deteriorates for $Z \gtrsim 10$ as higher order perturbation corrections to the $2^3P_1 - 2^1P_1$ mixing become more important. Second, the relativistic random phase calculations of Lin et al. (1977), and the relativistic perturbation calculations of Vainshtein and Safranova (1978) are accurate for high Z, but deteriorate for $Z \lesssim 20$ due to the inadequate treatment of correlation effects. The Drake (1979) results include both relativistic and correlation effects, and should therefore yield the most accurate values throughout the entire range of nuclear charge.

The total decay rate of the $1s2p\ ^3P_1$ state also contains a contribution from the allowed $1s2p\ ^3P_1 - 1s2s\ ^3S_1$ transiton. However, this contribution increases only in proportion to Z, is compared with a Z^{10} increase for the $1s2p\ ^3P_1 - 1s^2\ ^1S_0$ intercombination transition. The total decay rates are compared with experimental measurements in Table 20. With the possible exception of F^{7+} (Z=9), agreement between theory and experiment is quite satisfactory.

Table 19.   Comparison of calculated $1s2p\ {}^3P_1 - 1s^2\ {}^1S_0$ decay rates (sec^{-1})[a].

Z	Drake (1979)	Laughlin (1978)	Lin et al. (1977)	Vainshtein et al. (1977)
2	1.764(2)[b]	1.71(2)	2.33(2)	
4	3.993(5)	3.99(5)	4.21(5)	5.71(5)
6	2.825(7)	2.83(7)	2.89(7)	2.99(5)
8	5.499(8)	5.53(8)	5.56(8)	5.49(8)
10	5.356(9)	5.44(9)	5.40(9)	5.25(9)
12	3.376(10)	3.50(10)	3.40(10)	3.29(10)
14	1.571(11)	1.68(11)	1.58(11)	1.52(11)
16	5.822(11)	6.51(11)	5.87(11)	5.65(11)
18	1.800(12)	2.15(12)	1.82(12)	1.74(12)
20	4.786(12)	6.25(12)	4.85(12)	4.65(12)
30	1.252(14)	3.79(14)	1.22(14)	1.23(14)
40	7.017(14)		6.98(14)	
60	4.853(15)		4.86(14)	
80	1.672(16)		1.68(16)	
100	4.250(16)		4.27(16)	

[a] See Drake (1979) for a more extensive tabulation, including also the $1s2p\ {}^1P_1 - {}^1S_0$ transion rates.
[b] Numbers in brackets indicate powers of ten.

The methods discussed in this section and in IV.B   allow relativistic effects to be added to essentially nonrelativistic calculations with a minimum of additional computational effort. Since it is in general easier to find accurate solutions to the Schrödinger equation than it is to solve a many-particle generalization of the Dirac equation, the approach discussed here may turn out to be both more economical and more accurate. It can be extended to many electron systems to the extent that accurate nonrelativistic solutions are available.

ACKNOWLEDGEMENT

Research support by the National Science and Engineering Research Council of Canada is gratefully acknowledged. The author wishes to thank the Ford Motor Company for the hospitality and courtesy extended to him during the preparation of this contribution.

Table 20. Comparison of experimental and theoretical values for the total 1s2p $^3P_1$ decay rate (sec^{-1}).

	$2^3P_1-1^1S_0$	$2^3P_1-2^3S_1$	Total Theory	Experiment
6	2.825(7)	5.575(7)	8.400(7)	
7	1.394(8)	0.664(8)	2.058(8)	(1.7 ± 0.30) (8)[a]
8	5.499(8)	0.770(8)	6.269(8)	(5.80 ± 0.50)(8)[b]
				(6.01 ± 0.33)(8)[c]
				(6.01 ± 0.42)(8)[d]
9	1.834(9)	0.087(9)	1.921(9)	(1.77 ± 0.10)(9)[e]
				(1.77 ± 0.07)(9)[c]
10	5.356(9)	0.098(9)	5.454(9)	
11	1.406(10)	0.011(10)	1.417(10)	
12	3.376(10)	0.012(10)	3.388(10)	(3.45 ± 0.17)(10)[f]
13	7.523(10)	0.013(10)	7.536(10)	(7.81 ± 0.41)(10)[f]
14	1.571(11)	0.014(10)	1.572(11)	(1.57 ± 0.08)(11)[g]
				(1.55 ± 0.07)(11)[f]
15	3.101(11)	0.015(10)	3.103(11)	
16	5.822(11)	0.016(10)	5.824(11)	(6.37 ± 0.73)(11)[g]
17	1.045(12)	0.017(10)	1.045(12)	
18	1.800(12)	0.018(10)	1.800(12)	

[a] Sellin et al. (1968)
[b] Sellin et al. (1970)
[c] Richard et al. (1973)
[d] Moore et al. (1973)
[d] Mowat et al. (1973)
[e] Armour et al. (1981)
[f] Varghese et al. (1976)

REFERENCES

Aashamar, K. and Austvik, A. (1976). Phys. Norv. 8, 229.
Aashamar, K. and Hambro, L. (1977). J. Phys. B 10, 553.
Aashamar, K., Lyslo, G. and Midtdal, J. (1970). J. Chem. Phys. 52, 3324.
Accad, Y., Pekeris, C. L. and Schiff, B. (1971). Phys. Rev. A. 4, 516.

Akhiezer, A. I. and Berestetskii, V. B. (1965). "Quantum
        Electrodynamics", p. 703, Interscience, New York.
Araki, H. (1957). Prog. Theo. Phys. Japan, 17, 619.
Araki, G., Ohta, M., and Mano, K. (1959). Phys. Rev. 116, 651.
Armour, I. A., Myers, E. G., Silver, J. D. and Trabert, E.
        (1979). Phys. Lett. 75A, 45.
Armour, I. A., Silver, J. D. and Trabert, E. (1981) J. Phys. B
        14, 3563.
Barbieri, R., Caffo, M. and Remiddi, E. (1973). Lett. Nuovo
        Cimento, 7, 60.
Barker, W. A. and Glover, F. N. (1955). Phys. Rev. 99, 317.
Bayer, R., Kowalski, J., Neumann, R., Noehte, S., Suhr, H.,
        Winkler, K. and zu Putlitz, G. (1979). Z. Phys. A. 292,
        329.
Behring, W. E., Cohen, L., Feldman, U. and Doschek, G. A. (1976).
        Astrophys. J. 203, 521.
Berry, H. G. and Bacis, R. (1973). Phys. Rev. A 8, 26.
Berry, H. G., DeSerio, R. and Livingston, A. E. (1978). Phys.
        Rev. Lett. 41, 1652.
Berry, H. G., DeSerio, R. and Livingston, A. E. (1980). Phys.
        Rev. A. 22, 998.
Bertin, A., Carboni, G., Duclos, J., Grastaldi, U., Gorini, G.,
        Neri, G., Picard, J., Pitzurra, O., Placci, A., Polacco,
        E., Torelli, G., Vitale, A. and Zavattini, E. (1975).
        Phys. Lett. 55B, 411.
Bethe, H. A. (1947). Phys. Rev. 72, 339.
Bjorken, J. D. and Drell, S. D. (1964). "Relativistic Quantum
        Mechanics," McGraw-Hill, New York.
Blanchard, P. and Drake, G. W. F. (1973). J. Phys. B. 6, 2495.
Borie, E. (1975). Z. Phys. A. 275, 347.
Borie, E. (1976). Z. Phys. A. 278, 127.
Borie, E. (1981), Phys. Rev. Lett. 47, 568.
Borie, E. and Rinker, G. A. (1978). Phys. Rev. A. 18, 324.
Borie, E. and Rinker, G. A. (1980). Z. Phys. A 296, 111.
Breit, G. (1929). Phys. Rev. 34, 553.
Brodsky, S. J. and Mohr, P. J. (1978), in "Structue and
        Collisions of Ions and Atoms," (I. A. Sellin, ed.). Vol.
        5, p. 3. Topics in Current Physics, Springer, Berlin.
Brown, G. E. and Ravenhall, G. (1951), Proc. Roy. Soc. Lond.
        A208, 552.
Brown, R. T. and Cortez, J. L. M. (1971), J. Chem. Phys. 54,
        2657.
Buchet, J. P., Buchet-Poulizac, M. C., Denis, A., Desesquelles,
        J., Druetta, M., Grandin, J. P. and Husson, X. (1981).
        Phys. Rev. A. 23, 3354.
Campani, E. (1970). Lett. Nuov. Cim. 4, 982.
Chang, T. N. and Poe, R. T. (1976). Phys. Rev. A. 14, 11.
Cheng, K. T., Sepp, W. D., Johnson, W. R. and Fricke, B. (1978).
        Phys. Rev. A. 17, 489.

Cheng, K. T., Kim, Y. K. and Desclaux, J. P. (1979). At. Data
    Nucl. Data Tables, 24, 111.
Cok, D. R. and Lundeen, S. R. (1981), Phys. Rev. A. 23, 2488.
Cosens, B. L. (1968)., Phys. Rev. 173, 49. See Taylor et al.
    (1969) for discussion of this experimental value.
Curnutte, B., Cocke, C. L. and DuBois, R. D. (1981), in
    "Proceedings of the Sixth International Conference on
    Fast Ion Beam Spectroscopy," Laval. To be published in
    Nucl. Instr. Methods.
Daley, J., Douglas, M., Hambro, L. and Kroll, N. M. (1972),
    Phys. Rev. Lett. 29, 12.
Dalgarno, A. and Drake, G. W. F. (1969), Chem. Phys. Lett. 3,
    349.
Dalgarno, A. and Lewis, J. T. (1956), Proc. Roy. Soc. Lond. A233,
    70.
Dalgarno, A. and Stewart, A. L. (1957), Proc. Roy. Soc. Lond.
    A240, 274.
Dalgarno, A. and Stewart, A. L. (1960), Proc. Phys. Soc.
    (London), 75, 441.
Davis, W. A. and Marrus, R. (1977). Phys. Rev. A. 15, 1963.
de Jager, C. W., de Vries, H. and de Vries, C. (1974), At. Data
    Nucl. Data Tables, 14, 479.
Denne, B., Huldt, S., Pihl, J. and Hallin, R. (1980). Phys. scr.
    22, 45.
Dere, K. P. (1978), Astrophys. J. 221, 1062.
Desclaux, J. P. (1975), Comput. Phys. Commun. 9, 31.
Desclaux, J. P., Cheng, K. T. and Kim, Y. K. (1979). J. Phys. B
    12, 3819.
DeSerio, R., Berry, H. G., Brooks, R. L., Hardis, H., Livingston,
    A. E. and Hinterlong, S. (1981). Phys. Rev. A. 24, 1872.
Deutsch, C. (1976), Phys. Rev. A. 13, 2311.
Dietrich, D., Lebow, P., de Zafra, R. and Metcalf, H. (1976),
    Bull. Am. Phys. Soc. 21, 625.
DiGiacomo, A. (1969). Nucl. Phys. B11, 411. His eq. (22) for the
    vacuum polarization correction is too large by a factor
    of 2, although the final results are correct.
DiGiacomo, A. (1970), Nucl. Phys. B23, 641.
Douglas, M. H. (1971), PhD. Thesis, University of California, San
    Diego (unpublished, available from University Microfilms,
    Ann Arbor, Michigan).
Douglas, M. and Kroll, N. M. (1974), Ann. Phys. (N.Y.), 82, 89.
Doyle, H. T. (1969), Adv. At. Mol. Phys. 5, 377.
Doyle, H., Oppenheimer, M. and Drake, G. W. F. (1972), Phys. Rev.
    A. 5, 26.
Drake, G. W. F. (1976), J. Phys. B 9, L169.
Drake, G. W. F. (1979), Phys. Rev. A. 19, 1387. The signs in eq.
    (A5) of this reference should be corrected to agree with
    (43) of the text.

Drake, G. W. F. (1981), in "Proceedings of the Sixth
      International Conference on fasdt Ion Beam Spectroscopy,"
      Laval. To be published in Nucl. Instr. Methods.
Drake, G. W. F. (1982), Adv. At. Mol. Phys. 18, 399.
Drake, G. W. F. and Dalgarno, A. (1969), Astrophys. J. 157, 456.
Drake, G. W. F., Goldman, S. P. and van Wijngaarden, A. (1979),
      Phys. Rev. A. 20, 1299.
Edlén, B. (1934), Nova Acta Reg. Soc. Sci. Ups. (IV), 9, no. 6.
Edlén, B. (1979), Phys. Scr. 19, 255.
Edlén, B. and Lofstrand, B. (1970), J. Phys. B 3, 1380.
Eidelsberg, M. (1972), J. Phys. B 5, 1031.
Elton, L. R. B., Hofstadter, R. and Collard, H. R. (1967), in
      "Landolt-Bornstein Numerical Data and Functional
      Relationships in Science and Technology," (K. H. Hellwege
      and H. Schopper, eds.) New Series Group I, Vol. 2.
      Springer-Verlag, Berlin.
Engelhardt, W. and Sommer, J. (1971), Astrophys. J., 167, 201.
Engfer, R., Schneuwley, H., Vuillemier, J. L., Walter, H. K. and
      Zehnder, A. (1974), At. Data Nucl. Data Tables, 14, 509.
Erickson, G. W. (1971), Phys. Rev. Lett. 27, 780.
Erickson, G. W. (1977). J. Chem. Phys. Ref. Data, 6, 831.
Erickson, G. W. and Yennie, D. R. (1965), Ann. Phys. N. Y. 35,
      271, 447.
Ermolaev, A. M. (1973), Phys. Rev. A. 8, 1651.
Ermolaev, A. M. (1975). Phys. Rev. Lett. 34, 380.
Ermolaev, A. M. and Jones, M. J. (1974), J. Phys. B. 7, 199.
Farley, J. W., MacAdam, K. B. and Wing, W. H. (1979), Phys. Rev.
      A. 20 1754.
Fawcett, B. C. (1970), J. Phys. B 3, 1152.
Friar, J. L. and Negele, J. W. (1973), Phys. Lett. 46B, 5.
Friar, J. L. (1979), Ann. Phys. (N.Y.), 122, 151.
Gell-Mann, M. and Low, F. (1951), Phys. Rev. 84, 350.
Gould, H. and Marrus, R. (1978), Phys. Rev. Lett. 41, 1457. See
      Goldman and Drake (1981) for a correction to their value.
Grant, I. P. (1970), Adv. Phys. 19, 747.
Herzberg, G. (1958), Proc. Roy. Soc. Lond., A248, 309.
Holt, R. A., Rosner, S. D., Gaily, T. D. and Adam, A. G. (1980),
      Phys. Rev. A. 22, 1563.
Horak, Z. J., Lewis, M. N., Dalgarno, A. and Blanchard, P.
      (1969), Phys. Rev. 185, 21.
Ivanov, L. N., Ivanova, E. P. and Safranova, U. I. (1975), J.
      Quant. Spectrosc. Radiat. Transfer, 15, 553.
Ivanova, E. P. and Safranova, U. I. (1975), J. Phys. B 8, 1591.
Jette, A. N., Lee, T. and Das, T. P. (1974), Phys. Rev. A. 9,
      2337.
Kelsey, E. J. and Spruch, L. (1978). Phys. Rev. A. 18, 15, 845
      and 1055.
Klarsfeld, S. (1977). Nucl. Phys. A285, 493.
Klarsfeld, S. and Maquet, A. (1973), Phys. Lett. 43B, 201.

Klimchitskaya, G. L. and Labzovskii, L. M. (1971), Zh. Eksp.
        Teor. Fiz. 60, 2019; [(2971), Sov. Phys. JETP, 33, 1088].
Kpanou, A., Hughes, V. W., Johnson, C. E., Lewis, S. A. and
        Pichanick, F. M. J. (1971), Phys. Rev. Lett. 26, 1613.
Kramer, P. B. and Pipkin, F. M. (1978), Phys. Rev. A. 18, 212.
Kugel, H. W., Leventhal, M. and Murnick, D. E. (1972), Phys. Rev.
        A. 6, 1306.
Kugel, H. W., Leventhal, M., Murnick, D. E., Patel, C. K. N. and
        Wood, O. R. II. (1975), Phys. Rev. Lett. 35, 647.
Kugel, H. W. and Murnick, D. E. (1977), Rep. Prog. Phys. 40, 297.
Labzovskii, L. N. (1970), Zh. Eksp. Teor. Fiz. 59, 168. [(1971)
        Sov. Phys. JETP. 32, 94].
Lamb, W. E. Jr. and Retherford, R. C. (1947), Phys. Rev. 72, 241.
Larsson, S. (1968), Phys. Rev. 169, 49.
Laughlin, C. (1978), J. Phys. B 11, L391.
Lautrup, B. E., Peterman, A. and de Raphael, E. (1972), Physics
        Reports, 36, 193.
Lawrence, G. P., Fan, C. Y. and Bashkin, S. (1972), Phys. Rev.
        28, 1612.
Layzer, D. and Bahcall, J. (1962). Ann. Phys. (N.Y.), 17, 177.
Lepage, G. P., Yennie, D. R. and Erickson, G. W. (1981), Phys.
        Rev. Lett. 47, 1640.
Leventhal, M. (1975), Phys. Rev. A. 11, 427.
Leventhal, M., Murnick, D. E. and Kugel, H. W. (1972), Phys. Rev.
        Lett. 28, 1609.
Lewis, S. A., Pichanick, F. M. J. and Hughes, V. W. (1970), Phys.
        Rev. A. 2, 86.
Lewis, M. L. and Serafino, P. H. (1978), Phys. Rev. A. 18, 867.
Lin, C. D., Johnson, W. R. and Dalgarno, A. (1977), Phys. Rev. A
        15, 154.
Lipworth, E. and Novick, R. (1957), Phys. Rev. 108, 1434.
Lofstrand, B. (1972), Phys. Scr. 8, 57.
Lundeen, S. R. and Pipkin, F. M. (1981), Phys. Rev. Lett. 46,
        232.
Mann, J. B. and Johnson, W. R. (1971), Phys. Rev. A. 4, 41.
Martin, W. C. (1981), Phys. Scr. 24, 725.
McKibben, C. S. and Stewart, A. L. (1969), J. Phys. B 2, 24.
Meggers, W. F. (1935), J. Res. N.B.S., 14, 487.
Mittleman, M. H. (1971), Phys. Rev. A. 4, 897.
Mittleman, M. H. (1972), Phys. Rev. A. 5, 2395.
Mohr, P. J. (1976), in "Beam Foil Spectroscopy," (I. A. Sellin
        and D. J. Pegg, eds.), pp. 89-95. Plenum Press, New York.
Moore, C. F., Braithwaite, W. J. and Mathews, D. L. (1973), Phys.
        Lett. A 44, 199.
Mowat, J. R., Sellin, I. A., Peterson, R. S., Pegg, D. J., Brown,
        M. D. and McDonald, J. R. (1973), Phys. Rev. A 8, 145.
Murnick, D. E. (1981), in "Proceedings of the Workshop on the
        Foundations of the Relativistic Theory of Atomic Structure,
        ANL80-126," pp. 172-194. Argonne National Laboratory.

Murnick, D. E., Leventhal, M. and Kugel, H. W. (1972), in
        "Proceedings of the Third International Conference on
        Beam-Foil Spectroscopy," Boulder, Colorado.
Myers, E. G., Kuske, P., Andra, H. J., Armour I. A. Jelley, N. A.
        Klein, H. A., Silver, J. D. and Trabet, E. (1981), Phys.
        Rev. Lett. 47, 87.
Narasimham, M. and Strombotne, R. L. (1971), Phys. Rev. A. 4, 14.
Newton, G., Andrews, D. A. and Unsworth, P. J. (1979), Phil.
        Trans. Roy. Soc. Lond., 290, 373.
O'Brien, R., Silver, J. D., Jelley, N. A., Bashkin, S., Trabert,
        E. and Heckmann, P. H. (1979), J. Phys. B 12, L41.
Pekeris, C. L. (1959), Phys. Rev. 115, 1216.
Perkins, J. F. (1976), Phys. Rev. A 13, 915.
Rashid, K. and Fricke, B. (1980), Z. Phys. A 297, 279.
Richard, P., Kauffman, R. L., Hopkins, F. F., Woods, L. W. and
        Jamison, K. A. (1973), Phys. Rev. Lett. 30, 888.
Rinker, G. A. (1976), Phys. Rev. A 14, 18.
Rinker, G. A. and Steffen, R. M. (1977), At. Data Nucl. Data
        Tables, 20, 145.
Robiscoe, R. T. and Shyn, T. W. (1970), Phys. Rev. Lett. 24, 559.
Salpeter, E. E. (1952), Phys. Rev. 87, 328.
Salpeter, E. E. and Bethe, H. A. (1951), Phys. Rev. 84, 1232.
Sanders, F. C. and Scherr, C. W. (1969), Phys. Rev. 181, 84.
Sandlin, G. D., Brueckner, G. E., Scherrer, V. E. and Tousey, R.
        (1976), Astrophys. J. 205, L47.
Sapirstein, J. (1981), Phys. Rev. Lett. 47, 1723.
Scherr, C. W. and Sanders, F. C. (1968), Int. J. Quantum Chem. 2,
        29.
Schiff, B., Accad, Y. and Pekeris, C. L. (1973), Phys. Rev. A 8,
        2272.
Schiff, B., Pekeris, C. L. and Accad, Y. (1971), Phys. Rev. A 4,
        885.
Schwartz, C. (1961), Phys. Rev. 123, 1700.
Schwartz, C. (1964), Phys. Rev. 134, A1181.
Schwartz, C. and Tiemann, J. J. (1959), Ann. Phys. N.Y., 6, 178.
Schwinger, J. (1949), Phys. Rev. 75, 651.
Sellin, I. A., Donnally, B. L. and Fan, C. Y. (1968), Phys. Rev.
        Lett. 21, 717.
Sellin, I. A., Brown, M., Smith, W. W. and Donnally (1970), Phys.
        Rev. A 2, 1189.
Shestakov, A. F. (1979), Opt. and Spektrosk, 46, 209. [(1979).
        Opt. and Spectrosc. 46, 117.]
Sick, I., McCarthy, J. S. and Whitney, R. R. (1976), Phys. Lett.
        64B, 33.

Simon, G. G., Borkowski, F., Schmitt, Ch. and Walther, V. W.
        (1980). Z. Naturforsch. 35A, 1.
Sims, J. S. and Hagstrom, S. A. (1975), Phys. Rev. A 11, 418.
Sims, J. S., Parmer, D. R. and Reese, J. M. (1982). J. Phys.
        B 15, 327.
Snyder, R. (1971), J. Phys. B 4, 1150.
Snyder, R. (1974), J. Phys. B 7, 335.
Stamp, M. F., Armour, I. A., Peacock, N. J. and Silver, J. D.
        (1981), J. Phys. B 14, 3551.
Stone, A. P. (1961), Proc. Phys. Soc. 77, 786.
Stone, A. P. (1963), Proc. Phys. Soc. 81, 868.
Sucher, J. (1958), Ph.D. Thesis, Columbia University
        (unpublished; available from University Microfilms, Ann
        Arbor, Michigan) and Phys. Rev. 109, 1010.
Sucher, J. (1981), in "Proceedings of the Workshop on the
        Foundations of the Relativistic Theory of Atomic
        Structure, ANL80-126," pp. 1-26. Argonne National
        Laboratory.
Suh, K. S. and Zaidi, M. M. (1965), Proc. Roy. Soc. Lond. A296,
        94.
Tam, A. C. (1979), Phys. Rev. A 20, 1784.
Taylor, B. N., Parker, W. H. and Ladenberg, D. N. (1969), Rev.
        Mod. Phys. 41, 375.
Triebwasser, S., Dayhoff, E. S. and Lamb, W. E. Jr. (1953), Phys.
        Rev. 89, 98.
Uehling, E. A. (1935), Phys. Rev. 48, 55.
Vainshtein, L. A. and Safranova, U. I. (1978), At. Data Nucl.
        Data Tables, 21, 49.
van Wijngaarden, A. and Drake, G. W. F. (1978), Phys. Rev. A 17,
        1366.
Varghese, S. L., Cocke, C. L. and Curnutte, B. (1976) Phys. Rev.
        A 14, 1729.
von Edidy, T. and Desclaux, J. P. (1978), Z. Phys. A 288, 23.
Welton, T. A. (1948), Phys. Rev. 74, 1157.
Wichmann, E. H. and Kroll, N. M. (1956), Phys. Rev. 101, 843.
Widing, K. G. and Purcell, J. D. (1976), Astrophys. J. 204, L151.
Wood, O. R. II, Patel, C. K. N., Murnick, D. E., Nelson, E. T.,
        Leventhal, M., Kurgel, H. W. and Niv, Y. (1982), Phys.
        Rev. Lett. 48, 398.
Zaprayagaev, S. A. and Manakov, N. L. (1976), Yad. Fiz. 23, 917,
        [(1976), Sov. J. Nucl. Phys. 23, 482].
Zemach, A. C. (1956), Phys. Rev. 104, 1771.
Zon, B. A., Manakov, N. L. and Rapaport, L. P. (1972), Yad. Fiz.
        15, 508. [(1972), Sov. J. Nucl. Phys. 15, 282].

ELECTRON EXCITATION PROCESSES AND QUANTUM ELECTRODYNAMICS IN HIGH-Z SYSTEMS*

Gerhard Soff

Gesellschaft für Schwerionenforschung mbH
Planckstraße 1, Postfach 110541
D-6100 Darmstadt 11, West Germany
and
Udo Müller, Theo de Reus, Paul Schlüter, Andreas Schäfer,
Joachim Reinhardt, Berndt Müller, and Walter Greiner

Institut für Theoretische Physik der
Johann Wolfgang Goethe-Universität
Robert-Mayer-Straße 8-10, Postfach 111 932
D-6000 Frankfurt am Main, West Germany

1.   INTRODUCTION

Collisions of very heavy ions with bombarding energies close to the Coulomb barrier offer us the possibility to perform a spectroscopy of electronic states in transient superheavy systems within a charge range of $100 \leq Z \leq 190$ ($_{92}U + _{98}Cf$). The pivotal question to answer is whether the binding energy of the strongest bound electronic state can reach or exceed twice the electron rest mass. An important task in this connection is the investigation of electron excitation processes in superheavy systems. Two possible excitation mechanisms are visualized in Fig. 1. As an example we consider the schematic level structure for a head-on collision of Pb + Cm with a total charge Z = 178 at 5.9 MeV/u bombarding energy. Binding energies of the adiabatic $1s\sigma$ - and $2s\sigma$- states[1] are displayed versus the internuclear separation R. The turning point between incoming and outgoing trajectroy is given by R_{min} = 17 fm.

*This work was supported by the Bundesministerium für Forschung und Technologie (BMFT) and the Deutsche Forschungsgemeinschaft (DFG). One of us (G.S.) acknowledges the support of the DFG-Heisenberg Programm.

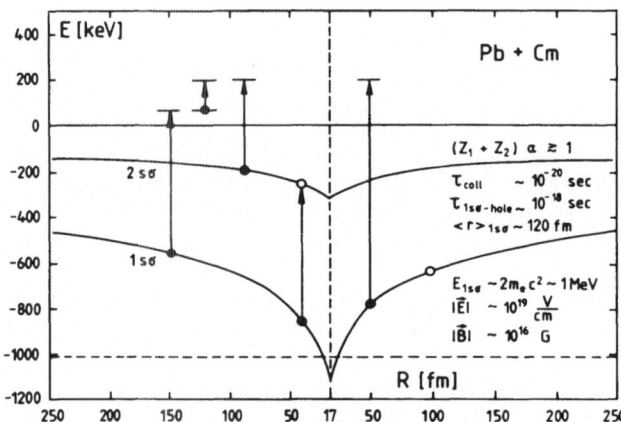

Fig. 1: Binding energies of the 1sσ and 2sσ state versus internuclear separation for the Pb+Cm system. One-step and two-step
electron excitation processes are indicated qualitatively. On the
right hand side characteristic numbers describing the collision
are given.

A typical one-step excitation process is the direct ionization
from the 1sσ- state to the positive continuum. In addition the
multi-step processes play an important role. Actually they form
the dominant part of inner shell vacancy production probabilities.
An example for a two-step process is depicted in Fig. 1, where a
2sσ- electron is ionized to the continuum. The remaining hole is
filled by the 1sσ- electron which leaves a vacancy in the K-shell.
Since initial and final state of both processes are indistinguishable, they have to be treated coherently. This is achieved by

solving coupled channel equations for the electron occupation amplitudes of the quasimolecular adiabatic states.

Strong relativistic effects are best visible in the wave-
functions. Fig. 2 shows the radial density $r^2|\psi(r)|^2$ of the 1s -
electron in the field of an extended nucleus. The Schrödinger hydrogenic 1s wavefunction is given by $\psi_{1s}(r) = 2(Z/a_B)^{3/2} e^{-rZ/a_B}$,
where $a_B = \lambda_e/\alpha = 52918$ fm is the Bohr radius. The axes in fig. 2
have been scaled appropriately in order to make the shape of the
nonrelativistic density independent of the charge Z. Deviations
therefore are solely due to relativistic effects. While the shape
of $r^2|\psi(r)|^2$ does not alter much in the region of the ordinary
periodic table, the wavefunction becomes severely contracted as $Z\alpha$
approaches or exceeds the value 1. Although a complete "collapse
of the wavefunction" is averted, the scaled Bohr radius a_B/Z loses
its importance if $Z\alpha > 1$ and the only relevant length scale is determined by the nuclear radius R. The inset in Fig. 2 gives the
electron density at the origin, $|\psi(0)|^2$, as a function of Z. Apart
from the Z^3-dependence known from the nonrelativistic solution the
density increases by additional three orders of magnitude going to
Z = 180.

The figure also displays the electron density at the origin
for the $2p_{1/2}$-wavefunction. For a nonrelativistic p-wave, characterized by the sharp angular momentum $\ell = 1$, this value is exactly
zero. The $2p_{1/2}$ Dirac bispinor, however, carries a mixture of $\ell = 0$
(upper component) and $\ell = 1$ (lower component orbital angular momentum. In atoms with $Z\alpha > 1$ the "small" and "large" components of the
wavefunction become comparable in magnitude. As a consequence the
distinction between $s_{1/2}$- and $p_{1/2}$-states is diminished. This ex-
plains the steep rise of the $2p_{1/2}$ density as well as the strong
increase of its binding energy as a function of Z.

In fig. 3 we compare the radial density distribution of the K-shell
electron in lead with that in the superheavy atom Z=164, where the
maximum of $|\psi|^2r^2$ is located at about r ~ 20 fm. The radial expec-
tation value $<r>_{1s}$ changes from 840 fm for Z=82 to 140 fm for Z=164.

The shrinkage of the 1s-wavefunction at shortest internuclear
distances is a typical feature in collisions of very heavy ions.
Sticking to our example Pb + Cm at E_{lab} = 5.9 MeV/u we find a radial
expectation value for the 1sσ orbital of about 120 fm at the distance of closest approach. Due to this extreme localization (within
the Compton wavelength of an electron, λ_e = 386 fm) transfer of
high momenta components to the strongly bound electrons resulting
from the nuclear motion becomes possible. Therefore δ-electrons up
to kinetic energies of 2 MeV can be observed experimentally[3], whereas the classical allowed maximum energy transfer to an electron at
rest is about 10 keV.

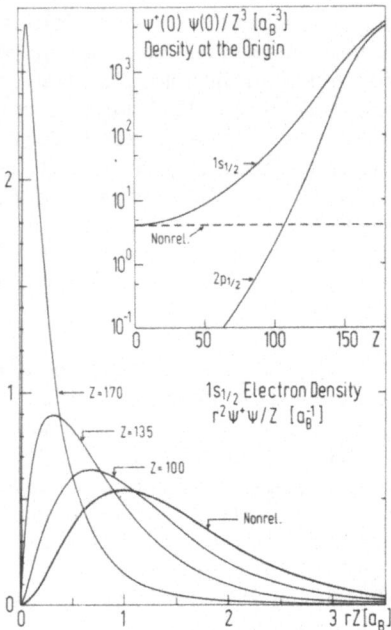

Fig. 2: The radial electron density of the $1s_{1/2}$ state in superheavy atoms with nuclear charges Z = 100, 135, and 170. For comparison the nonrelativistic Schrödinger wavefunction is drawn. The
axes are scaled with powers of Z in order to make the nonrelativistic density independent of the charge. The inset shows the
drastic increase of the electron density at the origin over its
nonrelativistic value. The wavefunctions of the $1s_{1/2}$ and $2p_{1/2}$
states become very similar in the region $Z\alpha > 1$.

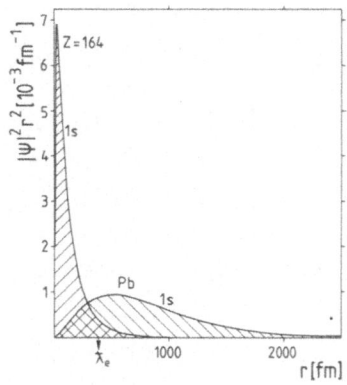

Fig. 3

After a short theoretical survey concerning electron excitation processes we discuss in chapter 3 as a special topic the bound state solution of the two-center Dirac equation. In the following we present our numerical results and compare them with some experimental data. The influence of strong collisional magnetic fields on spin polarization of electrons is treated in chapter 5. Section 6 and 7 deal with the selfenergy of electrons in critical fields and with some aspects of the Glashow-Weinberg-Salam interaction. After this we present a few new results concerning positron creation in deep inelastic nuclear collisions. Finally we discuss the possibility of internal electron-positron pair creation in supercritical compound systems.

2.   THEORETICAL DESCRIPTION OF ELECTRON EXCITATIONS

In this chapter we summarize the essential equations where our calculations are based on. We emphasize that in superheavy quasi-molecules a full relativistic treatment of the electron motion is absolutely necessary since the combined charge $Z = Z_1 + Z_2 \gtrsim 1/\alpha$ and the electron binding energies become comparable with the electron rest mass or even exceed it. If we want to describe the dynamical evolution of an electron involving two colliding nuclei we thus have to solve the time-dependent two-centre Dirac equation

$$i\partial/\partial t \; \Phi_i \; (\vec{R}(t)) = H_{TCD} \; (\vec{R}(t)) \; \Phi_i \; (\vec{R}(t)) \; . \qquad (2.1)$$

H_{TCD} is the relativistic two-centre Hamiltonian which depends sensitively on the internuclear separation $R(t)$. For this reason it is useful to expand the total wave function Φ_i into Born-Oppenheimer states ϕ_n which are represented by the stationary molecular states. The expansion reads

$$\Phi_i(t) = \sum_j a_{ij}(t) \; \phi_j \; (\vec{R}(t)) \; \exp(-i\chi_j(t)), \qquad (2.2)$$

where the sum includes an integration over continuum states with positive and negative energies. The phase factors χ_j are chosen as

$$\chi_j(t) = \int^t dt' <\phi_j(\vec{R}(t'))|H_{TCD}(\vec{R}(t'))|\phi_j(\vec{R}(t'))> \; . \qquad (2.3)$$

Inserting the expansion of eq. (2.2) into eq. (2.1) and projecting with stationary eigenfunctions we obtain the following set of first-order coupled differential equations for the occupation amplitudes $a_{ij}(t)$

$$\dot{a}_{ij}(t) = \sum_k a_{ik}(t)<\phi_j|\partial/\partial t|\phi_k> \exp\{+i(\chi_j-\chi_k)\} \; . \qquad (2.4)$$

The time derivative operator can be expressed in terms of a radial and rotational coupling

$$\partial/\partial t \rightarrow \dot{R} \; \partial/\partial R - i\vec{\omega}\cdot\vec{j} \; . \qquad (2.5)$$

Our calculations are restricted to radial couplings only.

The set $\Phi_i(t)$, already containing the dynamical excitations, can be used as a basis for solving the many-particle problem. To do this, one can expand the total wave function in a basis of many-electron configurations which may be represented by Slater determinants of the single particle basis functions Φ_i. The amplitude for exciting a final configuration starting from a given initial configuration turns out to be just the determinant of the corresponding single-particle amplitudes $a_{ij}(t)$. Therefore the number of physical particles p or holes q in a particular level can be deduced completely from the single-particle amplitudes[4,5]

$$N_p = \sum_{r>F} |a_{rp}|^2 \qquad \text{for } p>F \; , \tag{2.6}$$

$$N_q = \sum_{r>F} |a_{rq}|^2 \qquad \text{for } q<F \; . \tag{2.7}$$

F denotes the Fermi surface of occupied states. For the number of correlated particle-hole pairs $N_{p,q}$ one finds

$$N_{p,q} = N_p \cdot N_q + \left| \sum_{r<F} a_{rp}^* \, a_{rq} \right|^2 \; . \tag{2.8}$$

This formula should be applied to analyze experiments like those presently performed for coincidence between electrons and 1sσ-vacancy formation[3,6]. Eq. (2.8) holds also for particle-particle or hole-hole correlations if the sign of the second term is inverted.

3.   BOUND STATE SOLUTIONS OF THE TWO—CENTER DIRAC EQUATION

In this section we find solutions of the stationary Dirac equation for a two-center potential. We start with the representation of the Dirac equation in spherical coordinates:

$$[i\gamma_5\sigma_r(\frac{\partial}{\partial r} + \frac{1}{r} - \frac{\beta}{r}\hat{K}) - \frac{Z_1e^2}{|\vec{r} - \frac{1}{2}\vec{R}|}$$

$$- \frac{Z_2e^2}{|\vec{r} + \frac{1}{2}\vec{R}|} + \beta m_e] \phi_\mu(\vec{r}) = E\phi_\mu(\vec{r}). \tag{3.1}$$

The spin-orbit operator is defined through the relation

$$\hat{K} \equiv \beta(\vec{\sigma}\vec{\ell} + 1). \tag{3.2}$$

It is known that the two-center Dirac equation is more difficult to handle than its nonrelativistic counterpart, since it is not separable in any orthogonal coordinate system. The most accurate and flexible approach at present is based on a multipole expansion of the wave function[7,8]

$$\phi_\mu(\vec{r}) = \sum_\kappa \phi_{\mu\kappa}(\vec{r}) = \sum_\kappa \begin{pmatrix} g_\kappa(r) \, X_\kappa^\mu \\ if_\kappa(r) \, X_{-\kappa}^\mu \end{pmatrix}. \tag{3.3}$$

$f_\kappa(r)$ and $g_\kappa(r)$ are the radial wave functions, and the spinor spherical harmonics are given by

$$X_\kappa^\mu = \sum_{m=\pm 1/2} (\ell, \frac{1}{2}, j, \mu-m, m) Y_\ell^{\mu-m}(\theta,\phi) X^m. \tag{3.4}$$

$-\kappa$ is the eigenvalue of $\hat{K}$ and connected to the angular momentum through

$$\kappa = \begin{cases} \ell & \text{for } j=\ell-\frac{1}{2} \\ -\ell-1 & \text{for } j=\ell+\frac{1}{2}, \end{cases} \tag{3.5}$$

$j = |\kappa| - \frac{1}{2}$ being the total angular momentum. The magnetic quantum number μ is the projection of the total angular momentum on the axis connecting the two nuclei (z axis). After expanding the two center potential into multipoles

$$V(\vec{r},\vec{R}) = \sum_{\ell=0}^\infty V_\ell(r,R) P_\ell(\cos\theta), \tag{3.6}$$

where for $r < R/2$,

$$V_\ell(r,R) = \frac{-2Z_1 e^2}{R}\left(\frac{2r}{R}\right)^\ell - \frac{2Z_2 e^2}{R}\left(\frac{-2r}{R}\right)^\ell , \qquad (3.7)$$

and where for $r > R/2$,

$$V_\ell(r,R) = \frac{-Z_1 e^2}{r}\left(\frac{R}{2r}\right)^\ell - \frac{Z_2 e^2}{r}\left(\frac{-R}{2r}\right)^\ell , \qquad (3.8)$$

the coupled radial equations read

$$\frac{d}{dr} g_\kappa(r) = (E+m)f_\kappa(r) - \frac{\kappa+1}{r} g_\kappa(r)$$

$$- \sum_{\bar\kappa,\ell} f_{\bar\kappa}(r) V_\ell(r,R) A^\mu_{(-\kappa,\ell,-\bar\kappa)} ,$$

$$\frac{d}{dr} f_\kappa(r) = \frac{\kappa-1}{r} f_\kappa(r) - (E-m) g_\kappa(r) \qquad (3.9)$$

$$+ \sum_{\bar\kappa,\ell} g_{\bar\kappa}(r) V_\ell(r,R) A^\mu_{(\kappa,\ell,\bar\kappa)} .$$

The coefficients $A^\mu_{(\kappa,\ell,\bar\kappa)}$ are easily determined by angular momentum algebra

$$A^\mu_{(\kappa,\ell,\bar\kappa)} = \langle \chi^\mu_\kappa | P_\ell | \chi^\mu_{\bar\kappa} \rangle$$

$$= \sum_{m=\pm 1/2} (-1)^{\mu-m}(\ell_\kappa,\tfrac{1}{2},j_\kappa,\mu-m,m)(\ell_{\bar\kappa},\tfrac{1}{2},j_{\bar\kappa},\mu-m,m)$$

$$\qquad\qquad\qquad\qquad\qquad\qquad\qquad\qquad\qquad\qquad (3.10)$$

$$[(2\ell_\kappa+1)(2\ell_{\bar\kappa}+1)]^{1/2} \begin{bmatrix} \ell_\kappa & \ell & \ell_{\bar\kappa} \\ 0 & 0 & 0 \end{bmatrix} \begin{bmatrix} \ell_\kappa & \ell & \ell_{\bar\kappa} \\ m-\mu & 0 & \mu-m \end{bmatrix} .$$

The sum in Eq. (3.9) has been truncated at a sufficiently large angular momentum j_{max}. The remaining $2(2j_{max}+1)$ coupled differential equations can be solved by a five-point Adams integration code. The energy eigenvalues are determined by iteration, which is stopped when the change of the energy value is less than 10^{-14}. In this case the accuracy of the two-center wave function is better than $3 \cdot 10^{-3}$.

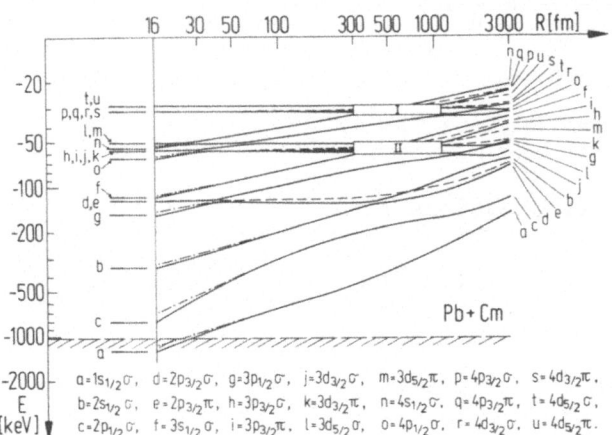

a=1s$_{1/2}$σ,   d=2p$_{3/2}$σ,   g=3p$_{1/2}$σ,   j=3d$_{3/2}$σ,   m=3d$_{5/2}$π,   p=4p$_{3/2}$σ,   s=4d$_{3/2}$π,
b=2s$_{1/2}$σ,   e=2p$_{3/2}$σ,   h=3p$_{3/2}$π,   k=3d$_{3/2}$π,   n=4s$_{1/2}$σ,   q=4p$_{3/2}$π,   t=4d$_{5/2}$σ,
c=2p$_{1/2}$σ,   f=3s$_{1/2}$σ,   i=3p$_{3/2}$π,   l=3d$_{5/2}$σ,   o=4p$_{1/2}$σ,   r=4d$_{3/2}$σ,   u=4d$_{5/2}$π.

Fig. 4a

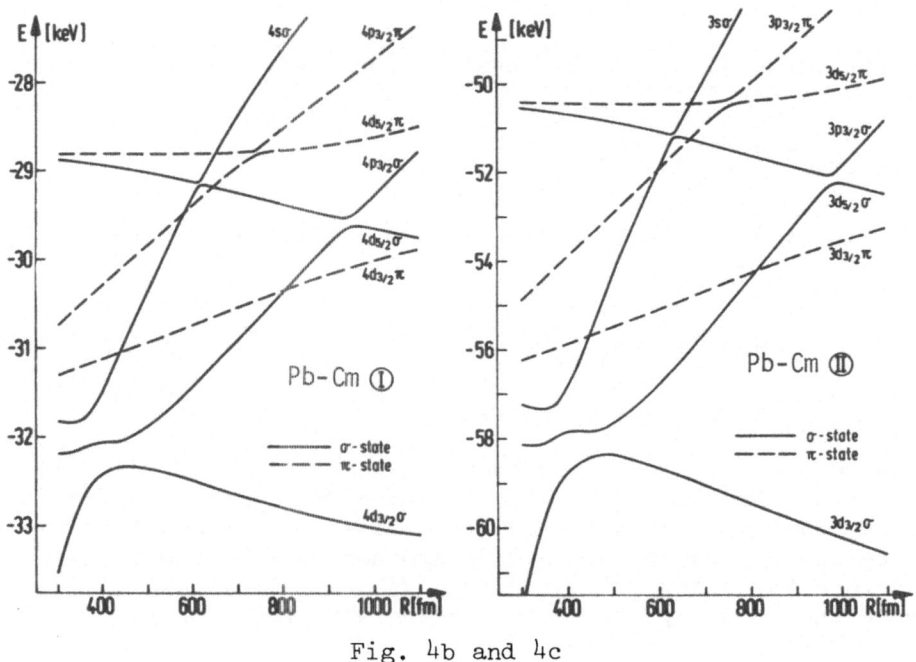

Fig. 4b and 4c

   The adiabatic correlation diagram for the symmetric Pb-Pb system
has been published elsewhere [1] . In the following we discuss a
few examples of relativistic correlation diagrams for asymmetric
systems. Fig. 4 shows the dependence of the binding energies on
the two-center distance R for the $_{82}$Pb - $_{96}$Cm system. For a more

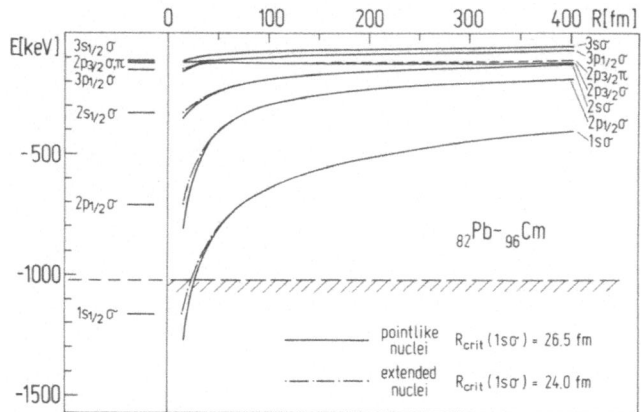

Fig. 4d : Correlation diagram for the superheavy quasimolecule
           Pb-Cm. For two-center distances R < 400 fm the
           energies of the 7 lowest bound states are presented
           on a linear scale.

accurate analysis various representations of the correlation diagram are chosen. In fig. 4a a we plot on a double logarithmic scale
the 21 lowest σ- (solid lines) and π-(dashed lines) states between
R=16 fm and R = 3000 fm. The dashed-dotted lines denote the energies
for extended nuclei. The relativistic splitting between the states
$2p_{3/2}\sigma$ - $2p_{1/2}\sigma$ is striking. It amounts to more than the electron
rest mass. At the critical distance R_{cr} = 24 fm the 1sσ state
dives into the negative energy continuum. The energy levels in
rectangles I and II between 300 and 1000 fm are presented separately
on a linear scale in figs. 4b and 4c . The sudden energy change
of the strongest bound states near the distance of closest approach
is most impressive if energies as a function of two-center distance
R are drawn on a linear scale (Fig. 4d).

The strongest energy change is found for the $2p_{1/2}\sigma$-state. Its
binding energy increases by 56.4% from R=100 fm to R=16 fm. The
variation of the $1s\sigma$ binding energy amounts to about the electron
rest mass in a two-center distance interval of $\Delta R=84$ fm. The
nuclear extension lowers the binding energy of the $1s\sigma$-state by
10% and that of the $2p_{1/2}\sigma$-state by 15%. The three following
asymmetric correlation diagrams $_{92}U-_{96}Cm$, $_{82}Pb - _{98}Cf$, $_{92}U-_{98}Cf$ show
an equivalent structure as the $_{82}Pb - _{96}Cm$ system.

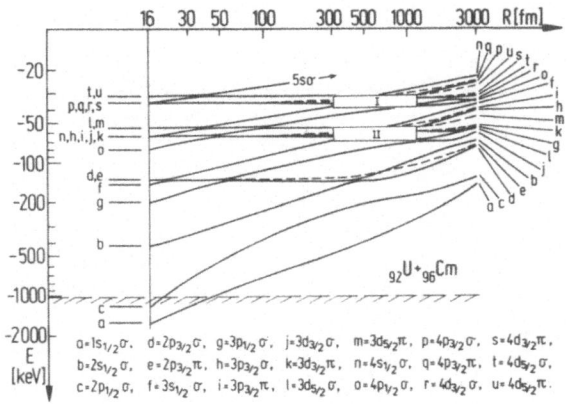

Fig. 5a

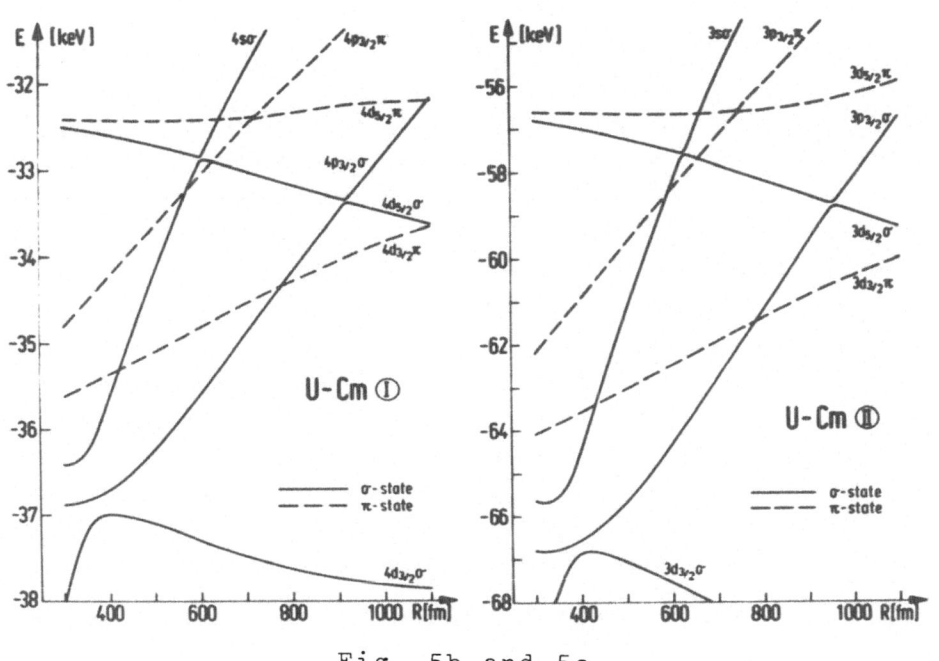

Fig. 5b and 5c

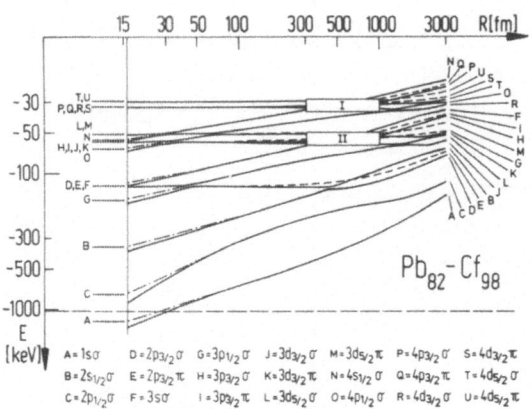

Fig. 6a

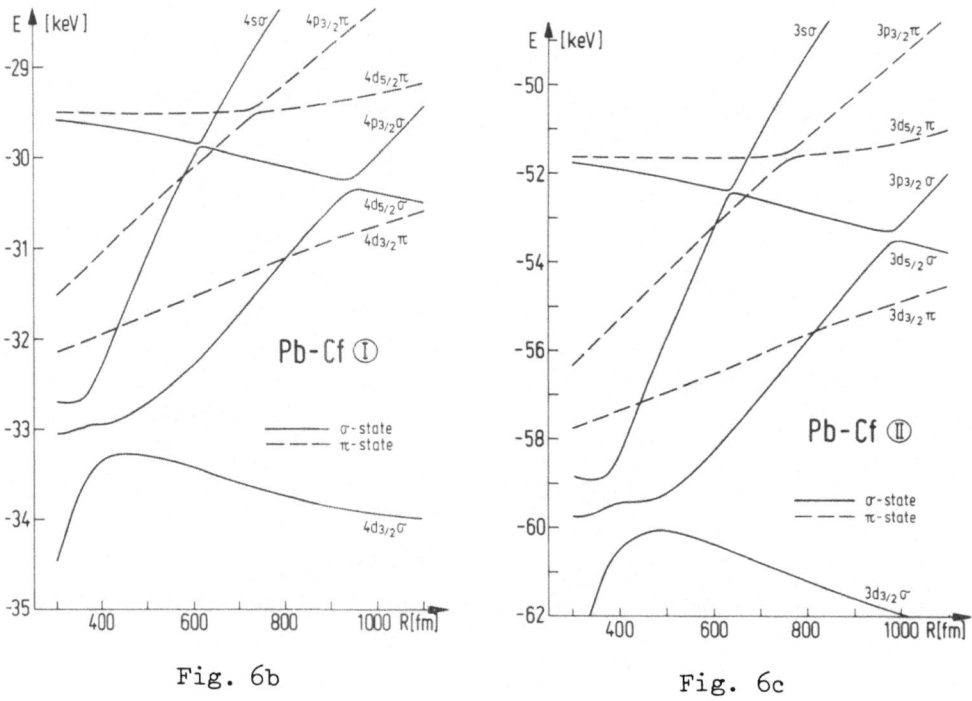

Fig. 6b   Fig. 6c

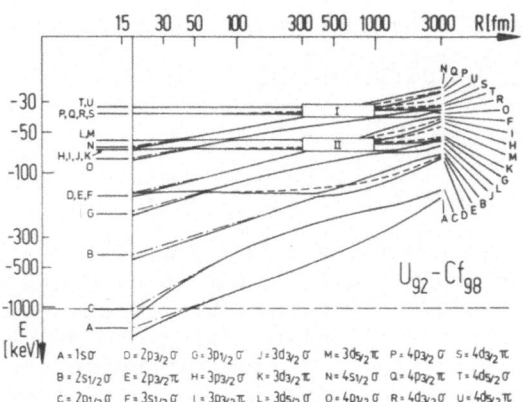

Fig. 7a

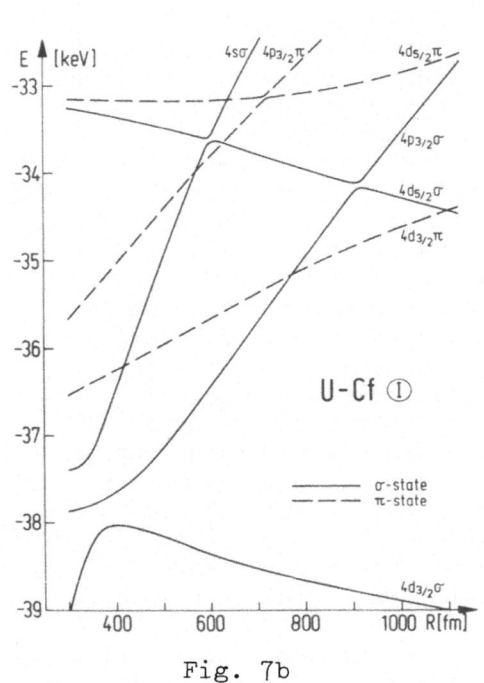

Fig. 7b

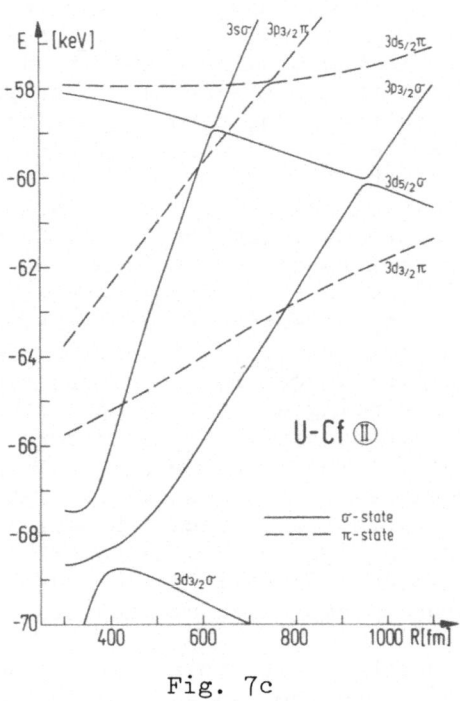

Fig. 7c

4.   RESULTS FOR THE VACANCY FORMATION AND δ-ELECTRON PRODUCTION
      IN SUPERHEAVY QUASIMOLECULES

    After expanding the two-centre potential into multipoles and
restricting ourselves to the dominant monopole part V_0 we obtain
two first-order coupled differential equations

$$d/dr\ u_1 = -\kappa/r\ u_1 + (E + m - V_0)\ u_2\ ,$$

$$d/dr\ u_2 = - (E - m - V_0)\ u_1 + \kappa/r\ u_2\ ,$$

$$(4.1)$$

where $u_1 = r \cdot g(r)$ and $u_2 = r \cdot f(r)$. The eqs. (4.1) are solved
numerically both for bound states and continuum states. With the
obtained wavefunctions we solved the coupled channel equations
(2.4) for the occupation amplitudes. Using the expressions (2.6),
(2.7) and (2.8) we calculated the vacancy production probabilities
and δ-electron emission rates. A large variety of results are
already published in refs. 4,5,9 and 10. For corresponding experimental results we refer to refs. 3,6,11 - 14. In the following we
thus restrict ourselves to the presentation of few selected examples
for electron excitation calculations.

    Binding energies as well as the radial expectation value
$<r_{1s\sigma}>$ play an important, but contrary role when calculating K-
hole formation and δ-electron production. Whereas a strong increase in the binding energy hinders, e.g., the ionization of an
electron from the K-shell - thus producing a K-hole and, eventually,
a δ-electron - the strong decrease of $<r_{1s\sigma}>$ leads to high Fourier
frequencies in the pulse spectrum - thus stimulating K-shell
ionization. Therefore one can predict a maximum for K-hole production
rates, $P_{1s\sigma}$, around $Z \sim 170$. Subsequently we investigate this effect
more systematically for K-hole formation as well as for δ-electron
production rates. For this purpose we evaluated K-vacancy production rates for various combined nuclear charges Z=134 up to
Z=184. We present coupled channel calculations stressing the Z-
dependence of $P_{1s\sigma}(b)$ in Fig. 9. Part a shows K-hole production
rates for charges Z≤164. We observe an increasing slope with
rising binding energy. Vacancy formation increases especially for
small impact parameters as a consequence of the sharp localized
1sσ-state. The latter effect allows the transfer of high Fourier
frequencies to the K-shell originating from the nuclear motion,
thus leading to increased K-hole production. In Fig. 9b combined
nuclear charges beyond Z=164 are considered In this region the
ionization probabilities decrease due to the extemely strong
binding energy which amounts nearly twice the electron rest mass.

    Another observable quantity where the influence of 1sσ-binding
energies and the radial expectation value $<r_{1s\sigma}>$ can be demonstrated
is the differential emission probability of δ-electrons as a func-

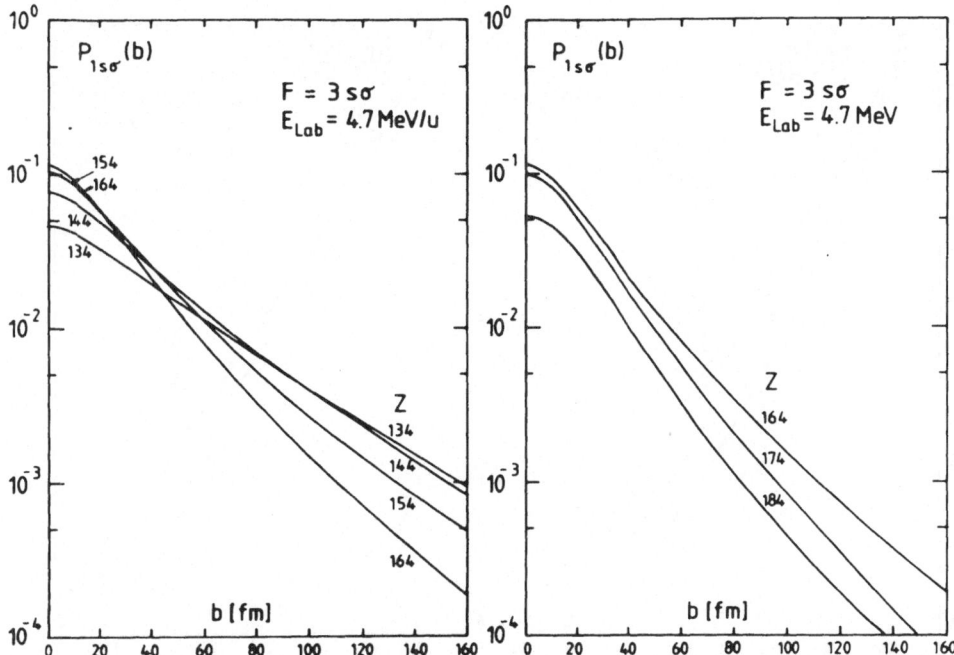

Fig. 9: Number of created 1sσ-vacancies versus impact para-
        meter b. The different curves belong to different
        total charges Z. Asymmetric systems are assumed.
        Part (a) for Z ≤ 164, part (b) for Z ≥ 164.

tion of the kinetic electron energy. In Fig. 10 δ-electron cross
sections are displayed for systems with united charge Z=134 up to
Z=184. The δ-electrons have to be measured in coincidence with
1sσ-vacancies. In all calculations we chose a Fermi level of
F=3sσ and a bombarding energy E_{lab}=4.7 MeV/u. Asymmetric systems
are considered. In the region of united charge Z<164 the δ-electron
cross section rises with increasing Z. The high-kinetic part with
E_{e^-}>500 keV grows faster than the low-kinetic one. This is due
to the stronger localization of the 1sσ-orbit which allows for
high momenta transfer as described above. But, if the united
charge exceeds Z~170 the influence of the rapidly increasing
1sσ-binding energy becomes dominant. Thus the emission cross
section even dimishes which is shown for Z=184. Therefore it would
be highly desirable to measure more systematically K-hole formation
and δ-electron cross sections in dependence on the united charge
Z in order to verify the interplay of strong binding and sharp
localization in ionization processes.

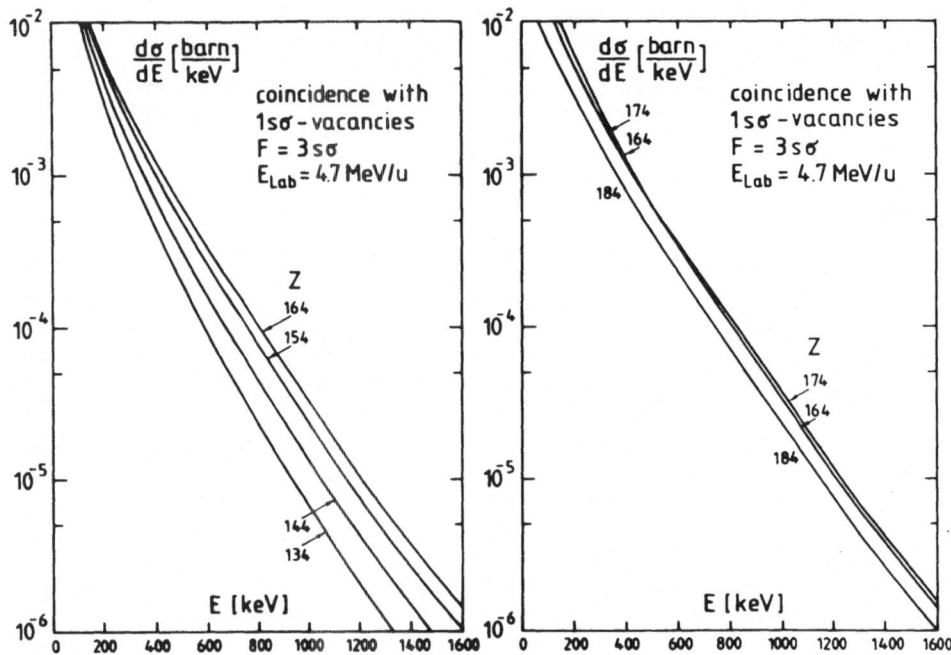

Fig. 10: δ-electron distribution in coincidence with 1sσ-
          vacancies versus the kinetic electron energy for
          different combined charges Z.

For the J+Pb system with $Z_1+Z_2=135$ the δ-electron spectra are
shown as a function of the kinetic electron energy in Fig. 11. The
dependence is of exponential form and less steep for more central
collisions. Part a shows a triple coincidence measurement (scattering
angle, δ-electron energy, and K-hole formation) by W. Koenig,
C. Kozhuharov et al.[3,6]. Good agreement is achieved for all cm-
kinetic electron energies and for all impact parameters. In part b
the double differential electron cross section $d^2\sigma/d\Omega dE$ is displayed.
The triangle denote the total spectrum of emitted δ-electrons. The
experimental data representing coincidence measurements with K-
vacancies are in overall agreement with our theoretical calculations.
Also for the lower Z-system Br+Pb fair agreement with the measured
data of ref. 6 is achieved[15].

The δ-electron distribution of the system Pb+Sn in coincidence
with 1sσ-vacancies is displayed in Fig. 12a. For comparison with
experimental data of Kozhuharov et al.[3] we divided our results by
4π assuming a spherical symmetric angular distribution of the
δ- ray emission. The latter assumption was verified experimentally[3].
We obtain a theoretical cross section which slightly underestimates
the measured data. The δ-electron distribution for the selected impact
parameters b=0,5,10 and 20 fm is shown in fig. 12b for the same system.

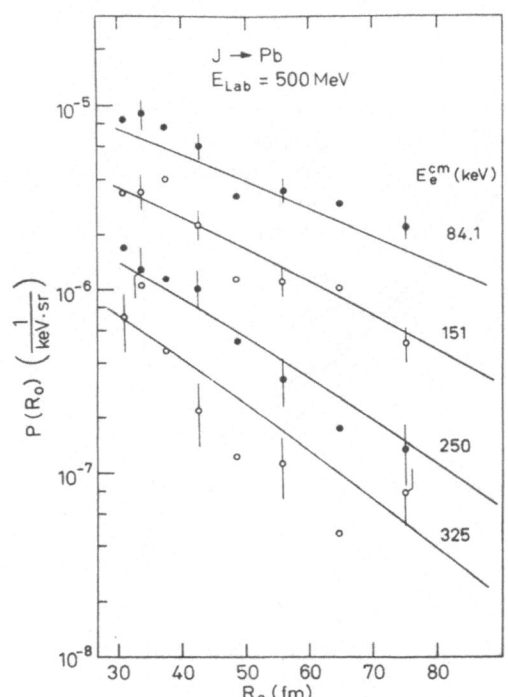

Fig. 11a: Triple coincidence measurements[3,6] (scattering angle, δ-electron energy, and K-hole formation) of δ-electrons versus $R_O=R_{min}$ in the system J+Pb at $E_{lab}=500$ MeV. Different curves belong to different kinetic electron energies.

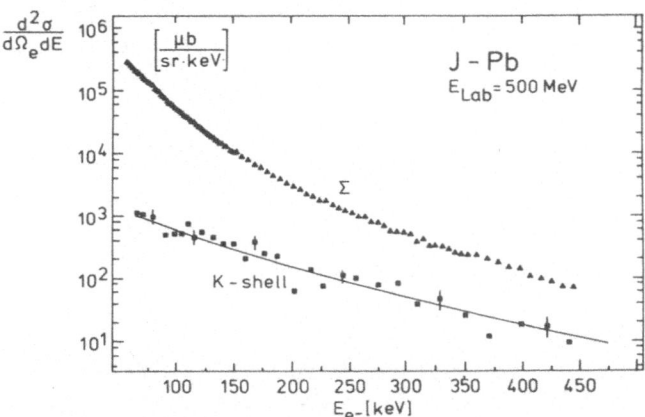

Fig. 11b: Total δ-electron distribution (triangles) and δ-electrons in coincidence with K-vacancies for the same system versus kinetic energies are results of Refs. 3,6. The latter data are compared with absolute values of coupled channel calculations.

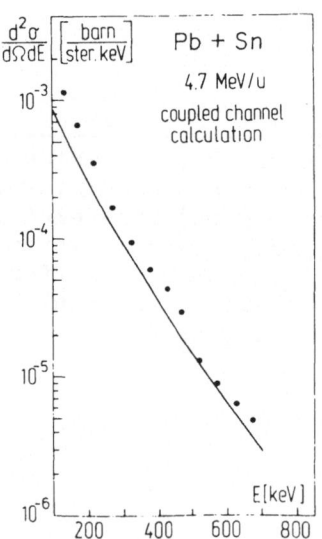 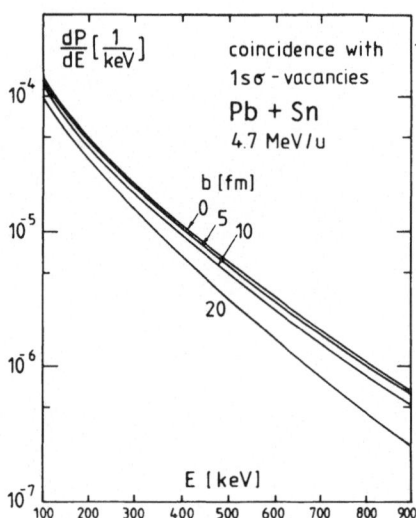

Fig. 12: a) δ-ray spectrum for the Pb+Sn system which should
be measured in coincidence with 1sσ-vacancy formation.
The data are divided by 4π and compared with measurements of Kozhuharov et al.[3]. The experimental error
bars are not drawn. b) The δ-electron distribution
for selected impact parameters for the same system.

The differential production probability of δ-electrons measured
in coincidence with a 1sσ-vacancy is displayed in Fig. 13 as a function of the classical impact parameter b. The superheavy systems
J+Pb and J+U at E_{lab}=3.94 MeV/u are under investigation. The various
curves correspond to different electron energies E_{e^-}=300,...,900 keV.
For b ≥ 15 fm the curves show an exponential fall-off, which is
steepest for the highest energy E. At smaller impact parameters the
curves pitch down. As initial bound states we take into account the
1sσ- up to the 3sσ-state.

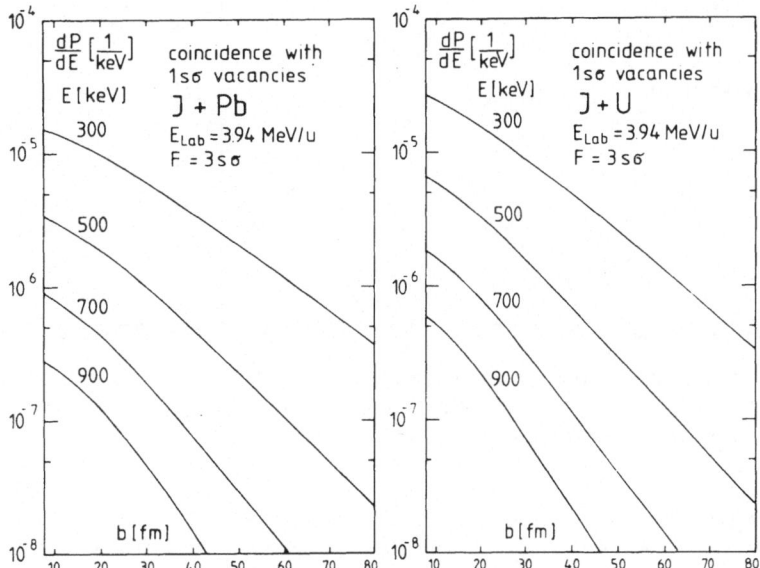

Fig. 13: a) Production probabilities for δ-electrons in coincidence
with 1sσ-vacancies versus impact parameter b, considering
a I + Pb collision at E_{lab} = 3.94 MeV/u. The different
curves belong to different kinetic electron energies.
b) The same for the I+U system.

In fig. 14 we compare for various superheavy quasimolecules the
differential cross section for δ-electron emission with respect
to the kinetic electron energy E. The systems Br+Pb and Br+U at
E_{lab} = 6.5 MeV/u, I+Pb and I+U at E_{lab} = 3.94 MeV/u and Au+U at
E_{lab} = 3.5 MeV/u are considered. A coincidence measurement with
1sσ vacancies is assumed. In each system we find the typical experimental shape. However, the exponential decline depends strongly
on the combined nuclear charge Z and the bombarding energy, which
reflects the different momentum components in the initial bound
state wavefunction.

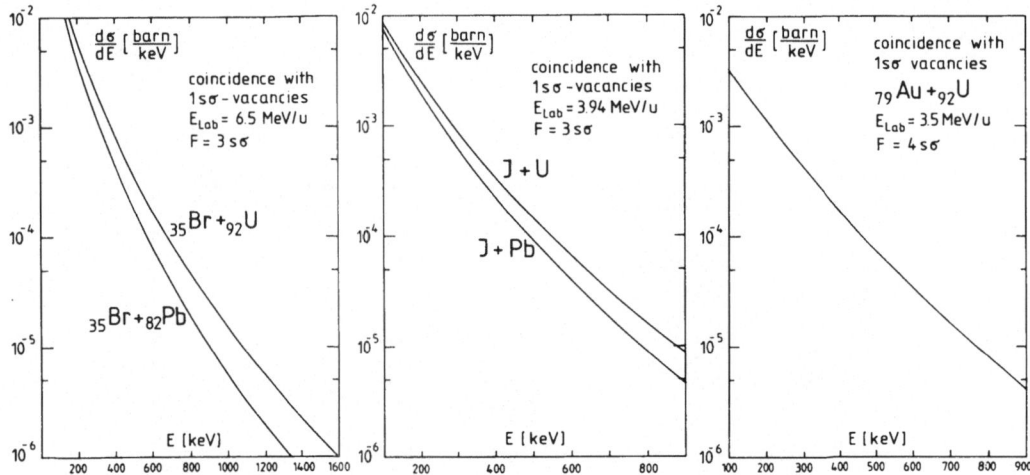

Fig. 14: Differential cross section for δ-electron emission with
respect to the kinetic electron energy E for various
superheavy quasimolecules.

As an example for $2p_{1/2}\sigma$ excitation we finally show in fig. 15
the number of created $2p_{1/2}\sigma$ vacancies per Xe+Pb collision. Different bombarding energies are considered. We have to stress that
rotational coupling contributions and electron screening corrections are neglected. Obviously the impact parameter dependence of
P(b) for the $2p_{1/2}\sigma$ state is less stepp than for the strongest
bound $1s\sigma$-state.

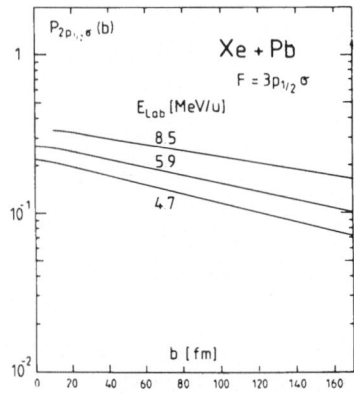

Fig. 15: Number of created $2p_{1/2}\sigma$-vacancies per collision versus
classical impact parameter b for the Xe+Pb system. Various
bombarding energies are considered.

5. SPIN POLARIZATION OF ELECTRONS BY STRONG COLLISIONAL MAGNETIC
    FIELDS

The strongest magnetic fields on a microscopic scale accessible
to experimental observation are created[7,16] in heavy ion collisions
with $(Z_1+Z_2)\alpha > 1$. The maximum magnetic field strength may reach
values in the order of $\vec{B}_{max} = 10^{15}$ G.

In superheavy quasimolecules the $1s\sigma$-electron moves almost
adiabatically close to the nuclei and is highly localized. Hence
it may serve as a test particle for the high-$\vec{B}$ limit of Quantum-
electrodynamics. The strong collisional magnetic fields cause a
spin polarization of electrons which will be discussed in this
chapter.

In the following discussion we neglect electron screening and
retardation effects which should have only small influence on
polarization. We now have to solve the coupled channel equations
taking into account the scalar and vector potentials in the
transvers Coulomb gauge. For this purpose we split the relativistic
Hamiltonian into two parts,

$$H = H_0 + H' \tag{5.1}$$

where H_0 is the two-center Dirac Hamiltonian. H' contains any
interaction responsible for electron excitations which is not
included in H_0. In the case of magnetic interactions it is given
by

$$H' = -\vec{\alpha} \cdot \vec{A} \tag{5.2}$$

with the vector potential $\vec{A}$ created by the current of both colliding
ions. This leads to a set of first-order coupled differential
equations for the occupation amplitudes $a_{ij}(t)$.

$$\dot{a}_{ij}(t) = -\sum_k a_{ik}(t)\{<\phi_j|\partial/\partial t|\phi_k> + i<\phi_j|H'(t)|\phi_k>\}\exp(i\chi_{jk}) \tag{5.3}$$

with the phases

$$\chi_{ik} = \chi_i - \chi_k. \tag{5.4}$$

In contrast to eq. (2.4) there occurs an additional magnetic
interaction term resulting from the vector potential $\vec{A}$ which reads
in the Coulomb gauge

$$\vec{A}_C(\vec{r},t) = - \sum_{i=1,2} \frac{z_i e^2 \vec{v}_i(t)}{2|\vec{r}-\vec{R}_i(t)|} -$$

$$- \sum_{i=1,2} \frac{z_i e^2 \vec{v}_i(t) \cdot [\vec{r}-\vec{R}_i(t)][\vec{r}-\vec{R}_i(t)]}{2|\vec{r}-\vec{R}_i(t)|^3}$$

$$= \vec{A}_C^{(1)} + \vec{A}_C^{(2)} \quad . \tag{5.5}$$

The corresponding result in the Lorentz-gauge $\vec{A}_L(\vec{r},t)$ becomes

$$\vec{A}_L(\vec{r},t) = 2\vec{A}_C^{(1)}(\vec{r},t) \quad . \tag{5.6}$$

For simplicity we restrict our derivations to symmetric systems with

$$Z_1 = Z_2 = Z \tag{5.7}$$

and use nuclear trajectories prescribed as Rutherford hyperbolas in the x-z plane, where the z-axis connects both nuclei. The matrixelements resulting from the interaction term in eq.(5.3)are evaluated in the monopole approximation since exact continuum solutions of the two-centre Dirac equation are unknown up to now. Due to selection rules or triange rules we find that all contributions to couplings between s-states with the same magnetic quantum number just cancel in the Coulomb gauge

$$\langle s' \pm \tfrac{1}{2}|H'_{Coulomb}|s \pm \tfrac{1}{2}\rangle = 0. \tag{5.8}$$

This is different from the Lorentz gauge, where we obtain

$$\langle s' \pm \tfrac{1}{2}|H'_{Lorentz}|s \pm \tfrac{1}{2}\rangle = -\frac{Ze^2}{3}iv_R\int_0^\infty r^2 dr \frac{r_<}{r_>^2}(fg'-f'g) \quad . \tag{5.9}$$

However, also in this case there is no contribution to the Zeeman splitting: Due to the minus sign in the integrand of eq. (5.9) the diagonal matrix elements (s'=s) vanish. For the spin flip transitions we find in summary the simple result

$$<s'\mp\tfrac{1}{2}|H'|s\mp\tfrac{1}{2}> = \mp\frac{Ze^2}{3}iv_\phi\int_0^\infty r^2 dr\frac{r_<}{r_>^2}(fg'+f'g) =$$

$$= \mp iA_{s'\mp,s\mp'},\qquad (5.10)$$

which is the same in the Coulomb as well as in the Lorentz gauge.
The only contributing rotational matrixelement in the monopole
approximation leads to

$$<s\pm1/2|-i\vec{\omega}\cdot\vec{j}|s\mp1/2> = \mp bv_\infty/2R^2 . \qquad (5.11)$$

Up to now we only discussed the fate of a single electron
influenced by the collision dynamics. In order to describe the
many-electron problem the field operator $\hat{\psi}$ is expanded   in terms
of the complete adiabatic basis states ϕ_q. This leads us to expressions for the number operators of particles N^y_{q+} with a given
spin projection along a marked axis y. The spin polarization of
holes q along the y axis is defined by

$$\eta_q = (N^y_{q+} - N^y_{q-})/(N^y_{q+} + N^y_{q-}) \qquad (5.12)$$

and correspondingly for particles

$$\eta_p = (N^y_{p+} - N^y_{p-})/(N^y_{p+} + N^y_{p-}) . \qquad (5.13)$$

Using symmetry, time-reversal-symmetry and completeness relations of the expansion coefficients $a_{i\pm j\pm}$ to a given spin projection
we can rewrite the two preceding equations into

$$\eta_q = \frac{2\text{Im}\left(\sum_{r<F} a^*_{q-,r+}a_{q-,r-}\right)}{1 - \sum_{r-<F}|a_{q-,r-}|^2 - \sum_{r+<F}|a_{q-,r+}|^2} , \qquad (5.14)$$

$$\eta_p = \frac{2\text{Im}\left(\sum_{r-<F} a^*_{r-,p+}a_{r-,p-}\right)}{\sum_{r-<F}|a_{r-,p-}|^2 + \sum_{r-<F}|a_{r-,p+}|^2} . \qquad (5.15)$$

The solutions for a modified set of coupled channel equations
containing expressions for $\dot{a}_{i+j+}$, $\dot{a}_{i+j-}$, $\dot{a}_{i-j+}$ also take into
account the rotational coupling eq. (5.11), the diagonal and off-
diagonal magnetic interaction and as usual the radial coupling.

We only consider results for the collision Pb+Cm ($Z_1+Z_2=178$)
hence being the heaviest system investigated with respect to $1s\sigma$-
ionization probabilities.

As a main result we found that the additional magnetic interaction does not change the total ionization probabilities but leads
to spin polarization of electron states. The created polarization
is preserved during the collision despite the importance of multi-[16]
step excitation processes. As dominant effect the Zeeman splitting
of the $1s\sigma$-state gives rise to a stronger ionization of one of the
usually degenerated spin states whereas the other becomes less
ionized. However, the total sum remains almost exactly the same as
obtained with the radial coupling only.

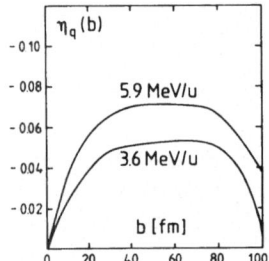

Fig. 16:   Spin polarizations of $1s\sigma$-vacancies versus impact
           parameter b. E_{lab}=3.6 MeV/u and 5.9 MeV/u,
           respectively. $Z_1 + Z_2 = 178$.

The spin polarization η_q of created $1s\sigma$-vacancies has been
calculated according to eg. (5.14). Its dependence on the impact
parameter b is presented in Fig. 16 . As can be seen, η_q=0 for
head-on collisions since the rotational velocity v_ϕ is zero in this
case. Its maximum is obtained for medium impact parameters between
b=30 and 80 fm. For larger impact parameters (distances) the magnetic
field is too small in order to influence ionization processes. This
theoretical prediction still needs experimental verification.
Ionization measurements of $1s\sigma$-electrons and also δ-electron spectroscopy therefore may yield direct information about the behaviour
of electrons in strong magnetic fields.

6. THE SELF-ENERGY OF ELECTRONS IN CRITICAL FIELDS

The K-electron binding energy E_{1s} increases strongly as a function of the nuclear charge Z. For Z = 150, E_{1s} amounts to about the electron rest mass and hence one enters the truly relativistic domain. For Z $\gtrsim$ 170 the binding energy exceeds twice the electron rest mass and the K-shell electron gets imbedded as resonance in the negative energy continuum, which opens the possibility of sponateous positron production[17].

The major motivation of our investigations was the question wether field theoretical corrections, such as vacuum-polarization and self-energy may prevent such an extraordinary strong binding. These processes are visualized by the Feynman-diagrams in fig. 17. The double lines indicate the exact propagators and wave functions in the Coulomb field of the nucleus. The dominant vacuum-polarization contribution is provided by the attractive Uehling potential.

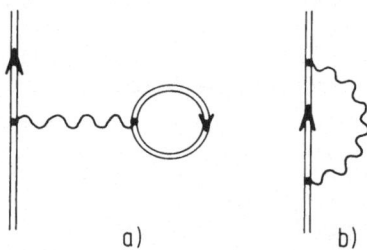

Fig. 17: Feynman diagrams for the lowest order vacuum-polarization (a) and self-energy (b). The double lines indicate the exact propagators and wave functions in the Coulomb field of a nucleus.

Its influence on electronic binding energies for superheavy systems has been calculated by various authors[18-20]. For the critical nuclear charge Z_{cr} the Uehling potential leads to an energy shift $\Delta E_{VP}^{(n=1)}$ = -11.8 keV[20], which decreases Z_{cr} by 1/3 of a unit. The remaining vacuum-polarization effects in lowest order of the fine-structure constant α but in all orders of $(Z\alpha)^n$ with n $\geq$ 1 were evaluated by M. Gyulassy[21,22] and by Rinker and Wilets[23]. These authors made use of the angular momentum decomposition of the electron propagator in spherically symmetric potentials that was developed by Wichmann and Kroll[24]. The obtained energy shift of $\Delta E_{VP}^{(n>1)}$ = +1.15 keV[21] is very small compared with the total K-shell binding energy of 1 MeV.

Electronic self-energy corrections for high-Z systems have been first
studied in the pioneering work of Brown, Langer and Schaefer[25-27]
In these theoretical investigations the traditional expansion[28] of
the Feynman diagrams in powers of the copling constant (Zα) of the
external field was avoided. This method was further refined and
successfully applied in computations of electron energy shifts in
high-Z elements by Desiderio and Johnson[29], who allowed for a
realistic nuclear charge distribution as well as the electron-
electron interaction in the Hartree-Fock approximation. The precise
analysis of self-energy corrections by P. Mohr[30] is based on the
Coulomb potential for point-like nuclei. Due to the singular nature
of the potential these calculations are restricted to nuclear
charges below $Z = \alpha^{-1} \sim 137$. Cheng and Johnson[31] continued the
calculations of ref. 29) up to Z = 160, where a repulsive energy shift
for K-shell electrons of ΔE_{SE} = +7.3 keV was found.

Recently Liesen et al.[14] measured the ionization probability P(b)
of the strongest bound electron states versus the classical impact
parameter b in collisions of Pb and Cm with a combined charge
$Z_1 + Z_2 = 178$. For almost central collisions deviations from an
empirical scaling law for P(b) were found. The authors speculated
that a strong self-energy shift of the quasimolecular 1sσ-state
could be responsible for the observed modification of the
ionization probability.

In our calculations we employed the methods developed by Desiderio
and Johnson[29], which may be slightly simplified by restriction
to K-shell electrons. We performed our calculations for hydrogen-
like systems. The external potential energy V(x) is determined by
the nuclear charge distribution, for which a homogeneously charged
sphere with a radius $R = 1.2 A^{1/3}$ fm has been assumed. To check our
computer code, we computed the self-energy contribution to the
K-shell binding energy in mercury (Z=80). Assuming a fictitous nuclear
mass number A=1 (R=1.2 fm) we obtained ΔE_{SE}=206.1 eV which is about
0.7 eV smaller than the result of Cheng and Johnson[31] for a point-
like nucleus. For the superheavy system Z = 130 we found ΔE = 2.537
keV for a nucleus with A=1, as compared with the point-nucleus value
(ref. 31) of ΔE_{SE} = 2.586 keV ±.156 keV. These numbers are drastically
reduced if one takes into account a realistic nuclear size
determined by A = 2.5 Z. This lowers the energy shift to ΔE_{SE} =
1.896 keV. The complementary result of Cheng and Johnson is
ΔE_{SE}^{HF} = 1.844 keV ±.029 keV, where in addition, electron screening
effects within a mean field Hartree-Fock (HF) potential were taken
into account. For Z = 150 the present calculation leads to ΔE_{SE} =
4.963 keV and for Z = 160 to ΔE_{SE} = 7.759 keV, respectively.
The latter number differs by 393 eV from the corresponding HF-values
of ref. 31), where the numerical error was estimated to ± 354 eV.
Presumably this slight disagreement is caused by the considerable
difference of a HF-potential from a Coulomb potential for finite

size nuclei. Our calculation for Z=169 yielded ΔE_{SE} = 10.839 keV.
For the critical nuclear charge Z = 170 we adjusted the nuclear
mass number and hence the nuclear radius such that the K-electron
energy eigenvalue differed only by 10^{-3} eV $\sim 10^{-9}$ E_{1s}^{b} from the

border line of the negative energy continuum. As the most important
result we found an energy shift of ΔE_{SE} = 10.989 keV, which still
represents only a 1% correction to the total K-electron binding
energy. Therefore it may safely be neglected in investigations of
ionization probabilities[14] in superheavy quasimolecular systems.
If one adds to this the vacuum-plarization calculations of ref. 20)
and 21) for critical external potentials, the total energy shift
due to radiative corrections of order α amounts only to 300 eV. This
tiny effect is at present far outside of any measurable consequences.
The various calculations for the self-energy correction of K-electrons
in high-Z atoms are summarized in fig. 18[32]. On a logarithmic scale
the energy shift is displayed versus the nuclear charge Z. For
$Z \geq 70$ it is well described by an exponential increase.

We conclude that radiative corrections as vacuum-polarization or
self-energy may not prevent the K-shell binding energy from
exceeding $2mc^2$ in superheavy systems with Z > $Z_{cr} \sim 170$.

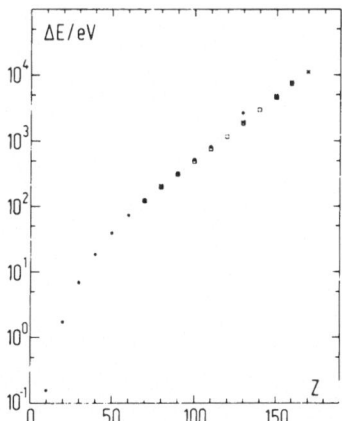

Fig. 18: The self-energy shift ΔE of K-shell electrons as a function
of the nuclear charge Z. The calculations are performed
to first order in α but to all orders in the coupling
constant (Zα) of the external field. The dots denote the
numerical results of P. Mohr[30] for 1s-electrons in the
Coulomb field of point-like nuclei. The squares represent
the values obtained by Cheng and Johnson for a Hartree-
Fock potential and extended nuclei. The results of the
present calculations for extended nuclei are indicated
by crosses.

7. THE GLASHOW-SALAM-WEINBERG MODEL AND ELECTRON POLARIZATIONS

IN HEAVY-ION COLLISIONS

In this section we roughly estimate the order of electron polarization phenomena in heavy-ion collision caused by the Glashow-Salam-Weinberg interaction. In this model the interaction Lagrangian for the coupling of a neutral Z-boson to leptons is given by

$$L_1 = - \frac{g}{4 \cos \theta_w} \, \bar{e}(x) \, \gamma_\mu \, (c_v + \gamma_5) e(x) Z^\mu(x) \tag{7.1}$$

with $c_v = 1-4 \sin^2 \theta_w$ and $\sin^2\theta_w = 0.22$. θ_w denotes the Weinberg angle and g the coupling constant being related with the charge by $e = g \sin\theta_w$. $e(x)$ is the field operator of the electron and $Z^\mu(x)$ described the field of the neutral Z-boson in analogy to the A^μ-field in the case of the electromagnetic interaction. On the other hand the coupling to quarks is determined by[33]

$$L_2 = \frac{g}{4 \cos\theta_w} \, \{\bar{u}(x) \, \gamma_\mu(c_v' + \gamma_5) \, u(x) + \bar{c}(x) \, \gamma_\mu(c_v' + \gamma_5) \, c(x)$$

$$- \bar{d}' \, \gamma_\mu(c_v'' + \gamma_5) \, d' - \bar{s}' \gamma_\mu(c_v'' + \gamma_5) \, s' - \ldots \} Z^\mu(x) \tag{7.2}$$

with

$$c_v' = 1 - \frac{8}{3} \sin^2 \theta_w$$

$$c_v'' = 1 - \frac{4}{3} \sin^2 \theta_w \tag{7.3}$$

$u(x)$ and $c(x)$ are the fields of the up- and charm-quarks, respectively.

$$\frac{G}{\sqrt{2}} = \frac{g^2}{8M_W^2} \tag{7.4}$$

is the Fermi coupling constant. The mass of the charged W-boson M_W and of the neutral vector boson M_Z are related through

$$M_Z \cos\theta_w = M_W \tag{7.5}$$

Furthermore we introduced the mixing

$$d' = d \cos\theta_c + s \sin\theta_c$$

$$s' = -d \sin\theta_c + s \cos\theta_c \tag{7.6}$$

for the fields of the down- and strange quarks. θ_C is the Cabbibo angle. In order to simplify the hadronis current (7.2) we perform some approximations:

(1) Only the up- and down-quarks are considered. Hence, the contribution of the "sea-quarks" is neglected. This immediately leads to

$$L_2 \rightarrow \frac{g}{4\cos\theta_W} \{\bar{u}(x)\ \gamma_\mu\ (c_v'+\gamma_5)\ u(x) - \bar{d}\ \gamma_\mu\ (c_v'' + \gamma_5)d\}z^\mu(x) \quad (7.7)$$

(2) We replace the quark currents by the nucleon currents

$$J_p^Z = 2\ J_u^Z + 1\ J_d^Z$$

$$\qquad\qquad\qquad\qquad\qquad\qquad\qquad\qquad (7.8)$$

$$J_n^Z = 1\ J_u^Z + 2\ J_d^Z$$

Thus we neglect the internal structure of the nucleons except for their quark content. This represents an approximation since for the charged currents eq. (7.8) is valid only for the vector part whereas the axial vector part has to be renormalized by a factor of about 1.25. With

$$J_u^Z = \bar{u}\ \gamma_\mu\ (c_v' + \gamma_5)u$$

$$\qquad\qquad\qquad\qquad\qquad\qquad\qquad\qquad (7.9)$$

$$J_d^Z = -\ \bar{d}\ \gamma_\mu\ (c_v'' + \gamma_5)d$$

we get

$$J_p^Z = \bar{p}\ \gamma_\mu\ (2\ c_v' - c_v'' + \gamma_5)p = \bar{p}\ \gamma_\mu\ (c_v + \gamma_5)p$$

$$\qquad\qquad\qquad\qquad\qquad\qquad\qquad\qquad (7.10)$$

$$J_n^Z = \bar{n}\ \gamma_\mu\ (c_v' - 2c_v'' - \gamma_5)n = -\ \bar{n}\ \gamma_\mu\ (1 + \gamma_5)n$$

In consequence it follows from equation (7.7)

$$L_2 \rightarrow \frac{g}{4\cos\theta_W}\ \frac{1}{3}\ \{\bar{p}\ \gamma_\mu\ (c_v + \gamma_5)p - \bar{n}\ \gamma_\mu(1 + \gamma_5)\ n\}\ z^\mu(x) \quad (7.11)$$

(3) We adopt the nonrelativistic limit for the protons and neutrons.

$$p = \begin{pmatrix} \phi_p \\ \chi_p \end{pmatrix} \sim \begin{pmatrix} \phi_p \\ 0 \end{pmatrix} \qquad\qquad\qquad (7.12)$$

With this approximation it results

$$\bar{p}\gamma_0 p \sim \phi_p^+ \phi_p$$

$$\bar{p}\gamma_5 p \sim 0$$

$$\vec{p}\vec{\gamma}p \sim 0$$

$$\vec{p}\vec{\gamma}\gamma_5 p \sim \phi_p^+ \sigma_p \phi_p$$

(4) In non-polarized nuclear matter the term $\sum_N \phi_N^+ \sigma_N \phi_N$ is much smaller smaller than $\sum \phi_N^+ \phi_N$. Furthermore, due to the specific value of the the Weinberg angle, we also neglect the contribution of the protons $(1-4\sin^2\theta_W \sim 0.1)$. Then the expression (7.11) simplifies considerably

$$L_2 \rightarrow - \frac{g}{4\cos\theta_W} \frac{1}{3} \phi_n^+(x) \phi_n(x) Z^0(x) \tag{7.12}$$

With $\rho_n(x) = \phi_{n,f}^+(x) \phi_{n,i}(x)$ the electron-nucleus interaction takes the form $\tilde{\rho}_n(\vec{x})e^{i x_0(E_f-E_i)}$. Due to the large mass of the Z-boson the corresponding propagator may be written as

$$\frac{g_{0\mu} - \frac{k_0 k_\mu}{M_z^2}}{k^2 - M_z^2} \sim \frac{g_{0\mu}}{M_z^2} = D_z(k)_{0\mu} \tag{7.13}$$

The interaction matrix element now may be transformed into

$$M = - \int d^4x \int d^4y\, j_e^\mu(x) \int \frac{d^4k}{(2\pi)^4} D_z(k)_{0\mu} e^{ik(x-y)} \frac{g}{4\cos\theta_W} \rho_n(y)$$

$$= - \frac{g}{4\cos\theta_W} \frac{1}{M_z^2} \int d^4x \int d^4y\, j_e^\mu(x) g_{0\mu} \rho_n(y)\delta^4(x-y)$$

$$= - \frac{g}{4\cos\theta_W M_z^2} \int d^4x\, j_e^0(x) \rho_n(x) \tag{7.14}$$

This integral contributes only inside the nuclear volume. The electron current is split off into helicity eigenstates e_R and e_L.

$$j_e^o = - \frac{g}{4 \cos\theta_w} \; e_f^+ \, (c_v + \gamma_5) e_i$$

$$= - \frac{g}{4 \cos\theta_w} \cdot \frac{1}{2} \left\{ e_f^+ \, \frac{1+\gamma_5}{2} \, e_i \, \frac{1+c_v}{2} - e_f^+ \, \frac{1-\gamma_5}{2} \, e_i \, \frac{1-c_v}{2} \right\}$$

$$= - \frac{g}{8 \cos\theta_w} \left\{ e_{R,f}^+ \, e_i \, (1-2\sin^2\theta_w) - e_{L,f}^+ \, e_i \, 2\sin^2\theta_w \right\} \quad (7.15)$$

$$\text{with} \quad e_{R,L}^+ = e^+ \, \frac{1 \mp \gamma_5}{2} \tag{7.16}$$

If we insert this into eq. (7.14) we obtain

$$M = \frac{G}{4\sqrt{2}} \int d^4x \left\{ (1-2\sin^2\theta_w) \, e_{L,f}^+ \, e_i - 2 \sin^2\theta_w \, e_{R,f}^+ \, e_i \right\} \rho_n(x) \tag{7.17}$$

Thus results a polarization of electrons which are scattered by neutral currents. It is given by

$$\frac{W(e_L) - W(e_R)}{W(e_L) + W(e_R)} = \frac{(1-2 \sin^2\theta_w)^2 - (2\sin^2\theta_w)^2}{(1+2 \sin^2\theta_w)^2 - (2\sin^2\theta_w)^2} \sim 0.16 \tag{7.18}$$

Provided that the initial states are unpolarized the final polarization amounts to about 16%.

Finally we compare the matrix elements of the electromagnetic interaction with that resulting from the neutral currents

$$M_n = \frac{g^2}{(4\cos\theta_w)^2 \, M_z^2} \int d^4x \frac{1}{2} \left\{ e_{L,f}^+ \, e_{L,i} \, (1-2 \sin^2\theta_w) \right.$$

$$\left. - e_{R,f}^+ \, e_{R,i} \, 2 \sin^2\theta_w \right\} \rho_n(x) \tag{7.19}$$

$$M_{el-mag} = e^2 \int d^4x \int d^4y \, \frac{1}{2} \left\{ e_{L,f}^+ \, e_{L,i} + e_{R,f}^+ \, e_{R,i} \right\} \cdot$$

$$\int \frac{d^4k}{(2\pi)^4} \, \frac{1}{k^2+i\varepsilon} \, e^{i k (x-y)} \, \rho_p(y) \tag{7.20}$$

Now we assume

$$\rho_p(y) = \tilde{\rho}_p (\vec{y}) \, e^{i(E_f-E_i)y_o}$$

It follows

$$|M_{el-mag}| = |e^2 \int d^4x \, \frac{1}{2} \{e^+_{L,f} e_{L,i} + e^+_{R,f} e_{R,i}\} \int \frac{d^4k}{(2\pi)^4} \int d^3y$$

$$\delta(k^0 - E_f + E_i) \cdot \frac{1}{k^2 + i\varepsilon} e^{i[k_0 x_0 - \vec{k}(\vec{x}-\vec{y})]} \tilde{\rho}_p(\vec{y})| =$$

$$|e^2 \int d^4x \, \frac{1}{2} \{e^+_{L,f} e_{L,i} + e^+_{R,f} e_{R,i}\} \int \frac{d^3\vec{k}}{(2\pi)^4} \int d^3\vec{y} \, \frac{1}{(E_f - E_i)^2 - \vec{k}^2 + i\varepsilon} \cdot$$

$$e^{-i\vec{k}(\vec{x}-\vec{y})} e^{i(E_f - E_i)x_0} \tilde{\rho}_p(\vec{y})| =$$

$$|\frac{e^2}{8\pi^2} \int d^4x \, \frac{1}{2} \{e^+_{L,f} e_{L,i} + e^+_{R,f} e_{R,i}\} e^{i(E_f - E_i)x_0}$$

$$\int d^3\vec{y} \, \frac{1}{|\vec{y}-\vec{x}|} e^{i|E_f - E_i||\vec{x}-\vec{y}|} \tilde{\rho}_p(\vec{y})| \qquad (7.21)$$

where we employed

$$\int \frac{d^3\vec{k}}{(2\pi)^4} \frac{e^{i\vec{k}\cdot(\vec{x}-\vec{y})}}{(E_f - E_i)^2 - \vec{k}^2 + i\varepsilon} = -\frac{1}{8\pi^2} \frac{1}{|\vec{y}-\vec{x}|} e^{i|E_f - E_i||\vec{x}-\vec{y}|} \qquad (7.22)$$

In general eq. (7.21) may not be further simplified. The comparison with eq. (7.19) clearly expresses the difference between the Coulomb – and the point-interaction. The weak interaction is proportional to the overlap of the electron density and nucleon density. A rough estimate is attainable from eq. (7.20)

$$|\int \frac{d^4k}{(2\pi)^4} \frac{1}{k^2 + i\varepsilon} e^{ik(x-y)}| \geq |\int \frac{d^4k}{(2\pi)^4} \frac{1}{|k^2|_{max}} e^{ik(x-y)}| =$$

$$\frac{1}{|k^2|_{max}} \delta^4(x-y) \qquad (7.23)$$

The maximum momentum transfer which is observed experimentally[3,6] for electron excitation process in superheavy quasimolecules is about 2 MeV. As upper limit we assume $|k^2|_{max} = (7 \text{ MeV})^2$. Then we find

$$\left| \frac{M_n}{M_{el\text{ -mag}}} \right| = \frac{g^2(7\text{MeV})^2}{(4\cos\theta_w)^2 \, M_Z^2 \, e^2} \, \frac{N}{Z} \sim \frac{1}{(4\sin\theta_w \cos\theta_w)^2} \left(\frac{7 \text{ MeV}}{70\text{GeV}}\right)^2 \frac{N}{Z}$$

$$\sim \frac{N}{3Z} \, 10^{-8} \sim 10^{-8} \qquad\qquad\qquad (7.24)$$

Thus the interaction via neutral currents is strongly suppressed and may be safely neglected in polarization measurements of electron excitation processes in superheavy quasimolecules. Still we have to emphasize that this does not exclude observable effects due to the neutral current interaction in superheavy systems. A corresponding measurement would be favourable whenever an electromagnetic transition is forbidden by angular momentum or parity selection rules.

8. POSITRON CREATION IN SUPERCRITICAL QUASIMOLECULES

At the nuclear charge value $Z \simeq 173$ the 1s-state enters the negative
energy continuum, $E_{1s} < -mc^2$; the binding energy has reached the
threshold where spontaneous pair production becomes possible. At
this point the spectrum of eigenstates of the Dirac equation is
subject to a characteristic change: The 1s-state becomes a resonance
in the lower energy continuum. In a Gedankenexperiment it is
possible to increase the charge of a bare nucleus from $Z < Z_{cr} \simeq$
173 to $Z > Z_{cr}$, the unoccupied 1s-state will then be filled under
emission of two (due to spin-degeneracy) positrons. The new stable
ground state of the system consists of the nucleus plus two electrons
in the K-shell; it is called the charged vacuum[34,35]. The
experimental exploration of this new phenomenon would constitute
an important test of the theory of Quantum Electrodynamics (QED)
in the region of strong fields.

At present, however, the unique way of extending the periodic table
of elements, where these atomic effects can be pursued, lies in
heavy ion collisions. In such scattering experiments, however, the
dynamics of the collision becomes extremely important. The
time-scale must be sufficiently long to allow the electron
(positron) to adjust to the variation of the combined Coulomb field
of the two nuclei. Since the typical velocities required to bring
the nuclei close together are about $v/c \simeq .1$, the adiabatic
picture is meaningful only for electrons in relativistic motion.

In this section we will concentrate on positron creation. In
particular, it is our aim to give an adequate description for
positron production in supercritical collision systems, where
$Z_T + Z_P$ exceeds 173. In supercritical collisions the resonance
property of the 1sσ-state must be handled with care. In ref. 17
a formalism is developed, which avoids those difficulties and
moreover has heuristic value for the interpretation of the
positron creation process. The method is based on the
observation that the continuum wavefunction of the supercritical
system at resonance energy $E_p = E_{res}$ is quite similar to the
discrete 1sσ-state in the subcritical case except for an
oscillating tail - small in amplitude - reaching out to infinity.
This structure reflects the occurrence of a tunneling process
through the gap separating the particle- and antiparticle solutions
of the Diracequation. Apart from the asymptotic behaviour the
1sσ-wavefunction retains much of its identity, e.g. the strongly
localized charge distribution having the extension of the atomic
K-shell. This fact can be used to develop a general method to
treat resonance scattering. In this context Wang and Shakin[36]
introduced a projection formalism for resonances in the nuclear
continuum shell model: After having defined a normalizable
quasibound wavefunction ϕ_R, a new continuum $\tilde{\mathscr{Y}}_{E_p}$ is constructed
which spans a subspace orthogonal to ϕ_R and replaces the old

continuum ζ_{Ep}. Here ζ_{Ep} satisfies the Dirac equation

$$(H_{TCD} - E_p) \; |\varphi_{Ep}> = 0 \tag{8.1}$$

and contains the resonance; E_p denotes the energy of the positron. The modified continuum wavefunction $\tilde{\mathfrak{Y}}_{Ep}$ are eigenstates of a projected Hamiltonian

$$(P \; H \; P - E_p) \; |\tilde{\varphi}_{Ep}> = 0 \; , \tag{8.2}$$

with

$$P = 1 - Q := \int dE_p \; |\tilde{\varphi}_{Ep}><\tilde{\varphi}_{Ep}| \; ,$$

$$\tag{8.3}$$

$$Q = |\phi_R><\phi_R| + \sum_\alpha |\varphi_\alpha><\varphi_\alpha| \; ,$$

where Q is an operator projecting on the resonance state ϕ_R and on all higher bound and electron continuum states $\tilde{\mathfrak{Y}}_\alpha$. If ϕ_R was chosen judiciously the newly defined modified continuum $\tilde{\mathfrak{Y}}_{Ep}$ will no longer show resonance behaviour. Using eqs. (8.3) and the orthogonality relations, one obtains

$$(H - E_p) \; |\tilde{\varphi}_{Ep}> = <\phi_R|H|\tilde{\varphi}_{Ep}>|\phi_R> \; . \tag{8.4}$$

The modified continuum states satisfy the original Dirac equation supplemented by an inhomogeneous term containing an integral over the solution $\tilde{\mathfrak{Y}}_{Ep}$.

If the states ϕ_R and $\tilde{\mathfrak{Y}}_{Ep}$ are used as part of the basis the 1sσ-state couples to the positron continuum by two separate coupling operators

$$\dot{R}<\tilde{\varphi}_{Ep}|\partial/\partial R|\phi_R> + i/\hbar<\tilde{\varphi}_{Ep}|H|\phi_R> \; . \tag{8.5}$$

The second matrixelement arises since ϕ_R and $\tilde{\mathfrak{Y}}_{Ep}$ are not exact eigenstates of the two-centre Hamiltonian H. It does not depend on the nuclear motion and leads, in the static limit R(t) = const, to an exponential decay of a hole prepared in ϕ_R with the width

$$\Gamma = 2\pi \; |<\tilde{\varphi}_{E_{res}}|H|\phi_R>|^2 \; . \tag{8.6}$$

The values of Γ deduced by this method agree well with the exact widths obtained by a phase shift analysis of the old continuum ψ_{Ep}.

The developed formalism thus has led to the emergence of "induced" and "spontaneous" positron coupling, the latter resulting from the presence of an unstable state ϕ_R in the expansion basis. In practice, however, this does not result in a threshold behaviour at the border of the supercritical region. Firstly both coupling matrixelements enter via their Fourier transforms depending on the time development of the heavy ion collision. Their contributions have to be added coherently so that in a given collision there is no physical way to distinguish between them. Secondly in Coulomb collisions the rapid variation of the quasimolecular potential causes significant contributions from the dynamical coupling, whereas the period of time for which the internuclear distance $R(t)$ is less than R_{cr} is usually very short ($\sim 10^{-21}$ s) compared to the decay time of the 1sσ-resonance ($\sim 10^{-19}$ s).

Therefore, the predicted production rates and energy spectra of positrons continue smoothly from the subcritical to the supercritical region. It turns out that the emergence of a new coupling is accompanied by a reduction of the radial matrixelement $<\psi_{Ep}|\partial/\partial R|\phi_R>$ and both effects nearly cancel in collisions on Coulomb trajectories. Qualitative deviations of the positron production rate in supercritical collision systems are expected only under favourable conditions, i.e. in encounters with a prolonged interaction time. Since the "spontaneous" and "dynamical" couplings exhibit a different functional dependence on the nuclear motion (the "induced" coupling is proportional to the velocity $\dot{R}$ while the "spontaneous" part depends on the distance R), an increase in collision time can be expected to provide a clear signature for supercritical collisions. Therefore Rafelski et al.[37] suggested the study of positron emission in heavy ion reaction at bombarding energies above the Coulomb barrier, where the formation of a dinuclear system or of a compound nucleus would eventually lead to a time delay within the bounds of the critical distance R_{cr}. During this delay time T the spontaneous decay of the 1sσ-resonance, by filling dynamically created K-shell holes under emission of positrons, might be strongly enhanced. For an extensive discussion concerning the influence of a nuclear time delay on positron spectra and δ-electron distributions we refer to refs. 38-40.

Since beams of very heavy ions at energies close to the Coulomb barrier or even above have become available at the Gesellschaft für Schwerionenforschung (GSI) in Darmstadt, a number of experiments has been performed to study positron emission in highly charged collision systems. At the same time, experiments measuring K-hole and δ-electron production have yielded independent information on the excitation mechanism and have served to strengthen confidence in the underlying theoretical framework, see refs. 3,6, 11-14.

Before we will reflect on experimental results for positron creation and their comparison with theory, we first want to discuss the main theoretical predictions. The first calculations had been performed in the framework of perturbation theory which describes much of the physics involved in the excitation processes. However, the growing importance of muli-step processes indicates that perturbative calculations are of limited validity. In the following, therefore, we will present the results of coupled channel calculations [17].

We have solved the system of differential equations using the monopole approximation including up to 8 bound states and ~15 states in the upper continuum, separately for the angular momentum channels $\kappa = +1$ and $\kappa = -1$.

Positron emission rates increase very fast with total nuclear charge, flattening somewhat for the highest Z-values. If parametrized by a power law $(Z_T+Z_P)^n$ the power takes values of 20 down to 13. Here a Fermi-level at $F = 3s\sigma$, $4p_{1/2}\sigma$ is assumed, i.e. the states above the $3s\sigma$- and $4p_{1/2}\sigma$-levels are empty, which should give an upper bound for positron production rates. If the distance of closest approach R_{min} is kept fixed instead of the impact velocity, n becomes still larger. In collisions of bare nuclei (F=0) positron production is increased by up to two orders of magnitude. Here also the $\dot{Z}$- dependence is extremely steep (n~29). Mainly responsible for this effect is the contribution of the 1s-state which in normal col- lisions (F>0) is suppressed due to the small K-vacancy probability. If the K-shell is empty it becomes the dominant final state for pair production. This clearly reflects the strong coupling between the 1s-state and the antiparticle continuum which it approaches and even enters in the supercritical region. In all cases investigated the channels $\kappa = -1$ and $\kappa = +1$ contribute about equally to the total result.

Now we turn to the discussion of experimental results and their comparison with the described predictions. But first we want to point out a major problem in analyzing the experimental data.

Already for bombarding energies well below the Coulomb barrier $(E \sim .8E_C)$ the nuclei can be excited by Coulomb excitation, the emitted photons with energy above 1022 keV can undergo pair con- version. Although this process takes place long after the collision $(\sim 10^{-13}$ sec, while "atomic" positrons are emitted within $\sim 10^{-20}$ sec), it cannot be distinguished experimentally from the quasimolecular mechanism by ordinary methods, resulting in a background which is difficult to handle. For nuclei with a simple level-structure (e.g., ^{208}Pb: $2^+,3^-$) Coulomb excitation can be calculated. The resulting pair creation is then deduced from the theoretically known conversion coefficient. Otherwise one has to measure simultaneously the γ-spectrum and fold it with the conversion coefficient. Here one has to know - or to assume - the γ-ray multipolarity.

In collisions with lighter targets, where the combined charge of projectile and target nucleus, $Z_u = Z_p + Z_T$, is well below the critical charge so that no significant contribution of atomic positrons is expected, the γ-ray spectra have been measured. The data can be fitted nicely by pair conversion if E1 multipolarity is assumed. Beginning in the region $Z_p+Z_T \gtrsim 160$, however, all experiments have found an increase which could not be explained by nuclear conversion alone (Fig. 19). This was the first hint for "atomic" positron production in heavy-ion collisions and, up to now, all conclusions had to rely on the described procedure for background subtraction.

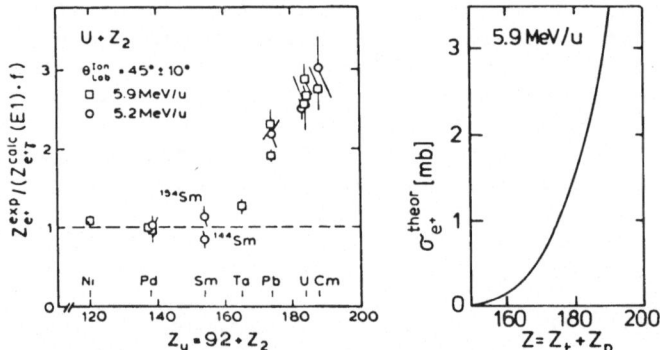

Fig. 19: In the left part the ratio between measured positron production rates and the expected pair conversion rates deduced from the nuclear γ-spectra is displayed as a function of the united charge. For $Z_u > 160$ an increase in positron emission is observed, which cannot be explained by nuclear excitation processes[41]. The right part shows theoretical predictions for the positron cross section, $\sigma_{e^+}(Z)$, in heavy ion collisions at E=5.9 MeV/u. A strong increase is expected.

The first generation of experiments further established the
dependence of positron excitation rates on the kinematic conditions
as well as on the combined charge Z_u. Fig. 20 shows results of
Kozhuharov et al.[42] for three collision systems, Pb-Pb, U-Pb, and
U-U, at 5.8 MeV/u bombarding energy, measured with an orange-type
β-spectrometer. The positron production probability in an energy
window $E(e^+) = 490\pm50$ keV is shown as a function of projectile
cm-scattering angle Θ_{cm}. The nuclear background is subtracted.
Because target and projectile cannot be distinguished in the
experiment, the theoretical values have been symmetrized with respect
to forward and recoil scattering.

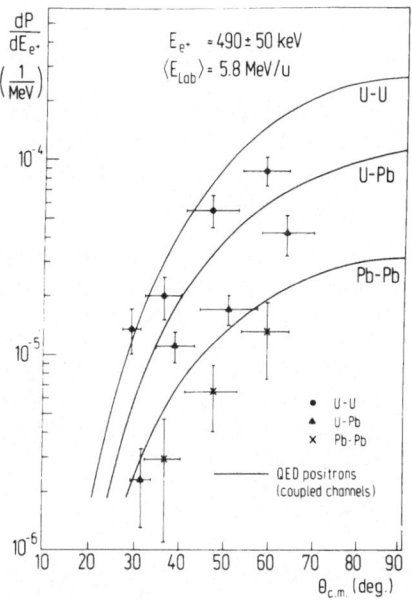

Fig. 20: For 5.8 MeV/u Pb-Pb, Pb-U, and U-U collisions positron
         production probability in an energy window $E_e+=490\pm50$ keV
         is displayed as a function of projectile scattering angle
         Θ_{cm}. The experimental data are taken from Kozhuharov et
         al (ref. 42). Here, the nuclear background is subtracted.

The shape of the theoretical curves is in quite good agreement with the experimental data. Also the Z-dependent increase is well described, which spans an order of a magnitude while $\Delta Z/Z$ is only 12%. On the other hand the absolute magnitude of the theoretical values is generally too high.

Using another type of experimental set-up, a solenoidal spectrometer, Backe et al.[44] obtained integrated and differential positron probabilities for various impact parameters for the heavy-ion collision systems Pb-Pb, Pb-U, U-U, and U-Cm. As in the experiment described above general agreement with theory is found. In the Pb-Pb case and, for smaller distances of closest approach, even in Pb-U and U-U collisions the data seem to agree also in absolute values, in contrast to the experiment discussed above. But in the heaviest accessible system U-Cm (Z_u=188) and for larger distances R_{min} theory again has a tendency to overestimate the measured data. The experimental slopes are somewhat steeper than predicted, while the Z-dependence is overestimated.

For given kinematic parameters {E,b} the Z-dependence of the total positron probabilities is given by $Z^{\sim 17}$, for lighter systems as well as for heavier ones [44]. From these data no qualitative signature for the "diving" of the 1s -state in U-U, U-Cm collisions can be extracted, in agreement with theoretical predictions.

More sensitive information can be obtained by the measurement of energy spectra of positrons detected in coincidence with the scattered ions. Their knowledge is most useful if one wants to find deviations hinting to the positron creation mechanism.

Fig. 21 shows the first published positron spectra of Backe et al.[44,45] for three collision systems, U-Pd, U-Pb, and U-U, at 5.9 MeV/u bombarding energy; the ions are detected in an angular window θ_{lab}=45°±10°. For U-Pd (Z=138) no atomic positrons are expected, the data can be fully accounted for by nuclear conversion (light curve). Extrapolating this procedure to the U-Pb system (light curve) the sum of background and calculated QED positron rates (full curve) is in excellent agreement with the observed emission rates. In the spectrum of the supercritical U-U system ($E_{1s} \approx 1200$ keV for $R_{min} \approx 21$ fm) some deviations are seen.

To obtain theoretical predictions for positron production it is essential to include the "spontaneous" coupling for collisions where the 1sσ-state joins the lower continuum. If it is left out of the calculation the resulting positron spectra would be strongly altered: The "induced" radial coupling is changed at the same time as the spontaneous coupling becomes important. Both contributions add up coherently and cannot be observed separately[17]. A promising strategy to get a clear qualitative signature for the

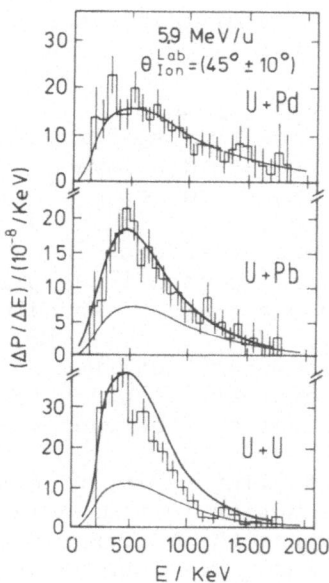

Fig. 21: Spectra of emitted positrons in 5.9 MeV/u collisions measured by Backe et al.[44,45] in coincidence with ions
scattered in the angular window $\theta_{lab}=45°\pm10°$. The spectrum
in the lightest system, U-Pd, is explained by nuclear pair
conversion alone (light line). In the U-Pb and U-U systems
the sum of nuclear and calculated atomic positron production rates (full lines) is displayed.

diving process will be to modify the time structure and to select
heavy-ion collisions with prolonged nuclear contact time [37]. Such
nuclear reactions are expected to occur at energies close to or
above Coulomb barrier. The nuclear delay time T should provide
a handle to distinguish supercritical systems.

Within the schematic trajectory model[39] we have performed
coupled channel calculations for the four heavy-ion collision
systems Pb-Pb, Pb-U, U-U, and U-Cm, correspondong to Z_{united} = 164,
174, 184, and 188, respectively. The contributions of s- and $p_{1/2}$-
partial waves ($\kappa=\pm1$), which are nearly equal for Coulomb-trajec-
tories [17], are taken into account.

Independently of assumptions on the incoming and outgoing

path, of dissipation of nuclear kinetic energy or angular momentum, and of the position of the Fermi level, all positron spectra exhibit the following features:

In subcritical collision systems ($Z_T + Z_p \lesssim 173$) a delay time causes modulations in the positron spectrum with a width $\Delta E = 2\pi\hbar/T$. In Fig. 22a positron spectra are displayed for a Pb-Pb collision, ($E_{lab} = 8.73$ MeV/u, b=7.11 fm, F=3sσ, $4p_{1/2}\sigma$) with delay times T=0 (pure Rutherford scattering), 3·, 6·, and $10 \cdot 10^{-21}$ sec. The modulations are due to interference effects in much the same way as predicted for the δ-electron spectra in deep inelastic heavy-ion collisions[40]. Fig. 22b shows the corresponding spectra for a Pb-U collision ($E_{lab} = 8.97$ MeV/u, b=7.27 fm, F=3), a system, which is just at the border to supercriticality. The spectra of Figs. 22a, b are qualitatively similar, although the heavier system is characterized by the presence of constructive interference.

In addition to the oscillatory interference patterns an enhancement of positron production in time-delayed supercritical collisions is observed, where the binding energy of the lowest bound states exceeds the value $2mc^2$. For long delay times a distinct peak in the positron spectrum is found at the location of the supercritical bound state resonance (binding energy minus $2mc^2$) due to the spontaneous pair-creation mechanism. A detailed analysis of the spectra reveals that this peak emerges gradually as $Z_u = Z_T + Z_p$ exceeds Z_{cr}.

Positron spectra for the supercritical system U-U ($E_{lab} = 7.35$ meV/u, b=3.72 fm, F=3) are shown in Fig. 22c. With increasing delay time the position of the maximum drifts slowly from the kinematic maximum to the "resonance energy", which depends on the combined charge, the separation of the two nuclei and on the nuclear charge distribution.

The integrated positron emission probability $P_{e}+$ as a function of T for these three collision systems and for U-Cm ($E_{lab} = 7.5$ MeV/u, b=3.74 fm, F=3) is displayed in Fig. 22d, where the solid lines denote s-state-, the dashed lines $p_{1/2}$-state contributions. The absolute values, $P_{e}+(T=0)$, should not be compared since the impact energies are chosen differently. However, the variation of $P_{e}+$ with delay time T is of particular interest. In the investigated subcritical systems, $P_{e}+ (T)$ remains roughly unchanged or is even reduced by up to a factor two due to destructive interference between incoming and outgoing path. In the supercritical systems the probability rises approximately in linear relation with T, which reflects the increasing contribution of the spontaneous decay. In the schematic model its slope depends on the separation of the two nuclei, R_{min}; in general it depends also on the time dependent charge distribution during nuclear contact. At small values of T interference

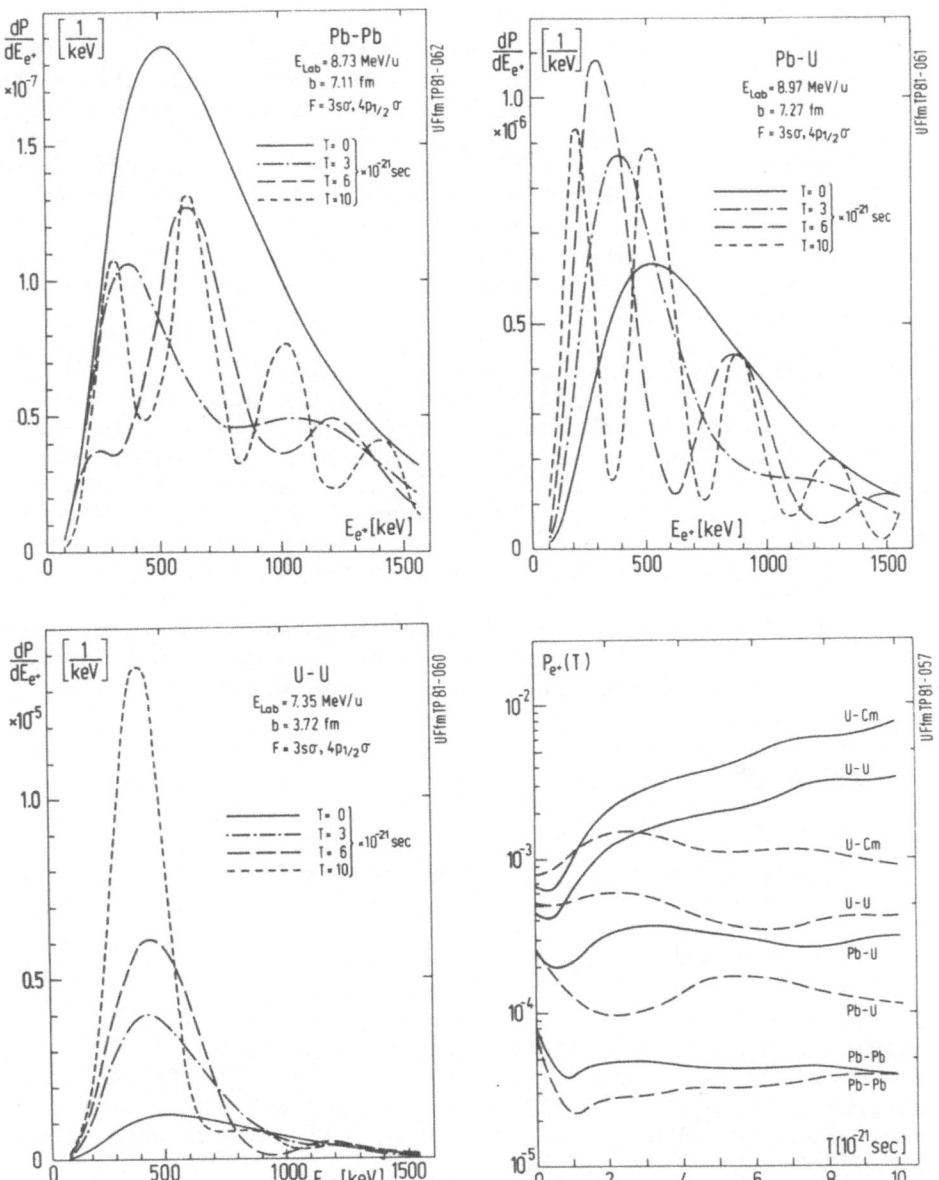

Fig. 22: Spectra of positrons created in subcritical (parts a,b)
and supercritical (part c) heavy-ion collisions assuming
grazing Coulomb trajectory (full lines) and nuclear reactions leading to delay times T=3·, 6·, and 10·10⁻²¹ sec,
resp., using a schematic model for the trajectory. The
kinematic parameters {E,b} chosen are displayed within the
figures, dissipation of energy or angular momentum is not

(continued)

(continued)
taken into account. A Fermi level F=3sσ, $4p_{1/2}σ$ is assumed.
Whereas for the lighter collision systems modulations in
the positron spectra are present, a distinct peak at the
"resonance" energy $E_{1sσ}(R_{min})$ builds up for systems with
$Z_u>173$.- Part d: Probability for the emission of positrons,
integrated over kinetic positron energy, as a function of
reaction time T. Full lines: contribution of s-partial
waves, dashed lines: $p_{1/2}$-partial waves' contribution. For
the supercritical systems U-U and U-Cm $P_{e^+,s}$ increases
strongly with T.

effects seem to prevail even in supercritical collisions. The results described so far were obtained within the schematic model for
the nuclear motion. It facilitates a systematic study of the time
delay effect and allows for an investigation of the conceptually
interesting limit of large sticking times. To analyze a given
experiment, however, the employed nuclear trajectories should be
consistent with the elastic and inelastic heavy-ion scattering
data. For a corresponding discussion we refer to ref. 38, 39.

However, for any chosen set of experimental parameters, the
nuclear reaction time T may (and will) not be a sharp but distributed over a certain range. Then the positron spectrum is determined by

$$\int dT \; \frac{dP(T)}{dE_{e^+}} \; f(T)$$

with the time distribution f(T). As an assumption we took a
Gaussian centered at $\overline{T}$

$$f(T) = \frac{1}{\sqrt{2\pi}\tau} \; \exp\left(-\frac{(T-\overline{T})^2}{2\tau^2}\right) \tag{8.7}$$

The resulting positron spectra for the parameters $\overline{T} = 16\cdot10^{-21}$s
and τ = 0 or τ = $2\cdot10^{-21}$s, are displayed in fig. 23 for a central
U-U collision at E_{lab} = 5.9 MeV/u. If we consider the p-states
(fig. 23a) only we observe that the oscillations disappear already
for τ = $2\cdot10^{-21}$s, except for the first peak.

Also in the total spectrum (fig. 23) the oscillations are
damped out for increasing τ. But most striking is the occurrence
of the dominant first peak, which originates from the spontaneous
part of the positron production mechanisms. Nevertheless we
may conclude that the appearence of several oscillations in the
lepton spectra can be expected only for sufficient sharp nuclear
reaction times.

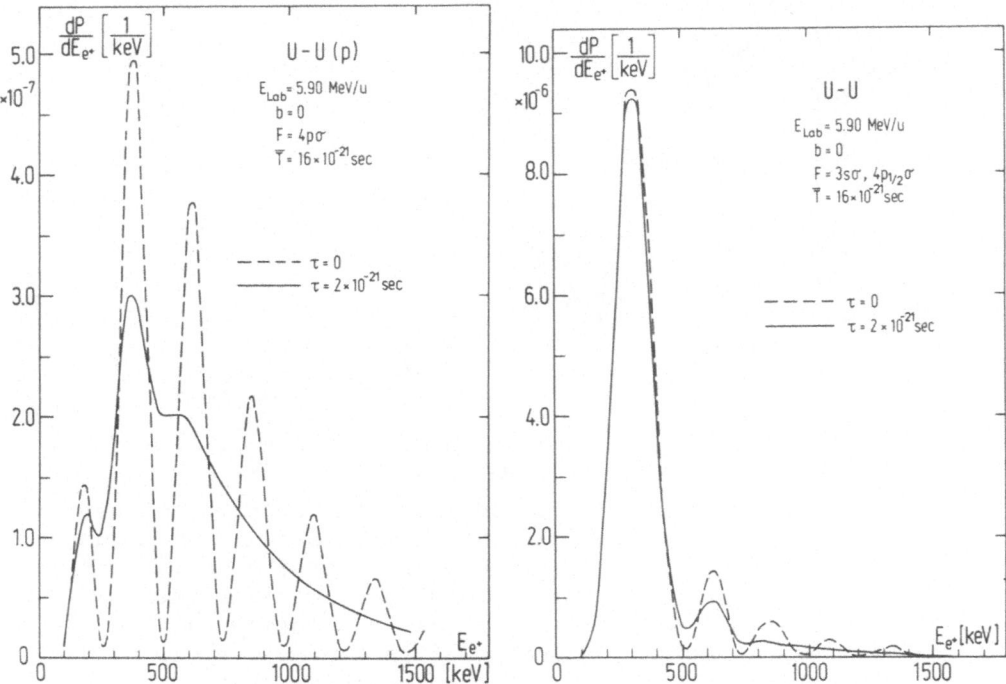

Fig. 23: a) Differential positron production probability versus
            kinetic positron energy E_{e^+} in a central U-U collision
            at E_{lab} = 5.9 MeV/u. It is assumed that the nuclear
            reaction time is distributed over a certain range which
            is determined by a Gaussian centered at $\overline{T}$. For the
            width τ we have chosen τ=0 and τ=2·10⁻²¹s, respectively.
        b) The same as in a) but including also the contribution
            of the s-states. Most striking is the appearance of the
            first pronounced peak which originates from spontaneous
            positron production.

The superheavy compound system may not be a static object. In the
framework of a simple model assumption we investigated[46] the influence of the internal dynamics of the dinuclear system on the
energy spectra of the emitted positrons. The internuclear separation
$R(t)$ is supposed to oscillate around the distance of closest approach
R_0 of the Coulomb trajectory for a fixed duration T

$$R = R_0 \ (1- \alpha_0 \ \sin(2\pi\nu t)) \tag{8.8}$$

In fig. 24 dP/dE is plotted for a central U-U collision at E_{Lab} =
6.2 MeV/u.

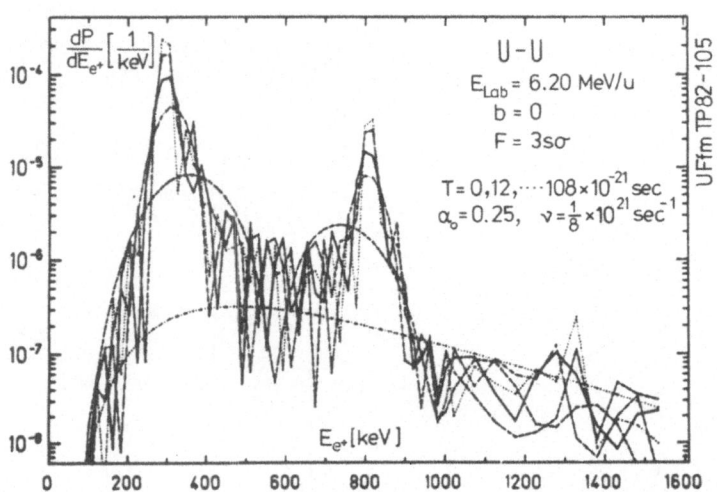

Fig. 24: Positron spectra calculated under the assumption that
distance between the nuclear centres oscillates around
the distance of closest approach R_O of the Coulomb trajectory (see eq. (8.8)). a) Nuclear reaction times
$T = 0,12,36,60,84, 108 \cdot 10^{-21}$s are considered. The
longest duration T corresponds to the most pronounced
"spontaneous peak", etc. Note the logarithmic scale.

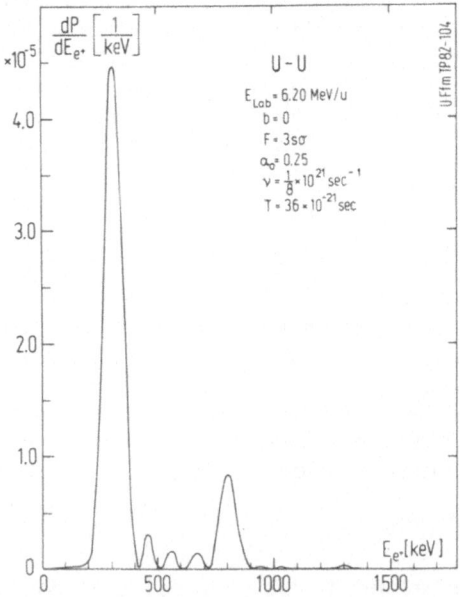

Fig. 24b: The same as in a) for the reaction time $T=36 \cdot 10^{-21}$s
on a linear scale.

In order to demonstrate the qualitative effect we have restricted the calculations to couplings between the s-states only. The parameters in the expression (8.8) are fixed by $\alpha_0 = 0.25$ and $\nu = 0.125 \cdot 10^{21} s^{-1}$. Various sticking periods between T=0 and $T=108 \cdot 10^{-21}s$ are considered in fig. 24 a). Fig. 24 b) shows on a linear scale the computed positron spectrum for $T=36 \cdot 10^{-21}s$. As in fig. 23 we find the dominant "spontaneous peak". But in addition a second pronounced peak appears at about $E_{e+} = 800$ keV. This reflects the fact, that part of the oscillation energy of the dinuclear system is transferred to the emitted lepton. The presented model calculation thus may be understood as a classical picture for the conversion process of internal pair creation in the supercritical quasimolecule.

We now turn again to the influence of a nuclear time delay in deep-inelastic heavy ion reactions on δ-electron spectra. To determine the excitation amplitudes in delayed collisions, we make here the simplest possible ansatz for the nuclear motion. The branches of two Rutherford hyperboles having the same distance of closest approach are joined by a section without any radial motion. Hence the effect of the reaction enters only via an additional phase[40) in the set of coupled differential equaitons for the occupation amplitudes $a_{ij}(t)$. We calculated δ- electron spectra in the central heavy-ion collision U+U, where the nuclei stick together for about 10^{-21} to $10^{-20}s$. Fig. 25 shows the spectra of emitted δ-electrons choosing a delay time T=0 (fig. 25a) and $T=10^{-20}s$ (fig. 25b). A coincidence with created $1s\sigma$-vacancies is assumed. Only the excitation of s-state electrons is considered. Since the p-states also contribute considerably the present data may not be compared quantitatively with experimental data. The spectra in fig. 25 are calculated according to eq. (2.8). The dashed lines show the separate contributions of both terms in eq. (2.8). The oscillatory pattern in fig. 25b is caused by the real correlation between the δ-electron and the $1s\sigma$-vacancy. The term $N_q \cdot N_q$ which describes accidental coincidences produces always a smooth distribution.

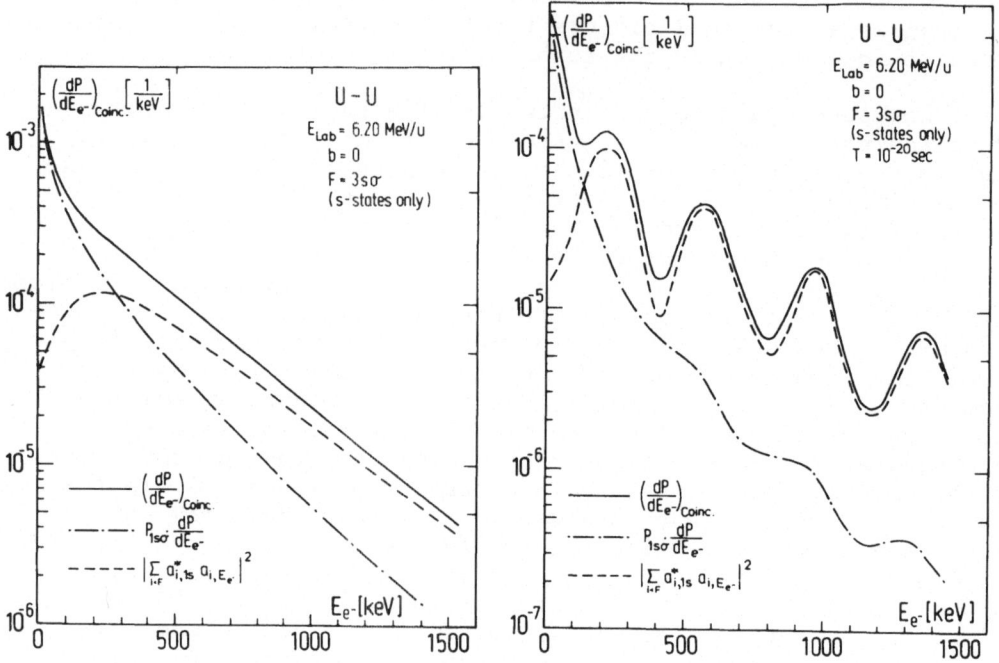

Fig. 25: Differential emission probability of δ-electrons in dependence on the kinetic electron energy for the system
U+U at E_{LAB} = 6.2 MeV/u and b=0. A coincidence with
1sσ-vacancies is assumed. Only the s-states are taken
into account. The dashed-dotted line shows the contribution
of the accidential coincidences $N_p \cdot N_q$ of eq. (2.8).
a) No nuclear sticking time   b) T = 10^{-20}s.

9.   CONVERSION PROCESSES IN SINGLE ATOMS AND IN SUPER-CRITICAL
      COMPOUND SYSTEMS

In collisions of very heavy ions such as U+U with $E_{LAB} > 3$ MeV/u
both nuclei are strongly Coulomb excited. For bombarding energies
at about the Coulomb barrier transfer reactions or even deep inelastic
nuclear reactions can take place which lead to additional excitations
of the nuclei. This internal excitation energy may be irradiated by
a photon or may be transferred to a bound electron or to an electron
of the negative energy continuum, which leads to ionization and
electron-positron pair creation, respectively. The latter process
requires nuclear transition energies ω larger than twice the electron
rest mass $\dot{m}$. Nuclear E0-transitions are characterized by the absence
of single photon emissions, because a photon must carry at least
one unit of angular momentum. Therefore E0-transitions predominantly
occur by bound state electron conversion or by internal electron-
positron pair creation.

In this section we investigate the internal electron-positron
pair creation accompanying the decay of excited nuclei. We mainly
focus our attention to the calculation of the pair conversion
coefficient (PCC)

$$\beta = P_{e^+,e^-} / P_\gamma .\tag{9.1}$$

β is defined as the ratio of the pair production probability compared
with that of photon emission for a specific nuclear transition with
energy ω. Since the energy of the electron and the positron takes
continuous values we may express β also as integral of the positron
spectrum $d\beta/dE$ with respect to the positron energy E,

$$\beta(\omega) = \int_1^{\omega-1} \frac{d\beta}{dE}(E)dE.\tag{9.2}$$

$d\beta/dE$ is called differential pair conversion coefficient (DPCC).
The lower bound of the integral is determined by the rest mass
of the electron, which corresponds to vanishing kinetic energy,
while the upper bound is given by the nuclear transition energy ω
minus m_e. We will use the units $h=m_e=c=1$. The investigation of β,
which is typically of the order of 10^{-4}-10^{-3}, instead of P_{e^+,e^-}
has several theoretical and experimental advantages. In contrast
to P_{e^+,e^-} the pair conversion coefficient, β, to a good approximation is independent of the nuclear wave function. Therefore only
the knowledge of the electron and positron wave function for a
given nuclear charge distribution is necessary for the evaluation
of β. Experimentally it is easier to measure the ratio of two
counting rates. The theory of internal pair creation is strongly
connected with conversion of bound state electrons. In the latter
process the excitation energy of a nucleus is transferred to a
bound state electron which becomes excited to the positive energy

continuum. Here the ratio of the probabilities of inner-shell vacancy formation and photon emission is defined as conversion coefficient,

$$\alpha = P_{e^-}/P_\gamma \; . \tag{9.3}$$

In particular this mechanism is important for low energy nuclear transitions. A major motivation for the investigation of internal pair creation results from the test of quantum electrodynamics in the presence of strong external fields. A large contribution to the total positron production rate in supercritical collisions stems from internal pair conversion following nuclear Coulomb excitation. For the experimental separation of both processes it is therefore necessary to know precisely the conversion coefficient β. To subtract the internal conversion process one has to measure the nuclear photon spectrum and multiply it with β. Our theoretical investigations are expecially motivated by the recently observed structures[47,48] in the energy distribution of emitted positrons which have been produced in collisions of very heavy ions like U-Th, U-U, and U-Cm. These systems are supercritical and in principle may lead to spontaneous positron production. A conventional explanation for this peculiar structure would be positron production via E0-conversion processes. Contrary to E0-processes nuclear transitions of multipolarity E1 or E2 should also be observable in the emitted photon spectra provided that proper Doppler shift corrections are performed.

The basic processes under investigation are depicted schematically in Fig. 26. The nucleus which undergoes E0-transition is labelled by its initial and final state angular momenta J_i, $J_f=J_i$ and eigenenergies E_i, $E_f=E_i-\omega$. Process a) describes the electron-positron pair creation. An electron of the negative energy continuum ($\epsilon=-E < mc^2$) with Dirac quantum number κ is lifted to the positive energy continuum. The final state energy obviously amounts to $E' = \epsilon + \hbar\omega$ whereas the angular momentum quantum number remains unchanged. Since neither the initial electron state energy nor the final state energy is fixed one expects a continuous energy distribution for the emitted positrons. Process b) indicates the conversion of a K-shell electron ($n=1$, $l=0$, $j=\frac{1}{2}$, $\kappa=-1$) with energy eigenvalue $E_{1s_{1/2}}$. Thus bound states with definite energies are involved. Energy conservation then simply causes monoenergetic lepton emission for a fixed nuclear transition energy $\hbar\omega$. Process c) symbolizes monoenergetic positron production. Here an electron of the negative energy continuum is excited to a bound state, e.g. to the $1s_{1/2}$-state. This represents a rather rare process since it requires a vacent bound state with a large overlap with the nuclear interior. If we investigate a realistic heavy-ion collision of $_{92}$U on $_{92}$U with a bombarding energy close to the Coulomb barrier, the number of created K-vacancies per collision with impact para-

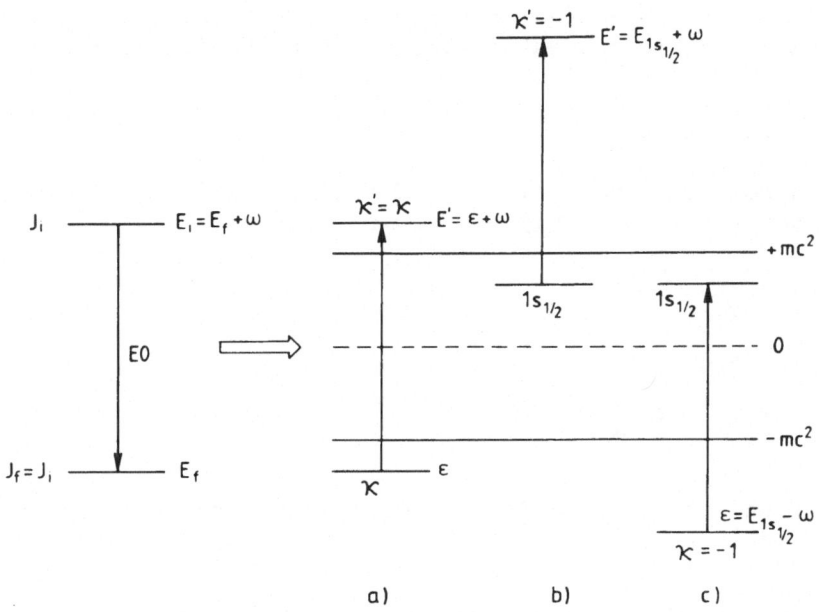

Fig. 26 : Schematic representation of electron conversion processes accompanying nuclear E0-transition from a state
$\{E_i, J_j\}$ to a state $\{E_f=E_i-\hbar\omega, J_f=J_i\}$. a) Electron-
positron pair production leading to a continuous energy
distribution of positrons and electrons. b) Conversion
of K-shell electrons – a monoenergetic electron-pro-
duction mechanism. c) Monoenergetic positron-production
– a negligible process.

meter $b \lesssim 20$ fm is typically of the order of 10% [17] . The
transition time (X-ray decay) of an electron from a higher shell
to the K-shell is [49] about 10^{-17} sec. This has to be compared with
nuclear transition times of about 10^{-13} sec, after which conversion
takes place. Hence the K-vacancy will be decayed before it could
eventually be filled again due to pair conversion of a nuclear state.

    Therefore the monoenergetic positron formation is suppressed
in contrast to ordinary K-shell conversion by at least 7 orders of
magnitude, because one can conclude [50] that the E0-conversion of a
K-shell electron has a probability larger by about 3 orders of
magnitude than the monoenergetic e^+-production where the electron
occupies the vacant K-shell. This ratio even increases by one order
of magnitude if one considers higher bound states in relation to
the positron production process. Thus process c) can be neglected
completely.

The ratio η of the two conversion probabilities P_{e^+,e^-} for electron-positron pair creation and P_{e^-} for the ionization of bound state electrons

$$\eta = \frac{P_{e^+,e^-}}{P_{e^-}} \qquad (9.4)$$

must be investigated more closely. The differential ratio with respect to the positron energy E obviously is

$$d\eta/dE = \frac{dP_{e^+,e^-}}{dE} \, / \, P_{e^-} \qquad (9.5)$$

and consequently

$$\eta = \int_{mc^2}^{\hbar\omega-mc^2} \frac{d\eta}{dE} \, dE \; . \qquad (9.6)$$

The final expressions for the conversion probabilities are to lowest order in r

$$\frac{dP_{e^+,e^-}}{dE} \propto |M|^2 \sum_{\kappa=-1}^{+} |C_{e^+,e^-}|^2 \; , \qquad (9.7)$$

$$P_{e^-} \propto |M|^2 \sum_{m}' |C_{e^-}|^2 \; , \qquad (9.8)$$

where M denotes the nuclear E0 matrix element

$$M = \int_0^\infty dV_n \rho_n(\vec{r}_n) r_n{}^2 \; , \qquad (9.9)$$

and C is a normalization factor

$$C = \lim_{r\to 0} \frac{ff' + gg'}{r^{2j-1}} \; . \qquad (9.10)$$

Here f, f' and g,g' are the radial Dirac wavefunctions belonging to the initial and final state of the electron, respectively. Σ' in eq.(9.8) denotes the sum over all occupied states (E_m, j_m) excluding the summation over magnetic substates; in eq. (9.7) only $j=1/2$ ($\kappa=\pm 1$) - states are taken into account.

Thus the ratio eq. (9.5) now reads

$$\frac{d\eta}{dE} = \sum_{\kappa=\pm 1} |C_{e^+,e^-}|^2 \, / \, \sum_{m}' |C_{e^-}|^2 \; , \qquad (9.11)$$

where the nuclear matrixelement drops out. Due to Eq. (9.10) the ratio η of pair conversion to bound state electron conversation is completely determined by the density of the electron wavefunctions at the nuclear origin.

It is also possible to derive an analytical expression for the ratio $d\eta/dE$. For this purpose we utilized electron wavefunctions that are solutions of the Dirac equation with a point-nucleus potential. It results

$$\frac{d\eta}{dE} = \frac{(EE' - \gamma^2)}{\pi(Z\alpha)^2(\omega+2\gamma)\Gamma(2\gamma+1)} \, e^{\pi(B'-B-\bar{B})} \cdot \left(\frac{pp'}{Z\alpha\bar{p}}\right)^{2\gamma-1} \left|\frac{\Gamma(\gamma+iB)\Gamma(\gamma+iB')}{\Gamma(\gamma+i\bar{B})}\right|^2 \cdot$$

$$(9.12)$$

The variables belonging to the final state electron are always indicated by a prime.

Here we used the abbreviations

$$E' = \omega-E$$

$$\bar{E} = \gamma+\omega$$

$$\gamma = [\kappa^2-(Z\alpha)^2]^{\frac{1}{2}}$$

$$(9.13)$$

$$B = \frac{Z\alpha E}{p}, \quad B' = \frac{Z\alpha E'}{p'},$$

$$p = [E^2-1]^{\frac{1}{2}}, \quad p' = [E'^2-1]^{\frac{1}{2}},$$

According to eqs. (9.10) and (9.11) we computed the differential conversion ratio $d\eta/dE$ with respect to the positron energy E for E0-transitions. As bound state only the atomic K-shell has been taken into account. The conversion probability of higher bound states is at least one order of magnitude smaller. For the nucleus $_{92}$U the energy distributions of emitted positrons is depicted in Fig. 27. Nuclear transition energies of 1323 keV, 1423 keV, 1523 keV, and 1623 keV are considered, which correspond to maximum positron energies of E^{kin}_{max} = 300 keV, 400 keV, 500 keV, and 600 keV, respectively.

In Fig. 28 we compare our numerical results for $d\eta/dE$ (solid lines) using Dirac wavefunction for finite size nuclei with the analytical formula (9.12) (dashed lines). Good agreement is achieved for low and medium Z nuclei. Even for Z = 92 the agreement is still remarkable. In conclusion we can state that nuclear E0-transitions with $\omega>2m \sim 1$ MeV may be identified by measuring the positron to K-shell vacancy production rate. The differential

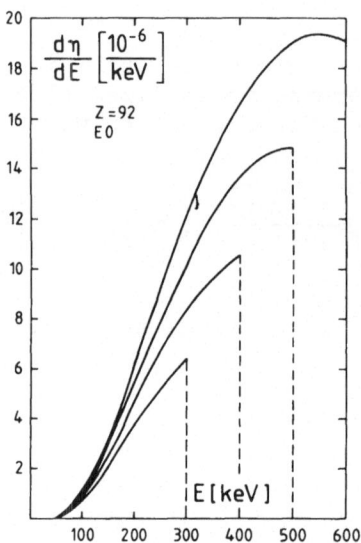

Fig. 27 : Differential conversion probability ratio dη/dE with
          respect to the kinetic positron energy E for nuclear E0-
          transitions in U. Nuclear transition energies ω= 1323 keV,
          1423 keV, 1523 keV, and 1623 keV are considered, corres-
          ponding to maximum kinetic positron energies of
          E_{max} = 300 keV, 400 keV, 500 keV, and 600 keV.

ratio $\frac{d\eta}{dE}$ as well as η are typical quantities for E0-transitions.
For the purpose of nuclear spectroscopy formula (9.12) can be
applied as a good approximation.

    The numerical integration of (9.11) yields the dimensionless
ratio η. For Z = 92 η is presented in Fig. 29 as function of the
nuclear transition energy.

    For ω = 1.423 keV we find that the probability for K-shell
conversion is 556 times larger than the e^+, e^- pair creation probability. In the considered range of transition energies η
varies from 10⁻⁴ to 2·10⁻². The dashed line represents results of
R. Thomas[51]   . Both calculations are in fair agreement.

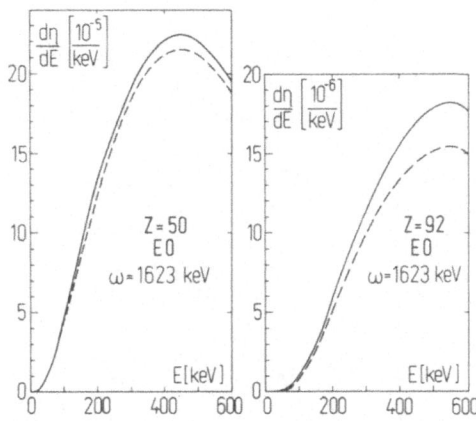

Fig. 28 : Differential conversion probability ratio dη/dE as
          function of the kinetic positron energy E. The nuclei
          Z = 50 and Z = 92 are considered. The nuclear transition
          energy was chosen to be ω = 1 623 keV. The solid line re-
          presents a numerical calculation utilizing Dirac wave-
          functions for finite size nuclei. The dashed line follows
          from the analytical expression.

        For comparison of the spectral shape of dη/dE in fig. 27
we show in fig. 30 the equivalent differential conversion coefficient dβ/dE for nuclear E1 and E2 transitions, which are defined
in eq. (9.1) and (9.2). Obviously the ratio β determines the
number of created electron-positron pairs divided by the number
of created photons. For details concerning the theoretical formalism and for the numerical numbers we refer to refs. 52-54.
Similar positron distributions as in fig. 27 are found.

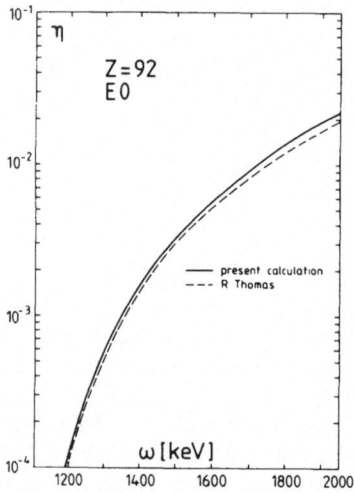

Fig. 29: The ratio η versus the nuclear transition energy ω for
         Z = 92. The dashed line is obtained by R. Thomas[51].
         and the solid line is the present calculation. Both
         calculations are in fair agreement.

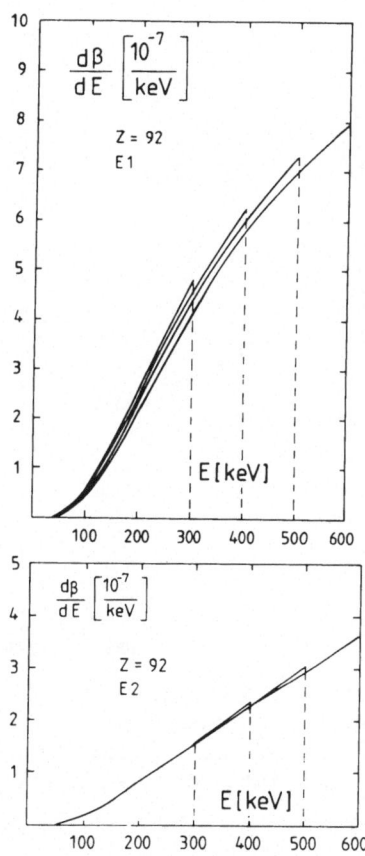

Fig. 30 : Differential conversion coefficient dβ/dE with respect to
the kinetic positron energy E for nuclear E1- and E2-
transitions in $_{92}$U. The same transition energies as in
fig. 27 .

For completeness we show in fig. 31 the K-shell conversion coefficient α_K (eq. 9.3) as function of the nuclear transition energy ω. Nuclear transitions of multipolarities E1, E2 and M1 in $_{92}U$ are considered. The data are taken from the table 55. Thus for $\omega > 600$ keV the ionization of the N-shell is suppressed typically by a factor of 100 compared with the emission of a single photon.

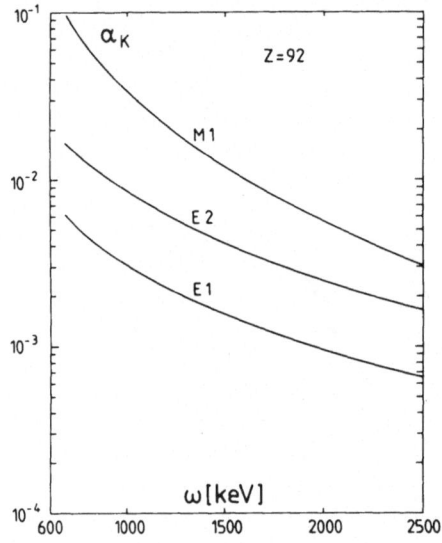

Fig. 31: The conversion coefficient $\alpha_K = P_{e^-_K}/P_\gamma$ as function of the nuclear transition energy ω. Nuclear transitions of multipolarity E1, E2 and M1 in $_{92}U$ are considered. The data are taken from the table 55.

Now we can discuss the probability that the recent observed structures in positron spectra originates from nuclear E0-transitions. One convincing argument against this interpretation is related to the shape of the e^+-energy distribution. According to Fig. 27 the halfwidth of the spectra should be at least $\Delta E \gtrsim 150$ keV. However, the observed structure is much narrower. The second argument is connected with the energy distribution of the emitted δ-electrons. The measured differential positron production probability dP_{e^+}/dE with respect to the kinetic positron energy at $E \sim 400$ keV for U-U collisions with a bombarding energy of

E_{lab} = 5.9 MeV/u and ion scattering angles of Θ_{ion}^{lab} = 45°±10° amounts to $dP_{e+}/dE \simeq 10^{-7}/$ keV [47].

Assuming that these positrons are of pure nuclear origin one can easily deduce the probability P_N for nuclear excitation

$$P_N \simeq \frac{dP_{e^+}}{dE} \; / \; \frac{d\eta}{dE} \simeq 10^{-2} \; . \tag{9.14}$$

In principle the competing K-shell conversions should lead to a sharp line in the electron spectra at $E_{e^-} \simeq 1300$ keV.

However, the experimental energy resolution for these high energy electrons is presently not better than $\Delta E_{e^-} = 25$ keV, which therefore yields a differential e^--production probability of

$$dP_{e^-}/dE_{e^-} \simeq P_N/\Delta E_{e^-} = 4 \cdot 10^{-4}/\text{keV} \; . \tag{9.15}$$

This is still about an order of magnitude larger than the expected δ -electron energy of $E_{e^-} \simeq 1300$ keV, which results from dynamical electron excitations during the collision. This factor can be increased by an order of magnitude if one measures the emitted electrons in coincidence with the decaying K-vacancy (e.g., K_α-radiation). Hence we can conclude: If the observed structure in a positron spectra is caused by nuclear E0-transitions one should also observe a distinct peak in the δ-electron distribution. Furthermore we emphasize that sharp peaks in the energy distribution of emitted positrons with a halfwidth of less than 100 keV may not originate from nuclear E0-conversion processes.

We now turn to the discussion of electron-positron pair [56] conversion in supercritical compound systems. We consider a collision of very heavy ions such as $_{92}U + _{92}U$ with a combined charge $Z=Z_1+Z_2=188$. According to the ideas of Reinhardt et al. [39] the projectile and target may come into nuclear contact even for bombarding energies slightly below the nuclear Coulomb barrier. These authors furthermore assume that for a period of $T \sim 4 \cdot 10^{-19}$ s a superheavy compound system is produced. In the superheavy system $Z=184$ the electron K-shell is imbedded as resonance in the negative energy continuum. A vacant K-shell thus may lead to spontaneous positron production [17]. Reinhardt et al. predicted a pronounced peak in the positron spectra due to the spontaneous part of the positron production mechanism thereby indicating the energetical location of the $1s_{1/2}$-resonance in the negative energy continuum.

Motivated by these theoretical investigations we considered the question whether positrons may be created due to conversion processes during the sticking period T. In this case pair conversion is not distinguishable from ordinary K-shell conversion. This situation is visualized in fig. 32 . The energy spectrum of of the Dirac equation is shown with the upper continuum and lower continuum, which includes the supercritical K-shell denoted by its energetical location E_{res}.

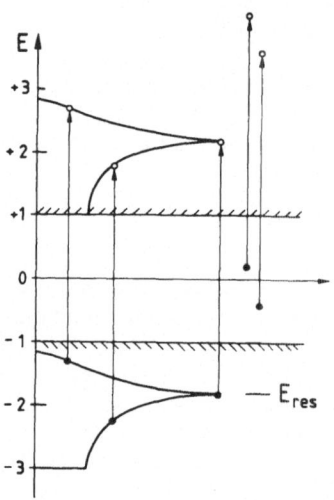

Fig. 32 : Conversion processes indicated by arrows in supercritical systems. The energy spectrum of the Dirac equation is shown with the upper continuum and lower continuum, which includes the supercritical K-shell denoted by the energetical location E_{res}. The curves in the lower and upper continuum display the spectral shape shape of the positron and electron distribution, respectively. The two arrows on the right-hand side denote the conversion of bound state electrons in high lying states.

A supercritical nucleus Z=184 which undergoes a transition with w > 2 during the sticking period T may transfer this excitation energy to one of the electrons in the negative energy continuum. The remaining hole is defined as positron. But also the K-shell electron can be lifted to the upper continuum. If T is longer than the spontaneous decay width of the K-shell resonance the K-vacancy will be filled again leading to spontaneous positron emission. The conversion processes are indicated by the arrows in fig.32 . The curves in the lower and upper continuum display the spectral shape of the positron and electron distribution, respectively. In the positron spectra a sharp peak appears at E = E_{res}. The two arrows on the right hand side denote the conversion of bound state electrons in high lying states. The conversion coefficients β and the differential distribution dβ/dE are calculated in precisely the same manner as described extensively in ref. 53. For the nuclear charge distribution a homogeneously charged sphere with a radius R_n = 10.88 fm has been assumed. In fig. 33 we show the

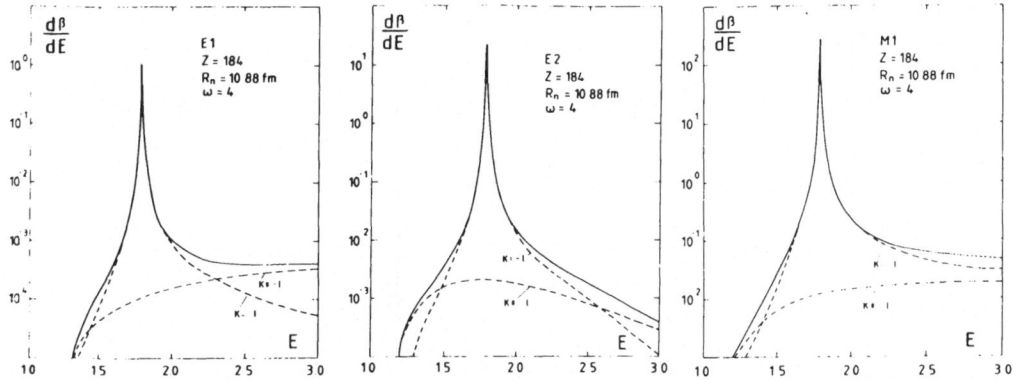

Fig. 33 : Differential ratio dβ/dE for the electron-positron pair
creation compared with the photon emission probability
as function of the positron energy E. Nuclear transitions
with w=4 and of multipolarity E1, E2 and M1 in a supercritical nucleus Z=184 with a radius R_n = 10.88 fm are
considered. The dashed curves signify the various portions
of the electron angular momentum states.

differential ratio dβ/dE for the electron-positron pair creation
compared with the photon emission probability as function of the
positron energy E. Nuclear transitions with w=4 and of multipolarity
E1, E2 and M1 are considered. Striking is the appearance of the
pronounced peak at E = E_{res}. The dashed curves signify the various
portions of the electron angular momentum states. The contributions
of the states with κ = -1 and the states with κ ≠ -1 are given separately. As expected the resonance shows up only in the (κ =-1)-
part.

In fig. 34 we can compare dβ/dE for w=4 and w=6. Again the
multipolarities E1, E2 and M1 are considered. In addition the
maximum values for dβ/dE for the multipolarities E3, M2 and M3
are given. Obviously lepton emission predominantly occurs for
nuclear transitions with magnetic multipolarities.

In the following table we present for the multipolarities E1,
E2, M1 the total conversion coefficient β following from a numerical
integration of dβ/dE. The second line always gives percentage
contribution of the κ = -1 electron angular momentum state, which
mainly correspond to K-shell conversion.

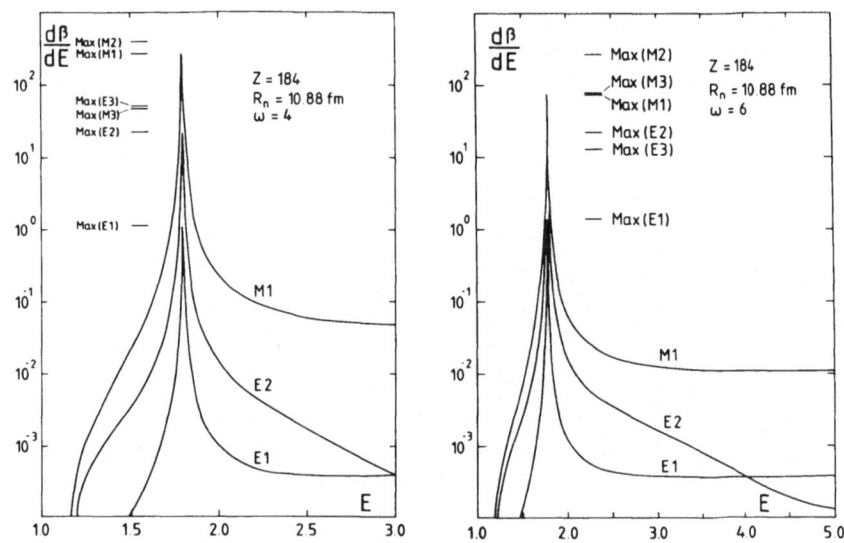

Fig. 34 : The same as in fig. 33 for w=4 and w=6. In addition the
         maximum values for dβ/dE for various multipolarities are
         given.

ω	E1	E2	M1
4	1.74E-2 95.4%	3.26E-1 97.1%	4.03 96.2%
6	2.17E-2 92.1%	3.21E-1 96.2%	1.11 94.2%
8	2.04E-2 88.7%	2.28E-1 95.3%	4.43E-1 92.0%

Thus especially for magnetic transitions the total probability of
lepton production may exceed the photon emission probability even
for w > 2. However, one has to emphasize that the contribution of
the discussed conversion processes to the positron yield in a
realistic heavy ion collision depends strongly on the specific
nuclear structure especially for the compound system. The observed
positron production probability per collision is about 10^{-5}. This
requires transition times shorter than 10^{-14}s in the compound
nucleus in order that internal pair conversion can contribute considerably. Furthermore we should stress that in contrast to the
measured structure in the positron spectra just one peak can be
expected due to the discussed pair production mechanism.

REFERENCES

1. G. Soff, W. Greiner, W. Betz, and B. Müller, Phys. Rev. A20 (1979) 169
2. J. Reinhardt, W. Greiner, Heavy Ion atomic physics (theoretical) preprint
3. C. Kozhuharov, in: Physics of Electronic and Atomic Collisions, 1982, p. 179, ed.: S. Datz, North-Holland
4. J. Reinhardt, B. Müller, W. Greiner, G. Soff, Phys. Rev. Lett. 43 (1979) 1307
5. G. Soff, J. Reinhardt, B. Müller, and W. Greiner, Z. Physik, A294 (1980) 137
6. F. Güttner, W. Koenig, B. Martin, B. Povh, H. Skapa, J. Soltani, Th. Walcher, F. Bosch, C. Kozhuharov, Z. Physik A304 (1982) 207
7. J. Rafelski, B. Müller, Phys. Rev. Lett. 36 (1976) 517
8. W. Betz, Thesis, Universität Frankfurt, 1980
9. G. Soff, B. Müller, W. Greiner, Z.Physik A299 (1981) 189
10. T. de Reus, U. Müller, J. Reinhardt, P. Schlüter, K.H. Wietschorke, B. Müller, W. Greiner, G. Soff, preprint, GSI-81-38
11. D. Liesen, P. Armbruster, H.-H. Behncke, and S. Hagmann, Z. Physik A288 (1978) 417
12. D. Liesen, P. Armbruster, H.-H. Behncke, F. Bosch, S. Hagmann, P.H. Mokler, H. Schmidt-Böcking, and R. Schuch, in: 'Electronic and Atomic Collisions', N. Oda and K. Takayanagi, eds., North Holland, Amsterdam (1980), p. 337.
13. R. Anholt, W.E. Meyerhof, and C. Stoller, Z. Physik A291 (1979) 287
14. D. Liesen, P. Armbruster, F. Bosch, S. Hagmann, P.H. Mokler, H.J. Wollersheim, H. Schmidt-Böcking, R. Schuch, and J.B. Wilhelmy, Phys. Rev. Lett. 44 (1980) 983
15. G. Soff, T. de Reus, B. Müller, W. Greiner, Phys. Lett. 88A (1982) 398
16. G. Soff, J. Reinhardt, W. Greiner, Phys. Rev. A23 (1981) 701
17. J. Reinhardt, B. Müller, W. Greiner, Phys. Rev. A24 (1981) 103
18. F.G. Werner, J.A. Wheeler, Phys. Rev. 109 (1958) 126
19. W. Pieper, W. Greiner, Z. Physik 218 (1969) 327
20. G. Soff, B. Müller, J. Rafelski, Z. Naturforsch. 29a (1974) 1267
21. M. Gyulassy, Phys. Rev. Lett. 33 (1974) 921
22. M. Gyulassy, Nucl. Phys. A244 (1975) 497
23. G.A. Rinker, L. Wilets. Phys. Rev. A12 (1975) 748
24. E.H. Wichmann, N.M. Kroll, Phys. Rev. 101 (1956) 843
25. G.E. Brown, G.W. Schaefer, Proc. Roy. Soc. (London) A 233 (1956) 527
26. G.E. Brown, J.S. Langer, G.W. Schaefer, Proc. Roy. Soc. (London) A 251 (1959) 92
27. G.E. Brown, D.F. Mayers, Proc. Roy. Soc. (London) A 251 (1959) 105
28. G.W. Erickson, D.R. Jennie, Ann. Phys. 35 (1965) 271 and 447
29. A.M. Desiderio, W.R. Johnson, Phys. Rev. A3 (1971) 1267

30. P.J. Mohr, Ann. Phys. 88 (1974) 26 and 52
31. K.T. Cheng, W.R. Johnson, Phys. Rev. A14 (1976) 1943
32. G. Soff, P. Schlüter, B. Müller, W. Greiner, Phys. Rev. Lett. 48 (1982) 1465
33. C. Itzykson, J.-B. Zuber, Quantum Field Theory, McGraw-Hill, 1980.
34. J. Rafelski, B. Müller, and W. Greiner, Nucl. Phys. B68 (1974) 585
35. L. Fulcher and A. Klein, Phys. Rev. D8 (1973) 2455
36. W.L. Wang and C.M. Shakin, Phys. Lett. 32B (1970) 422
37. J. Rafelski, B. Müller, and W. Greiner, Z. Physik A285 (1978) 49
38. U. Müller, J. Reinhardt, T. de Reus, P. Schlüter, G. Soff, K.H. Wietschorke, B. Müller, W. Greiner, preprint GSI-81-39
39. J. Reinhardt, U. Müller, B. Müller, W. Greiner, Z. Physik A303 (1981) 173
40. G. Soff, J. Reinhardt, B. Müller, W. Greiner, Phys. Rev. Lett. 43 (1979) 1981
41. H. Backe, in: "Present Status and Aims in Quantum Electrodynamics", Lecture Notes in Physics, vol. 143, G. Gräff, E. Klempf, and G. Werth, eds., Springer, Berlin, Heidelberg, New York, p. 277 (1981).
42. C. Kozhuharov, P. Kienle, E. Berdermann, H. Bokemeyer, J.S. Greenberg, Y. Nakayama, P. Vincent, H. Backe, L. Handschug, and E. Kankeleit, Phys. Rev. Lett. 42 (1979) 376
43. H. Backe, L. Handschug, F. Hessberger, E. Kankeleit, L. Richter, F. Weik, R. Willwater, H. Bokemeyer, P. Vincent, Y. Nakayama, and J.S. Greenberg, Phys. Rev. Lett. 40 (1978) 1443
44. H. Backe, W. Bonin, W. Engelhardt, E. Kankeleit, M. Mutterer, P. Senger, F. Weik, R. Willwater, V. Metag, and J.B. Wilhelmy, GSI Scientific Report 1979, GSI 80-3 (1980) 101
45. H. Backe, W. Bonin, W. Engelhardt, E. Kankeleit, M. Mutterer, P. Senger, F. Weik, R. Willwater, V. Metag, and J.B. Wilhelmy, "Positron Production in Heavy Ion Collisions", preprint (1979)
46. U. Müller. thesis, Universität Frankfurt
47. E. Berdermann, F. Bosch, M. Clemente, F. Güttner, P. Kienle, W. Koenig, C. Kozhuharov, B. Martin, B. Povh, H. Tsertos, W. Wagner, Th. Walcher: GSI-Sientific Report 1980, GSI 81-2, p. 128
48. H. Bokemeyer, H. Folger, H. Grein, S. Ito, D. Schwalm, P. Vincent, K. Bethge, A. Gruppe, M. Waldschmidt, R. Schulé, J.S. Greenberg, J. Schweppe, N. Trautmann: GSI Scientific Report 1980, GSI 81-2, p. 127
49. R. Anholt and J.O. Rasmussen, Phys. Rev. A9 (1974) 585
50. G. Soff, P. Schlüter, W. Greiner, Z. Physik A303 (1981) 189
51. R. Thomas, Phys. Rev. 58 (1940) 714
52. P. Schlüter, G. Soff, W. Greiner, Z. Physik A 286 (1978) 149
53. P. Schlüter, G. Soff, W. Greiner, Phys. Rep. 75 (1981) 327
54. P. Schlüter and G. Soff, Atomic Data and Nuclear Data Tables 24 (1979) 509

55. F. Rösel, H.M. Fries, K. Alder, H.C. Pauli, At. Data Nucl.
    Data Tables 21 (1978) 292
56. P. Schlüter, thesis, Universität Frankfurt

ACKNOWLEDGEMENTS

We acknowledge very fruitful discussions with P. Armbruster,
H. Backe, H. Bokemeyer, F. Bosch, J.S. Greenberg, W. Koenig,
C. Kozhuharov, D. Liesen, P. Mokler, and P. Senger concerning
their experiments. We are grateful to W. Betz for making his calculations on the two-centre Dirac equation available to us.

POSITRON PRODUCTION IN HEAVY ION-ATOM COLLISIONS[+]

H. Backe

Institut für Physik der Universität Mainz
Postfach 3980, D-6500 Mainz, West-Germany

INTRODUCTION

The question of pair creation in strong electrical fields is a very old problem. Already in 1929, within the framework of the new Dirac theory of the electron[1], Klein considered the behaviour of an electron wave with a total energy $E = E_T + m_e c^2$ impinging on a one-dimensional step barrier potential $V_o > m_e c^2/e$ (fig. 1a). If the kinetic energy E_T of an electron travelling on the z-axis from left to right is less than eV_o, then for $z > 0$ an exponentially damped wave function is expected as a solution of the Schrödinger equation. According to Klein's calculations with the Dirac equation, this behaviour was also found for $eV_o - 2m_e c^2 < E_T < eV_o$, but for a kinetic energy $E_T < eV_o - 2m_e c^2$ the wave function starts oscillating again. This means that, in a forbidden region, electrons behave as free particles with negative energy (and also negative momentum). This anomaly was called the Klein paradox.

To resolve the Klein paradox it was necessary to understand the solutions of the Dirac equation with negative energies. This was achieved with Dirac's hole theory. According to this, all states with negative energy can be thought to be occupied with electrons. A missing electron can then be regarded as a positron. This interpretation resulted in the solution of another problem at the same time, namely electrons of an atom are therefore not allowed to undergo transitions to the negative

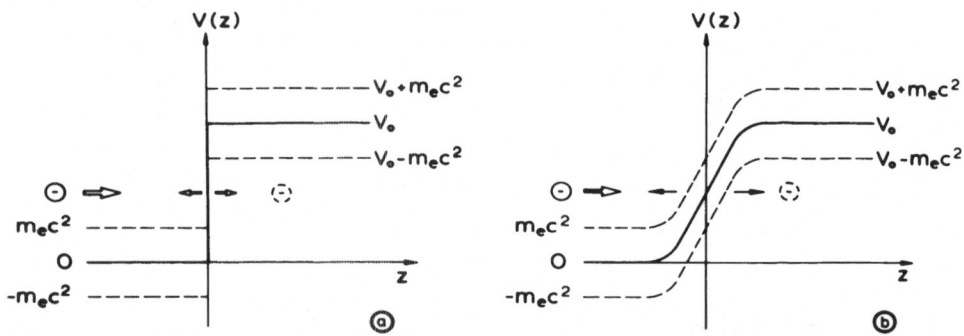

Fig. 1. (a) The step barrier potential for which
Klein [1] calculated the reflection behaviour
of electron waves using the Dirac equation.
(b) A more realistic potential for which
Sauter [2,3] performed corresponding calculations.

energy continuum states because of the Pauli exclusion
principle. Consequently, in its original form the Klein
paradox can be interpreted in the following way: A positron wave approaching from z>0 toward the step barrier
at z=0 results in radiationless annihilation with an
electron wave coming from z<0. While the electron wave
is reflected partially, the positron wave is not. With
properly chosen initial conditions, the continuous creation of electron positron pairs at z=0 takes place. In
today's terminology this is called spontaneous positron
creation.

It is well-known that potential differences having
$eV_0 > 2m_e c^2$ are easily attainable. Electrostatic accelerators operate routinely with a terminal voltage of 12 MV
or greater. Nevertheless, a vacuum break through which
results in spontaneous electron positron pair creation
has never been observed. The reason for this was recognized by Sauter [2,3] only shortly after Klein's paper was
published. Klein's assumption of a step barrier implies
an infintely high electric field strength at z=0 which
is of course, unphysical. In a more realistic potential
like that presented in Fig. 1b, the electrons have to
tunnel through a barrier. Assuming a homogeneous electric
field the transmission D of the electrons through this
barrier is approximately given by

$$D \simeq \exp(-\pi m_e c^2 / ((\hbar/m_e c^2)e(dV/dz)))$$ (1)

This transmission is unobservably small for field
strengths which can be realized in the laboratory. D
only gets large for critical fields

$$E_{crit} = (dV/dz)_{crit} \simeq m_e c^2/(\hbar e/m_e c) = 1.33 \text{ kV/fm}$$

$$(2)$$

exhibiting a potential jump of $m_e c^2/e$ over a distance of
the Compton wave lengths of the electron $\hbar/m_e c = 386$ fm.
Such fields are called strong fields.

   Strong fields can in principle be realized near
an atomic nucleus but for the known elements there are
no bound states with $E_B > 2m_e c^2$ which could be occupied
by electrons of the negative energy continuum accompanied
by the emission of positrons. However, in the last decade
the predicted existence of superheavy elements[4,5] and
their feasible production by fusion or transfer reactions with new heavy ion accelerators has led to renewed
interest in the theoretical aspects of this subject.
This has resulted in a rapidly growing study of the
"Quantumelectrodynamics of strong fields" which has been
recently reviewed[6,7] (see also the contribution of G.Soff
to this school).

   Superheavy atoms have not been found up to now
but superheavy collisional quasi atoms can be formed
at these new heavy ion facilities. In this contribution
I would like to review the progress in the search for
positron emission after close collisions of very heavy ions with atoms (for example U+U). These experiments
for the most part have been performed in the last 5 years
at GSI in Darmstadt. However, before I discuss these
experiments I would like to present a few more details
which might explain why they are interesting. In doing
this we follow closely Ref.6,7.

POSITRONS FROM SUPERHEAVY ATOMIC SYSTEMS

   The famous Dirac-Sommerfeld fine-structure formula
is well-known:

$$E = m_e c^2 [1-(Z\alpha)^2]^{1/2}$$

$$(3)$$

From this the binding energy $E_B = m_e c^2 - E$ of the ls ($\varkappa = -1$)
electron in a one electron system can be obtained (see
fig.2). According to (3) the total energy of the ls electron becomes imaginary for $Z\alpha > 1$, (i.e. $Z > 137$). However,
even as early as 1945, it was realized that this property of relation (3) is caused by the singularity of the

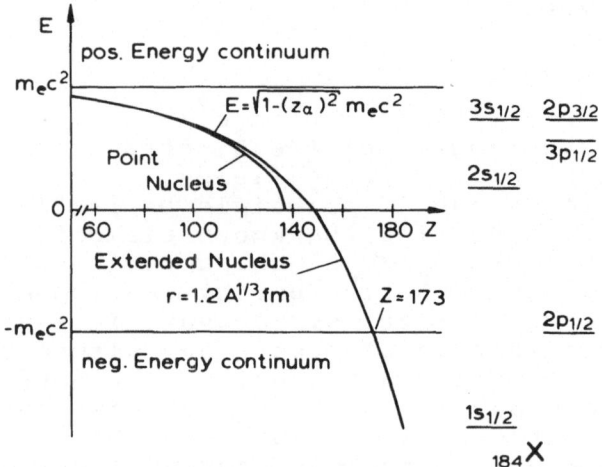

Fig.   2.   Binding energies of a 1s electron in the
            Coulomb potential of a point nucleus and
            an extended nucleus with the nuclear radius
            r=1.2 A$^{1/3}$ fm, as a function of the nuclear
            charge number Z. The right hand side presents
            the level scheme for a hypothetical hydrogen-
            like atom with Z=184.

Coulomb potential at the origin[8]. Assuming a realistic
charge distribution of the nucleus, recent calculations
show the binding energy exceeds $2m_e c^2$ =1.022 MeV at a
critical charge Z_{cr}>173. This state "dives" into the
negative energy Dirac sea. The level scheme of such a
supercritical atom with Z=184 is shown in Fig.2. Such
an atom stripped of all electrons would behave in a quite
extraordinary way. The 1s shell would be spontaneously
filled by an electron of the Dirac sea resulting in a
hole which escapes as a positron with kinetic energy
$E_T = E_B - 2m_e c^2$. This process may happen a second time de-

pending on whether the binding energy of the two electron
system is below 4 $m_e c^2$ or not. An important feature of
such an atom is that the 1s ($\varkappa=-1$) state is not anymore
a stationary state but a decaying resonance embedded
in the negative energy continuum. The decay width for
a Z=184 one electron system having a binding energy of
about 1540 keV has been calculated to be 8 keV[9], corresponding to a lifetime $\tau=\hbar/\Gamma$ of 8.2 10^{-20}s. This is what
makes positron spectroscopy after close collisions of
very heavy ions with atoms so attractive. If we assume
that somehow under certain kinematical conditions a <u>nu-</u>
<u>clear</u> molecule consisting of two uranium nuclei may be
formed having a lifetime of "only" a few times 10^{-20}s,
then the emission of positrons with discrete energies
is expected. The study of the spectroscopy of such positrons would be a unique way to investigate both atomic
and nuclear aspects of such hypothetical nuclear species.
       How realistic is such a study? At typical energies
below about 5 MeV/u the colliding uranium nuclei in nearly all cases follow pure Rutherford trajectories. Nevertheless, even in these cases a quite interesting process
occurs: the formation of a so-called quasi atom. The
electrons are exposed for the very short time (about
$2 \cdot 10^{-21}$s) to the two center nuclear charge of the colliding nuclei. If the two centers are much closer together than the Compton wave length of the electron this
two center potential cannot be "resolved" by the electron. It has been shown that the feature of the nuclear
charge distribution which effects the electron is essentially the monopole part of the two center multipole
expansion (see Ref. 8, 37). If we assume that the 1s
electron in a very high Z system is localized to its
Compton wave length then in a time of only $2 \cdot 10^{-21}$s it
may cover a distance of about 600 fm and may pass this
region only twice. While this brings into question the
justification for speaking about quasi "atoms", it is
possible in any case to describe the electrons quantum-
mechanically in an adiabatically varying two center
potential of the scattered ions. It is assumed that at
every distance the electrons have time enough to adjust
to the two center potential. The validity of this assumption can be tested by determining whether physical phenomena such as δ-electron or positron emission can be
described adequately by lowest order time dependent adiabatic perturbation theory. If this is a bad approximation, we will be able to obtain physically significant
results only with a coupled channel calculation.
       The current theoretical and experimental invest-
igations on these quasi atoms give credance to the use

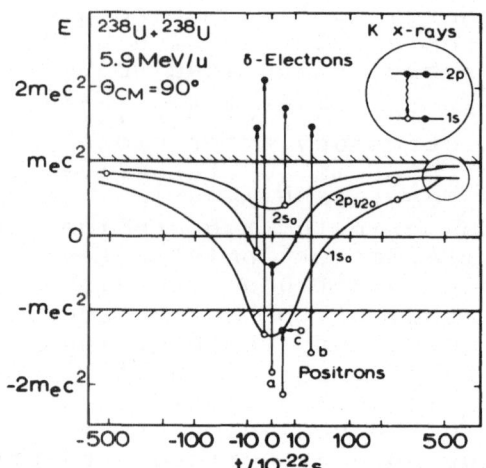

Fig. 3.   Binding energies of the 1s, $2p_{1/2}$ - and 2s
           levels as a function of time for the $^{238}U+^{238}U$
           collision system. Indicated are the inner-shell
           ionization processes with δ-electron emission;
           the three positron creation processes, induced
           (a), direct (b) and spontaneous (c); and
           the most probable process by which an inner
           shell vacancy is filled - the K X-ray emission
           in the separated atoms (taken from ref. 10).

of this adiabatic basis. In the example presented in
fig. 3 the binding energy of the 1s state exceeds, for
the very short time of 2×10^{-21} s, twice the rest mass
of the electron. The experimental question is: If this
situation actually occurs how can it be observed? In
principle there are four possibilities for investigating
these quasi atoms. All of them are based on the presump-
tion that a very strong perturbation acts on the electrons,
resulting in δ-electron emission. Experimentally either
the δ-electrons or processes associated with the result-

ant holes can be investigated. These holes can be observed by measurement of: emission of quasimolecular
X-rays, emission of characteristic X-rays from the sep-
arated atoms or emission of positrons[10,11,12,13,14,15,16,17,18] (see also the contribution of F.Bosch to this
school). Experiments to detect these emissions have yield-
ed evidence of the existance of quasi atoms. The wave
functions for the electrons at the origin are highly
relativisticly enhanced. The binding energies of inner
shell electrons are certainly very large but it is at
the moment an open question whether they exceed twice
the rest mass of the electron[10]. The most direct way
to investigate this question experimentally should be
with positron spectroscopy, since a preionized diving
level may decay by spontaneous positron emission. The
Fourier frequency spectrum of the collision process,
however, results in a dynamically induced contribution
to the positron spectrum. This is expected to be the
most important positron production process for supercritical scattering systems such as U+U with Z_u=184 and
can be investigated experimentally in subcritical systems
for which no level diving is expected. One objective
of experiments in the past has been to establish level
diving by comparison of the difference of positron spectra
or integrated positron production probabilities for subcritical and supercritical systems. In more recent experiments the main interest was focused on the search
for peaks or structure in the positron spectra.
        There is a major difficulty in performing these
experiments. Nuclear levels of the collision partners
which decay by transitions with energies greater than
$2m_e c^2$, may be exited by Coulomb excitation. Positrons
due to internal pair conversion of these transitions
cannot be easily distinguished from the atomic contribution because the time delay is only of the order of
10^{-16} to 10^{-13} s, and therefore these events cannot be
distinguished using timing or recoil techniques. The
procedure for estimating the nuclear positron background
is based on converting the observed γ-ray spectra into
positron spectra. This procedure was first proposed by
Meyerhof[20] and will be discussed in more detail later
in this paper.

DEFINITION OF PHYSICAL QUANTITIES

        The objective of positron spectroscopy in heavy
ion collisions is, in principle, to investigate a triple
differential cross section

$$d^3\sigma_{e^+} /(dE_{e^+}\, d\Omega_{e^+}\, d\Omega_p)(E_{e^+}\,,\; \theta_{e^+},\; \Phi_{e^+},\; \theta_p,\; E_1,\; M_1,\; Z_1,\; M_2, Z_2$$

$$(5)$$

which is differential with respect to the kinetic positron energy E_{e^+}, the solid angle element $d\Omega_{e^+}$ of the
emitted positron and the solid angle element $d\Omega_p$ of
scattered projectiles or recoils. The quantity θ_{e^+} and
Φ_{e^+} describe the emission angle of the positron relative to the beam axis and the scattering plane, respectively. θ_R is the scattering angle relative to the beam
direction of scattered projectiles or recoils. All quantities are measured in the laboratory system. The cross
section is determined as a function of the charge numbers
Z_1, Z_2 and masses M_1, M_2 of the projectile and target nuclei
respectively.

Since the positron production cross sections in
heavy ion collisions are fairly small, positron spectrometers for these experiments have to accept a large solid
angle to get a reasonable counting rate. Therefore what
is measured is a cross section

$$d^2\sigma_e'/(d\Omega_p\, dE_{e^+}) =$$

$$= \iint d\Omega_{e^+}\, \tilde{\epsilon}(E_{e^+},\theta_{e^+},\Phi_{e^+})[\, d^3\sigma_{e^+} / (dE_{e^+}\;\; d\Omega_{e^+} d\Omega_p)] \qquad (6)$$

whereby $\tilde{\epsilon}\,(E_{e^+},\theta_{e^+},\Phi_{e^+})$ is the positron detection efficiency and Ω_{e^+} the acceptance solid angle of the spectro-
meter (1.9 to 9.3 sr for the various spectrometers in
use).

From an experimental stand point it is more convenient, however, to measure the energy differential probability for positron production

$$\frac{dP_{e^+}^{exp}}{dE_{e^+}} = \frac{d^2\sigma_{e^+}/(d\Omega_p\, dE_{e^+})}{d\sigma_R/\, d\Omega_P} \simeq \frac{N_{e^+}^{exp}/(\epsilon_{e^+}(E_{e^+})h)}{\Delta E_e \cdot N_p} \qquad (7)$$

which is obtained by normalizing the cross section in
eq.(6) with respect to the Rutherford cross section for
projectiles and target recoils:

$$d\sigma_R /d\Omega_p =(d\sigma_R /d\Omega_p)^{proj} + (d\sigma_R /d\Omega_p)^{rec} \qquad (8)$$

Experimentally, the number of positrons $N_{e^+}^{exp}$ in an energy interval E_{e^+} (at energy E_{e^+}) must be obtained in
coincidence with the number N^e_P of scattered particles.
To get $dP_{e^+}^{exp}/dE_{e^+}$, the right hand side of eq.(5) must
be evaluated taking into account the detection efficiency

for positrons $\varepsilon_e+(E_e+)$ and the fraction of time h the apparatus is ready for data taking. Eq.(5) is only approximately valid because the particle counter has a finite acceptance angle $\theta_p \pm \Delta\theta_p$.

If dP_e^{exp}/dE_e+ is integrated over the positron energy, the total positron production probability

$$P_e^{exp} = \frac{d\sigma_e+/d\Omega_P}{d\sigma_R/d\Omega_P} = \frac{N_e^{exp}/(\bar{\varepsilon}_e+ h)}{N_P} \qquad (9)$$

is obtained. In this case a detector which simply counts the number of positrons is sufficient. For positrons with a spectral distribution $F(E_e+)$ the mean detection efficiency is

$$\bar{\varepsilon}_e+ = (\int \varepsilon_e+(E_e+)F(E_e+) \ dE_e+)/(\int F(E_e+) \ dE_e+) \qquad (10)$$

Finally, we can determine the total positron production cross section

$$\sigma_e^{tot} = \int P_e^{exp} (\theta_P)(d\sigma_R /d\Omega_P) \ d\Omega_P = \frac{N_e^{exp}/(\bar{\varepsilon}_e+ h)}{(N_T/A) \ N_B} \qquad (11)$$

The product $(N_T/A)N_B$ is the number of target nuclei N_T per target area A, times the number of projectiles N_B. This product is related to the number of particles N_M detected in a monitor counter of solid angle $\Delta\Omega$, and the known Rutherford cross section by the expression

$$(N_T/A)N_B = \frac{N_M/h}{\int (d\sigma_R /d\Omega_P) \ d\Omega_P} \qquad (12)$$

A few remarks should be added concerning the transformation of the positron production probabilities or cross sections to the CM system. This is a complication which arises due to the fact that positrons are emitted from fast moving systems. Unfortunately, there is a further complication because atomic and nuclear positrons are emitted from different systems. Atomic positrons originate from a quasi "compound" system moving with the velocity $v = \beta c$ along the beam direction, while nuclear positrons originate from scattered particles and recoils which have velocity components perpendicular to the beam direction. Even if positrons with discrete energies are emitted in the CM system, large Doppler

broadening effects are expected in the laboratory system
depending on the angle θ_e+ observed by the positron spectrometer. For example, the energy shift can be calculated
using the transformation formula to first order in β

$$E_e+ = E'_e+ (1 + \beta' \cos \theta_e+) \qquad (13a)$$

with

$$\beta' = \beta (1 + 2m_e c^2/E'_e+)^{1/2} \qquad (13b)$$

where E'_e+ is the kinetic energy of a positron line in
the CM system. More details on the transformation procedure will not be presented here except to mention that
additional assumptions concerning the angular distribution of emitted positrons in the CM system are necessary to perform the correction. Atomic positrons are
usually assumed to be emitted isotropically, as is suggested from the monopole approximation.

In a consideration of the positron emission probabilities in a Rutherford scattering process, some kinematical quantities which characterize the trajectory
are of importance. These are the distance of closed approach of the projectile and target nuclei

$$R_o = a(1 + \epsilon) \qquad (14)$$

and the impact parameter

$$b = a \, ctg \, (\theta_{CM}/2) = a(\epsilon^2 - 1)^{1/2} \qquad (15)$$

Here 2a is the minimum distance of closest approach for
a head-on-collision

$$2a = \mu[Z_1 Z_2 e^2/(E_1/M_1)] \qquad (16)$$

with $\mu = M_1 M_2/(M_1 M_2)$ the reduced mass and ϵ the excentric-
ity for the scattering angle θ_{CM} in the center-of-mass
system

$$\epsilon = 1/\sin (\theta_{CM}/2) \qquad (17)$$

It has been shown by Kankeleit[11,14] that in addition
to these quantities the scattering time

$$2\hat{t} = (2a/v)(\epsilon + 1.6\epsilon + 0.45/\epsilon) \qquad (18)$$

($v = (2E_1/M_1)^{1/2}$, the projectile velocity at infinity)

is an important quantity for understanding the positron
production process. This is the time between the extrema
of $\ddot{R}(t)/R(t)$, where $R(t)$ is the relative distance and
$\dot{R}(t)$ the relative radial velocity in the scattering process at the time t. For subcritical systems t can be used
to adequately describe the dynamically induced positron
production. For a symmetrical system such as U+U all
these quantities are defined unambigiously only for θ_p =45°
in the laboratory system. This is therefore a preferred
angle for experiments. For a typical U+U experiment at
E_1 =5.9 MeV/u and θ_p =45° the kinematical quantities are
2a=17.36 fm, ϵ=1.41, R_o =20.95 fm, 2t=1.71x10^{-21} s, b=8.68 fm.
The Rutherford cross section for the projectile in the
laboratory system is $d\sigma/d\Omega$ = 2.13 b/sr and the integrated positron production probability is P_e+ $\simeq$ 2 x 10^{-4}.
The total positron production cross section at this energy is $\simeq$3mb.

EXPERIMENTAL CONFIGURATIONS FOR IN-BEAM POSITRON SPECTROS-
COPY

Spectrometers for use in in-beam positron spectroscopy
have to have a large positron collection efficiency and
a broad energy acceptance band, while at the same time
effectively suppressing other background radiation such
as electrons, γ-rays and neutrons. These requirements
are fairly well met by the two different types of spectrometers which will be described in this section. These
are the magnetic orange type ß-spectrometer and the positron transport systems having Si(Li) detectors for energy
determination. I will not describe the first generation
spectrometers here again. Details of these can be found
in Ref.21,22. Since the time these spectrometers were
in use, positron detection techniques have been refined
considerably and I would like to concentrate on recent
progress on the second generation and the just emerging
third generation spectrometers.

THE ORANGE TYPE ß-SPECTROMETER

In fig.4 the improved set up for measurements of positrons[23,29] with the "orange" type ß-spectrometer[25] is
shown. Positrons emitted from a target between 30° and
70° relative to the beam direction can be focused by
the toroidal magnetic field produced by 60 coils, onto
a special detection system which accepts a momentum bite
of 14%. The positron detection system, shown in fig.4,
consists of a 12 cm long conical ($\emptyset$=12-16mm) plastic

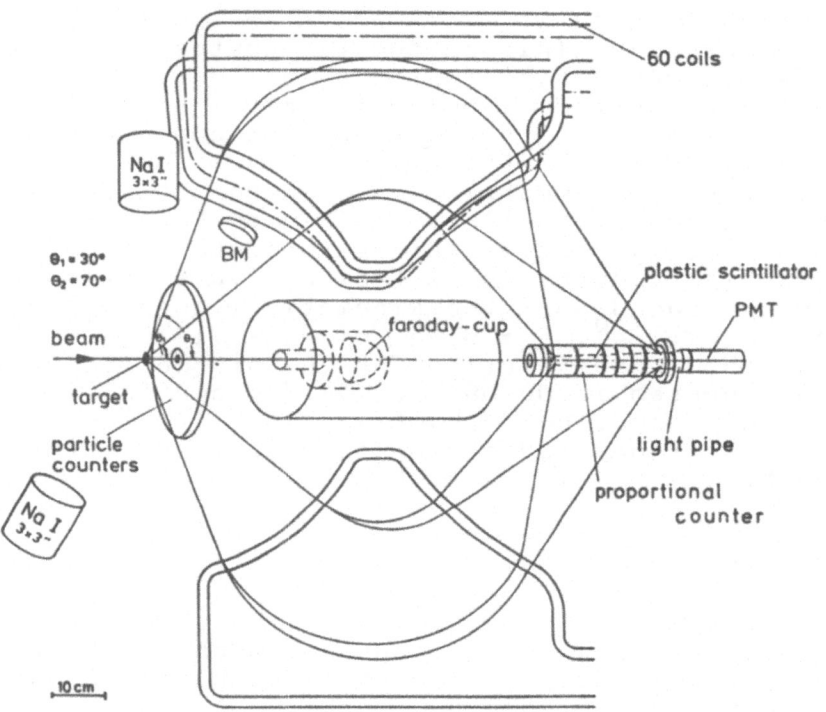

Fig. 4.   Schematic drawing of an experimental set up
          to measure positron production in heavy ion-
          atom collisions in coincidence with scattered
          particles. Positrons are focused by the toroid-
          al field of an "Orange" type ß-spectrometer
          on a conically shaped plastic-scintillator
          surrounded by a position sensitive proportion-
          al counter. Scattered ions are detected by
          an anular parallel plate avalanche counter,
          with 16 concentric anode rings. BM indicates
          a Si-beam monitor. γ-rays are detected by
          NaI-scintillation counters (taken from ref. 23,
          24).

scintillator coupled with optimized light collection
efficiency to a photomultipliertube, placed along the
axis of the spectrometer in the region of the momentum
focus. The positrons are stopped in the scintillator
and produce an energy proportional signal with an energy resolution of 17% at 500 keV. To suppress the low

energy background which originates mainly from Compton
scattered γ-rays, an annular proportional counter sur-
rounds the plastic scintillator. This proportional counter is made position sensitive by subdivision of the outer
cathode foil. Thus the accepted momentum bite of 14%
can be subdivided into several bins for better momentum
resolution. The total positron detection efficiency is
up to 15% of 4π (dependent on the angular range accepted)
in a momentum bite of 14%. The response function is independent of energy. A positron spectrum is measured
by varying the magnetic field up and down in a selected
number of steps within a certain momentum range. The
scanning is controlled by counting to a preselected number of the ions scattered into the avalanche detector.

Scattered ions and recoils from the target are detected by an annular parallel plate avalanche counter,
the anode of which is subdivided into 8 to 16 concentric
rings making it angle sensitive. In order to handle high
count rates (up to 10^7/s instantaneous) each ring is read
out independently. Best results have been obtained with
a counter subtending in angular range from 12° to 51°
in the laboratory system. Therefore the positron acceptance angle had to be reduced to values between 52.4°
and 70°, lowering the detection efficiency to 8%. Also
kinematic coincidences could be registered only in a
limited angular range (± 5%) around θ_p = 45°.

In this axially symmetric arrangement it should
be possible to reduce the Doppler broadening of the energy
of the positrons emitted from a quasi "compound" system
to nearly the intrinsic resolution of the spectrometer
($\Delta p/p \approx 0.5$%). This was shown[26] for conversion electrons
emitted from a compound nuclear system moving with a
velocity of 7,5%c. This feature may be of great importance in the search of positron lines. However, what may
be of even more importance is the fact that experimentalists are able to decide whether positron lines originate from a quasi "compound" system or from excited
nuclear levels of projectile or target like species.
This could be done by introducing a φ-sensitivity in
the particle- and positron-detectors[24]. In the case
of quasimolecular positron "lines" there would be no
φ-dependence in the spectrum whereas one should find
such a dependence in the case of scattered ions or of
recoils.

Of great importance were Si-detectors, mounted a-
round 45° relative to the beam, which monitored the energy spectra of the scattered ions. In this way the degree of target detoriation could be continously monitored.
For a detoriated target low energy tails below the peaks
of scattered particles or recoils were observed.

The γ-spectra in coincidence with scattered particles were recorded by two 3"x3"NaI-scintillation counters placed at 90° and 150° relative to the beam.

SOLENOIDAL POSITRON TRANSPORT SYSTEMS

Another type of spectrometer used for positron spectroscopy at GSI in Darmstadt is the soleniod transport system
with normal conducting coils and Si(Li) detectors for
energy determination. The improved version of the solenoid used for the pioneering positron experiments[21] is
shown schematically in fig.5. Beam and solenoid axis
are perpendicular to each other. Positrons created in
the target spiral in the magnetic field to a Si(Li) detector assembly 88 cm away from the target. The magnetic
mirror field configuration focuses onto the detector
all positrons with an emission angle

$$\theta < \theta_{max} = \pi/2 + \arcsin \, (B_T/B_{max})^{1/2} \qquad (19)$$

with respect to the ζ-axis. B_T is the magnetic field
at the target position and B_{max} its maximum value at
ζ=6.5cm. With the magnetic field configuration shown
in fig. 5 the acceptance angle is θ_{max} = 148.7°, corre-

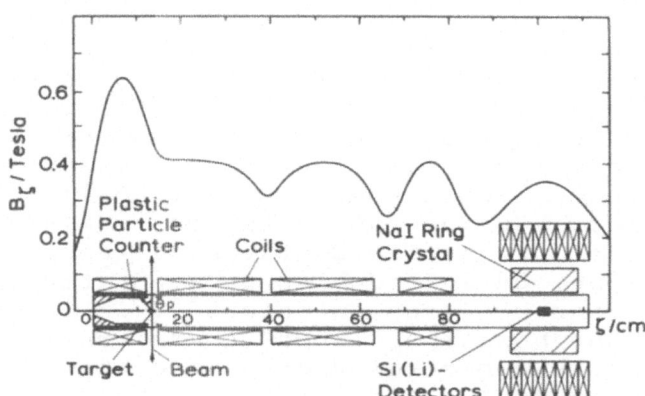

Fig. 5. Schematic drawing of the solenoid spectro-
          meter. Also shown is the magnetic field on
          the ζ-axis. The heavy ion beam enters the
          vacuum chamber perpendicularly to the solenoid
          axis[27].

sponding to a solid angle of $\Omega/4\pi=0.74$. Unfortunately,
all electrons resulting from Coulombic projectile target interactions or internal conversion transitions of
excited nuclear levels are also focused onto the detector.
To exclude the detection of this very high electron background without appreciably decreasing the positron detection efficiency, the detector system shown in fig.6 has
been used[27]. It consists of 2 Si(Li) diodes with a diameter of 19.5 mm and a thickness of 3mm mounted with
their surfaces parallel to the solenoid axis. One of
these is placed above the axis while the other is below. The sensitive face of the upward counter looks to
the left, that of the downward counter to the right.
With this configuration positrons spiraling in a right
hand manner have a good probability of entering the sensitive area of one of the Si(Li)-diodes. Electrons spiraling in the opposite direction are absorbed or scattered
in the 3mm thick aluminium back cover of the detector.
Only the very few electrons with an energy greater than
about 1.4 MeV can penetrate the aluminium absorber. A
suppression factor of about 150 was measured for 365 keV
electrons. In spite of this strong suppression of electrons, the Si(Li) counting rate observed for in-beam
positron experiments is still caused nearly entirely
by electrons. In order to select the very few positrons,
the Si(Li) diodes were surrounded by a 4-fold segmented
NaI(Tl) ring crystal which permitted detection of the
511 keV annihilation radiation from the positron. To
characterize a positron, a coincidence was required bet-

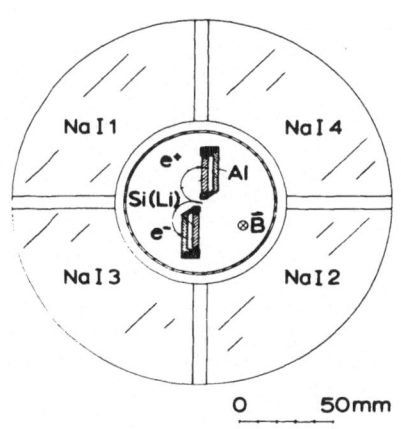

Fig. 6.

The NaI-Si(Li) detector
assembly for the detection
of positrons in a view perpendicular to the solenoid
axis at J =102 cm. The
circular Si(Li) detectors
have a diameter of 20 mm
and a thickness of 3 mm.
The fourfold segment-
ed NaI ring crystal has
the following dimensions:
inner diameter 90 mm,
outer diameter 204 mm,
length 150 mm [27].

ween an event registered in the 511 keV region of one
of the NaI(Tl) segments and in the total energy region
of the opposite one. Both Si(Li) diodes operate essentially independent of each other. This feature is useful
in the search for peaks in the positron spectra[28]. For
example, suppose that a hypothetical superheavy "compound" system formed after a $^{238}U+^{238}U$ collision at 5.9
MeV/u emits positrons with a discrete energy of 300 keV.
Such a superheavy system moves with about 5.6% of the
velocity of light. Positrons accepted by the magnetic
field of the solenoid would appear broadened in energy
by the Doppler effect (eq.13). The corresponding energy
distribution is nearly rectangular with a width of 71 keV.
The Si(Li) counters at the top and at the bottom are
sensitive primarily to positrons emitted parallel or
antiparallel respectively relative to the beam direction.
Therefore, for these counters, positron distributions
are expected to result in two peaks separated by about
35 keV. Such an observation would be a very characteristic
fingerprint for a positron peak.
        The positron efficiency obtained by unfolding the
Si(Li)-detector response function is shown in fig.7,
and was measured with intensity calibrated ^{22}Na and
^{68}Ge/Ga sources. The calibration determination was performed in the same manner as in the in-beam experiment.
The total positron yield $P_{e}+$ can be obtained by integration of the energy spectrum. In this procedure only
the fraction of positrons which strikes the Si(Li)-de-
tector is used. Obviously, the detection efficiency can
be increased by approximately a factor of 3 if the complete Si(Li)-detector assembly is used as a passive positron catcher. In this case the detection efficiency shown

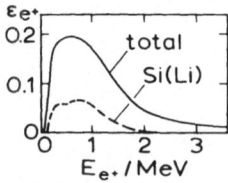

Fig.   7.   Positron detection efficiency of the solenoid
           spectrometer. The lower dashed curve labeled
           "Si(Li)" corresponds to energy analyzed
           positrons. For the upper curve labeled "total"
           the Si(Li) detector assembly serves as a
           passive positron catcher only [27].

in fig.7 drops off for low positron energies because
of target selfabsorbtion effects and for high energies
because positrons get absorbed in the vacuum chamber
or miss the catcher. Nevertheless, the shape of the
detection efficiency is sufficiently flat in the relevant
region between 0.2 and 1 MeV. As shown below only very
few positrons have energies beyond this region.
     Positrons are measured in coincidence with scatter-
ed particles or target recoils using a plastic scintillat-

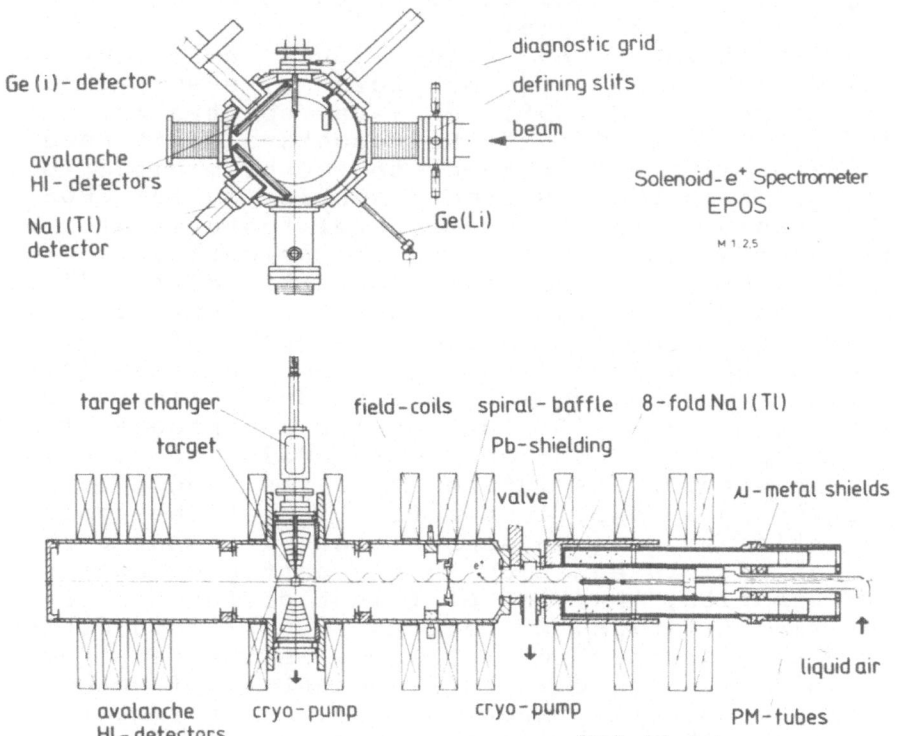

Fig. 8.   A schematic view of the solenoidal spectrometer
           EPOS[29]. The heavy ion beam axis is perpendicular
           to the plane of the drawing. The four coils to
           the left produce a magnetic mirror field con-
           figuration similar to that shown in fig. 5.
           Target electrons are excluded from detection
           by two pencillike positron Si(Li)-detectors
           (outer diameter 10 mm, sensitive thickness
           2.5 mm, length 50 mm) and an Al spiral baffle.
           The 8-fold NaI detector has the following
           dimensions: inner diameter 100 mm, outer
           diameter 200 mm, length 200 mm).

ion counter with a thickness of 50 μm, shown in fig.5.
Recoils of light target contaminants such as carbon
or oxygen are excluded from detection electronically
by pulse hight discrimination. Projectile fission products resulting from fusion reactions with light target
contaminants were kinematically excluded from entering
the sensitive region of the plastic counter at angles
between 35° and 55° degrees. For U beam energies below
5.9 MeV/u, a small fraction of fission fragments from
target projectile interactions were also detected.
However, positrons associated with these events were
estimated to be negligible.

The condition of the target as well as the product
$N_p(N_t/A)$ (see eq.(12)) were monitored by a surface
barrier detector at θ_B =45°. The γ-ray spectra were
recorded by one or two 3"x3" NaI counters positioned
at 45°,90° or 135° with respect to the beam direction.

The new solenoid positron detection system EPOS
at GSI[29], shown in fig.8, has several features of the
device just described. These include primarily the
principle of operation using the magnetic mirror effect.
However in addition this spectrometer has several improvements. For example, the tube diameter of 260mm
is large enough to insert two position sensitive parallel plate avalanche detectors with delay line readout
for particle detection[30]. The angular acceptance of
each of the particle counters is 20°≤θ≤70° at constant
ΔΦ=60°. The angular resolution is better than 1°.
This results in an excellent definition of the scattering kinematics and a good suppression of unwanted reaction channels. In addition to the particle scattering
angle information, the time of flight difference between the two particles, as well as rough Z-information
via the energy loss of the particles in the detector
gas, is obtained. The coincident detection of both
particles is especially advantageous. For asymmetric
systems such as U+Pb, the distance of closest approach
R_0 can be determined uniquely under the assumption
of Rutherford scattering conditions from the kinematic
(θ_1, θ_2)-correlation. For symmetric systems such as
U+U, the distinguishing of close from distant collisions
is, in principal, impossible. However, information
on the energy-loss during the collision (Q-value) can
be obtained from the sum of the two scattering angles
(θ_1, θ_2), which should be 90° for elastic and <90° for
inelastic events. Futhermore, the kinematic coincidence
can be used together with the rough Z-identification
of the detected ions to select desired reaction channels and discriminate against background from the target
backing and/or covering foils. Although the set-up

allows for the selection of binary, so-called quasielastic events, it should be noted that separation
of reaction channels involving the transfer of a few
nucleons is not possible as long as both resulting
nuclei are stable against fission.

Electrons are suppressed by a spiral baffle between target and positron detector which makes use of
the different helicity of positrons and electrons.
The baffle was optimized for highest positron transmission with minimal limitation of the momentum band
passed (see fig.9). Background counts in the e^+-detector
due to the leakage of scattered electrons through the
baffle and due to other possible background sources
outside the target are reduced by using a pencil- like
e^+-detector located along the solenoid axis. The coaxial
geometry of the e^+-detector combines maximum efficiency
for target produced positrons with minimum sensitivity
for positrons and electrons which have been scattered
on their way from the target or which do not originate
from the target. Positrons are identified by demanding
coincidences between the e^+-detector and at least one

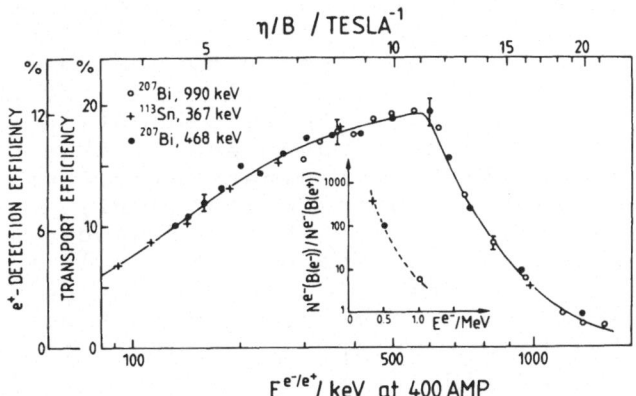

Fig. 9. Transport efficiency of the solenoid-system
as a function of the electron/positron kinetic
energy $E^{e+/e-}$ and the scaling parameter η/B,
the magnetic field at the target and the e^+/e^--
momentum in units of $m_e c$. Insert: e^--suppres-
sion factor (ratio of total e^--transmission
for opposite field settings) as a function
of initial kinetic e^--energy [29].

of the two 511 keV annihilation γ-quanta of the positron
as registered in an eight-fold cylindrical NaI-detector
array surrounding the e^+-counter. By summing up the
signals of the 8 separate NaI-crystals electronically
and requiring a total γ-sum $E_\gamma \geq 440$ keV (to exclude
low energy γ-rays) the NaI-efficiency reaches ≈58%
per positron detected in the e^+-counter.

The pencil-like positron detector consists of
two coaxially drifted Si(Li)-detectors. The detectors
are mounted closely one behind the other on a cooling-
rod. The energy resolution of the e^+-detector is better
than 10 keV for the 662 keV conversion electron line
of ^{137}Cs. This resolution is much better than the expected width of an e^+-line measured in-beam, a width
which will be determined mainly by the maximum Doppler-
broadening (eq.(13)). A reduction of the Doppler-broadening should be possible by adjusting the beam direction
with the solenoid axis and by accepting only positrons
emitted at small angles with respect to the solenoid
axis. As in the other experiments described above,the
γ-ray spectrum is measured using a 3"x3" NaI detector
operating in coincidence with the heavy ion detectors.

THE TORI SPECTROMETER - AN S-SHAPED POSITRON TRANSPORT
SYSTEM

Positron experiments with the solenoid spectrometer
shown in fig.5 have now been terminated. In the middle
of 1981, this solenoid was replaced by the so-called
"Tori" spectrometer[31] shown in fig.10; an instrument
which probably represents the third generation of positron spectrometers. The main advantage of this new
instrument is the ability to separate electrons and
positrons and to measure them simultaneously in different detectors. This is achieved by an S-shaped magnetic
field configuration produced by pancake coils surrounding two quarters of a toroidal tube. In the bent magnetic
field of the first quarter torus, electrons and positrons leaving the target experience a drift in their
guiding center perpendicular to the spectrometer plane.
This drift is opposite for electrons and positrons
and can give complete separation of the two. The electrons are absorbed in a detector, while the positrons
are left free to pass into the second quarter torus
in which the drift is reversed. This spectrometer has
image characteristics for positrons which are very
similar to those of the solenoidal transport systems.
However, because the rejection of electrons is achieved
in a way that does not effect the transmission of posi-

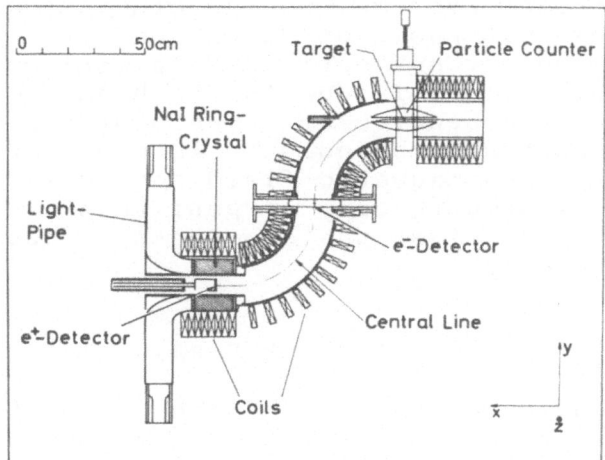

Fig. 10. The Torispectrometer. Beam direction and
z-drift is perpendicular to the plane of
the drawing[31]. The e+ detector is a Si(Li)
detector with 40 mm diameter and 5 mm thickness. The NaI ring crystal is the same as
shown in fig. 6.

trons, the broad band acceptence of positrons is improved. According to recent calculations[32] an efficiency of 22% for positrons in the energy range from
100 keV to 800 keV is expected. Scattered particles
are registered by two parallel plate avelanche counters
using the delay line principle. This set-up allows
the separation of two body reactions from reactions
involving three or more particles. The scattering angles which can be observed range from 20° to 70° with
an angular resolution of 2°.

In concluding this discussion on spectrometers,
the relative advantages of the two types of positron
spectrometers will be summarized. The advantage of
the orange-ß-spectrometer is that certain interesting
parts of the positron spectrum can be investigated
with very good momentum resolution. This is primarily

due to the coincidence of beam and spectrometer axis. The orange spectrometer is therefore well suited to the investigation of structures in the positron spectrum. The transport systems, on the other hand, are designed for highest positron detection efficiencies but are at the moment not optimized to minimum Doppler broadening. This is mainly due to the beam and spectrometer axis being perpendicular to each other. If desired these spectrometers could be turned by 90° and by choosing a proper magnetic field configuration this could also allow spectroscopy of positrons with good energy resolution, but at the expense of a loss in detection efficiency.

EVALUATION OF ATOMIC POSITRON SPECTRA

The evaluation procedure for atomic positron spectra will be described only for the experiments performed with the solenoidal spectrometer[28] as shown in fig.5. For details of the evaluation procedure for other experiments the reader is referred to ref.24, 30.

To calculate the atomic positron spectrum $\Delta P_{e^+}^{atom}/\Delta E_{e^+}$ using equ.(7), the number of measured positrons $\Delta N_{e^+}^{exp}$ has to be corrected for several contributions

$$\Delta N_{e^+}^{atom} = \Delta N_{e^+}^{exp} - \Delta N_{e^+}^{int}(Z_1,M_1) - \Delta N_{e^+}^{int}(Z_2,M_2)$$
$$- \Delta N_{e^+}^{ext}(T) \quad - \Delta N_{e^+}^{ext}(S) \qquad (20)$$

These contributions include positrons produced in the internal pair decay of excited levels in the projectile and target nuclei ($\Delta N_{e^+}^{int}(Z_1,M_1)$ and $\Delta N_{e^+}^{int}(Z_2,M_2)$) as well as positrons from external pair conversion of γ-rays in the target $\Delta N_{e^+}^{ext}(T)$ and in the solenoid $\Delta N_{e^+}^{ext}(S)$. From test experiments and calculations it was concluded that

$$\Delta N_{e^+}^{ext}(S) \simeq 0 \qquad (21)$$

On the other hand the external pair creation in a $1mg/cm^2$ Pb or U target requires corrections of the order of 10% to the positron yield. Similar corrections have to be applied for the determination of the integral number of positrons $N_{e^+}^{atom}$.

As already mentioned, the internal pair conversion contribution of positrons is desired from γ-ray spectra. For the ^{208}Pb + ^{208}Pb scattering system the γ-ray spectrum is very simple because only the 3$^-$ level at 2.614 MeV and the 2$^+$ level at 4.09 MeV are populated by Coulomb excitation. In this case the correction procedure for nuclear positrons is straight forward[21]. However, to obtain the highest possible $Z_u = Z_1 + Z_2$ ^{238}U beams have been used, resulting in the excitation of a large number of nuclear levels having unknown decay characteristics. In this case, the fraction of atomic and nuclear positrons are determined by observing the positron- and γ-ray production for various targets. To calculate the fractions the following assumptions were made

(i)   In low $Z_u = Z_1 + Z_2$ systems only nuclear positrons are emitted.

(ii)   The multipolarity of the observed γ-ray spectrum is target independent.

(iii)   The angular distribution of emitted positrons and γ-rays in the laboratory system is roughly the same for the different scattering systems investigated.

(iv)   Nuclear excited states will not decay by internal pair creation during the existence of the quasi atom. (Such an effect could be expected to be strongly Z_u dependent).

Corrections for the nuclear positrons are carried out as follows. A typical γ-spectrum is shown in fig.11. After unfolding the response of the NaI(Tl) detector from this spectrum the resultant γ-distribution $dN_\gamma/dE_\gamma(E_\gamma)$ can be transformed into a positron spectrum using theoretical pair conversion coefficients[34,35,36] $d\beta_{M\lambda}/dE_{e^+}(E_\gamma, E_{e^+}, Z)$ of multipolarity $M\lambda$

$$\frac{dN_{e^+}}{dE_{e^+}}(E_{e^+}, M\lambda) = \int_{2m_ec^2}^{\infty} dE_\gamma \frac{dN_\gamma}{dE_\gamma}(E_\gamma) \frac{d\beta_{M\lambda}}{dE_{e^+}}(E_\gamma, E_{e^+}, Z) \tag{22}$$

Fig.11 shows two positron spectra in the form

$$\Delta P_{e^+}/\Delta E_{e^+} = (\Delta N_{e^+}/\Delta E_{e^+}(E_{e^+}, M\lambda))/N_p \tag{23}$$

calculated from the γ-distribution assuming E1 and E2 multipolarities. The remarkable difference between these two spectra emphasizes the difficulty in correcting for nuclear background positrons when the multipolarity of the γ-ray transition is unknown. However,

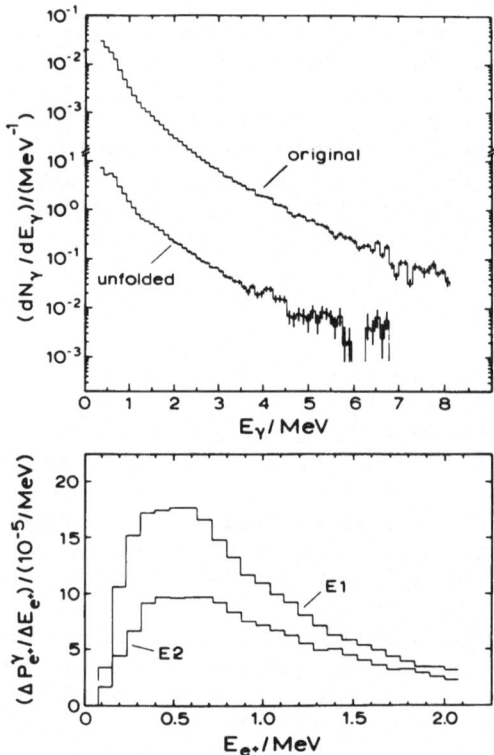

Fig. 11.Target γ-ray spectrum taken with the 7.5 x 7.5 cm
NaI detector. A ^{108}Pd target was bombarded with
^{238}U projectiles at an energy of 5.9 MeV/u. The
γ-ray spectra were recorded in coincidence with
the ^{108}Pd recoils, uniquely detected in the particle counter at θ_p = 45°±10°. The spectrum labeled "unfolded" was obtained from the original spectrum by unfolding the detector response and applying the efficiency correction. The lower part of
this figure exhibits positron spectra calculated
from the unfolded γ-ray spectrum, assuming E1 and
E2 multipolarity (taken from ref. 33).

the experimentally measured positron spectrum shown
below in fig.16 strongly suggests an E1 multipolarity
if we make the plausible assumption that there are
negligible amounts of atomic positrons created in the
U + Pd scattering system.

Further information on the γ-ray multipolarity
can be gained from energy integrated positron yields.
These can be determined with much less beam time than
positron spectra due to a higher total efficiency (see
fig.7). We calculate

$$N_{e^+\gamma}^{calc} (M\lambda) = \int_0^\infty dE_{e^+} \; \varepsilon_{e^+}^{tot}(E_{e^+}) \; \frac{dN_{e^+}}{dE_{e^+}} \; (E_{e^+}, M\lambda) \qquad (24)$$

and compare this number with the experimental value
$N_{e^+}^{exp}$ forming the ratio $N_{e^+}^{exp}/N_{e^+\gamma}^{calc}(M\lambda)$. The result is
shown in fig.12. The data have been normalized to unity
in the region $Z_u < 160$ by introduction of a factor f
which is itself close to unity.

The displayed ratio clearly exceeds unity for
the U+Pb, U+U and U+Cm scattering systems. This fraction of excess positrons is believed to originate from
atomic processes for which we can write

$$N_{e^+}^{atom} = N_{e^+}^{exp} - N_{e^+}^{ext}(T) - N_{e^+\gamma}^{calc}(E1) \cdot f \qquad (25)$$

From equation (20) and (21) we can identify the last
term of (25) as the nuclear contribution of positrons:

$$N_{e^+\gamma}^{calc} (E1) \; f = N_{e^+}^{int} (Z_1, M_1) + N_{e^+}^{int} (Z_2, M_2) \qquad (26)$$

It should be emphasized that this procedure is only
correct if the assumptions (i) to (iv) given above
are valid. The approximations introduced by these assumptions may lead to an error of up to 30%.

RESULTS AND DISCUSSION

The experimental results obtained up to now can
be grouped into four categories:
(i)   measurement of the total positron production
       cross sections $\sigma_{e^+}^{tot}$,
(ii)   measurement of energy differential positron pro-
       duction probabilities P_{e^+} ,
(iii) measurement of energy differential positron pro-
       duction probabilities $\Delta P_{e^+}/\Delta E_{e^+}$ at a fixed posi-
       tron energy $E_{e^+} = (478 \pm 54)keV$ and

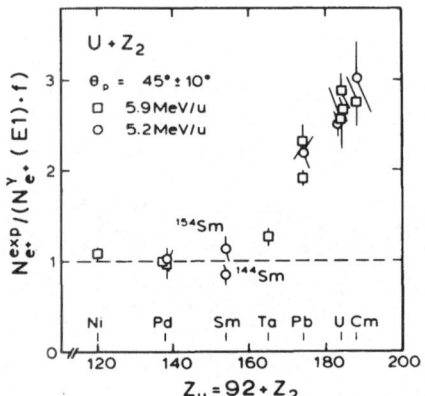

Fig. 12.   Ratio of measured to calculated positron
           yields $N_{e+}^{exp}/N_{e+}^{\gamma}$ (E1)·f   as a function of the
           united nuclear charge Z_u, assuming El multi-
           polarity for the nuclear target γ-rays.
           (θ_p = 45°±10°, U-beam □ 5.9 MeV/u, ○ 5.2 MeV/u).
           For Z_u >174 there is a clear indication of
           a large number of excess positrons which
           are attributed to atomic processes (taken
           from ref. 33).

(iv)   measurement of positron spectra $\Delta P_e+/\Delta E_e+$.

       Various scattering systems with various kinemati-
cal conditions have been investigated in each group,
including systems with low $Z_u = Z_1 + Z_2$ for which only
emission of nuclear positrons is expected. A discussion
of the four groups follows.
(i)   The total positron cross section σ_{e+}^{tot} is shown
in fig.13 as a function of distance of closest approach
in a head on collision[28], (see eq.(16)). This parameter
is particularly advantageous since the interaction
radius

$$R_{int} = 1.16 \; (A_1^{1/3} + A_2^{1/3} + 2 \;) \; fm \qquad (27)$$

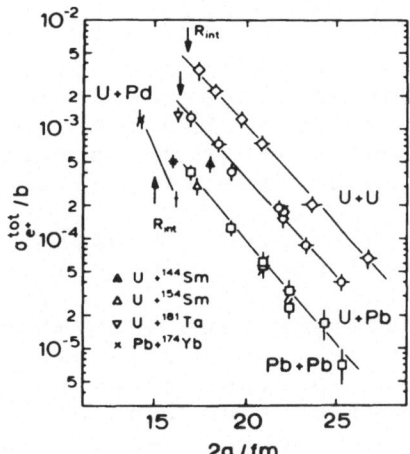

Fig. 13. Total positron production cross section (not
         corrected for nuclear positrons) for different
         scattering systems as a function of the closest
         approach in a head on collision 2a, see equ.(16).
         Also indicated is the interaction radius R_{int}
         (see equ.(27)).

defines two regions. Roughly speaking, in the region
$2a \gtrsim R_{int}$ the collision partners follow essentially pure
Rutherford trajectories and nuclear background positrons originate mainly from mutual Coulomb excitation
of projectile and target nuclei. In the region $2a \lesssim R_{int}$
nuclear reactions may contribute. A strong target dependence of $\sigma_{e^+}^{tot}$ for the nuclear systems
$^{238}U + ^{181}Ta$, $^{238}U + ^{144,154}Sm$ and $^{238}U + ^{108}Pd$ for 2a=const
can clearly be recognized. This reflects the Z and
v dependence of Coulomb excitation of pair decaying

nuclear levels in ^{238}U and, possibly, a target nuclear structure dependence. A nuclear structure effect may be responsible for the cross section difference observed in the ^{238}U+^{144,154}Sm scattering systems, although it can not be ruled out that this effect is caused by different contaminations of oxygen in the metallic ^{144}Sm and ^{154}Sm targets. The ^{238}U +^{238}U, ^{238}U +^{208}Pb and ^{208}Pb systems have been investigated at different beam energies. In the semilogarithmic representation of fig.13 the total cross sections follow straight lines. It is worth noting that cross sections have been measured in a range from several μb to several mb.

Up to now the total atomic positron cross section (corrected for nuclear background positrons) could only be determined for the Pb + Pb scattering system (see ref. 21). No attempts have been made to correct the U+U and U+Pb systems since a γ-ray background continuum resulting from spurious γ-ray sources (slit scattering of the beam or reactions of the beam particles with light target contaminants) can hardly be distinguished from the target contribution itself. Therefore, only the uncorrected positron cross sections are shown in fig.13.

(ii) Fig.14 shows the total atomic positron production probability P_{e^+} as a function of the scattering time $2\hat{t}$ (eq.18). As already mentioned in a preceeding section, this time characterizes the dynamically induced positron creation for subcritical systems. We present this following a development by Kankeleit[11,14]. He showed that the first order adiabatic transition amplitude

$$a(t=\infty) = - \int_0^\infty dt' <E_{e^-}|\partial/\partial t|E_{e^+}>\exp(i(E_{e^-} + E_{e^+} +2m_ec^2)t'/\hbar) \quad (28)$$

can be written as

$$a(t=\infty)=(f(E_{e^+})g(E_{e^-}))^{1/2}\exp(-\hat{t}(E_{e^-} + E_{e^+} +2m_ec^2)/\hbar)/(E_{e^-} + E_{e^+} +2m_ec^2) \quad (29)$$

The energy denominator $(E_{e^-} + E_{e^+} +2m_ec^2)$ originates from the application of the Hellmann-Feynman relation. The functions $f(E_{e^+})$, $g(E_{e^-})$ stem from the Coulomb repulsion of positrons or Coulomb attraction of electrons respectively, in analogy to the well-known Fermi function in β-decay. The double differential probability for pair creation can then be expressed as:

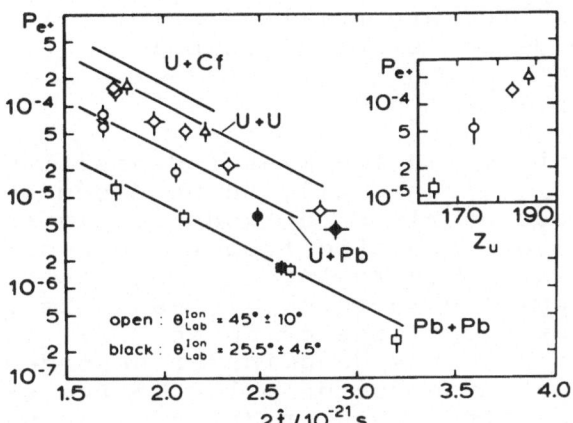

Fig. 14. The total atomic positron production probability P_{e+} for different systems as a function of scattering time $2\hat{t}$. The following designations are used: △ U+Cm, ◇ U+U, ○ U+Pb, □ Pb+Pb. Open points represent $\theta_{lab}=45°\ \pm10°$, black points $\theta_{lab}=25.5°\pm4.5°$. The full lines represent theoretical calculations[37]. The insert shows the Z_u dependence of P_{e+} for the fixed scattering time $2\hat{t}=1.75 \times 10^{-21}$ s.

$$d^2P_{e+}/(dE_{e}+dE_{e}-)=f(E_{e}+)g(E_{e}-)\exp(-2\hat{t}(E_{e}-\ +E_{e}+$$

$$+2m_ec^2)/\hbar)/(E_{e}-\ +E_{e}+\ +2m_ec^2)^2 \qquad\qquad (30)$$

We see directly that the transition amplitude factors into an exponential which contains the dynamical aspects of positron creation, and functions $f(E_{e}+)$ and $g(E_{e}-)$ which are connected to the electron and positron wave functions in the initial and final state. When electrons are not detected in the experiment we have to integrate over $E_{e}-$ and obtain the positron spectrum:

$$dP_{e}+/dE_{e}+ = h(E_{e}+,\hat{t})\exp(-2\hat{t}(E_{e}+\ +2m_ec^2)/\hbar) \qquad (31a)$$

The function $h(E_{e^+}, \hat{t})$ is responsible for the experimentally observed decrease of the positron intensity at
low energy (see fig.16). If we are only interested
in the total positron probability, a further integration
over E_{e^+} has to be performed yielding:

$$P_{e^+} = i(\hat{t}) \exp(-2\hat{t}(2m_e c^2)/\hbar) \tag{32a}$$

The high energy tails in the measured positron spectra
$\Delta P_{e^+}/\Delta E_{e^+}$ (see fig. 16 and 21), to be discussed later
in this paper, are nearly pure exponentials exhibiting
the steeper fall off expected from the exponential
in eq.(31a). This is due to the E_{e^+} dependence of $h(E_{e^+}, \hat{t})$
which at high positron energies, is essentially deter-
mined by the energy denominator in eq.(30). To a good
approximation this energy denominator dependence is
taken into account by a factor $m \simeq 1.4$ in the exponential
of eq.(31) which, after this modification, becomes

$$dP_{e^+}/dE_{e^+} = h'(E_{e^+}, \hat{t}) \exp(-m\, 2\hat{t}(E_{e^+} + 2m_e c^2)/\hbar \tag{31b}$$

The function $h'(E_{e^+}, \hat{t})$ now becomes a constant for large
E_{e^+}. After integration over the positron energy we
obtain for the total positron production probability

$$P_{e^+} = i'(\hat{t}) \exp(-m\, 2\hat{t}(2m_e c^2)/\hbar \tag{32b}$$

Experimentally we again observe the functional relationship between P_{e^+} and $2\hat{t}$ as an exponential fall off
which can be approximated by

$$P_{e^+}^{exp} \propto \exp(-m\, 2\hat{t}\, \Delta E/\hbar) \tag{33}$$

with $\Delta E/(m_e c^2) = 2.66(.72)$ for U+Cm, $2.86(.17)$ for U+U,
$2.91(.30)$ for U+Pb and $2.40(.19)$ for Pb+Pb, scattering
systems. It sounds resonable therefore to interprete
ΔE as the mean energy for creating an electron-positron
pair. As expected, this energy is above the threshold
of $2m_e c^2$. However, the assumption that ΔE does not
depend on the scattering time $2\hat{t}$ is only approximately
valid. Therefore the numbers for ΔE as given above
are mean values for scattering times in the range
$1.5\ 10^{-21}$ s $\leq 2\hat{t} \leq 3\ 10^{-21}$ s. These simple considerations
therefore lead to the explanation that the nearly
exponential slopes observed in both dP_{e^+}/dE_{e^+} (at
sufficiently high positron energies E_{e^+} (see below))and
in $P_{e^+}(\hat{t})$ are the results of the simple dynamical
aspects of collisional positron creation. On the other
hand, the absolute positron production yield and/or
deviations from the exponential slope can provide de-

tailed physical information about the wave function
in the matrix element $\langle E_e-|\partial/\partial t|E_e+\rangle$ of equ. (28).
For this to be true, however, there must be no such
deviations caused by interferencies with higher order
terms in the perturbation expansion. (Theoretical calculations[37] imply that these may not be negligible.)
Because of this, conclusions drawn from this model
must be accepted with caution.

As already pointed out, P_e+ in fig. 14 can be
approximated by an exponential function for every scattering system. No evidence of level diving is observed
in P_e+ for the U+Cm and U+U systems at lowest scattering time, even though these systems are expected
to be supercritical. This is also true for the Z_u dependence of P_e+, as shown in the insert of fig.15 for
the fixed scattering time $2t= 1.75 \times 10^{-21}$ s. In fact
the positron probability can be approximated by the
simple relation:

$$P_e+ \propto Z_u^{20.3} \tag{34}$$

A similar result was also reported in ref.24. It reflects the strong relativistic contraction of the wave
functions at the origin at the time of closed approach
of the colliding nuclei.

(iii) Next I would like to discuss the measurements of energy differential positron production
$\Delta P_e+/\Delta E_e+$ at a fixed positron energy $E_e+=(478\pm54)$keV.
These are shown in fig.15. However, before doing this
a few remarks concerning scaling laws seem to be appropriate. Bang and Hansteen[38] first proposed a scaling
law for Coulomb ionization by light ions. Recently
it was extended by the same authors to heavy-ion collision systems[39,40]. For the case of positron production
it can be written in the form

$$dP_e+/dE_e+ \propto \exp(-2q_0 b) \tag{35}$$

where q_0 is the minimum momentum transfer required
to produce an electron-positron pair with total energy

$$\Delta E=E_e+ +E_e- +2m_e c^2 \tag{36}$$

From

$$\Delta E = \frac{\vec{p}^2-\vec{p}'^2}{2M} = \frac{(\vec{p}+\vec{p}')(\vec{p}-\vec{p}')}{2M} \simeq \frac{2p}{2M} \hbar\, q_0 \cos(\vec{p},\vec{q}_0) \tag{37}$$

we obtain for the minimum q_0 with $\cos(\vec{p}, \vec{q}_0) = 1$

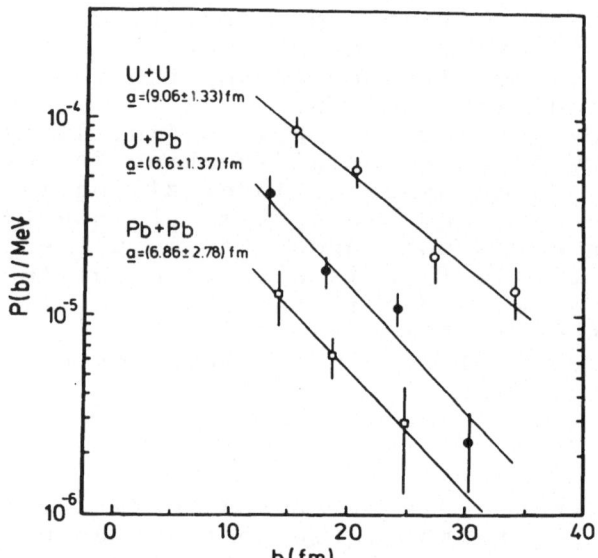

Fig. 15.   Impact parameter dependence of the production
            probability of positrons per MeV, P(b), at
            (478 ± 54) keV positron energy for (U+U),
            (U+Pb), and (Pb+Pb) - collisions at 5.9 MeV/u
            bombarding energy. The solid lines represent
            fits of the data to the relation
            P(b)=P(o)exp(-b/$\underline{a}$). The values for the decay
            constant $\underline{a}$ are given for various systems
            (taken from ref. 24,25).

$$q_o = \frac{\Delta E}{\hbar v} \qquad (38)$$

($\vec{p} = M\vec{v}$, $\vec{p}' = M\vec{v}'$ are the momenta of the ions before
and after the collision.)
      The exponentials in (35) and (31a) appear to be
quite different. However, after substitution of eq.(38)
and (15) into (35) we see that, for $\varepsilon \gg 1$ (i.e. small
scattering angles), both exponentials are the same.
It is for $\varepsilon \lesssim 1$ that there are considerable discrepancies.

These discrepancies are the source of some differences
of opinions concerning the correct scaling law (see
also ref.41 where still another scaling law is presented). As I have already mentioned, I believe that
conclusions drawn from such models have to be accepted
with caution. Because of this, I will not judge this
controversy but will present the data in the manner
preferred by the authors of the original work.
In fig.15 the data were fitted[42] with a function

$$P(b) = P(o) \exp(-b/\underline{a}) \qquad (39)$$

According to eq.(35) and (38) the constant $\underline{a}$ is related
to the energy transfer ΔE (eq.(36)) by

$$\underline{a} = 1/(2q_o) = \hbar v/(2 \Delta E) \qquad (40)$$

If we assume that $<E_e->$, the average energy transfer
to the δ-ray which is associated with the positron,
is small compared with $2m_e c^2 + E_e+$, then ΔE is 1.51 MeV.
From eq.(40) it then follows that $\underline{a}=7.2$ fm at 5.9 MeV/u.
This value is in good agreement with the experimental
observation (see fig. 15). This result again reflects
the dynamical aspects of positron creation in collisions.
(iv) The information providing the deepest insight
into the positron production process is expected to
come from the study of the complete positron spectrum.
There are now carefull investigations underway and
in the following I will discuss a few interesting aspects
of these experiments.
The first positron spectra measurements[28] have
been made with the solenoid spectrometer set up shown
in fig.5. Fig.16 presents the measured spectra corrected
for the detector response function and efficiency.
The spectrum of the low Z_u=138 system U+Pd, for which
only nuclear positrons are expected, can be reproduced
by converting the target γ-ray distribution into a
positron spectrum assuming E1 multipolarity. The nuclear
background contributions for the U+Pb and U+U systems
have similarly been determined. Comparing the supercritical U+U spectrum with the subcritical U+Pb spectrum, we can see evidence of small deviations in the
U+U system in the low energy part of the spectrum.
However, preliminary results of a repeat of the experiment performed by the some group indicate that
this effect could not be reproduced (cf. fig.21).
In this context the following may be interesting.
It was reported recently[43] that, for the U+Cm scattering system at 5.8 MeV/u, the low energy part of the
atomic positron spectrum ($E_e+<500$ keV) increases more

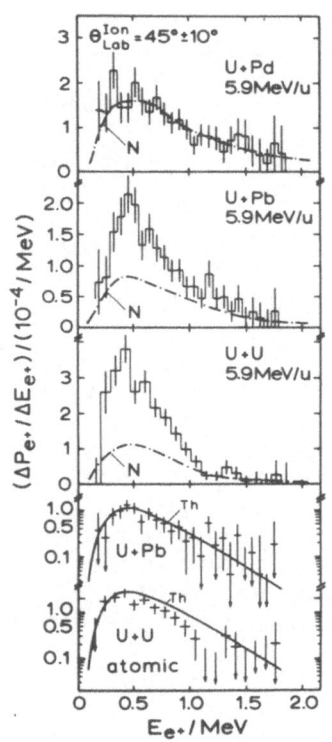

Fig. 16.

Positron spectra $\Delta P_{e+}/\Delta E_{e+}$ for (a) the nuclear U+Pd, (b) the subcritical U+Pb, (c) the supercritical U+U systems at a U beam energy of 5.9 MeV/u and $\theta_{lab}=45°±10°$. The dashed dotted curves "N" indicate nuclear positrons as derived from γ-ray spectra, assuming El multipolarity. (d) is a semilogarithmic presentation of the spectra of atomic positrons compared to theoretical calculations[37] indicated by "Th" (taken from ref. 28).

rapidly than the high energy part ($E_{e+}>600$ keV) when the minimum distance of closest approach decreases to near $R_o=22$ fm. It is especially interesting that this effect occurs as well at 4.3 MeV/u where the distance of closest approach is greater than 30 fm, while it is present neither in the U+Pb collision system nor in the U+Cm collision system for which the R_o values exclude overcritical binding.

Returning to the atomic positron spectra shown in fig.16d in a semilogarithmic representation, we

see that there is good qualitative agreement with the
dynamical model discussed above. We observe, as expected, the high energy exponential tails and the drop
off at low energies caused by the Coulomb repulsion
of the positrons from the nuclear charges.

Finally, we compare the positron production probabilities with theoretical calculations[37] based on
a semiclassical collision model. In this model, pair
creation induced by the time changing monopole Coulomb
field is determined by a coupled channel calculation.
The functional form of the results of these calculations
as shown in fig.16 and 14 agrees quite well with the
experiment. It is worth noting that these calculations
predict no significant effect from a diving $1s\sigma$ level.
Obviously the diving time is too short when the colliding nuclei follow pure Rutherford trajectories. There
is some indication that the theoretical calculations
overestimate the positron production probability, but
it should be stressed that the absolute experimental
errors are at present about 30%. However, if the deviations persist with improved experimental results, further
investigation will be necessary. Among the possible
causes of such deviations could be the absence of level
diving, the presence of nuclear delay time effects,
or that the Fermi level for the filled quasiatomic
states before collision is higher than assumed in the
theoretical calculations.

After this detailed discussion of gross properties
of positron production I will turn to the question:
Is there perhaps some fine structure in the positron
spectra? Fig.17 presents positron spectra resulting
from bombardment of 1 mg/cm thick ^{238}U foils with 5.9
MeV/u ^{238}U ions[44] as recorded by the orange spectrometer set-up shown in fig.4. Nuclear background, as
determined from the simultaneously measured γ-spectra
has <u>not</u> been subtracted. The positron spectra in coincidence with ions scattered to angles smaller than
37° are broad and mostly structureless, with maxima
between 450 and 500 keV positron energy and exponentially decaying high energy tails. However, the positron
spectrum in coincidence with ions scattered between
41.9° and 48.5° in the laboratory system shows a "line"
like structure peaked at about 370 keV and possibly
weaker ones around 720 and 950 keV. A subdivision of
the positron spectrum in the momentum band around 370
keV into six parts using the positron information from
the proportional counter confirms the existence of
a relatively sharp "line" with a width of 90 keV and
an energy integrated probability of about 10^{-5}.

Recently, the orange-spectrometer has acquired

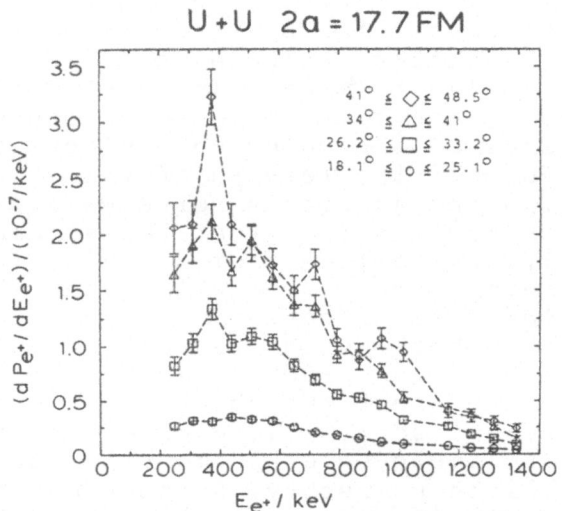

Fig. 17.   Positron production probability dP_{e^+}/dE_{e^+} per keV
            as a function of the positron kinetic energy
            in the c.m. system for (U+U)-collisions at
            5.9 MeV/u bombarding energy in coincidence
            with ions scattered into 4 angular ranges
            indicated (taken from ref. 44).

new U+U data. The momentum resolution has been improved
to $\Delta p/p \approx 3\%$ by making use of the positron sensitive
proportional detector for positrons[24,45]. Fig.18 shows
such a spectrum taken at a bombarding energy of 5.73
MeV/u. One notes a remarkable structure on this spectrum
with maxima at about 320 keV, 520 keV and possibly
700 keV; there may be another maximum at lower energies.
        With the new solenoid system EPOS shown in fig.8
the U+U scattering system was also investigated[30].
Fig.19 shows the total positron spectra observed for
the two systems $^{238}U + ^{238}U$ and $^{238}U + ^{248}Cm$ at a ^{238}U
ion energy of 5.9 MeV/u. The spectra are integrated

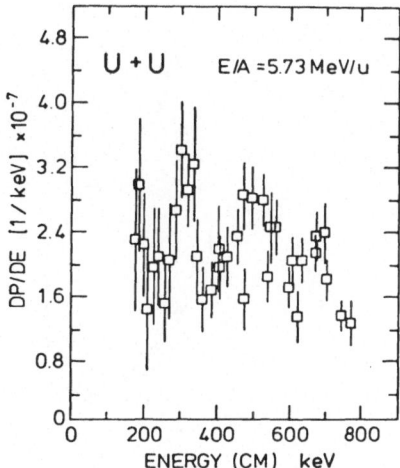

Fig. 18.   Positron spectra in coincidence with particles
           scattered between 41.5° and 50.3° in the
           laboratory system. The momentum resolution
           was Δp/p≈3% (taken from ref. 45).

over all observed scattering angles (20°≤θ_1≤70°)   and
over all elastic and quasieleastic two-body events.
These spectra are not corrected for nuclear e[+] contributions nor for the solenoid transport-efficiency
which partly determines the shape of the spectra at
the lowest and highest e[+]- energies. These spectra
seem to also indicate the possibility of peak-like
structures. This possibility led the EPOS group to
a detailed analysis of the positron spectra under special
kinematic conditions. In fig.20 the result for the

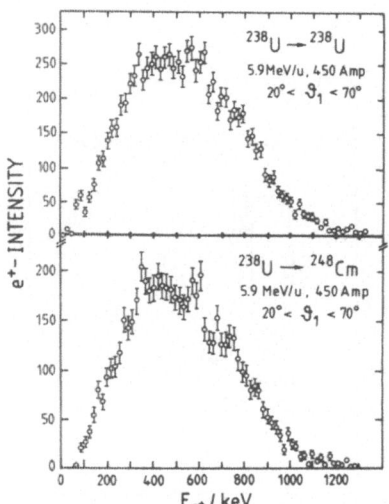

Fig. 19. e$^+$-energy spectra for the system ^{238}U+^{238}U and
^{238}U+^{248}Cm at 5.9 MeV/u projectile energy
integrated over particle scattering angles
between 20° and 70°. The spectra are not
corrected for the solenoid transport efficiency
and for the line-shape of the Si(Li)-detector.
The contribution of nuclear e$^+$-production
is not subtracted (taken from ref. 30).

^{238}U +^{238}U measurement at 5.9 MeV is shown. The spectrum was obtained by selecting the time of flight difference and the angle pulse-height corresponding to
binary events, by subtracting random events, and by
requiring $11° \leq |\Delta\theta| \leq 19°$ and $89.2° \leq \Sigma\theta \leq 89.8°$, where $\Delta\theta = \theta_1 - \theta_2$
and $\Sigma\theta = \theta_1 + \theta_2$. If elastic scattering is assumed, the
kinematic window on $\Delta\theta$ corresponds to selecting events
with $35.5° \leq \theta_1 \leq 39.5°$ or $50.5° \leq \theta_1 \leq 54.5°$. These two θ-regions
can be considered together because of the symmetry
of the experimental set-up around $\Delta\theta = 0°$. The window

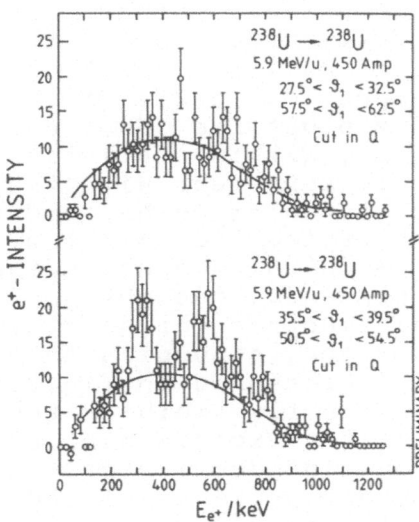

Fig. 20. Two selected e$^+$-energy spectra observed in
^{238}U+^{238}U-collisions at 5.9 MeV/u and kinematic conditions as explained in the text.
The spectra are not corrected for the solenoid
transport-efficiency, the Doppler-broadening
of the Si(Li)-detector line-shape and nuclear
background. The solid lines represent the
theoretical spectra normalized to the
upper spectrum (taken from ref. 30).

on $\Sigma\theta$ corresponds to slightly inelastic events (note
that $\Sigma\theta=90°$ for elastic scattering), and the selected
width is equivalent to a Q-value window of about $\Delta Q=15$
MeV (assuming inelastic scattering with no mass-trans-
fer). With these conditions, two peak-like structures
centered around positron energies of ∿320 keV and ∿590
keV are observed. In connection with the rather narrow
Q-window amounting to only approximately one third
of the experimental Q-resolution, it should be noted

that these structures are still present, although less
pronounced, if a very broad θ-window of approximately
100 MeV centered around the maximum of the observed
$\Sigma\theta$-distribution is selected. The structures are obviously
not produced by the Q-cut but rather are found to be
related to events of slight inelasticity.

Fig.20a shows the positron spectrum obtained for
$25°\leq|\Delta\theta|\leq35°$ (corresponding to $27.5°\leq\theta_1\leq32.5°$ and
$57.5°\leq\theta_1\leq62.5°$) under otherwise identical conditions
to the data in fig.20b. Within statistics, the shape
of this spectrum is in agreement with the bell-shaped
curve one expects by assuming Rutherford trajectories
(solid curve). By adjusting the height of this spectrum
to the experimental spectrum and using the theoretical
impact parameter dependence, one obtains the corresponding theoretical spectrum for scattering angles
satisfying $11°\leq\Delta\theta\leq19°$ (corresponding to $35.5°\leq\theta_1\leq31.5°$
and $50.5°\leq\theta_1\leq54.5°$) represented in fig.20b by the solid
line. The comparison clearly exhibits the additional
structure observed in the experimental spectrum for
$11°\leq|\Delta\theta|\leq19°$. This $\Delta\theta$-dependence itself excludes the
possibility that the structure is caused by instrumental
effects connected with the e^+-detection. (However,
I would suggest suggest a careful investigation of
the possibility that the lines may be caused by ns-delayed
conversion electron-target γ-ray coincidences. For
electrons, emitted out of the solenoid axis the spiral
baffle should have "leaks". The electron could therefore
be detected in the Si(Li) detector and, at the same
time, a high energy γ-ray from the target could be
detected in the NaI-ring crystal. A large $\Delta\theta$-dependence
should exist for such an effect).

Now the exciting possibility exists that two groups,
working completely independently with completely different experimental set-ups, have observed structures
at about the same energies in the $^{238}U + ^{238}U$ scattering
system. (The energies are $E_{e^+}=320$ keV for both groups
and $E_{e^+}=520$ keV for the orange-spectrometer group and
$E_{e^+}=590$ keV for EPOS.) It should be noted, however,
that the EPOS data were taken at 5.9 MeV/u while the
orange-spectrometer data were taken at 5.73 MeV/u.
It should also be noted that a third group, working
with the old solenoid spectrometer, has not seen such
structures. I would like now to discuss the possible
reasons of peak-structures in the positron spectra.
I think that, at the moment, experiments are just beginning to establish the existence of such structures
and no final conclusions can be drawn. However, I will
assume that these structures are not caused by some

unknown background effects in the apparatus and that
in future experiments they will be clearly reproducable.
The reproducibility itself is a difficult problem.
It was pointed out by the orange-spectrometer group[24,45]
that the observed positron line pattern shifts markedly with small changes of the bombarding energy. These
shifts impose a very severe requirement on maintaining
a constant effective bombarding energy for otherwise
structures may smear out. Such an effect could probably
explain why the evidence of a possible structure first
seen by the group working with the old solenoid spectrometer was not reproducable. The spectra as shown in
fig. 16 were taken with a 200 $\mu g/m^2$ sputtered ^{238}U
target but for the later experiments evaporated targets
with a thickness of about 500 $\mu q/m^2$ were used. No special
attention was paid to deterioration of these targets
during heavy ion bombardment.

   If the structures in the positron spectra have
a strong dependence on the beam energy and the scattering angle, then this could indicate nuclear transfer
reactions. In particular, the pair productions in the
E0-transitions can not be ruled out by inspection of
the target γ-ray spectrum. For E0-transitions a continuum positron distribution is expected with a sharp
cut-off at the threshold energy. Such a cut-off could
easily be misinterpreted as a line if statistics are
inadequate. If future experiments (as, for example,
the investigation of conversion electron spectra[46])
rule out the possibilities of a nuclear origin to the
structure in the positron spectra, one could think
about interference phenomena in positron creation.
This possibility has been discussed[24]. It is suggested
that there should be an interference of the transition
amplitudes for spontaneous positron creation between
the incoming and the outgoing parts of the collision.
The interference effect of spontaneous positron decay
from the entrance and exit channel may be enhanced
by additional interference with the amplitudes of induced pair-creation, but this by itself could not cause
a line structure.

   It might also be possible that nuclear and atomic
levels mix in the quasiatom and cause additional interference effects.It has also been suggested[9,30,47] that
the peaks signify the production of spontaneous positrons. If the times for which the collision systems
remain critical are prolonged due to nuclear interactions, an enhancement of the spontaneous e^+-produc-
tion probability is expected at focused positron kinetic energies[19], although rather long and unexpected
time delays of up to 10^{-20} sec would be required to

explain the observed line-width of the structure in
this case.

To check this idea, the group working with the
old solenoid spectrometer (fig.5) searched for sig-
natures of a time delay due to nuclear contact in deep
inelastic collisions of the $^{238}U + ^{238}U$ system[48]. These
reactions were distinguished from quasielastic collisions by the occurence of fission in at least one
of the reaction partners. In a plastic scintillator
ring counter accepting scattering angles $35° < \theta < 55°$,
the full energy peak from quasielastic events allowed
a separation from the broad distribution of fission
reactions. At projectile energies of 5.9, 7.5 and 8.4
MeV/u the quasieleastic contributions varied from about
100%, to 30%, to 0%, respectively.

It is well-known that the separation of quasiatomic and nuclear positrons is a major problem. The
technique applied was a measurement of γ-rays in coincidence with reaction products, a conversion of this
γ-ray spectrum into a positron spectrum with theoretical pair conversion coefficients, and a comparison
of the resulting spectrum with the U+Pd system for
which only nuclear positrons are expected. In recent
experiments this technique was complemented by a measurement of electron spectra coincident with the positrons. It turned out that their high energy tails are
of nuclear origin and this supports the correctness
of the γ-ray procedure.

A severe problem was presented by γ-rays produced
by neutrons in the solenoid material. These γ-rays
could be disentangled by observing their time delay
with regard to prompt nuclear γ-rays. Preliminary positron spectra corrected for background and detection
efficiency, are shown in fig.21.

With the exception of a possible fluctuation at
750 keV in the spectrum taken at 7.5 MeV/u, no statistically significant structures have been observed
in this experiment. Theoretical calculations on the
basis of pure Rutherford trajectories are also shown
in fig.21. For U+U at 5.9 MeV/u, the calculations over-
estimate the production by a factor of (1.5±0.3), but
reproduce the shape of the observed spectrum. Comparing all three spectra shown, one can see a remarkable
feature. Above an energy of about 600 keV all spectra drop off with the same shape, while from calculations based on pure Rutherford trajectories this
is not expected. This effect is in accord with a small
time delay of only about 10^{-21} s associated with
changed kinematical conditions due to energy dissipation in the outgoing channel. A time delay of this

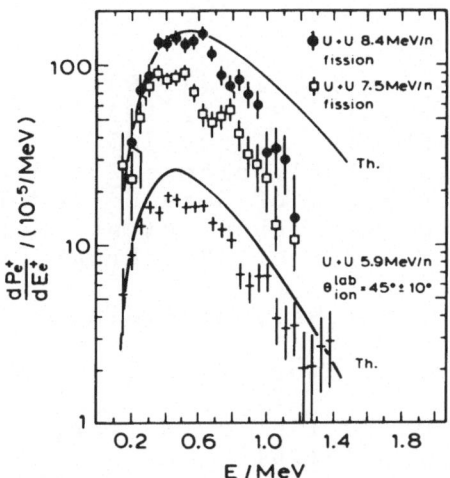

Fig. 21. Preliminary atomic positron spectra for U+U
          scattering systems. Spectra at 8.4
          and 7.5 MeV/u have been taken in coincidence
          with fission fragments. Pure Rutherford
          trajectories have been assumed in theoretical
          calculations Th. with θ_{CM}=85.2° at
          8.4 MeV/u and θ_{CM}=90.0°±20° at 5.9 MeV/u
          (taken from ref. 47).

order of magnitude is expected also from various theoretical models[49,50,51,52]. It can therefore be concluded that a time delay of about 10^{-20}s (or even more)
in a ^{238}U + ^{238}U collision with nuclear contact but
without any significant energy dissipation (i.e. no
fission!) would be a very great surprise.

Acknowledgements
        I would like to acknowledge fruitfull discussions

with P.Armbruster, H.Bokemeyer, F.Bosch, J.S.Green-
berg, E.Kankeleit, P.Kienle, W.Koenig, and C.Kozhuharov
as well as with W.Greiner, B.Müller, J.Reinhardt, and
G.Soff. I thank R.Moore for his efforts to make my
English understandable.

REFERENCES

1. O. Klein, Z.Physik 53, 157(1929)
2. F. Sauter, Z. Physik 69, 742(1931)
3. F. Sauter, Z. Physik 73, 547(1932)
4. J.R. Nix, Ann.Rev.Nucl.Science 22, 65(1972)
5. G. Herrmann, Yearbook of Science and Technology 1980,
   Mc Graw-Hill and references cited therein
6. J. Reinhardt, W. Greiner, Rep.Progr.Phys. 40, 219
   (1977)
7. J. Rafelski, L. Fulcher, A. Klein, Phys.Rep. 38C,
   227(1978)
8. I. Pomeranchuk, J. Smorodinsky, Journ. of Phys. IX,
   97(1945)
9. J. Reinhardt, U. Müller, B. Müller, W. Greiner,
   Z. Physik A303, 173(1981)
10. H. Backe, in: Trends in Physics 1978, ed. M.M.Woolf-
   son, Adam Hilger Ltd., Bristol 1979, p. 445
11. E. Kankeleit, in: Nuclear Interactions, ed. by
   B.A. Robson, Lecture Notes in Physics, Vol.92,
   Springer Verlag, New York 1979, p. 306
12. H. Bokemeyer, Atomic Physics with Very Heavy Ions,
   GSI-Report 79-4 and Heavy Ion Physics, eds.: A. Be-
   rinde, V.Ceausescu, I.A. Dorobantu, Proc. Predeal
   International School 1978, p. 489
13. P. Kienle, Progress in Particle and Nuclear Physics
   Vol.4, Heavy Ion Interactions p., ed.: D.Wilkinson,
   Plenum Press 1980
14. E. Kankeleit; in Proceedings of the 12th Summer
   School of Nuclear Physics, Mikolajki, Poland, 1979,
   Nukleonika 25, 253(1980)
15. D. Liesen, P.Armbruster, H.-H. Behnke, F. Bosch,
   S. Hagmann, P.H. Mokler, H. Schmidt-Böcking, R. Schuch,
   Electronic and Atomic Collisions, eds.: N. Oda
   and N. Takayanagi, North Holland Publ. Comp., Amster-
   dam 1980, p. 337.
16. J.S. Greenberg, Electronic and Atomic Collisions,
   p.351, eds.: N. Oda, T. Takayanagi, North Holland
   Publ. Comp., Amsterdam 1980
17. P. Kienle, Atomic Physics 7, p. 1, Plenum Press,
   New York 1981, eds.: D.Kleppner, F.M.Pipkin
18. C. Kozhuharov, Proceedings of the NATO Advanced
   Studies Institute "Quantumelectrodynamics of Strong
   Fields", Lahnstein 1981, Plenum Press, to be published

19. D. Liesen. P. Armbruster, F. Bosch, S. Hagmann,
    P.H. Mokler, H.J. Wollersheim, H. Schmidt-Böcking,
    R. Schuch, J.B. Wilhelmy, Phys.Rev.Lett. 44, 983(1980)
20. W.E. Meyerhof, R. Anholt, Y. El Masri, D. Cline,
    F.S. Stephens, R. Diamond, Phys.Lett. B69, 41 (1977)
21. H. Backe, L. Handschug, F. Hessberger, E. Kankeleit,
    L. Richter, F. Weik, R. Willwater, H. Bokemeyer,
    P. Vincent, Y. Nakayama, J.S. Greenberg, Phys.Rev.Lett.
    40, 1443 (1978)
22. G. Kozhuharov, P. Kienle, E. Berdermann, H. Bokemeyer,
    J.S. Greenberg, Y. Nakayama, P. Vincent, H. Backe,
    L. Handschug, E. Kankeleit, Phys.Rev.Lett. 42,
    376 (1979)
23. E. Berdermann, F. Bosch, M. Clemente, F. Güttner,
    P. Kienle, W. Koenig, C. Kozhuharov, B. Martin,
    W. Potzel, E. Povh, C. Tsertos, W. Wagner, Th. Walcher,
    GSI Scient. Rep. 80-3, p. 103
24. P. Kienle, Proceedings of the NATO Advanced Studies
    Institute "Quantumelectrodynamics of Strong Fields",
    Lahnstein 1981, Plenum Press, to be published
25. E. Moll, E. Kankeleit, Nukleonik 7, 180 (1965)
26. L. Handschug, H. Backe, H. Bokemeyer, NIM 161,
    117 (1979)
27. H. Backe, W. Bonin, E. Kankeleit, W. Patzner, P. Sen-
    ger, F. Weik, GSI Scient.Rep. 80-3, 168 (1980)
28. H. Backe, W. Bonin, E. Kankeleit, M. Krämer, K. Krieg,
    V. Metag, P. Senger, N. Trautmann, F. Weik, J.B. Wil-
    helmy, Proceedings of the NATO Advanced Studies
    Institute "Quantumelectrodynamics of Strong Fields",
    Lahnstein 1981, Plenum Press, to be published
29. A. Balanda, H.J. Beeskow, K. Bethge, H. Bokemeyer,
    H. Folger, J.S. Greenberg, H. Grein, A. Gruppe,
    S. Ito, S. Matsuki, R. Schulte, R. Schultz, D.
    Schwalm, J. Schweppe, R. Steiner, P. Vincent, M.
    Waldschmidt, GSI Scient. Rep. 80-3, p. 161
30. H. Bokemeyer, K. Bethge, H. Folger, J.S. Greenberg,
    H. Grein, A. Gruppe, S, Ito, R. Schule, D. Schwalm,
    J. Schweppe, N. Trautmann, P. Vincent, M. Waldschmidt,
    Proceedings of the NATO Advanced Studies Institute
    "Quantumelectrodynamics of Strong Fields", Lahnstein
    1981, Plenum Press, to be published
31. E. Kankeleit, R. Köhler, M. Kollatz, M. Krämer,
    R. Krieg, P. Senger, H. Backe, GSI Scient. Rep.
    81-2,p.195 (1981)
32. E. Kankeleit, R. Köhler, M. Kollatz, M. Krämer,
    R. Krieg, U. Meyer, H. Oeschler, P. Senger, GSI
    Scient. Rep. 82-1, p. 231 (1982)
33. H. Backe, Lecture Notes in Physics, Bd. 143, Present
    Status and Aims of Quantum Electrodynamics, Springer
    Verlag Berlin-Heidelberg-New York, 1981, p. 277

34. P. Schlüter, G. Soff, W. Greiner, Z.Physik A 286,
    149 (1978)
35. P. Schlüter, G. Soff, ADANDT 24, 509 (1979)
36. P. Schlüter, G. Soff, W. Greiner, Phys.Rep. 75(6),
    329 (1981)
37. J. Reinhardt, B. Müller, W. Greiner, Phys.Rev. A24,
    103 (1981)
38. J. Bang, J.M. Hansteen, Kgl.Dan.Vid.Sesk., Mat.Fys.
    Medd. 31, no 13 (1959)
39. J. Bang, J.M. Hansteen, Phys.Lett 72A, 218 (1979)
40. J. Bang, J.M. Hansteen, Physica Scripta 22, 609 (1981)
41. F. Bosch, D. Liesen, P. Armbruster, D. Maor, P.H.
    Mokler, H. Schmidt-Böcking, R. Schuch, Z.Physik
    A296, 11 (1980)
42. P. Armbruster, P. Kienle, Z.Physik A 291, 399 (1979)
43. H. Bokemeyer, H. Folger, H. Grein, S. Ito, D. Schwalm,
    P. Vincent, K. Bethge, A. Gruppe, R. Schulé, M. Wald-
    schmidt, J.S. Greenberg, J. Schweppe, N. Trautmann,
    GSI Scient. Rep. 81-2, 127 (1981)
44. E. Berdermann, F. Bosch, M. Clemente, F. Güttner,
    P. Kienle, W. Koenig, C. Kozhuharov, B. Martin,
    B. Povh, H. Tsertos, W. Wagner, Th. Walcher, GSI
    Scient.Rep. 80-3, 128 (1981)
45. E. Berdermann, F. Bosch, M. Clemente, P. Kienle,
    W. Koenig, C. Kozhuharov, H. Tsertos, W. Wagner,
    GSI Scient.Rep. 82-1, 138 (1982)
46. G. Soff, P. Schlüter, W. Greiner, Z.Physik A 303,
    189 (1981)
47. J. Rafelski, B. Müller, W. Greiner, Z.Physik A 285,
    49 (1978)
48. H. Backe, W. Bonin, E. Kankeleit, M. Krämer, R. Krieg,
    V. Metag, P. Senger, J.B. Wilhelmy, GSI Scient. Rep.
    82-1, 140 (1982)
49. G. Wolschin, Nukleonika 22, 1165 (1977)
50. J. Blocki, J. Randrup, W.J. Swiatecki, C.F. Tsang,
    Ann.Phys. 105, 427 (1977)
51. D.H.E. Gross, H. Kalinowski, Phys.Rep. 45, 175 (1978)
52. R. Schmidt, V.D. Toneev, G. Wolschin, Nucl.Phys.
    A 311, 247 (1978)

Supported in part by the Bundesministerium für Forschung
und Technologie (BMFT) and Gesellschaft für Schwerionenforschung (GSI) Darmstadt.

INNER SHELL ELECTRONS IN SUPERHEAVY QUASIATOMS

Fritz Bosch*

Gesellschaft für Schwerionenforschung mbH

D-6100 Darmstadt, Fed. Rep. of Germany

1. INTRODUCTION

In spring 1976, when the heavy ion accelerator UNILAC, developed by Christoph Schmelzer, worked successfully, it became possible for the first time to accelerate *all* ions up to several millions of electronvolts per nukleon and, thus, to reach the Coulombbarrier even of the heaviest target atoms. An old physicists' dream, to investigate nuclear reactions for all elements of the periodic system, to understand fusion, fission, and nuclear structure upon a broad experimental base, was converted into reality and - at least in principle - also the door was opened to the unknown realm of superheavy elements.

Nature, however, defending very obstinately this still uncultivated garden, has not revealed up to now more than element $Z = 107$ despite all the continuous strain of physicists. Much easier, however, seems to be the access to superheavy atoms, when looking not onto the nuclei but onto the electron clouds rather of a heavy ion-atom collision system.

Let us consider a central collision of 5 MeV/N Pb ions onto a Cm-target - that is scarely below the Coulombbarrier (Fig. 1): The smallest internuclear distance R in such an encounter amounts to approximately 20 fm. Looking now onto the innermost electrons of Cm and Pb one finds :

(i) At internuclear distances R, which are still very large as compared to the mean 1s-radius $<r>_{1s}$ (for Cm $<r>_{1s}$ is ≈ 600 fm), the inner shells are still completely separated;

binding energies, wave functions, and quantum numbers follow
from the Dirac equation of the *separated* systems Cm and Pb,
respectively.

(ii)   If R becomes comparable to $<r>_{1s}$, one already needs a *molecular* picture. Now, the inner electrons have a nonzero
probability distribution at both central charges Z_i; the
stronger bound 1s electrons of Cm populate the deepest
molecular 1sσ level, those of Pb the nearest-lying 2pσ niveau.
The binding energies increase appreciably due to the enhanced
effective charge.

(iii)   Finally, when R $<<$ $<r>_{1s}$ (e.g. at R = 20 fm), the 1sσ and
2pσ electrons "see" the practically *combined* charge
Z = Z_1+Z_2 = 178 with $\alpha Z > 1$ ($\alpha = e^2/\hbar c$ = fine structure
constant). Their orbitals can be described asymptotically
by the quantum numbers of a *superheavy atom* with charge
Z = 178.

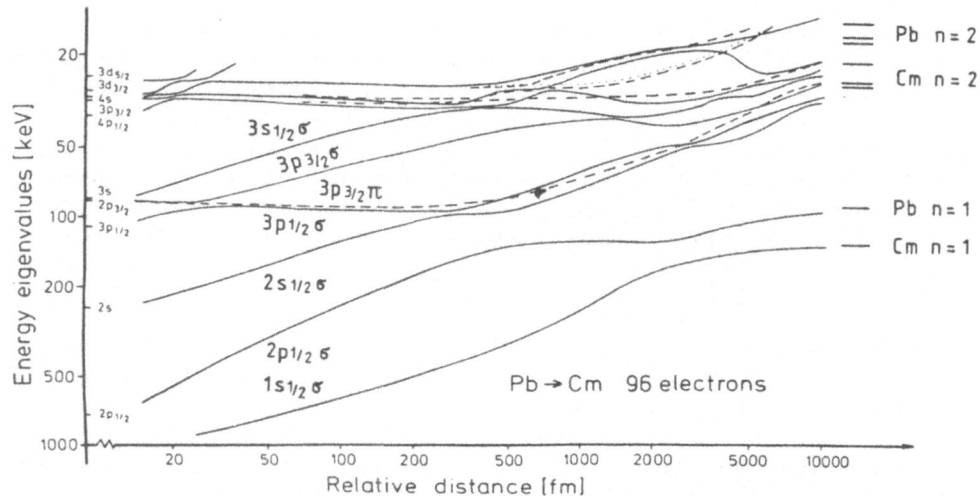

Fig. 1.   Correlation diagram for the system Pb+Cm (Z = 178) as a
function of the internuclear distance R according to
B. Fricke et al.[1] The curves are the result of a relativistic selfconsistent Dirac-Fock-Slater calculation.

    Hence, for a short time a superheavy quasiatomic system with
$\alpha Z > 1$ is built in such a collision supposed, the relative velocity
of the nuclei is, on one hand, sufficient to bring them closely together, but, one the other hand, distinctly smaller than the Bohr
velocity of the 1s electrons, so that these can adjust themselves
adiabatically onto the time changing electromagnetic field.

These fields, extended over atomic dimensions, reach in close collisions of heavy atoms an order of magnitude which is nearly incredible: The electric field strength $|\vec{E}| \gtrsim 10^{19}$ Volt/Meter, the magnetic induction $|\vec{B}| \gtrsim 10^{12}$ Tesla. Their Fourier frequencies can now take the electrons into empty bound states or into the continuum. A measurement of their excitation probability as a function of the internuclear distance and a spectroscopy of the emitted electrons then could reveal immediately information about the momentum distribution of electrons bound in a superheavy atom – supposed it would become possible to investigate these fleeting, collision created things during their hasty existence. Hence, the decisive question we have to clear up is how transitory superheavy quasiatoms can become the objects of real physical investigation and not only of "gedankenexperiments". We shall discuss this point extensively in section 4.

First of all, some of the quite fascinating predictions of theory concerning atoms $\alpha Z > 1$ will be presented shortly (section 2) – Gerhard Soff will explain them in detail in his talk –, then the experimentally accessible "messengers" of superheavy atoms will be enumerated (section 3), and, finally, both the experimental methods and results are described together with all the questions not yet answered (section 4).

2.   SOME PREDICTIONS OF THEORY FOR ATOMS $\alpha Z > 1$

For atoms with pointlike nuclei the electromagnetic coupling strength can never overstep the magic value $\alpha Z = 1$: The binding energy of the 1s orbital becomes complex for $\alpha Z > 1$, according to the Sommerfeld formula :

$$E_{1s} = m_0 c^2 \sqrt{1-(\alpha Z)^2} \qquad (1)$$

Going over to a more realistic extended nucleus, this singularity is removed, nevertheless, for $Z \gtrsim 150$ the 1s binding energy is predicted[2] to reach the rest mass $m_0 c^2$ of the electron and, for $Z \gtrsim 173$, even ther fermi level of the negative continuum (Fig. 2).

A "naked" charge $Z = 173$ would bind, in a spontaneous "abiogenesis", two electrons from the negative continuum into its 1s-shell, leaving there two empty states, i.e. positrons. The vacuum would change from the neutral ground state into a charged one.[3] Quite similarly, a collision generated quasiatom with $Z > 173$ could fill a 1s-hole with an electron from the negative continuum without energy transfer and, thus, produce "spontaneously" a positron.

In accordance to the drastically increasing binding energy, theory predicts a pronounced shrinking of the 1s wave function in configuration space. For an atom $Z = 164$, the K-shell radius should

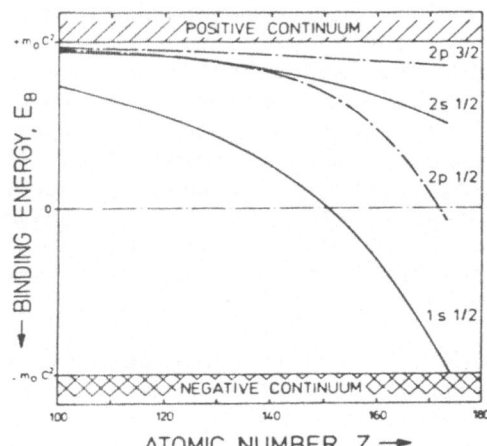

Fig. 2. Binding energies for existing
         (Z ≤ 107) and superheavy atoms
         as a function of the atomic
         number Z according to Ref. 2.
         At Z ≈ 150 the 1s orbital is
         predicted to reach the electron
         rest mass, and at Z ≈ 173 the
         negative continuum.

be less than 140 fm - the 1s radius of Pb, for comparison, is about
700 fm (Fig. 3) - that is, the mean value of the radial probability
distribution should fall considerably below the Compton wave length
λ_c ($\lambda_c = \hbar/mc = 386$ fm) of the electron.

In the thirties, shortly after the discovery of the Dirac
equation and the assertion of Klein's paradoxon, these predictions
were discussed violently. Born and Infeld proposed an additive
quadratic term of the electric field strength in the Maxwell
equations, which should lead to a finite maximum field strength
E_{MAX}. Meanwhile, however, this hypothesis was disproved unambiguously by several precision measurements of 1s binding energies in
heavy atoms. In the last decades a central point was whether the
interaction of the electron with the radiation field - which, of
course, is not incorporated into the one particle Dirac theory -

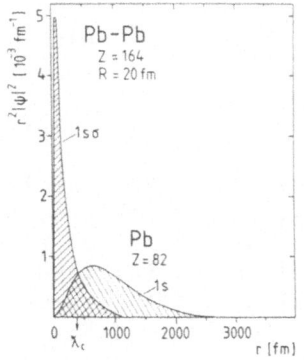

Fig. 3. Probability distribution of the 1s wave
function of Pb and of
the 1sσ orbital of the
Pb+Pb system at an
internuclear distance
R = 20 fm. The Pb+Pb
values were calculated
by G. Soff. λ_c indicates the Compton wave
length of the electron
(386 fm).

could prevent the drastic increase of binding energy and, thus, the
"diving" of the 1s level into the negative continuum. Very recent
calculations,[4] however, yielded a value of only 15 keV for the electron self energy at Z = 170, and for the vacuum polarisation a value
which even enhances the binding of the 1s electron.

At this present state of the art, we are interested especially
in three groups of problems, which probably could be elucidated by
experimental results :

(i)   Can the K-shell radius of a superheavy atom become signifi-
        cantly smaller than the electron Compton wave length ?

(ii)   Can the binding energy of an electron exceed its own rest
        mass ? If so, the electron would be the first known elemen-
        tary particle with such a property.

(iii)   Is the 1sσ orbital "diving" into the negative continuum for
        Z $\gtrsim$ 173 ?

We should mind, however, that the final answer on all these
questions lies on the knees of the gods, at least as far as we have
no *real* superheavy atoms at our disposal. For the time being, as
we are restricted to investigate collision induced quasiatoms, the
solution of the questions mentioned above, should rather be understood as an asymptotic goal. Before all, it is time to ask by which

experimental methods we hope to receive information about the
properties of superheavy quasiatoms.

3.   EXPERIMENTAL APPROACHES TO SUPERHEAVY QUASIATOMS

     For all collision systems discussed here, below the Coulomb
barrier, the Sommerfeld parameter κ

$$\kappa = \frac{2\ Z_1 Z_2 e^2}{\hbar v_p} \qquad (2)$$

(where the projectile velocity v_p is in the order of 2 % to 10 %
of the light velocity c) is much larger than 1, and, therefore, the
motion of the nuclei can be described by classical Rutherford trajectories. For a given projectile velocity, they are determined
unambiguously by the CM-scattering angle Θ_{CM}. The impact parameter
b and the smallest internuclear distance in the collision, R_0,
respectively, are connected with Θ_{CM} and the minimum distance 2a
in a head-on collision, by the relations of equation (3) :

$$b = a\ ctg\ (\Theta_{CM}/2);\quad R_0 = at\ \sqrt{a^2+b^2}$$
$$a = \frac{Z_1 Z_2 e^2}{2\ E_{CM}} \qquad (3)$$

     The most important atomic excitation processes during the
collision are shown in Fig. 4 :

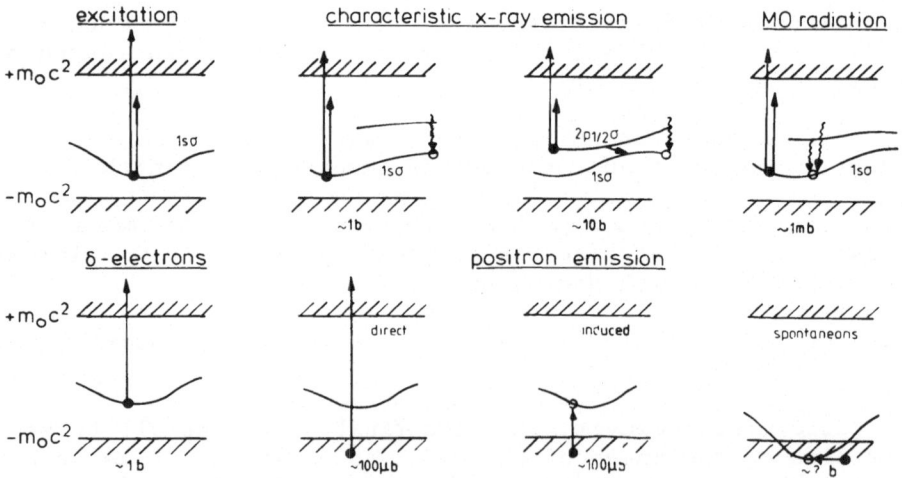

Fig. 4.   Atomic excitation processes induced by the time changing
          Coulomb field in a heavy ion-atom collision. The numbers
          indicate typical cross sections (b = barn).

(i) The time changing Coulombfield takes an electron from the
deepest lying 1sσ orbital into the continuum; the 1sσ-hole
asymptotically goes in the 1s shell of the *heavier* collision
partner, where it decays emitting characteristic K-x radiation. Similarly, a hole of the 2pσ-orbital finally becomes
a 1s-hole in the *lighter* atom.
When recording the particles scattered into an angle ϑ in
coincidence with characteristic K-radiation, one gets the
excitation probability of the 1sσ- and 2pσ-orbital, respectively, as a function of the impact parameter b according to
the relations of equation (3). A typical set-up for such an
experiment is shown in Fig. 5. Usually, the particles are
detected by a position sensitive, gasfilled parallel plate
avalanche counter, which even works at very high event
rates, but at the expense of a poor energy resolution.
Therefore, at large scattering angles, where a separation
of projectiles and recoils becomes indispensable, one has
to use either *two* counters or a surface barrier detector
with good energy resolution. The characteristic K-x rays of
both atoms are recorded by an intrinsic germanium diode. In
the third column of the first row of Fig. 4 the possibility
that a hole produced at first in the 2pσ-shell goes into
the 1sσ-orbital by virtue of dynamical coupling of the
molecular shells is indicated. The more symmetric the collision system is, the more this "vacancy sharing" grows due to
the narrowing gap between the shells, obscuring, thus, the
interpretation of the experiment. Hence, in order to obtain
a result as unambiguous as possible, one has to restrict to
rather asymmetric collision systems.

(ii) More detailed information is to be received by measuring
additionally the *energy distribution of the electrons* which
are kicked out from an inner shell into continuum states
(δ-electrons). A *coincidence* between δ-electrons and K-x
rays delivers the momentum distribution of the 1sσ and 2pσ
electrons, respectively, a *triple coincidence* (δ-electron,
K-x ray, scattered particle) their momentum distribution at
a fixed internuclear distance. Now, as we shall see and explain later, the previously bound electrons can reach an
energy of several hundred keV in the final state. Two detectors, which are especially well suited for the measurement
of such high energetic electrons, are shown in Figs. 6a and
6b. The first one, designed as an achromatic magnetic electron channel, separates a momentum byte of $\Delta p/p = 0.24$ and
focusses it onto a cooled Si(Li)-counter with an energy
resolution of 3 keV at 1 MeV. This spectrometer can be ro-
tated around the target, thus allowing a measurement of the
angular distribution of the δ-electrons. Due to the special
geometry, the electron spectrum is almost free of background
and, therefore, very reliable single spectra can be obtained.

Concerning coincidence measurements, however, the use of this
detector is very much restricted owing to its poor transmission (4 msr, and, in an improved version, 50 msr). More
appropriate for this purpose is the second detector shown in
Fig. 6b, an "orange"-β-spectrometer, developed by Moll and
Kankeleit: It produces by 60 current coils a toroidal magnetic field, which focusses - corresponding to the current
direction - electrons or positrons emitted from the target
onto a cone-shaped plastic detector. The "orange" connects
a variable momentum acceptance $\Delta p/p$, a complete suppression
of the "wrong" charge with a high transmission (≈ 1 sr).
Thus, the spectral distribution of the electrons or positrons
can be recorded in coincidence with both the scattered
particles and characteristic x rays.

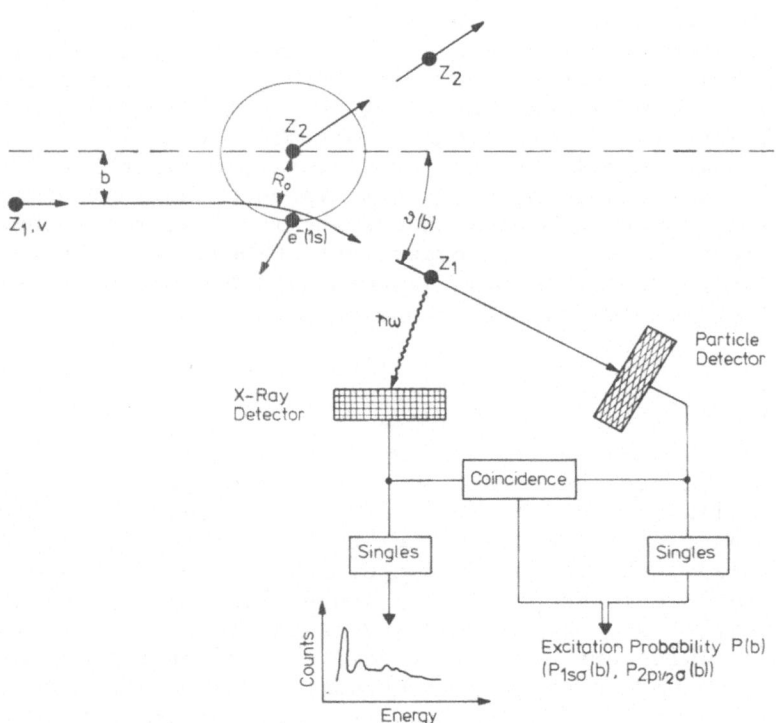

Fig. 5.   Experimental set-up for the determination of the $1s\sigma$ and
          $2p\sigma$ excitation probabilities by recording coincidences
          between scattered particles and characteristic K-x rays.
          Usually, the particles are detected with a position
          sensitive parallel plate counter, the K-x rays with a
          Ge(i) diode.

(iii)   If the time changing Coulombfield contains very high Fourier
        frequencies (Z ⪎ 160), *electrons of the negative continuum*
        can be excited into empty orbitals or the positive continuum,
        too. By this process, so called "induced" positrons are
        generated. If, finally, an electron of the negative continuum
        occupies an empty state of a "dived" 1sσ-orbital without
        energy transfer, a "spontaneous" positron is created. An un-
        ambiguous detection of such positrons would be the proof that
        the total energy of a strongly bound electron can become less
        than $-m_0c^2$. In principle, the experimental set-up for the
        detection of atomic positrons is the same as for δ-electrons;
        only the polarity of the magnetic field of the β-spectrometer
        has to be reversed. Both kinds of positrons shall be the ob-
        ject of Hartmut Backe's talk.

(iv)   Finally, it is indicated in Fig. 4 that an inner shell hole
        decays still during the existence of the collision system by
        the emission of quasimolecular x rays. Of course, the spectral
        distribution of this radiation contains the properties of the
        strongly bound electrons, too; however, for reasons of time
        I cannot discuss in detail these experiments concerning quasi-
        molecular radiation.

   Inner shell holes, electrons, and positrons, together with the
following x radiation, are the only "messengers" of the atomic processes arising in heavy ion collisions. Now, a quasimolecular system
is built for some 10^{-19} s, whereas the quasiatom, in its real
meaning, with a coupling strength $\alpha Z > 1$, exists only for times
$T_{QA} \gtrsim 10^{-20}$ s. (T_{QA} can be calculated from the condition $R << <r>_{1s}$).
Therefore, it seems completely mysterious, how one could get any
reliable experimental information within this extremely short time
where the quasiatom exists. Nature, however, is helping us for this
purpose with a simple, but far-reaching "trick", which we shall dis-
cover and discuss in the following section.

4.   THE "TRICK" OF NATURE

   Let us have a look to Figs. 7a and 7b. In part (a) we show an
x-ray spectrum recorded in the bombardment of a pollentube with a
2.5 MeV proton beam at the Heidelberg microprobe (beam width ⪎ 3 μm).
In part (b) the δ-electron spectrum of a Pb+Pb collision system is
to be seen. In a first and probably also in a second glance we cannot
find any reason to present these two pictures together. Nevertheless,
the underlying physics is quite similar and quite different at a
time, as we shall see in a third glance.

   Beside the characteristic K-x rays of the trace elements we
show in (a) a broad background stemming from the bremsstrahlung of
knocked out secondary electrons. This background ends rather sharply

$\Delta p/p = 12\%$

$\Delta \Omega = 4.1$ msr

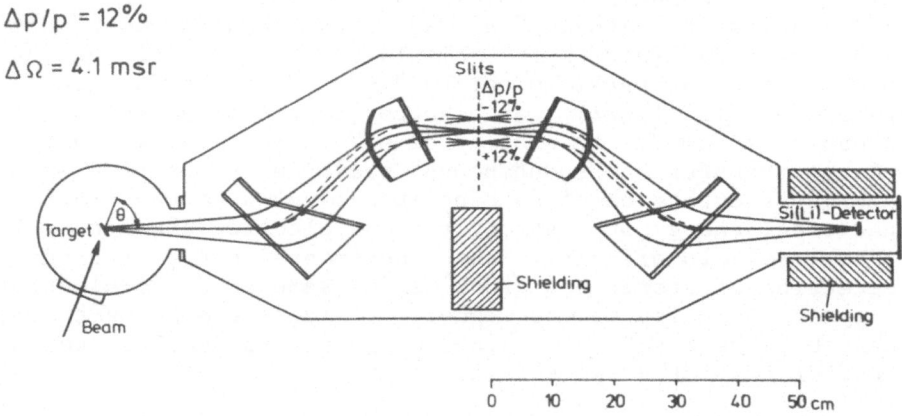

Fig. 6a.   Electron spectrometer consisting of an achromatic electron channel with four dipole magnets, which focusses
electrons from the target onto a cooled Si(Li) detector.
Momentum byte $\Delta p/p = 0.24$, solid angle $\Delta \Omega = 4$ msr. The
detector was constructed by B. Povh, K. Traxel, and
Th. Walcher at the MPI Heidelberg (Ref. 5).

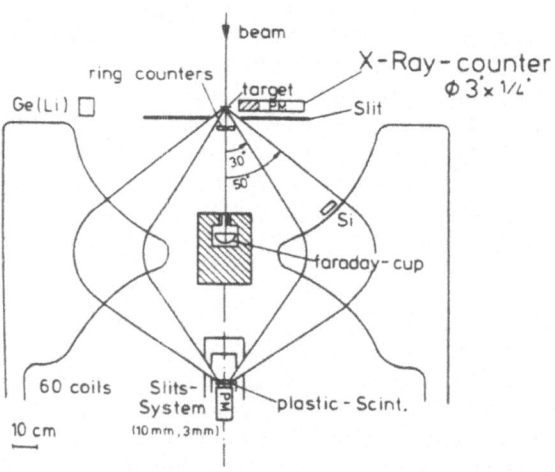

Fig. 6b.   Schematical picture of the iron free
"orange" β-spectrometer, developed by
Moll and Kankeleit. Momentum resolu-
tion $\Delta p/p \approx 0.08$ (can be changed by
entrance slits), transmission ≈ 1 sr.
The detector is used at the UNILAC
accelerator, Darmstadt (Ref. 5).

at 5 keV ($\hat{=}$ channel 200), which marks simultaneously the maximum energy of the secondary electrons. Classically, onto a *free* electron only an energy

$$E_{MAX} = 4 \frac{me}{mp} E_p \quad (\hat{=} 5 \text{ keV for 2.5 MeV p}) \qquad (4)$$

can be transferred.

In part (b) electrons of more than 1600 keV kinetic energy are measured with a cross section in the order of 10 μb/keV, that is more than 300 times the maximum energy of part (a).

Now, this difference has an explanation as easy as far-reaching: Considering the ionization of a *bound* electron with the aid of energy and momentum conservation only, we find for its initial momentum $\hbar\vec{k}_i$, momentum transfer $\hbar\vec{q}_0$, its final momentum $\hbar\vec{k}_f$, and energy transfer ΔE the relations:

$$\hbar\vec{k}_i + \hbar\vec{q}_0 = \hbar\vec{k}_f; \quad \hbar\vec{q}_0 = \frac{\Delta E}{v} \qquad (5)$$

(where the energy transfer $\Delta E = |E_B| + E_f$, with binding and final energy of the electron E_B and E_f, respectively).

Therefore, a δ-electron of 1600 keV kinetic energy got a momentum transfer of at least 16 MeV/c (at a projectile velocity v/c = 0.1). Compared therewith the final momentum $|\hbar\vec{k}_f| \, \stackrel{\sim}{} \, 2$ MeV/c is negligibly small, so that a value of about 14 MeV/c results for the initial momentum of the bound electron. This number again has to be compared with the Bohr-momentum of a Pb 1s-electron, for which we get the value

$$<\hbar\vec{k}_i>_{1s}^{Pb} \, \stackrel{\sim}{} \, 0.3 \text{ MeV/c} \qquad (6)$$

from the Virial-theorem.

Hence, high energetic δ-electrons necessarily have had an initial momentum almost two orders of magnitudes larger than the mean momentum in the Pb K-shell. They are, therefore, a "trigger" for the quasiatom, while such extremely high momentum components are not yet available in the separated atoms (imagine a Fourier-transformation of the Pb 1s and the Pb+Pb 1σ wave functions of Fig. 3 into the momentum space).

This property, being very clear for electrons of high final energy, surprisingly holds almost unchanged also for low energetic electrons. Supposed, we are looking to electrons ionized from *inner shells* of collision systems with a *high combined charge Z*. The reason for that is contained implicitly in equation (5): Looking to the relation between energy- and momentum transfer,

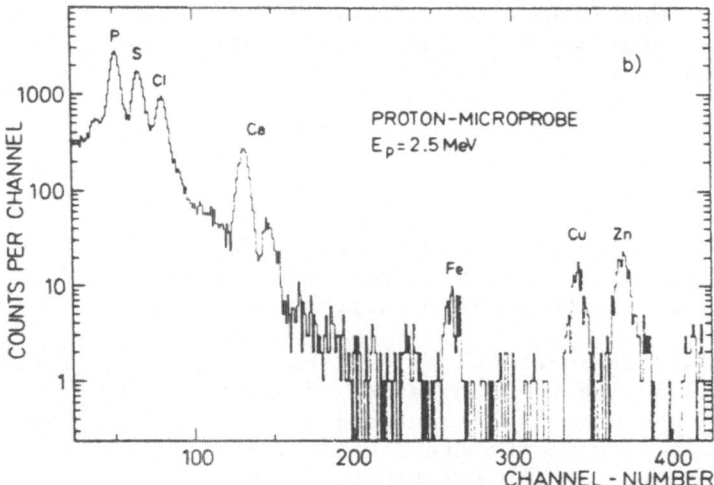

Fig. 7a.   X-ray spectrum of a pollentube recorded
           at the Heidelberg microprobe (Ref. 7)
           (beam with ≈ 3 μm). The broad background
           due to secondary electron bremsstrahlung
           sharply ends at channel 200 = 5 keV (Ref. 6).

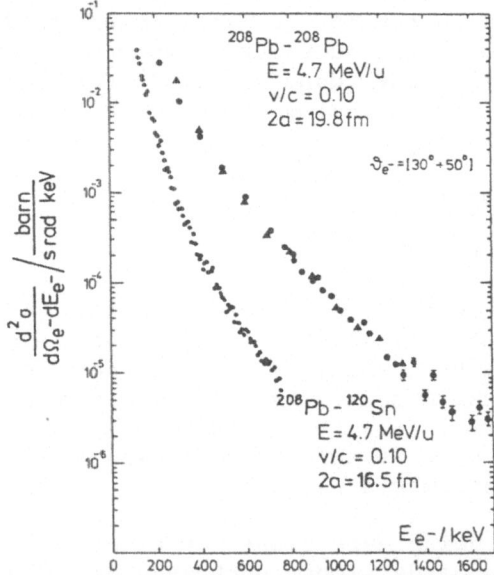

Fig. 7b.   δ-electron spectrum for the systems Pb+Pb
           and Pb+Sn recorded with the "orange"-β-
           spectrometer at the UNILAC, Darmstadt
           (Ref. 7).

ΔE and $\hbar\vec{q}_0$, respectively, keeping also in mind that for an inner electron of a very heavy collision system ΔE necessarily has to be in the order of some 100 keV, we can easily conclude that the needed momentum transfer reaches an order of several MeV/c. Now, for all practical cases, this number is much larger than the final electron momentum $|\hbar\vec{k}_f|$. Hence, the initial electron momentum has to be in the order of several MeV/c too, which, in turn, by far exceeds the mean K-shell momentum of any atom. These fundamental differences in the inner shell excitation of a "normal" ($\alpha Z \ll 1$) and a super-heavy ($\alpha Z > 1$) collision system are distinctly reflected in the experimental results presented in Fig. 8.

For the system Ne+Ni ($Z = 38$), for example, the momentum transfer onto a K-electron to be ionized is ≈ 0.4 MeV/c (for $E_f = 0$ and $v/c = 0.1$), the Bohr momentum of the K-shell ≈ 0.2 MeV/c, that is in the same order of magnitude. At an impact parameter b ≈ 1000 fm the $1s\sigma$ excitation probability $P_{1s\sigma}(b)$ grows steeply and remains practically constant up to very small b.

Now, this "excitation-threshold" at ≈ 1000 fm corresponds fairly well to the K-shell radius of an atom with $Z = 38$. Since the momentum transfer needed is not much larger than the Bohr momentum, one can find yet at internuclear distances in the order of the K-shell electrons with sufficient initial momentum, and, thus, K-shell excitation happens within a large impact parameter

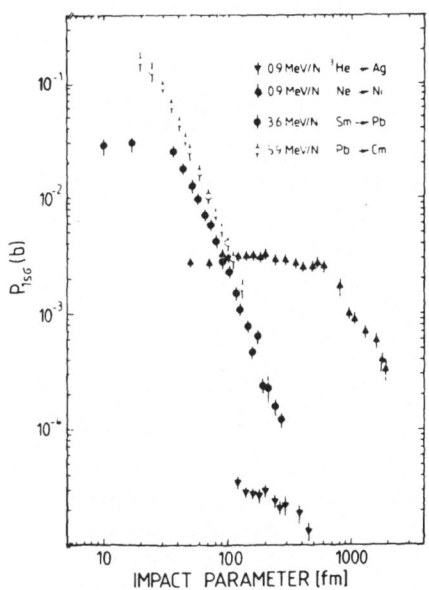

Fig. 8.
1s excitation probabilities $P_{1s\sigma}(b)$ as a function of the impact parameter b for collision systems $\alpha(Z_1+Z_2)$ = $\alpha Z \ll 1$ (^{3}He+Ag ($Z = 49$), Ne+Ni ($Z = 38$), and $\alpha Z > 1$ (Sm+Pb ($Z = 144$), Pb+Cm ($Z = 178$), revailing the principal difference between "normal" and superheavy collision systems. The measurements were performed at the MPI Heidelberg (Ref. 8) and at the UNILAC, Darmstadt (Ref. 9).

region, that is to say, during the hole collision time. In sharp
contrast therewith are the systems Sm+Pb (Z = 144), or Pb+Cm
(Z = 178). Noteworthy $1s\sigma$ excitation can be found for very small b
only (b $\lesssim$ 150 fm); the excitation probability here changes up to
a factor of 10 within some 70 fm, reaching a value of more than
10 % at small b. Once more the key for an understanding is the
relation of momentum transfer and Bohr momentum: $|\vec{\hbar q_0}| \approx 4$ MeV/c
(for E_f = 0, v/c = 0.1) $<\vec{\hbar k_i}>^{Pb}_{1s} \approx 0.5$ MeV/c (calculated for the
Pb K-shell).

The "trick" of Nature, which enables us to convert the investigation of superheavy quasiatoms from a gedankenexperiment to
a real one, can be summarized by the following statements:

(i)   In order to become ionized, the momenta of the innermost
         electrons have to be much larger than their mean momentum
         in the K-shell of the separated atoms.

(ii)   Such high momentum components are not available but next to
         the charge center of the combined system and for a very
         short time ($\lesssim 10^{-20}$ s).

(iii)   Hence, $1s\sigma$ excitation in systems $\alpha Z > 1$ does not happen but
         at very small internuclear distances R (R $<<$ $<r>_{1s}$), i.e.
         only as far as the innermost electrons feel the practically
         combined charge of both nuclei.

5.   RESULTS OF EXPERIMENTS

5.1   1sσ Excitation for αZ > 1

An excitation of a $1s\sigma$ or a $2p\sigma$-orbital leads to an exit
channel with a 1s vacancy in the heavier or the lighter atom,
which is filled by characteristic K-x radiation, as I have mentioned
previously. In an experiment, where coincidences between the
scattered particles and K-x rays are recorded, the measured quantities are, therefore,

$$P^H(\Theta) = \frac{N_c^H(\Theta)}{N(\Theta)\ \epsilon(K^H)} \quad \text{and} \quad P^L(\Theta) = \frac{N_c^L(\Theta)}{N(\Theta)\ \epsilon(K^L)} \tag{5}$$

where $N(\Theta)$ is the number of particles scattered into an angle (Θ),
$N_c^H(\Theta)$ and $N_c^L(\Theta)$ are the particles which are coincident with K_x
radiation of the heavier and lighter atom, respectively, and
$\epsilon(K^H)$ and $\epsilon(K^C)$ are the x-ray counter efficiencies at the corresponding energies.

$P^H(\Theta)$, however, is not yet the desired excitation probability
of the $1s\sigma$ orbital, $P_{1s\sigma}$. The reason is the already mentioned
"vacancy sharing" between the $2p\sigma$- and $1s\sigma$ orbitals. In order to
obtain $P_{1s\sigma}$, one has to correct P^H for this sharing according to

equation (6) :

$$P_{1s\sigma} = \frac{(1-w)P^H - wP_L}{1 - 2w} \qquad (6)$$

where w is the sharing probability $2p\sigma \to 1s\sigma$ in the model of Demkov and Meyerhof[10] For the Sm+Pb system we show in Fig. 9 at the left side the directly measured coincidence rates $P^H(\Theta)$ as a function of the impact parameter $b = a \, ctz \, (\Theta/2)$ and on the right side the same data but corrected for vacancy sharing. Although the sharing probability w is a few percent only for this rather asymmetric system ($\Delta Z = 20$); the correction is very pronounced at large b, whereas at small be ($b \lesssim 100$) it may be neglected.

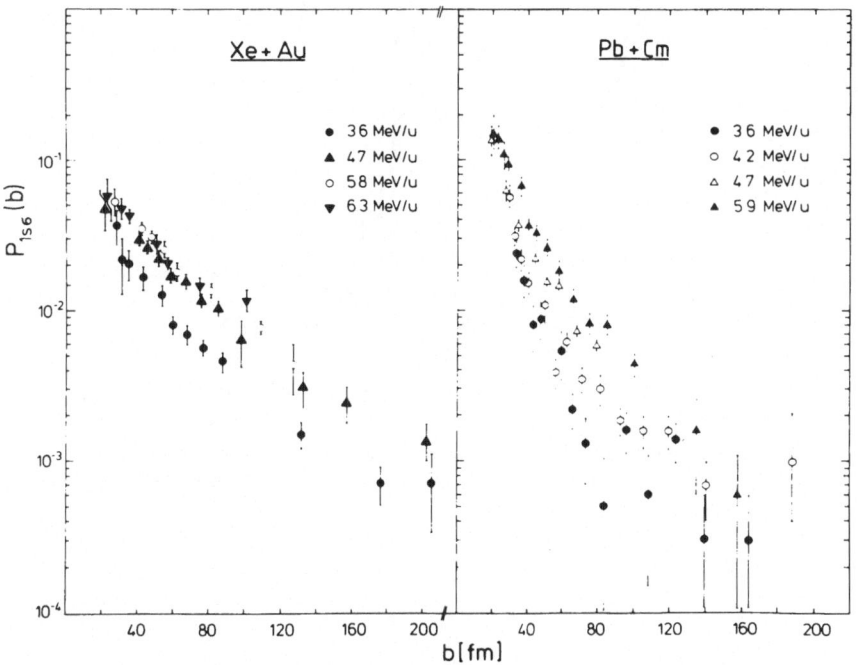

Fig. 9.   The system Sm+Pb (Z = 144) (Ref. 11).
          Left side: Scattered Sm particles in
          coincidence with Pb K-x radiation,
          yielding $P^H(\Theta)$, equation (5).
          Right side: The same spectrum corrected
          for vacancy sharing yields $P_{1s\sigma}$,
          equation (6).

Hence, in order to minimize this correction, one should choose
the collision system as asymmetric as possible. In the last four
years several systems between Z = 133 and Z = 178 have been investigated. These 1sσ excitation data show, without any exception, the
same typical features: A steep, exponential increase of $P_{1s\sigma}$ as a
function of b (an order of magnitude within 30 - 80 fm, depending
on Z) up to a maximum value about 10 % (cf. Fig. 10).

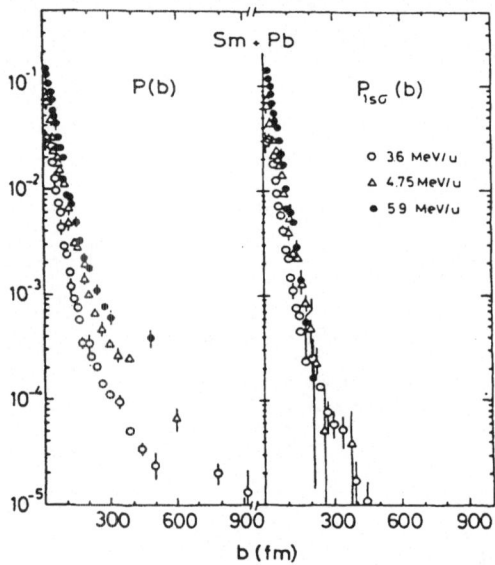

Fig. 10.   1sσ excitation data as a function of the impact parameter
            b for the systems Xe+Au (Z = 133) and Pb+Cm (Z = 178)
            recorded at several projectile energies. The data are
            corrected for vacancy sharing and nuclear Coulomb exci-
            tation.

In first order perturbation theory, taking into account only
the monopole part of the time dependent two center potential (a
reasonable approximation for *small* internuclear distances R), we
get for $P_{1s\sigma}$ the expression

$$P_{1s\sigma} \quad \alpha f(Z)\exp\{-mR_0 \frac{\Delta E(R_0)}{\hbar v}\} \quad \equiv \quad f(Z)\exp\{-mR_0 q_0(R_0)\} \qquad (7)$$

where f(Z) describes the coupling strength to the continuum, and the number m is closely connected to the electron density at the charge center. Hence, the wave function at the origin and the energy transfer to the electron should be reflected directly in the absolute value and the decay constant of $P_{1s\sigma}$, respectively. (Equation (7) is in close analogy to the classic Bang Hansteen scaling law $P_{1s\sigma} \propto \exp\{-2bq_0\}$ which was derived in the framework of the SCA theory.)[13]

The $P_{1s\sigma}$ data of the "light" collision systems Xe+Au (cf. Fig. 10), Xe+Pb, and Sm+Pb (cf. Fig. 9), perfectly fit on a straightline when plotted logarithmically versus the internuclear distance R_0; hence, the energy transfer ΔE should be constant for all R_0; the measured decay constant, therefore, is a direct measure for ΔE, i.e. the *binding energy of the 1sσ-orbital*, if we assume the final energy E_f of the electron to be zero (the electron goes into empty bound states with $E_f < 0$ *and* into continuum states with $E_f > 0$). Theory predicts m = 2 for $\alpha Z > 1$, nearly independent of Z.[14] Inserting now this value in equation (7) we get from the experimental decay constant directly the 1sσ binding energy. The 1sσ-excitation of the system Xe+Au (Z = 133), for example, was measured at few projectile velocities v_p. From the experimental decay constants $\hbar v_p/\Delta E \approx \hbar v_p/|E_B|$ we deduce a mean value of the binding energy $|E_B| = 250 \pm 35$ keV; this has to be compared with the theoretical 1s-binding energy of an atom Z = 133, which yields 290 keV.[2]

The 1sσ-excitation of the "heavy" systems (e.g. Sm+U Z = 154, Pb+Cm Z = 178), however, show a nonlinear dependence on R_0, that is, the energy transfer ΔE here changes sensitively as a function of R_0. This fact agrees with theory, which for Z $\gtrsim$ 150 predicts a pronounced dependence of the 1sσ binding energy on R_0, whereas for Z about 137 it should change less than 15 % within 100 fm.

For these systems the 1sσ binding energy can be determined by measuring at a fixed internuclear distance R_0 the excitation probability at *different* projectile velocities, by that cancelling the unknown prefactor F(Z) of equation (7). Some 1sσ binding energies at fixed R_0, which have been obtained in this way, are shown in Fig. 11.

If equation (7) describes correctly the 1sσ-excitation of superheavy systems, then in reverse, *all* experimental data (normalized in their absolute height) should fit on one *straight line with the slope 2*, drawn versus R_0q_0, whereby the binding energies $E_{1s\sigma}(R_0)$ are inserted from theory. In Fig. 12 this plot is shown for practically all data obtained up to now (more than 150 points). Obviously, most of the data fit very well on this universal straight line. The significant deviations at large values of R_0q_0, however, are rather probably due to the large errors of the vacancy sharing correction at such large internuclear distances.

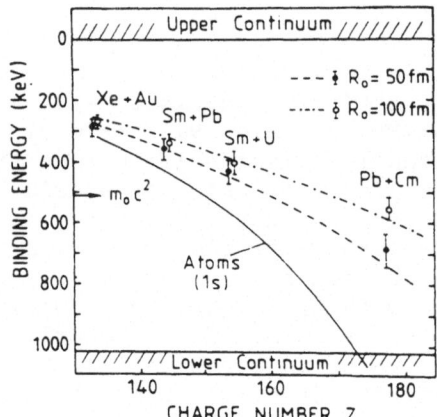

Fig. 11.   1sσ-binding energies at different internuclear
            distances R for collision systems between Z =
            133 and Z = 178 extracted from measured excita-
            tion probabilities according to equation (7)
            with m = 2. The dashed and dashed-dotted lines
            are the theoretical values of G. Soff, the full
            line indicates the 1s-binding energies for
            superheavy *atoms* according to Ref. 2.

        The application of equation (7) is by no means restricted on
the excitation of bound electrons. In Fig. 13 we show the yield of
atomic positrons versus the impact parameter with 423 keV $\leq E_{e+} \leq$
532 keV, obtained in very heavy ion collisions.[15] The data are
fitted according to P(b) = P(0) exp{-b/a}. In contrast to the
1sσ-data, however, the energy transfer ΔE is *known* in this case
(ΔE = $2mc^2 + E_{e+}$). The obtained decay constants "a" are, for all the
three collision systems investigated, in excellent agreement with
the decay constant ħv/2ΔE as extracted from equation (7). We may
conclude, therefore, that equation (7) describes correctly the
central point of excitation in heavy ion-atom collisions, both for
strongly bound electron states as for the negative continuum states.

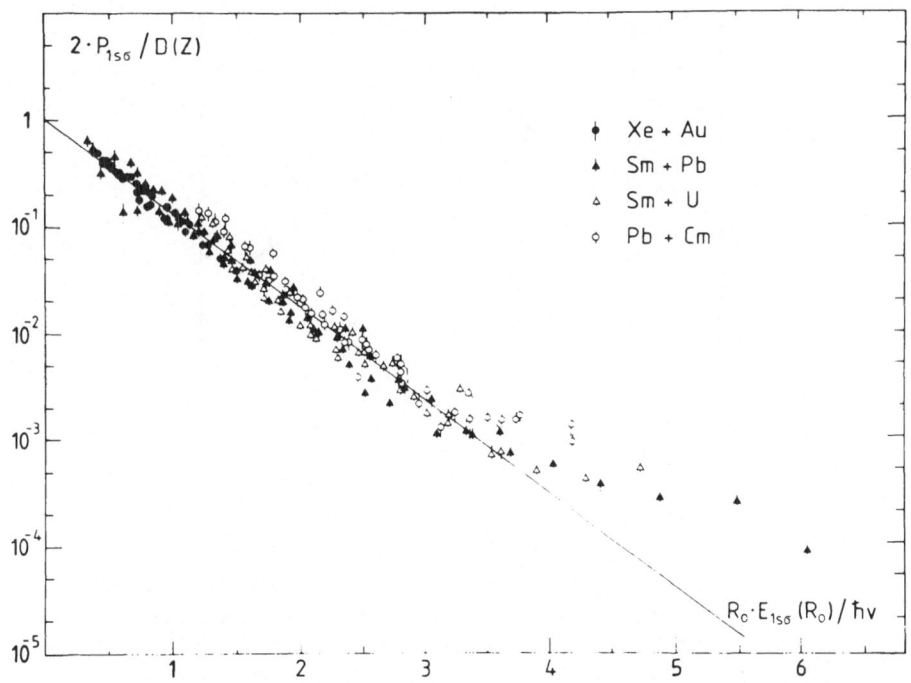

Fig. 12.   Universal plot of normalized 1sσ-excitation data in the
region $133 \leq Z \leq 178$ versus $R_0 \, E_{1s\sigma}(R_0)/\hbar v$ with *theoretical* values of $E_{1s\sigma}(R_0)$.

Although the agreement with theory seems to be well, the unsatisfactory aspects of this 1sσ-"spectroscopy", have to be pointed
out. The method is *indirect*, because the binding energy is not extracted from transitions between well defined levels; it is *not
selfconsistent*, because the factor "m" is inferred from theory; it
is, finally, *not unambiguous*, because the final energy E_f of the
electron is not observed. Therefore, the measurement of 1sσ-exci-
tation probabilities can never give a definite answer if the 1sσ-
orbital is "diving" or not into the negative continuum.

On the other hand, however, from the very similar structure of
so many experimental data the clear statement can be deduced, that
the binding energies of the innermost electrons in superheavy quasiatoms reach an order of several hundred keV.

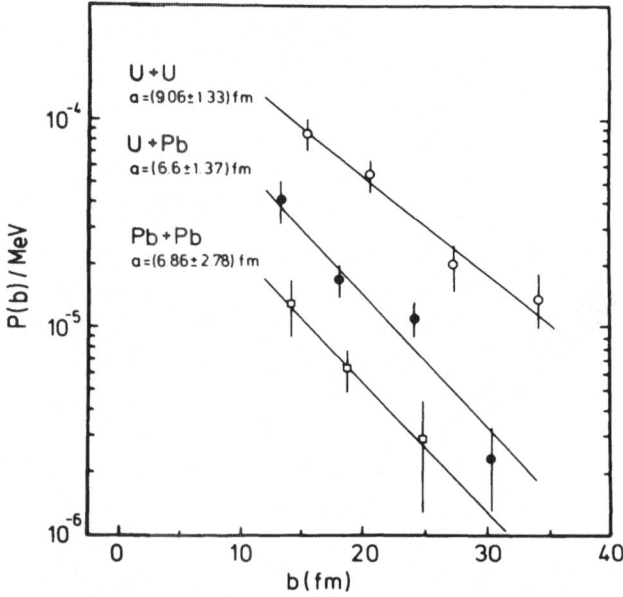

Fig. 13. Impact parameter dependence of the pro-
          duction probability for positrons, P(b)
          per MeV, of (478±54) keV energy at 5.9
          MeV/N projectile energies. The solid
          lines represent fits of the data to the
          relation P(b) = P(0) exp{-b/a}.

5.2 $P_{1s\sigma}$ at Very Small Impact Parameters

K-holes which are produced by internal conversion following
nuclear Coulomb excitation "simulate" the same final state as the
holes directly produced in the collision by the time changing
Coulomb field. And the smaller the impact parameters are, the more
important becomes this process of internal conversion (IC). Usually,
one disentangles the two types of K-hole excitation by the following
method: One measures the intensity of "all" γ-lines and takes into
account the angular distribution of the γ-rays and the appropriate
K-conversion coefficients. At *very* small impact parameters (b $\lesssim$ 20
fm), however, this procedure fails, because there are generated
even more holes by an IC-process than be direct atomic excitation
(AE) and, therefore, one has to find a more suitable method.

Is there any *physical* difference between IC- and AE-produced
K-holes ? Yes, there is ! By an IC-process namely the K-hole
originates just after the mean lifetime of the correspoding

excited nuclear state, whereas an AE-process generates the K-hole immediately, that is in the collision itself.

Let us discuss that in more detail, for instance at the collision system Pb+Cm. The lifetimes of all members of the ground state band of Cm are in the order of a picosecond or longer ($\tau^{30+} \approx 0.5$ ps). Therefore, no characteristic Cm K-radiation which fills a K-hole due to an IC-process can occur earlier than $\approx 10^{-12}$ s after the collision. The Cm K-rays following an AE-process, on the other hand, already appear $\approx 10^{-17}$ s (that is the mean lifetime of a Cm K-hole) after the collision.

How one could transform now these different time scales of 10^{-12} s and 10^{-17} s, respectively, into physically distinct observables ? The answer is: Via the Doppler effect. We use the 10^{-12} s for decelerating in a backing the recoiling Cm ions, then all Cm K-x rays following IC have as their noninterchangeable signature a *smaller* Doppler shift as compared with those K-x rays, which are due to an AE-process.

The details of the experimental set-up are shown in Fig. 14. A lead beam of 5.4 MeV/u (v/c = 0.107) hits the Cm target (≈ 400 µg/cm^2 thick), which is backed with 4 mg/cm^2 Ti. Target and backing are tilted on 30° with respect to the beam direction. The scattered particles are recorded in two position sensitive parallel plate avanlanche counters (with delay-line output); the γ- and x-radiation is detected by an intrinsic Ge-diode, placed behind the particle counter at right. The small impact parameters 3 fm $\leq$ b $\leq$ 11 fm corresponds to Cm recoils scattered into "Peter" between 17° and 47°. Now, the Cm K-x rays following AE are emitted from a Cm nucleus with the *full* initial velocity <v/c> = 0.07, the K-x rays due to an IC-process, on the other hand, always proceed from a Cm-nucleus which has traversed already the backing (passage time $\leq 0.4 \cdot 10^{-12}$ s) and, therefore, is *slowed down* (<v/c> = 0.047). Hence, for the two kinds of Cm-K x radiation ($E_{K\alpha 1}$ = 109 keV) a difference in energy arises in the order of ≈ 2.5 keV. Furthermore, the tilt angle of the target was chosen in such a way that this difference is approximately the same for all Cm-recoil reaching the particle counter at right (17° $\leq \Theta \leq$ 47°).

We see some conditions have to be fulfilled to make the method described working successfully :

(i)   The lifetime τ_γ of all nuclear states involved should be in the order of picoseconds or longer (because otherwise no sufficient slowing down on the recoil can be obtained)

$$\tau_\gamma \approx 10^{-12} \text{ s}$$

(ii)   The passage time Δt through the decelerating backing has to
        be smaller than the nuclear lifetimes (because otherwise the
        "IC"-lines are smeared out)

        $\Delta t < \tau_\gamma$

(iii)   The Doppler shifted K_x-lines should not overlap with other
        lines of the spectrum, and

(iv)   The backing should not influence too much the essential
        spectrum (in our case the grazing angle Cm → Ti was 13°).

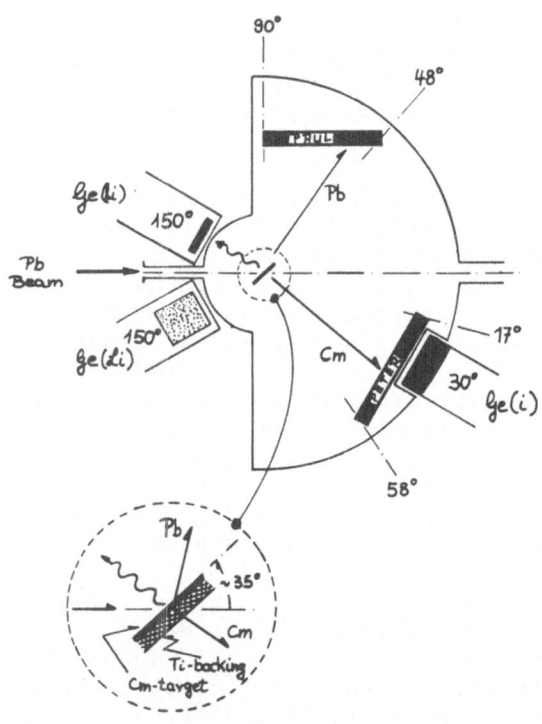

Fig. 14.   Experimental set-up for the measurement
           of the 1sσ excitation in the Pb+Cm system
           via a Doppler shift technique.

        Finally, we point out that it would be not at all convenient
to stop completely the recoil in the backing, because one needs in
any case clean kinematic coicidences between both particle
counters.

In the upper part of Fig. 15 the complete spectrum is drawn, as recorded by the intrinsic Ge-counter, in the lower part the region of the Cm-K_α lines is shown enlarged. Clearly, the $K^{AE}(\alpha 1)$- and the $K^{IC}(\alpha 2)$-lines, respectively, are well separated from all other lines, whereas the $K^{AE}(\alpha 2)$- and $K^{IC}(\alpha 2)$-lines are overlapping strongly. Using the literature-values, however, for the branching ratio of $K^{IC}(\alpha 2)$ and $K^{IC}(\alpha 2)$, respectively, this line can be easily unfolded, too. (The branching ratio should not deviate from the literature, because the inner shells are most probably rearranged after 10^{-12} s).

Fig. 15 demonstrates for the first time the feasibility to separate "hardware-like" atomic processes, which take place at time scales of 10^{-17} s and 10^{-12} s, respectively.

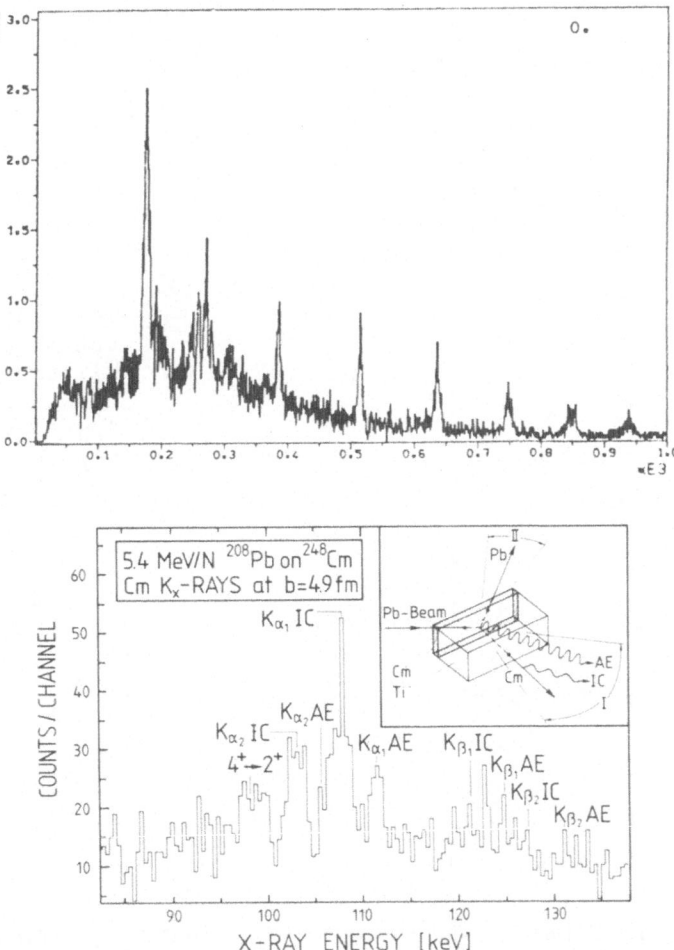

Fig. 15. Cm K spectrum for the impact parameter region $2.9 \leq b \leq 11$ fm.

   In Fig. 16 all directly measured data are presented (open
symbols) together with the IC-contributions which are extrapolated
in the usual way from the yield of all measured γ-lines using theoretical conversion coefficients (black triangles). Now, we are able
for the first time to compare the directly measured IC-yields with
the corresponding extrapolated values (open rectangles and black
triangles, respectively, at the two smallest b, where the Doppler
separation could be applied). Obviously, the extrapolated values
underestimate the directly measured ones by $\approx$ 40 % which indicates
strongly the request of a direct method at those small b. Taking
into account, however, the strong decrease of Coulomb excitation
as function of b, one can argue, that for large b (b $\approx$ 20 fm) the
error done by the indirect determination of IC should be tolerable
($\approx$ 20 %).

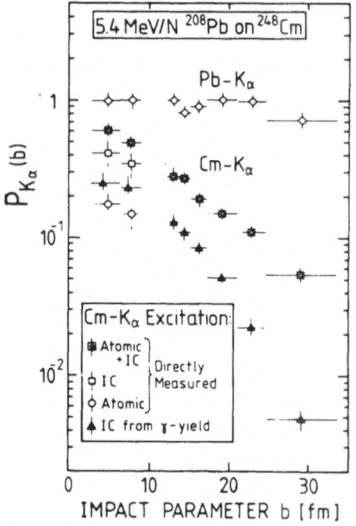

Fig. 16. Directly
measured exci-
tation proba-
bilities for
Cm K_α and Pb K_α
(open symbols)
together with
IC-values,
extrapolated
from the mea-
sured Coulomb
excitation
yield of Cm.

   In Fig. 17, finally, we show the $P_{1s\sigma}$ and $P_{2p\sigma}$ probabilities,
which were obtained from the direct data of Fig. 16 by scaling up
the K_α-yields to total K-yields (using the literature value of

K_α/K) and correcting for vacancy sharing. Thereby, for the large b
(b $\geq$ 13 fm) the values extrapolated from the γ-yield have been used
as IC-contribution. For comparison we also show the data of a previous measurement [9] (open symbols). Whereas the slope of the present
and previous 1sσ data seems to be very similar, the absolute values
differ by $\approx$ 25 % (1sσ) and $\approx$ 70 % (2pσ). This big difference of the
2pσ yields is not yet understood.

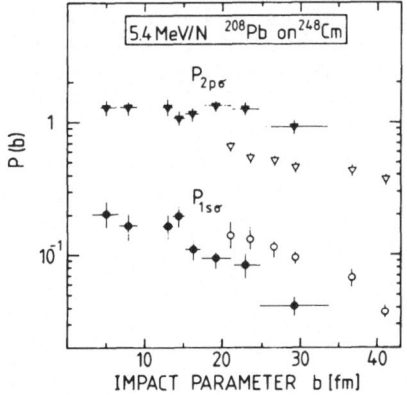

Fig. 17.   Black symbols: $P_{1s\sigma}$ and $P_{2p\sigma}$ excitation probabi-
            lities of this experiment. Open symbols: Previous
            data of Ref. 9. (P(b) is normalized on 2, due to
            the 2 electrons in the shells.)

5.3 High Energy δ-Electrons

From a measurement of the 1sσ excitation probability $P_{1s\sigma}$, as
described in the previous part, no information is obtained about
the final state of the excited electron. Above all, it remains completely unknown, whether the electron is still bound (in a higher
lying orbital) or it is ionized into the positive continuum. In

applying equation (7) for 1sσ-data, we needed an assumption concerning the mean final energy E_f. Obviously, a detection of the
emitted electrons simultaneously with both the scattered particle
and the characteristic K-radiation would give the maximum possible
information about the quasiatomic system. That kind of triple coincidence, however, demands very high detection efficiencies of all
counters involved.

For that reason δ-electron investigations have started first
with the measurement of the spectral distribution of the δ-electrons,
either in coincidence with K-radiation of the heavier atom[16] or without any coincidence condition at all.[5]

The pioneering experiments in this field have been done by
Ch. Kozhuharov in Munich and W. Koenig in Heidelberg. Fig. 18 shows
the electron spectrum recorded in the collision S on Pb (Z = 98).
The spectral distribution can be described well by the properly
weighted electron form factors of Z = 98 (solid line), confirming
that high energy electrons stem from the *combined system*.

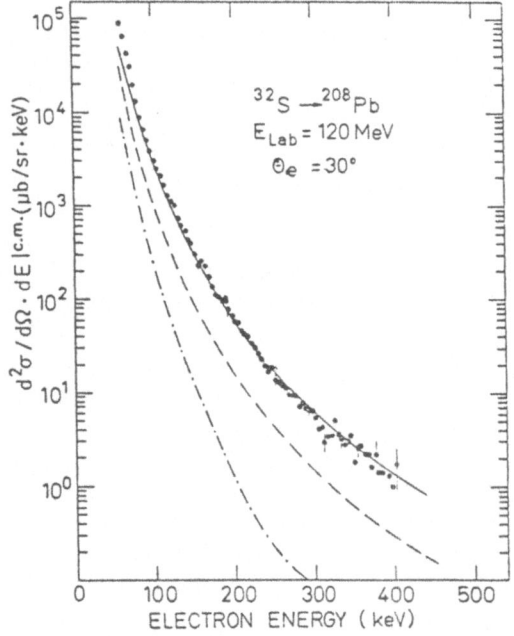

Fig. 18. δ-electron spectrum
(Ref. 5) of the S
on Pb collision
system ($Z_1 + Z_2 = 98$).
The solid line is
the properly
weighted sum of the
1s, 2s, and $2p_{1/2}$-
form factors of
Z = 98, without any
normalization, using
relativistic hydro-
genic wave functions.
The dashed line is
the same sum for
Pb, the dashed-
dotted line a non-
relativistic cal-
culation.

Going to heavier systems ($\alpha Z \gtrsim 1$), first it should be mentioned
that the separation of the K-shell contribution to the δ-electron
spectrum could be performed successfully, using the coincidence
technique as described previously. Fig. 19 shows the δ-spectrum of

the system J on Pb (Z = 135). In the upper part the single spectrum is presented (with contributions of several shells). In the lower part a coincidence with characteristic Pb K-radiation was required, hence, fixing the 1sσ orbital as initial state of the detected electrons.

Additionally, the same system was investigated by a triple coincidence experiment (Figs. 20,21), too (δ-electron, Pb K-radiation, scattered projectile). Such a measurement improves, indeed, the $P_{1s\sigma}$-experiments because now the final state E_f of the electron is known. Moreover, by varying E_f at a given internuclear distance R_0, or changing R_0 at fixed E_f, one can check very sensitively the concept of a "spectroscopy" as it was derived from equation (7).

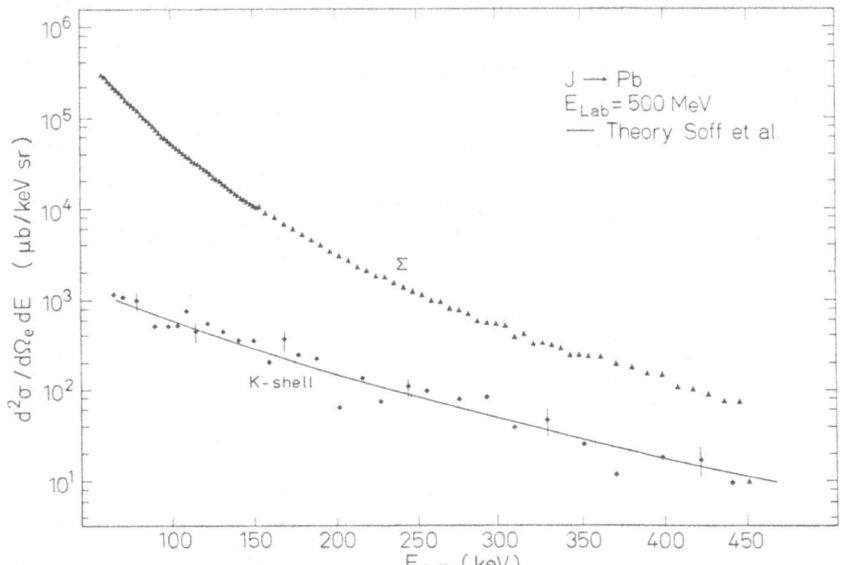

Fig. 19. δ-electron spectrum (Ref. 17) for the I+Pb system (Z = 135). In the lower part electrons in coincidence with Pb-K x rays are shown. The solid line indicates a coupled channel calculation of G. Soff.

The dependence of the decay constant on the energy transfer ΔE, as postulated in equation (7), is confirmed by the results, shown in Fig. 21. A variation of the electron energy from 84 keV to 325 keV leads to a significant change in the decay constant from 25 fm^{-1} to 15 fm^{-1}, respectively; from that, a value of 260 keV for the 1s-binding energy of Z = 135 can be derived, independently of m (Fig. 22). (The theoretical value for the binding energy of the quasiatom Z = 135 at R_0 = 40 fm yields to 292 keV.)

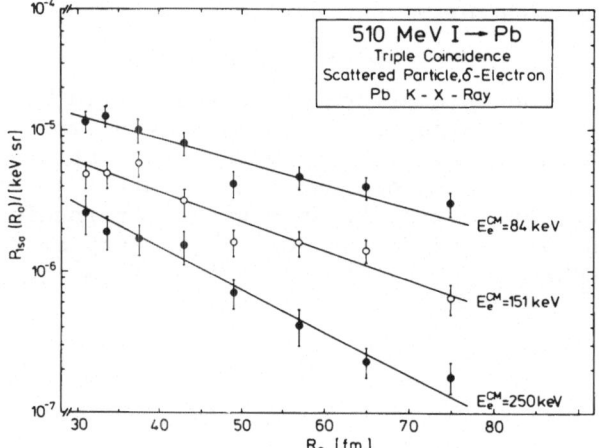

Fig. 20. Triple coincidence measurement (Ref. 17) (scattered
          particle, δ-electron, Pb K-x ray) for the I+Pb
          system. Drawn are the yields for a fixed final
          energy of the electrons versus the internuclear
          distance R_0.

Of course, also "m" can be extracted independently from the
measured decay constants. The value m = 2.1 ± 0.2 has been obtained.
In Fig. 22 also the I+Pb-data, without the Pb-x ray condition are
shown (open circles). Whereas m, quite similar to the triple coincidence data, is very close to m = 2, the binding energy (inter-
section on the negative x-axis) is only about 40 keV, indicating
that the main part of the δ-electrons stems from the L-shell of the
united system.[17] Also shown are the data for the Ni+Pb (Z = 110)
system with an obviously different m of about 1.2. And, indeed,
theory predicts an m-value of 1.1 for Z = 110 due to the drastically
different electron distribution at the origin, as compared with
Z = 135 !

     Concluding it should be stressed that the experiments on high
energy δ-electrons are in an early stage yet. It is planned to study
very heavy collision systems (e.g. Pb+Cm) in a systematic manner,

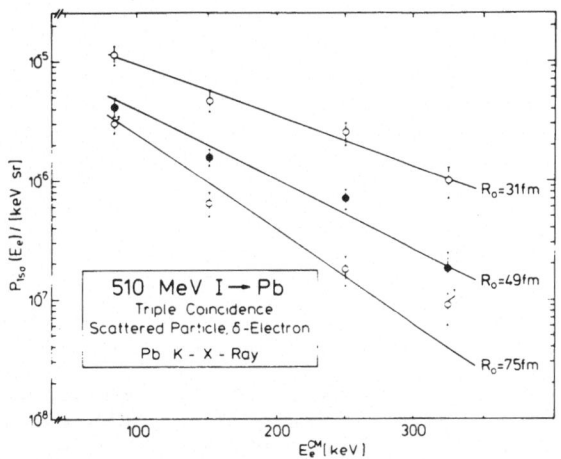

Fig. 21. Triple coincidence measurement (Ref. 17) for the
          I+Pb-system at fixed internuclear distances R_0
          versus the final kinetic energy of the δ-electrons.

that means varying all the relevant parameters as projectile energy,
impact parameter, final energy of the electrons, and so on. From
that we hope to obtain a much more precise understanding of the
dynamics in inner shell excitation of superheavy collision systems.
In particular, we expect to gain a sure information about binding
energies and wave functions of the most deeply bound orbitals in
quasiatoms – even more reliable than it was obtained up to now
from $P_{1s\sigma}$-data.

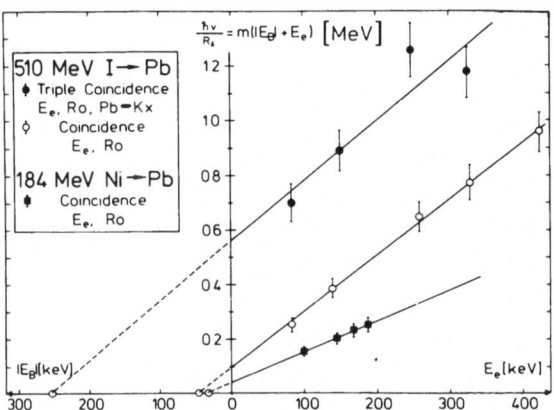

Fig. 22.   "Decay constants" $\hbar v/R_A$ [MeV] (cf. equation (7)) versus
the kinetic energy of δ-electrons. Black circles: triple
coincidence data of the I+Pb system (Z = 135) extracted
from the data of Fig. 21. Open circle: Coincidence data
(δ-electron, R_0) for I+Pb. Black squares: Coincidence
data for Ni+Pb (Z = 110). The intersections of the curves
on the negative x-axis deliver the corresponding binding
energies.

*This talk was prepared in collaboration with D. Liesen, Ch. Kozhu-
harov, W. Koenig, D. Schwalm, P. Armbruster, and P.H. Mokler.

REFERENCES

1.   B. Fricke, private communication
2.   B. Fricke and G. Soff, At. and Nucl. Data Tabl. 19, 83 (1977)
3.   J. Reinhardt, B. Müller, W. Greiner, Phys. Rev. A24, 103 (1981)
4.   G. Soff, Phys. Rev. Lett., to be published
5.   F. Bosch, H. Krimm, B. Martin, B. Povh, U. Traxel, R. Walcher,
       Phys. Lett. 78B, 568 (1978);
          C. Kozhuharov, Physics of Electronic and Atomic Collisions,
       p. 179, ed. S. Datz, North Holland (1982)
6.   F. Bosch, A. El Goresy, W. Herth, B. Martin, R. Nobiling,
       B. Povh, H.D. Reiss, U. Traxel, Nucl. Sc. Appl. 1, 33 (1980)

7. C. Kozhuharov, P. Kienle, GSI Scientific Report, 122 (1980)
8. H. Schmidt-Böcking, R. Schuch, I. Tserruya, MPI Heidelberg,
    Scientific Report, 112 (1979)
9. D. Liesen, P. Armbruster, F. Bosch, S. Hagmann, P.H. Mokler,
    H. Schmidt-Böcking, R. Schuch, H.J. Wollersheim,
    J.B. Wilhelmy, Phys. Rev. Lett. 44, 983 (1980)
10. W.E. Meyerhof, Phys. Rev. Lett. 31, 1341 (1973)
11. D. Liesen, P. Armbruster, F. Bosch, D. Maor, P.H. Mokler,
    H. Schmidt-Böcking, R. Schuch, A. Warczak, Abstracts to
    the XIIth Int. Conf. on the Physics of Electronic and
    Atomic Collisions, Gatlinburg (1981, pp).
12. F. Bosch, D. Liesen, P. Armbruster, D. Maor, P.H. Mokler,
    H. Schmidt-Böcking, R. Schuch, Z. Phys. A296, 11 (1980);
    B. Müller, G. Soff, W. Greiner, V. Ceaucescu, Z. Phys.
    A285, 27 (1978)
13. J. Bang, J.M. Hansteen, Kgl. Danske Vid. Selsk. Mat.-Fys.
    Medd. 31, 13 (1959)
14. J. Rafelski, B. Müller, W. Greiner, Z. Phys. A285, 49 (1978)
15. P. Armbruster, P. Kienle, Z. Phys. A291, 399 (1970)
16. C. Kozhuharov, P. Kienle, D. Jakubaßa, M. Kleber, Phys. Rev.
    Lett. 39, 540 (1977)
17. F. Güttner, W. Koenig, B. Martin, B. Povh, H. Skapa, J. Soltani,
    T. Walcher, F. Bosch, C. Kozhuharov, Z. Phys. A304,
    207 (1982)

INNER SHELL CAPTURE IN THE INTERMEDIATE VELOCITY RANGE

C.L. Cocke

Department of Physics, James R. Macdonald Laboratory

Kansas State University, Manhattan, Kansas   66506 USA

I. Introduction

Over the past decade a great deal of effort has been ex-
pended on and progress been made in the understanding of the capture of inner shell electrons by projectiles of intermediate velocity. Reasons for the interest in the problem are numerous,
including of course the continuing practical need for understanding
the capture process because of the role it plays in determining the
state of hot plasmas encountered in MFE and astrophysical problems.
More intriguing perhaps is the feature that even after more than
50 years of work on the elemental single electron capture problem,[1,2]
certain aspects remained unsatisfactorily treated theoretically,
at least until very recently. Over the last ten years or so a
great deal of experimental information has come forth of both a
volume and type heretofore unavailable. The new data include
measurements differential in angle or impact parameter and from
selected target shells and challenge the theory more severely than
do total cross section data. Theory has reacted vigorously to the
challenge, with a great deal of progress being realized especially
very recently.

In these two talks I will review some of the progress of

the last decade with an emphasis on the experimental data. The
elemental process of interest is $Z_A + (Z_B + e) \rightarrow (Z_A + e) + Z_B$, a
single electron process. I am discussing <u>inner</u>-shell capture
only, almost always <u>K</u> shell. The study of K capture from inner
K shells in systems with $Z_B \gg 1$ has at least two advantages:
1) The ratio Z_A/Z_B can be varied experimentally over a wide range
(not possible for p on H), and 2) For larger Z_B, the K-shell electrons still behave more as independent electrons than do outer-
shell electrons, and indeed screened hydrogenic wave functions are
usually adequate.

In order to establish what types of collision are to be
considered, I show in Fig. 1, a Madison-Merzbacher[3] map, applied to
capture, showing Z_A/Z_B and v/v_K on the two axes (Z_A and Z_B are
projectile and target nuclear charge, v and v_K are projectile
and target-electron velocities). Why are these parameters significant? Refer to Fig. 2, where the schematic of a collision involving nuclear charges Z_A and Z_B and a single electron is given.
In all cases which I will be discussing the de Broglie wavelength
of the projectile is very small compared to that of the electron,
and thus the nuclear motion can, and will consistently, be treated
classically, giving a classical $\vec{R}(t)$. The electronic wave function
$\psi(\vec{r},t)$ then evolves according to

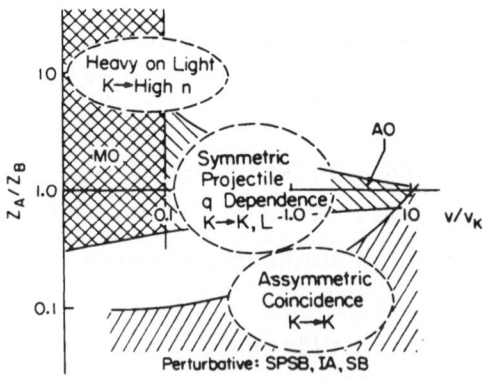

Fig. 1. M-M map for k-electron capture.

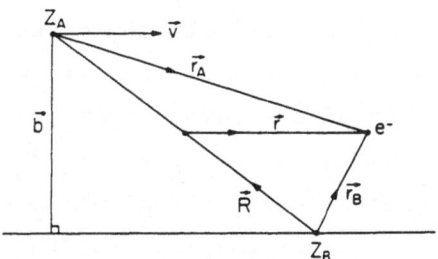

Fig. 2. Collision schematic.

$$\left(-\frac{\nabla^2}{2} - \frac{Z_A}{|\vec{r}-\vec{R}/2|} - \frac{Z_B}{|\vec{r}+\vec{R}/2|} - i\frac{\partial}{\partial t} \right) \psi(\vec{r},t) = 0 \qquad . \qquad (1)$$

Defining $\vec{\rho} = \vec{r}\, Z_B$ and $\tau = t\, Z_B^2$, this can be written as

$$\left(-\frac{\nabla^2}{2} - \frac{Z_A/Z_B}{|\vec{\rho}-\vec{R}'/2|} - \frac{1}{|\vec{\rho}+\vec{R}'/2|} - \frac{\partial}{\partial \tau} \right) \phi(\vec{\rho},\tau) = 0. \qquad (2)$$

If the scattering is very forward peaked, true for all cases discussed here, a very good approximation is that Z_A follows a straight
line: $\vec{R} = \vec{b} + \vec{v}t$, in which case $\vec{R}'=\vec{b}'+\vec{v}'\tau$ where $\vec{b}'$ $\vec{b}Z_B$ and
$v' = v/Z_B$. Thus by a simple length and velocity scaling, the
equation of motion for the electron is universally represented by
that for a projectile of charge Z_A/Z_B on atomic hydrogen. The
resulting scaling properties of transition probabilities P_{if} and
cross section σ_{if} between initial and final states i and f are
given by

$$P_{if}(\vec{b},\ \vec{v},\ Z_A,\ Z_B) = P_{if}\ (\vec{b}\, Z_B,\ \vec{v}/Z_B,\ Z_A/Z_B,\ 1) \qquad (3)$$

and

$$\sigma_{if}(v, Z_A, Z_B) = \sigma_{if}(v/Z_B, Z_A/Z_B, 1) / Z_B^2 \qquad (4).$$

This scaling holds for capture, excitation or ionization, but for a strictly hydrogenic system only. To the extent that inner and outer "screening parameters" differ, the scaling is only approximate. Since the velocity of the target K-electron, v_K, is Z_B, the important parameters from (4) are v/v_K and Z_A/Z_B.

The behavior of the charge cloud during transfer is quite different in different parts of the map, and thus differing are the theoretical approaches used to describe the process.

(a) In the lower right-hand part of the map the target wave function is only slightly disturbed by the projectile. Capture is weak compared to K ionization to the target continuum in this region, and that capture which does take place is dominantly to the projectile K shell. A perturbative treatment of the process is appropriate and quite successful for the case of ionization. For the case of capture, it is now widely recognized that an essential feature of this region is that both target and projectile potentials must act to enable the transfer, and thus an expansion, though the "Second Born" is necessary. An alternative way to say it is that appropriate intermediate states centered on the target must be included to describe the process.

(b) In the central region of the map the charge cloud is distorted in a major way and the process is best described in terms of an eigenfunction expansion (molecular orbital, MO, or atomic orbital, AO). The capture competes strongly with direct ionization in this region, often winning as a K-vacancy production mechanism. Near symmetry, a small number of basis functions (often two) is adequate, with K-shell to K-shell transfer being the dominant reaction channel.

(c) In the upper left hand region lies the area of capture from H and He by highly charged projectiles. Capture is the major reaction channel, and population of excited states on the projectile dominates. Eigenfunction expansions are appropriate,

although a large number of basis states are usually necessary.
In this region lies the low energy capture by high-q projectiles,
a subject to be discussed by other speakers at this school and
which I therefore will not discuss further here.

The data I will discuss fall into the symmetric, (b),
and asymmetric, (a), parts of the map. In both cases v/v_K
extends into the vicinity of one, often called the "intermediate"
velocity range. Two types of data isolate the K-K electron
transfer process:

(a) <u>Asymmetric Coincidence Experiments</u>: For light projectiles (p, He^{2+}, Li^{3+}) on heavy targets much of the total capture
comes from outer shells. The capture of specifically K-shell
electrons is signalled experimentally by the time coincidence of
a single charge reduction of the projectile and a K x-ray or K
Auger-electron from the filling of the target K vacancy produced
by the capture. (See Fig. 3) The interpretation of results of such
coincidence measurements is clean only for relatively weakly
ionizing/capturing projectiles, since the simultaneous capture of
an outer shell electron and, accidently in the same collision,

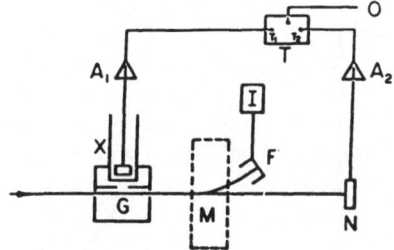

Fig. 3. Schematic of apparatus for coincidence experiment. Some
components are gas cell (G), Si [Li] detector (X), Magnet
(M), surface barrier detector (N), time-to-amplitude
converter (T). A proton beam enters from the left and
neutral H-atoms are detected in N, in coincidence with
K-X-rays from the target. (From Macdonald et al., ref. 4).

ionization or excitation of a K electron, is indistinguishable from
the K capture. Thus a system with higher Z_A/Z_B, such as the F + Ar
system, <u>cannot</u> be easily studied by this method, although p + Ar
<u>can</u> be. The experiment gives the ratio σ_{CK}/σ_{VK} which can be com-
bined with independent measurements of σ_{VK} to give σ_{CK} (σ_{CK} and
σ_{VK} are cross sections for K capture and for K-vacancy production,
respectively. σ_{IK} is the cross section for electron ejection to
the target continuum). Results from Macdonald et al.[4] are shown in
Fig. 4.

  b) <u>Symmetric Projectile Charge State Dependence of Target
K-Vacancy Production</u>: Fig. 5 shows the dependence of the cross
section for Ar K-vacancy production, signaled by the detection of
Ar K-x rays, as a function of the charge state of the F projectiles, data taken by Macdonald et al.[5] in 1975. At that time a

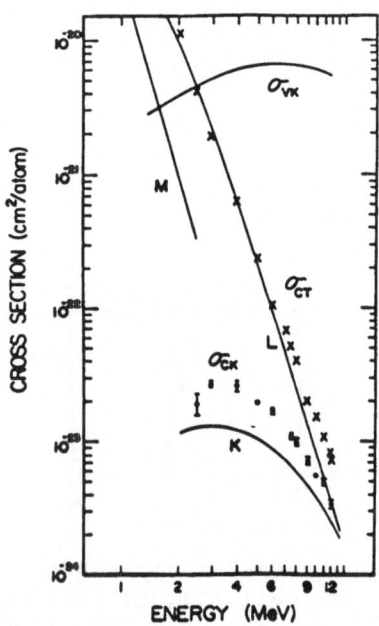

Fig. 4. Results from coincidence experiment for p on Ar. σ_{CT} is
         total capture. (From Macdonald et al. ref. 4).

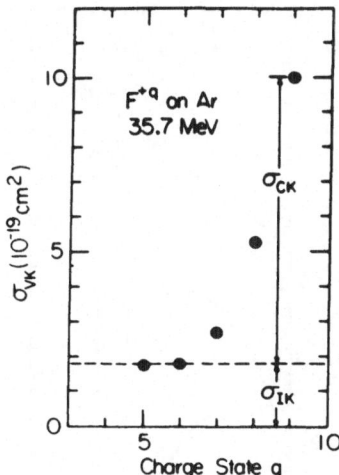

Fig. 5.   Projectile charge state dependence of Ar K-vacancy pro-
          duction, deduced from Ar K-X-ray yields, for F on Ar
          (ref. 5).

great deal of work was being done on Coulomb ionization of heavy
targets by lighter projectiles, for which the nuclear charge Z_A
does all the work, and it was somewhat startling that the electronic structure of the projectile could have such a dramatic
effect on the K-vacancy production.   The circumstantial evidence
is now overwhelming that the dramatic rise in σ_{VK} for projectiles
bearing K vacancies is due to the opening of the K-K electron
capture channel.   Using this interpretation, σ_{CK} can be obtained
from the vacancy production cross sections as $\sigma_{VK}^2 - \sigma_{VK}^0$, where the
superscript indicates the number of K vacancies on the projectile.

     Differential measurements using both a) and b) have now
been made as well, and are discussed later.

II.   Total Cross Section Results

     In Fig. 6 are shown results for the cases of p on Ne and
F on Ne.   The ionization data are from measurements of Ne-K-Auger-

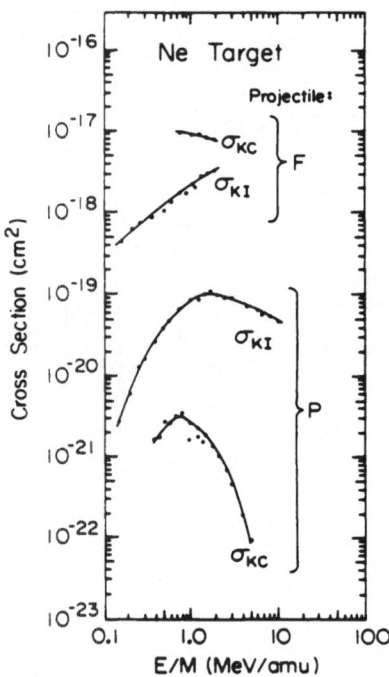

Fig. 6. Target K-capture and K-ionization cross sections for p
on Ne and F on Ne. The p data are from Rødbro et al.,
(ref. 6): the F data are from Woods (ref. 7).

electron production cross sections, using low F charge states
where a K-K transfer channel is closed. The capture data are
from coincidence experiments[6] and projectile-q dependence data[7]
for the p and F cases, respectively. The systematics are quite
different in the two cases, ionization dominating capture for p
and vice versa for F projectiles. In Fig. 7 is shown a plot, at
fixed v/v_K, of the capture and ionization cross sections versus
projectile charge, where again is seen the rapid rise of the importance of K-K capture relative to K ionization with increasing Z_A.

A.   Some Formalism:

In order to discuss the interpretation of data such as
these, a brief review of two elemental theoretical approaches will
be given, one following a perturbative and one eigenfunction expansion approach.

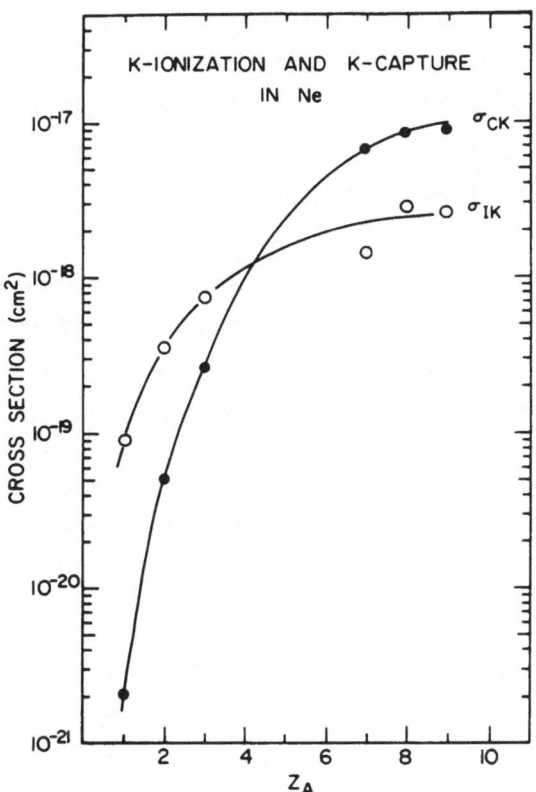

Fig. 7.   Dependence on Z_A of σ_{IK} for Z_A on Ne.   Data are from
ref. 6 ($Z_A < 4$) and ref. 7 ($Z_A < 4$).

1) <u>First order treatment: OBK</u> - The earliest attempts
to deal with data such as that shown in Fig. 5 was in terms of a
perturbative treatment, specifically the first order Born
approximation description of Oppenheimer[8] and Brinkman and Kramers[9]
(OBK).   There are several objections to such an approach, includ-

ing: 1) For F + Ar, Z_A is not $\ll Z_B$. 2) OBK is inadequate in principle, since it neglects the intrinsically "double"- (or "multiple") scattering nature of capture. (Discussed later.) Nevertheless the OBK is not totally devoid of worth, for reasons including: 1) the form of the transition matrix element has a physically appealing transparency, 2) for hydrogenic wave functions, a closed analytic form for the cross sections is obtained, and 3) the experimental scaling properties of σ_{CK} with Z_A and Z_B are apparently very close to those of σ_{OBK}.

     With these latter reasons in mind, we briefly describe the first order perturbation theory treatment, in a time-dependent formalism. With no loss of physics and some simplification of algebra, we take the target to be infinitely heavy. The projectile proceeds along a trajectory given by $\vec{R} = \vec{v}\, t + \vec{b}$, causing the electron to undergo a transition from an initial state $\psi_i = \phi_i(\vec{r}_B)e^{-i\,\omega_i t}$ to a final state $\psi_f = \phi_f(\vec{r}_A)e^{-i(\omega_f t - \vec{v}\cdot\vec{r}_B)}$. In first order perturbation theory,

$$a_{if}(b) = -i \int dt \int d\vec{r}\; \phi_f^*(\vec{r}_A)e^{i(\omega_f t - \vec{v}\cdot\vec{r}_B)} \frac{Z_A}{r_A} \phi_i(r_B)e^{-i\omega_i t} \tag{5}$$

Using the fact that $\phi_f(r_A)$ satisfies[10]

$$\left(-\frac{\nabla^2}{2} - \frac{Z_A}{r_A} + \frac{Z_A^2}{2n_A^2} \right) \phi_f(r_A) = 0$$

the Fourier-Transform of $\phi_f^* \cdot Z_A/r_A$ is found to be

$$\int \phi_f^*(r_A) \frac{Z_A}{r_A} e^{-i\,\vec{q}\cdot\vec{r}_A} d\vec{r}_A = \tfrac{1}{2}\left(q^2 + \frac{Z_A^2}{n_A^2} \right) G_f^*(\vec{q})$$

where $G_f^*(q) \equiv \int \phi_f^*(r) e^{-i\,\vec{q}\cdot\vec{r}} d\vec{r}$, so that

$$a_{if}(b) \propto \int dt \int d\vec{r} \int d\vec{q}\; \left(q^2 + \frac{Z_A^2}{n_A^2} \right) G_f^*(\vec{q})e^{i\,\vec{q}\cdot\vec{r}_A} \phi_i(\vec{r}_B)e^{-i\vec{v}\cdot\vec{r}_B} e^{i(\omega_f - \omega_i)t}.$$

setting $\vec{r}_A = \vec{r}_B - \vec{v}t - \vec{b}$ in the exponential and doing the t-integral leads to $\delta(\omega_f - \omega_i - \vec{q} \cdot \vec{v})$ in the integrand, so that the integral over the longitudinal component of q ($\parallel$ to $\vec{v}$) can be done to yield

$$a_{if}(b) \propto \int d\vec{q}_\perp \, e^{-i \, \vec{q}_\perp \cdot \vec{b}} \, \underline{(q^2 + \frac{Z_A^2}{n_A^2}) \, G_f^*(\vec{q}) \, G_i(\vec{q} - \vec{v})} \tag{6}$$

(The underlined part would be the transition amplitude, as a function of momentum transfer $\vec{q}$ to the projectile, had we done a PWBA treatment rather than the time dependent one.)

The δ-function has fixed $q_\parallel$ at $(\omega_f - \omega_i)/v$. If U_i and U_f are the target and projectile binding energies, respectively, $\omega_i = -U_i$, $\omega_f = -U_f + \frac{1}{2} v^2$, so $q_\parallel = (U_i - U_f)/v + v/2$. The integral over $\vec{q}$ is over a "surface" in $\vec{q}$-space dictated by $q_\parallel$ (see fig. 8). In the very asymmetric case where $U_i \simeq + v_K^2/2$ and $U_f \simeq 0$, $q_\parallel = (v_K^2/v + v)/2$ which puts the surface through the heart of G_i for $v = v_K$, near the maximum in the cross section.

This particularly simple result emphasizes that transfer in the OBK approximation only occurs if there is enough momentum overlap in the initial and/or final state to allow the electron to "jump" to its new home. If the necessary momentum components are not present in ψ_i or ψ_f, the potential V_A cannot provide them in the collision.

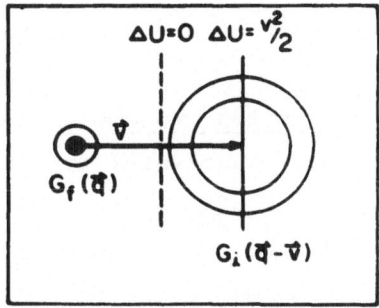

Fig. 8. $\vec{q}$-space picture of OBK amplitude (eq. 6).

For hydrogenic ψ_i and ψ_f, the total cross section can be evaluated analytically, yielding, for 1s to 1s transfer,[10]

$$\sigma_{OBK} = \frac{2}{5}\left(\frac{Z_A}{Z_B}\right)^5 \; \pi\left(\frac{a_o}{Z_B}\right)^2 \; \frac{F(s)}{V^2} \quad , \tag{7}$$

where $V = v/v_K = v/Z_B v_o$, $s = V\{1-[1-(Z_A/Z_B)^2]/v^2\}$ and $F(s) = (1 + s^2/4)^{-5}$. For $Z_A \ll Z_B$, $s \sim V\{1 - 1/V^2\}$, nearly independent of Z_A/Z_B and $F(s)$ becomes a universal function of V maximizing near $V = 1$. The corresponding first order PWBA cross section for ionization can be written[11] as

$$\sigma_{PWBA} = 8\left(\frac{Z_A}{Z_B}\right)^2 \; \pi\left(\frac{a_o}{Z_B}\right)^2 \; \frac{f(V,\theta)}{V}, \tag{8}$$

where θ is the ratio of experimental to hydrogenic-model binding energies and $f(V,\theta)/V$ maximizes near $V = 1$. A comparison of these expressions illuminates the relative importances of direct ionization and electron capture as competing K-vacancy production mechanisms. The ratio $\sigma_{OBK}/\sigma_{PWBA}$ goes as $(Z_A/Z_B)^3$ for fixed v. This ratio is rather weakly v-dependent at low v, since both σ_{OBK} and σ_{PWBA} maximize at $v/v_K \simeq 1$ and decrease together at lower v. For higher v, σ_{OBK} falls off faster with v than does σ_{PWBA} since the capture is very sensitive to having a momentum match between initial and final states while the ionization is not.

It should be reemphasized here that, while the OBK has pedagogical value, it fails in fundamental ways to describe the capture. Further discussion of more valid perturbation treatments will follow later.

2) Eigenfunction expansion treatment: Atomic Orbitals – Probably the most consistently successful treatment of the symmetric region has been in terms of the atomic expansion model. The formalism has been used often for light systems, and especially applied to heavy systems by Lin and collaborators.[12,13]

The internuclear distance R is assumed to follow a trajectory
given by classical equations of motion. The electronic wave
function $\psi(r,t)$ is expanded in terms of a small number (we take
two for example) bound states centered on the projectile and
target:

$$\psi(r,t) = \sum_m A_m(t)\phi_m(r_A)\exp[-i(\varepsilon_m t - \tfrac{1}{2}\vec{v}\cdot\vec{r} + \tfrac{1}{8}v^2 t)]$$

$$+ \sum_n B_n(t)\phi(r_B)\exp[-i(\varepsilon_n t + \tfrac{1}{2}v\cdot r + \tfrac{1}{8}v^2 t)]. \tag{9}$$

The coefficients A_m and B_n are found by numerical integration of
Schrödinger's equation. The approach has at least two advantages
of principle: 1) Translation factors are correctly treated, since in
the AE they are only plane wave factors. 2) The non-orthogonality
of states centered on target and projectile is correctly handled,
as discussed by Bates.[14]

The method is especially useful if the number of basis
states needed to describe what the wave function is actually doing
in the collision is small. For example, for $Z_A \sim Z_B$, K-K transfer
is rather well described by a two-state approximation (TSAE) in
which 1s wavefunctions only on target and projectile are retained.

In the TSAE, the resulting equations[12] are

$$i\dot{A} = \frac{h_{AB} - S_{AB}h_{BB}}{1-S^2} e^{i(\omega t + \delta)}\bar{B} \tag{10}$$

$$i\dot{B} = \frac{h_{BA} - S_{BA}h_{BB}}{1-S^2} e^{i(\omega t - \delta)}\bar{A}$$

where $h_{AB} = \int \phi_A^*(-Z_A/r_A)\phi_B\, e^{i(\vec{v}\cdot\vec{r})}\, d\tau$,

$$S_{AB} = \int \phi_A^*\phi_B\, e^{(i\vec{v}\cdot\vec{r})}\, d\tau, \quad h_{BB} = \int \phi_B^*(-Z_A/r_A)\phi_B\, d\tau$$

$$\omega = \varepsilon_A - \varepsilon_B.$$

The $\bar{A}$ and $\bar{B}$ are related, by a simple phase transformation, to A
and B via phase angles discussed by Lin et al.[12] If the overlap
terms S are neglected and $\bar{B} \simeq 1$ (perturbation treatment) one re-
trieves the OBK treatment in a time-dependent form since h_{AB} is
just the OBK transition matrix element.

B. A First Comparison with Experiment:

How well the OBK picture does at explaining the data for
the symmetric case is illustrated in Fig. 9, where OBK is too
high by a factor of about 10. A systematic survey of the data
shows OBK to be above the data for near-symmetric cases by factors
of 3 to 100, and there is thus ample reason on both theoretical
and experimental grounds to drop discussion of the OBK as applied
to this region. Why has this not happened? Nikolaev[16] pointed out
in 1956 that, even for asymmetric cases, the OBK result must be
divided by about 3 to bring it into agreement with data. Once
this was done, however, the correct v-dependence, Z_A/Z_B scaling
and distribution of σ among target/projectile shells seemed to be

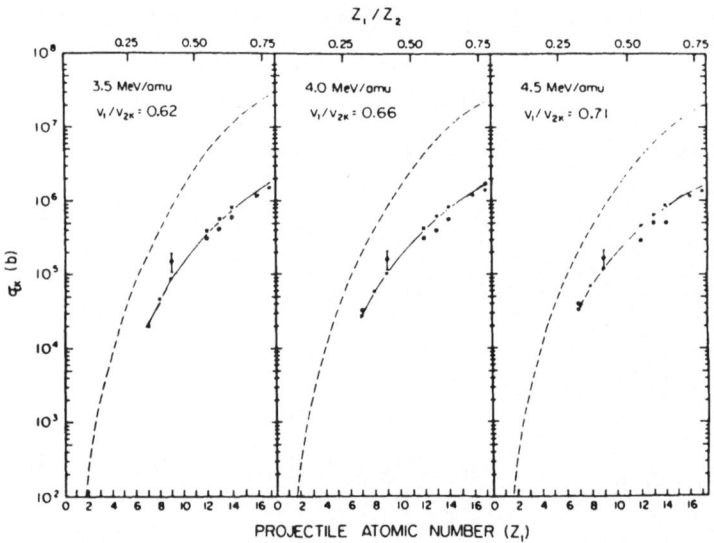

Fig. 9.   K-capture cross sections for various projectiles on a Ti
          target. The dashed lines are an OBK calculation. The
          solid lines are a TSAE calculation. (From J. Hall,
          ref. 15).

correctly given. Thus when data for the near-symmetric region
became available for heavy systems, a natural extension of this
procedure by replacing 3 by $3(v,Z_A,Z_B)$ occurred. Few results from
alternative formalisms were available then. This is no longer
true and use of OBK is in some sense obsolete. Nevertheless, the
closed form of σ_{OBK} will probably keep it in use by experimentalists who need a quick-and-dirty way of estimating capture cross
sections.

The use of OBK in the asymmetric case is more promising
than for the symmetric case, since this is clearly a perturbative
case. If one cannot understand the asymmetric case for p bombardment, the outlook for the case of capture by fast, highly charged
projectiles seems bleak. Considerable study of the asymmetric
case at KSU and Aarhus, using the coincidence technique, has
been made.[6] Results of some of the experiments on K capture by bare

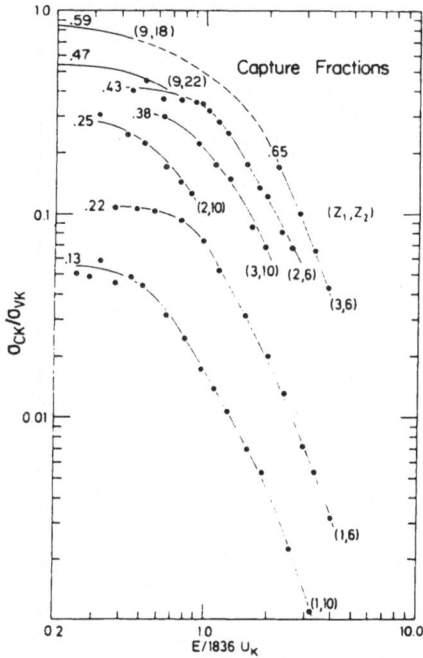

Fig. 10. Capture as a fraction of total vacancy production, versus
energy (approximately $(v/v_K)^2$). Except for F on Ti and
F on Ar, the data are from refs. 6 and 17.

H^+, He^{+2} and Li^{+3} projectiles are shown in Figs. 10-12. These data
are all taken by measuring coincidences between capture and target
K Auger electrons, and were usually done with sufficient resolution
to isolate "diagram lines" in the Auger spectrum. Thus only true
K-capture events are counted, and the multiple process discussed
earlier is excluded.

In Fig. 10 are shown the capture fractions σ_{CK}/σ_{VK}, where
the ratio is seen to go from 0.1% for p on Ne at 5 MeV to more than
40% for He^{++} on C at low v. For comparison, in the upper left hand
corner are shown data from projectile q-dependences for F on Ti and
F on Ar. Although there is a large velocity gap, the F + Ar data
are seen to appear consistent with the Li + C results at much
higher v. In the upper left hand corner the K-K transfer is clearly
much stronger than direct ionization.

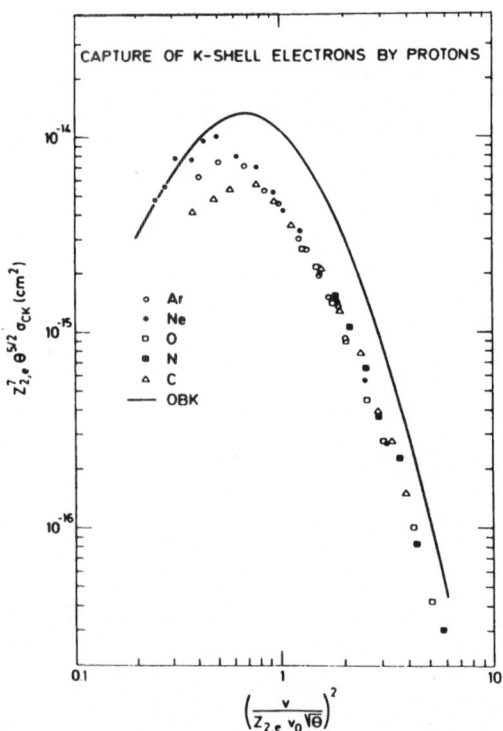

Fig. 11.   Target scaling properties of σ_{CK} for p on various
targets. Data from refs. 6 and 17.

In Fig. 11 the target Z_B scaling for protons capturing from various targets is tested. The form of the OBK used here is that of Nikolaev,[16] who introduced inner and outer screening parameters to account for the non-hydrogenic nature of the target wave functions. In the figure $Z_{2,e} = Z_B - 0.3$ and $\theta = U_K/2Z_{2,e}^2$, as in the PWBA. For high v, the OBK scaling approximately as Z_B^{-7}, is remarkably good. The OBK curve shown is Nikolaev's form of eq. 7, multiplied by 1.17 to convert a 1s→1s calculation into a 1s→ (all shells) result. This is done simply by noting[10] that the OBK 1s→ n result goes as $1/n^3$. The absolute value of the OBK result lies a consistent factor of about 3 above the data, and the scaling properties fail below v ≃ 1.

In Fig. 12 the projectile scaling laws for the case of a $C(H_4)$ target is tested. The ionization is seen to follow Z_A^2 scaling and to be in good agreement with PWBA calculations. The

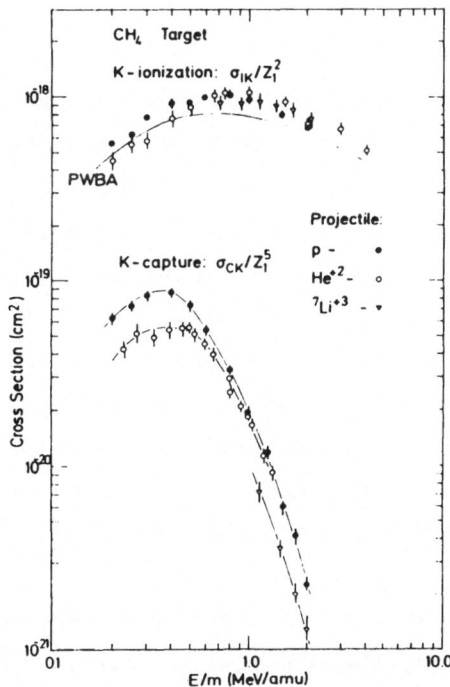

Fig. 12. Projectile scaling properties, for Z_A on the K-shell of C (in CH_4), for both ionization and capture.

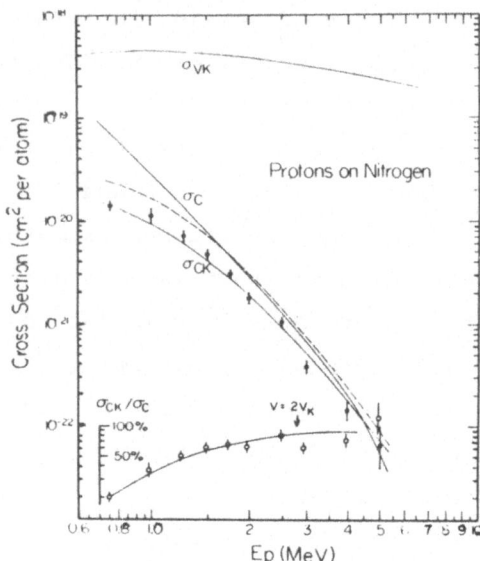

Fig. 13. Cross sections for K-capture and K-vacancy production in
          N by protons. σ_c is total capture from ref. 1. Other
          data are from ref. 17. The dashed line is σ_{OBK} divided
          by 3.9.

Z_A^5 scaling for capture is not perfect but remarkably good when
it is recalled that this factor goes over a dynamic range of 243.
A similar degree of success was found for Ne.[6]

      The coincidence data allow one to now assign experimentally
relative contributions to the total capture to various contributing
target shells. In Fig. 13 the case of p on N[(2)] is shown, where
it is seen that above v ~ $2v_k$, more than 75% of the capture is
from the K shell. The dashed line in that figure is OBK divided
by 3.9. In the case of argon, coincidence data is available to
allow capture from all three target shells to be separated experimentally, as is shown in Fig. 14. The solid curves are OBK
divided by 3.

      Two simple conclusions emerge from these, and other, data:
in the asymmetric region, the OBK scaling rules are quite good,
but the absolute OBK cross sections are too large by a factor of
three.

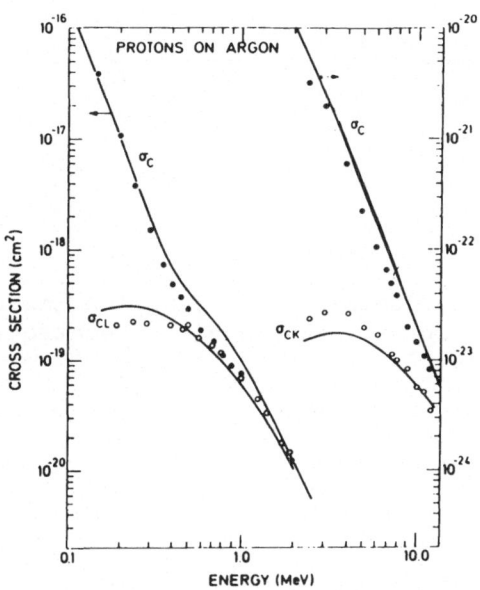

Fig. 14.   Target-shell distributions from ref. 6 for shell capture
from Ar L- and K-shells by protons. Solid lines are
σ_{OBK} divided by 3.

C.   Better Treatments of the Asymmetric Regime:

The failure of the OBK even in the asymmetric case is presumably due to its failure in principle to describe the basic
process.   Classically, the projectile can never capture a free
electron via the projectile-electron potential alone, regardless
of what velocity the initial orbit about the target may present.
This result follows from conservation of energy and momentum alone:
some energy must be released (e.g. by radiation) for capture to
occur.   There appears to be no real classical analogy to the OBK
result, in contrast to the case for ionization, where the PWBA[11]
and classical binary encounter approximation treatments[18] are very
close.   Thomas, in 1927,[19] described capture classically as a two-
step process in which the slow target electron is first scattered
from the projectile at 60° so that v_e v after this encounter
(see Fig. 15) and then scattered again from the target nucleus to
direct it nearly parallel to the projectile.   If the resulting

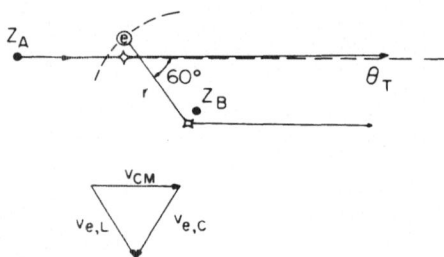

Fig. 15.   Schematic of double-scattering leading to classical
           electron capture. The angle of scattering of the pro-
           jectile is a proton.

kinetic energy of the electron relative to the projectile is less
than the corresponding potential energy, the electron is captured.
Thomas's calculation from straightforward classical mechanics, gave

$$\sigma_T = \frac{64\sqrt{2}}{3} \pi a_o^2 (Z_A)^{7/2} (Z_B)^2 (\frac{a_o}{r})^{7/2} (\frac{v_o}{v})^{11}$$

where a_o and v_o are the Bohr radius and velocity and r is the
radius of the initial electron with respect to Z_B. Note that the
asymptotic v-dependence is v^{-11}, as opposed to v^{-12} for OBK.

It is now widely understood that any correct treatment of
capture in the intermediate to high-v range must take account in
some way of double or multiple scattering, which, in perturbation
expansions, means including second and higher order Born terms.
As early as 1955 Drisko found,[20] from a second Born calculation,
an asymptotic result

$$\sigma_{B2} = \sigma_{OBK}[.295 + (5\pi v/2^{11})/(Z_A + Z_B)]$$

Shakeshaft and Spruch[21] showed that a modified treatment of the
Thomas double scattering, incorporating some quantal aspects of the
problem, led to

$$\sigma_{SS}^{D} = \left[(5\pi v/2^{11})/(z_A + z_B) \right] \sigma_{OBK}$$

and thus the second term of Drisko's result corresponds to the on-shell part of the 2nd Born with a direct classical interpretation. This term is only important at very high v and no experimental evidence for its importance has been reported. The 0.295 term is widely taken as an excuse for dividing σ_{OBK} by 3 to compare with experiment, although this is usually unjustified, since the derivation is only for asymptotically high v.

The last decade has seen much progress in both finding the correct perturbation expansion to use and in illuminating the relationships among the various approaches. A recent review of approaches[22-30] which incorporate in some way the multiple scattering nature of the problem is given by Briggs, Macek and Taulbjerg.[31] Especially important is the recent use of the Coulomb Green's' function to propagate the electron wave between scatterings.[32]

These theories are generally capable of giving excellent agreement with experiment for very asymmetric cases and high v. In Fig. 16 we show the p + Ne results compared to several theoretical curves. The TSAE result of Lin et al.[12], in this asymmetric case, nearly reduced to the OBK treatment except that the non-orthogonality of ϕ_i and ϕ_f is correctly treated. The terms containing S in equation 10 are important, leading to the reduction in σ needed to agree with data for high b. The failure of the TSAE at low v is expected, since no provision is made in the model for intermediate states through which the double scattering might proceed. A considerable improvement is realized in the coupled-channels (CC) calculation of Ford, Reading and Becker,[30] in which a large target centered basis is used to provide intermediate states through which the capture may proceed, including continuum states represented by pseudostates. The impulse approximation (IA) treatment of Jakubassa-Amundsen and

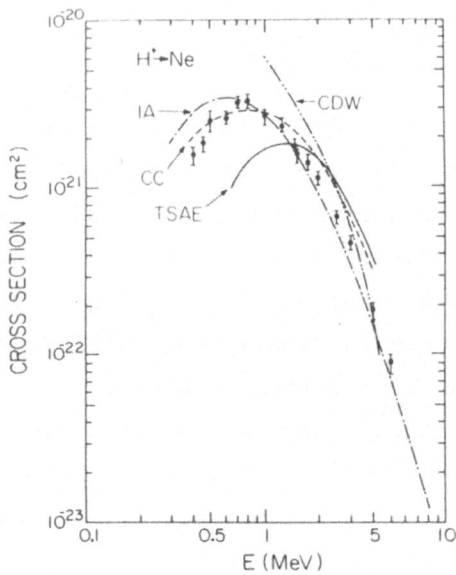

Fig. 16.   K-capture cross sections for p on Ne.   Theoretical curves
           are: TSAE, ref. 12; IA, ref. 24; CC, ref. 30; CDW, ref. 33.
           Data are from refs. 6 and 17.

Amundsen,[24] which uses an expansion which includes target-electron
scattering beyond second order, but projectile-electron scattering
only to first order. Again, rather good agreement with the data is
obtained. The continuum distorted wave treatment of Belkic, Gayet
and Salin[33] works well in its advertised region of validity, namely
high-v.

     In Fig. 17 the more symmetric case of Li^{+3} on Ne is dis-
played. As is the pattern generally, the TSAE does better in this
less asymmetric case. The Ford, Reading and Becker CC result is of
particular interest here, since they have directly addressed the
problem of the multi-electron nature of the target. The curves CCb
and CCa are calculated with and without inclusion of the all
processes which can lead to a target K-hole and a single electron
capture by the projectile. The experiment measured the "with"

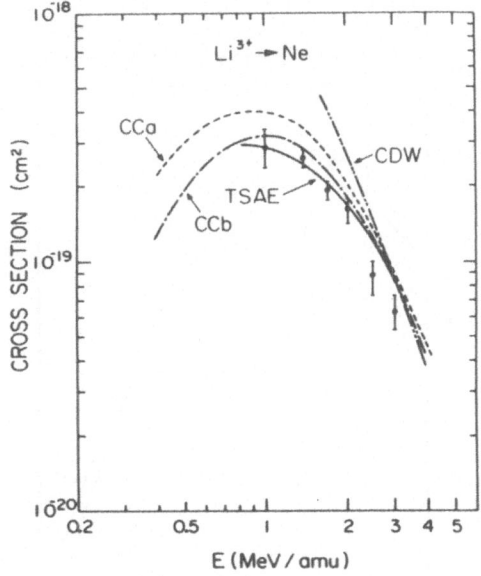

Fig. 17. Same as fig. 16, but for Li^{+3} projectiles.

case, and agreement is good.

In Fig. 18 the case of p on Ar is revisited. The strong potential second Born (SPSB) treatment of Macek and Alston[22] represents an expansion in Z_A/Z_B which is claimed valid at all v so long as $Z_A \ll Z_B$. A crucial aspect of the treatment is that scattering from the strong target potential is done in using Coulomb, rather than plane wave, Green's function, thus going above second order in interaction with the target. The impulse approximation (IA) can be obtained from the SPSB by neglecting "off-energy-shell" intermediate states, and which introduces errors of order $(Z_B/v)^2$. The IA and SPSB converge to each other at high v. The SPSB is seen to agree rather well for high v, but to diverge at low v. However, recent data of Horsdal-Pedersen[35] suggest that recent results of Adriamondje[34] et al. and the lowest

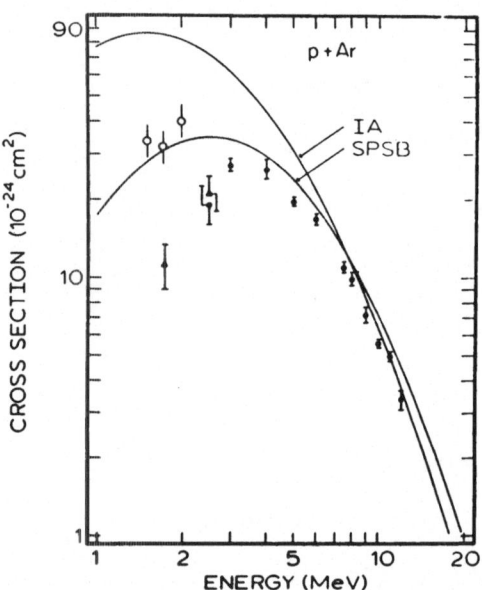

Fig. 18. K-capture cross sections for p on Ar, compared with SPSB and
          and IA calculations from ref. 22. Data points are:
          ● , ref. 4; ▲ , ref. 34; ○ , ref. 35.

data point of Macdonald et al.[6] may be too low, and are in better
agreement with experiment.

    While the SPSB cross section cannot be given in a simple
analytic form as could the OBK, it possesses similar scaling properties. For 1s to ns capture,

$$\sigma_{SPSB} = (Z_A^5/Z_B^7) \ M_n(\nu)^{\ 2} \ F(\nu)/n^3, \tag{11}$$

where $F(\nu)$ is a universal function tabulated by Macek and Alston,[22]
$\nu = Z_B/v$, and the "off-shell-factor"

$$|M_n(\nu)|^2 = 2n^2/(1 + e^{-2\pi\nu})(n^2 + \nu^2) \ .$$

The Z-scaling properties are the same as those of the OBK, which perhaps gives some alternative explanation for why the OBK scaling rules work as well as they do. Setting $|M(\nu)|^2 = 1$ leads to the IA of Briggs.[23] Since the factor $(Z_A{}^5/Z_B{}^7)/n^3$ is common to the OBK, SPSB and IA calculations, the difference between these results is exhibited simply as a difference in the universal function of v/v_K which multiplies this factor to give the K-capture cross section. A comparison of these three functions is shown in Fig. 19, where $Z_A/Z_B \to 0$ has been assumed in extracting the OBK curve from eq. 7.

D.   Symmetric Region:

In spite of the availability of a great deal of data on the q-dependence of K-vacancy-producing cross sections, relatively few calculations have been made in the symmetric region. One reason for this is probably that much of the data have fallen into the "no-man's land" between MO and perturbative regions in Fig. 1, a

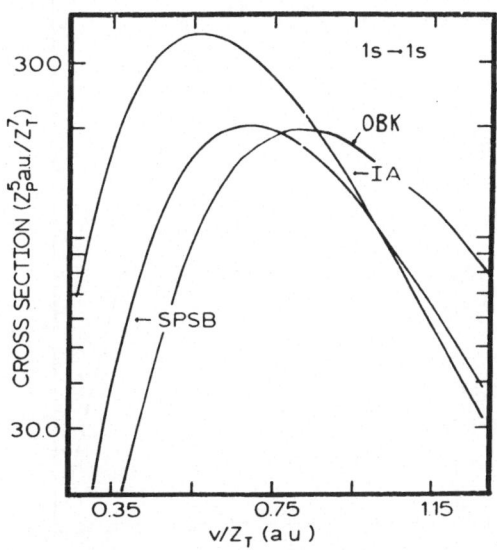

Fig. 19.   Cross sections for 1s to 1s charge transfer for asymmetric collisions in three different theoretical formulations: SPSB and IA, ref. 26; OBK, eq. 7.

region reluctantly visited by theorists. The most consistently
successful treatment of this region has been in terms of the AE
model. To set the stage, we show, in Fig. 20, a comparison between two-state AE calculations and experiment for the case of p on
He. For $Z_A/Z_B \overset{\sim}{>} 0.5$, it appears quite promising to try to represent
the evolution of the target K-electron wave function in terms of 1s
functions centered on target and projectile. For low velocities,
this procedure should fail, as the wave function has an important
part looking like the united atom K shell. Either an MO treatment
or at least the addition of a UA 1s state becomes appropriate.

In Figs. 9 and 21 are shown cross sections for K-K transfer only, deduced from data such as are shown in Fig. 3. The OBK
result is clearly so far off as to be useless. The TSAE result
agrees nicely with the data, improving as Z_A approaches Z_B. A
summary of the comparison of this data with the TSAE is shown in

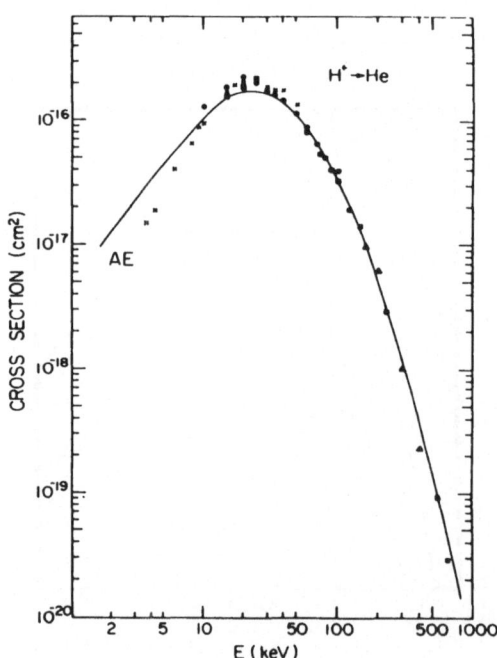

Fig. 20.   Cross sections for k-capture by p on He, compared with a
            TSAE Calculation. Theory: refs. 36, Expt: see ref. 1.

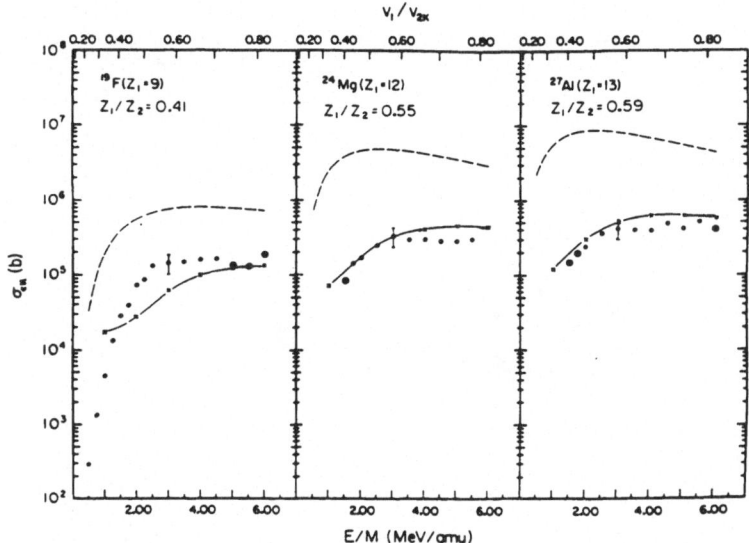

Fig. 21.   Cross sections for K-capture in the symmetric region
            from a Ti target.   The dashed line is OBK; the solid,
            TSAE, ref. 12, 37.   Data are from Hall, ref. 15.

Fig. 22, where two qualitatively different behaviors are noted for
symmetric and asymmetric situations.   The former shows excellent
agreement at all energies, while the later shows fair agreement
at high energy, but increasingly worse agreement as v is lowered.
This is presumably again due to the inadequacy of a theory which
does not provide for intermediate states in the target continuum
for very asymmetric cases.

III.   Differential Cross Sections

    A.   Introduction

        The recent availability of data on differential cross
sections for K-shell capture has presented the theory with substantially more challenge than does total cross section data.   In
these experiments the scattered projectile is detected at a
definite scattering angle in time coincidence with the K x ray or
K Auger electron from the filling of the K hole (see Fig. 23).
As for the total cross section case, in the asymmetric region

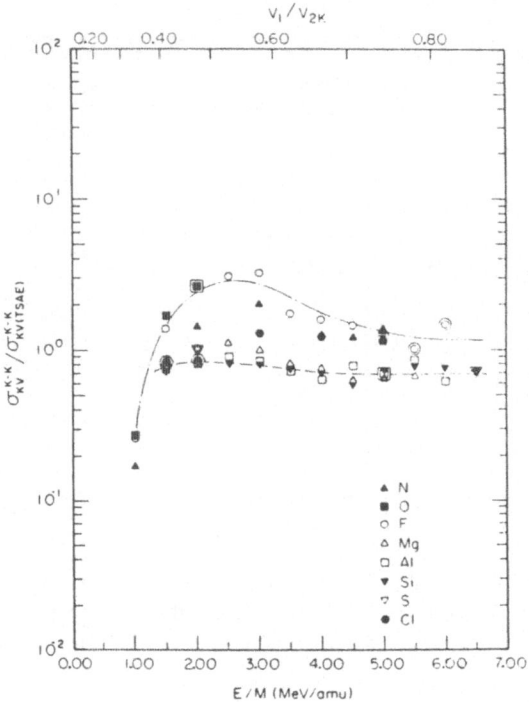

Fig. 22.   A comparison of experimental K to K electron transfer
cross sections on Ti targets with the TSAE calculation
of Lin, from ref. 15.   The agreement is clearly worse
for asymmetric cases.

charge capture is identified by charge analysis of the projectiles,
while for the near symmetric region the capture contribution is
identified by the projectile q-dependence of the K-hole production

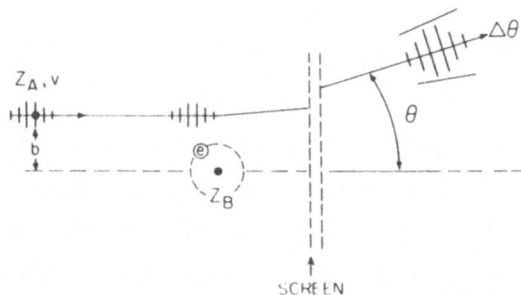

Fig. 23.   Scattering schematic for differential measurements.

probability.   The data are taken as θ -dependent capture probabilities $P(\theta)$, where P_{CK} is the ratio of K-capturing (coincidence)
events to the total number of scattered particles.   The results
are often either presented or analyzed as impact-parameter (b)
dependent probabilities $P(b)$, however, and a short discussion of
this relationship is in order.

It was shown by Bohr[38] in 1948 that the classical relationship between θ and b is expected to hold only if the scattering angle is large compared to the diffraction angle, that is if
$\kappa = 2Z_1 Z_2 e^2/\hbar v$ is large.   The argument is that a wave packet confined
to a small impact parameter range Δb near b will give rise to a
spread in scattering angle $\Delta\theta$ due to both diffraction and to the
classical variation of θ with b.   If one quadratically combines
these uncertainties and optimizes Δb, one finds $\Delta b/b \sim [\lambdabar/d_o]^{1/2} = \kappa^{-1/2}$, where λ is the projectile deBroglie wavelength and d_o is
the collision diameter $Z_1 Z_2 e^2/E$.   This conclusion holds if the
scattering is Rutherford, which is quite good for the penetrating
collisions which result in capture of K-electrons

The near-symmetric collision systems usually satisfy this
condition easily, and experimental values of $P(\theta)$ can be translated
into $P(b)$ using classical trajectories for comparison with impact-
parameter theoretical treatments.   If κ is not large, as is often
the case for asymmetric systems, the comparison requires more work
as was discussed by McCarroll and Salin.[39] The scattering amplitude then must take into account that scattering from different
impact parameters may contribute coherently at any angle θ.   The
simplest treatment of the problem involves erecting a diffraction
screen perpendicular to $\vec{v}$, just downstream of the scattering region (e.g., dashed lines in Fig. 23).   If $\vec{K}$ and $\vec{K}'$ are taken as
wave vectors of incident and exiting waves from this screen, with
$\vec{q} \equiv \vec{K}' - \vec{K}$, then the scattering amplitude

$$f(\theta) \propto \int a(\vec{b})\, e^{i\vec{q}_\perp \cdot \vec{b}}\, d\vec{b} \qquad (12)$$

where $\vec{b}$ is a two-dimensional vector lying in the plane of the
screen. If a(b) is calculated neglecting the inter-nuclear
potential W(R) (= $Z_A Z_B/R$ for a pure Coulomb scattering) then f(θ)
calculated from (12) will correspond to scattering only from the
electron, not to reality. Belkic and Salin[40] pointed out that this
problem could be solved in the Eikonal approximation by inserting
into (12) an additional factor to account for the phase accumulation
due to W(R):

$$f(\theta) \propto \int a(\vec{b})\, e^{-i\int_{-\infty}^{\infty} W(R)dt}\, e^{i\vec{q}_\perp \cdot \vec{b}}\, d\vec{b} \qquad (13)$$

For the case of azimuthally symmetric scattering, (13) reduces to

$$f(\theta) \propto \int_0^\infty a(b)\, e^{-i\int_{-\infty}^{\infty} W(R)dt}\, J_0(q_\perp b)\, b\, db \qquad (14)$$

Comparison with experiment in such cases should ideally be done in
terms of differential scattering cross sections $d\sigma/d\Omega \propto |f(\theta)|^2$,
rather than P(b).

Comparison of Theory and Experiment

B. Perturbative Regime

At least partial motivation for measuring differential
cross sections in the asymmetric case came from the apparent predictions of several theoretical treatments of the process that a
node could arise in the differential cross section from an interference between the OBK amplitude and one arising by the inclusion
of the internuclear potential in the calculation of the matrix
element in Eq. (5). That is, $-Z_A/r_A$ is replaced by $-Z_A/r_A + Z_A Z_B/R$.
Experimental searches showed no evidence for such a node in p
+ Ar or p + He.[41-43]

The role of the internuclear potential was discussed by Bates[14]
in 1954 and is made transparent in an impact-parameter formalism
in eqs. (10). If a coupled channel expansion of the rearrangement

process is used to obtain a first order transition amplitude, by
setting $\bar{B} = 1$ and $\bar{A} = 0$ in eq. (10), the amplitude is found
to be

$$a(b) = \frac{i\int_{-\infty}^{\infty} dt \langle \phi_A^* | (-Z_A/r_A + \langle \frac{Z_A}{r_A} \rangle_B) | \phi \rangle e^{i(\omega t + \delta)}}{1 - s^2} \qquad (14)$$

where $\langle -Z_A/r_A \rangle$ $(=h_{BB})$ is the projectile electron interaction
averaged over ϕ_B. The replacement of $-Z_A/r_A$ by $-Z_A/r_A + Z_A Z_B/R$
now has no effect on the matrix element. The term $\langle Z_A/r_A \rangle$, which
plays a role in some ways similar to a "nuclear potential" term,
is a function of R alone, but has a physical origin which is quite
different. Note that, as $R \to \infty$, $\langle Z_A/r_A \rangle \twoheadrightarrow Z_A/R$. The inclusion
of $Z_A Z_B/R$ in the calculation in eq. 5, which gives rise to
the strange angular structure, is an incorrect use of perturbation
theory. However, the inclusion of $\langle Z_A/r_A \rangle$ in the AE approach can
also give rise to structure in $d\sigma/d\Omega$.

Belkic and Salin[40] pointed out that a rather good description of the angular distribution for p on Ar was obtained by using
a b-dependent OBK transition amplitude a(b) and calculating $P(\theta)$
using eq. 13, with the nuclear potential coming in only through the
eikonal phase integral $\delta(b)$. The most recent results of Rivarola
et al.,[44] still show structure unobserved experimentally, however.
Also shown in Fig. 24 TSAE results which are in rather good
agreement with experiment. The TSAE does produce spurious nodal
structure at lower v, but its failure to predict the correct $d\sigma/d\Omega$
occurs where it already fails to give correct total cross sections,
so comes as no surprise.

In Fig. 25 are shown recent data of Horsdal-Pedersen et al.
for p + Ne, compared with the SCIA (Semi-Classical Impulse Aproxi-
mation) calculations of Jakubassa-Amundsen and Amundsen. These
results are presented in impact parameter form using classical
trajectories to convert θ to b. Although $\kappa \simeq 4.3$ here (not >> 1),

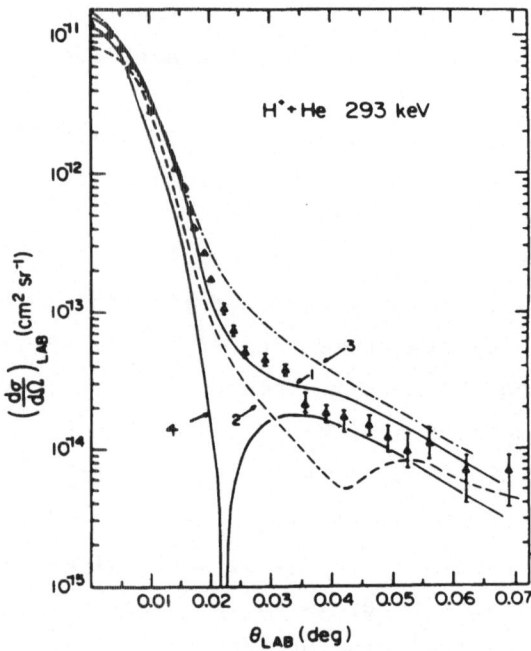

Fig. 24.   Angular distributions for capture from He.   Data: ref. 42.
          Theories are: TSAE, Lin et al., with (3) and without (1)
          screening in W(R) (ref. 45);   DW, Rivarola et al.,
          (ref. 44); straight Born approximation without using
          eq. 12, (ref. 46).

this procedure is argued to be adequate, since there is no sharp
structure in a(b) which would be washed out by calculating $d\sigma/d\Omega$
through eq. 12.   The SCIA does include the rescattering but only
includes on-the-energy shell intermediate states.   The results for
p on Ne are quite close, and give remarkable agreement with experiment.

     The scaling properties of the SCIA P(b) are remarkably simple
and very near those of an OBK treatment:

$$P(b, v, Z_A, Z_B) = (Z_A/Z_B)^5 \, \hat{P}(bZ_B, v/v_K) \tag{15}$$

where $\hat{P}$ is a universal function of the scaled impact parameter bZ_B

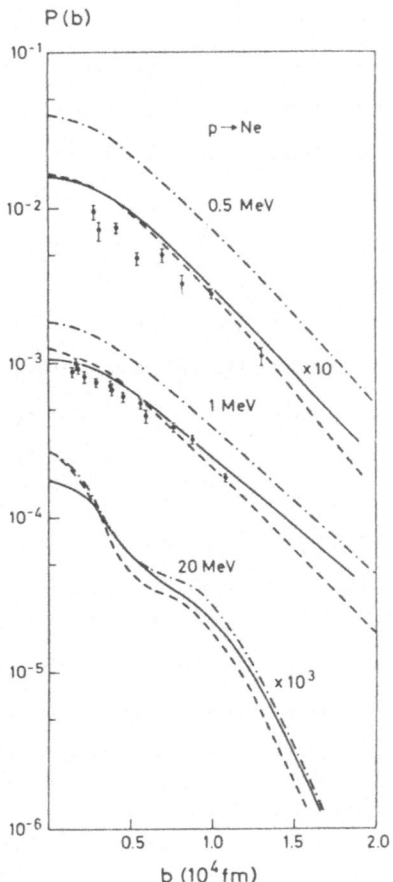

Fig. 25.   K-capture probabilities for p on Ne as a function of
impact parameter. Curves are SCIA calculations of
Jakubassa-Amundsen and Amundsen, full; ---- SCIA peaking;
-·-·, Briggs[23] Peaking (from ref. 24). Data are from
Horsdal-Pedersen et al., (ref. 47).

and scaled velocity parameter v/v_K. (In the OBK treatment[48] for
1s – 1s transfer and $Z_B \gg Z_A$,

$$P\ (b) = (Z_A/Z_B)^5\ [x^2\ K_2(x)]^2\ (4/(1 + s^2/4))(v/v_K)^2$$

where $x = b\ Z_B\ (1 + s^2/4)^{1/2} \approx b\ Z_B$ for $v \approx v_K$. K_2 is the second
order modified Bessel function and s is defined in eq. 7.) In

Fig. 25 are shown data for p, He^{+2} and Li^{+3} on C, Ne and Ar,[49], scaled according to the above prescription. The scaling law, as well as the absolute probabilities, are in excellent agreement with the data, even though the probabilities cover a dynamic range of one to 640. Recent coupled channels calculations by Ford, Reading and Becker[50] show also excellent agreement with this data for differential scattering cross sections.

      Although the circumstantial evidence that the Thomas double scattering plays an essential role in high energy capture, is there any direct experimental signature of this process which would clearly isolate the process? Perhaps. It is probably hopeless to detect the difference between a v^{-11} and v^{-12} behavior in σ_{CK} at high v. However, calculations of $d\sigma/d\Omega$ in the 2nd Born approximation for p on H show that, at extremely high v, a peak appears in the angular distribution at an angle of 0.03° corresponding to the scattering with a free electron which sends the electron off at 60°. (Fig. 27) Although the second born amplitudes are quite important for smaller v, this peak, which corresponds to the on-shell part

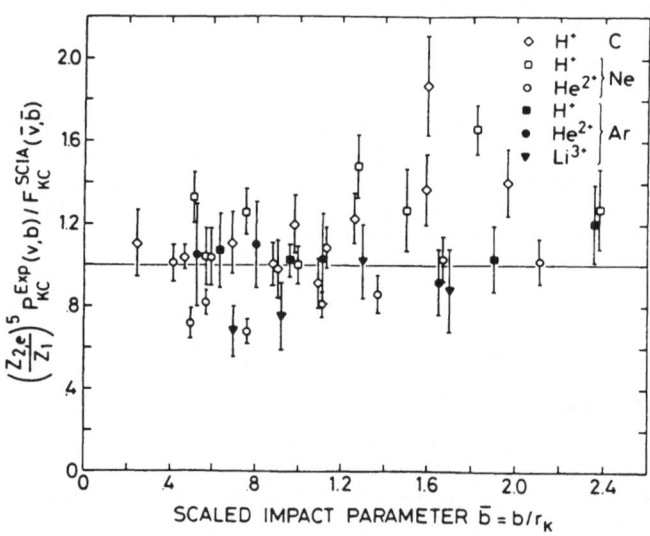

Fig. 26. Scaled probabilities for K-electron capture from Horsdal-Pedersen and Pedersen, (ref. 49). The scaling law is that for the SCIA (ref. 24).

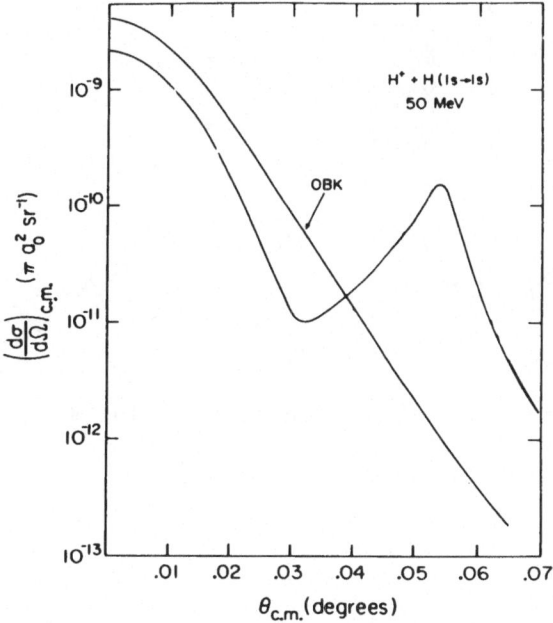

Fig. 27. Differential cross sections for electron capture by p
          from H, calculated in the second Born approximation by
          Simony and McGuire (ref. 28).

of the double scattering amplitude only, does not appear clearly
before 10 MeV, at which point $d\sigma/d\Omega$ is quite small. The experiment to detect this peak will be, at best, difficult. Nevertheless, such a peak would be a rather unambiguous signature of the
Thomas process.

     C. The Symmetric Case

     In 1962 Everhart and collaborators found that if one
fixes the impact parameter of the collision, the K-capture prob-
ility for the symmetric case of H$^+$ on H oscillates as a function
of velocity.[51] Qualitatively, this behavior has a simple explanation in terms of a two-quantum state double well model shown
schematically in Fig. 28. If the electron is initially centered
in the left hand well, its wave function will be a linear combination of the even (1sσ_g) and odd (1sσ_u) MO's. As the collision

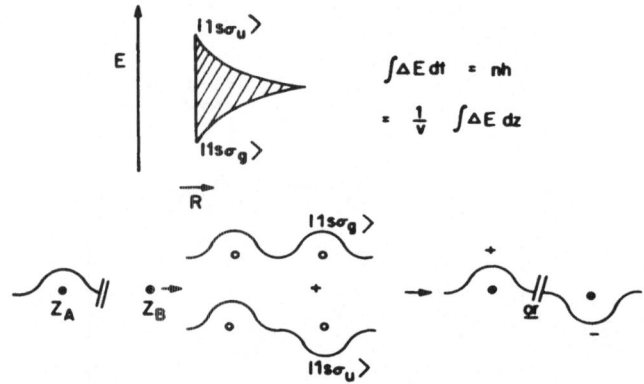

Fig. 28. Schematic representation of resonant K- to K- electron
          transfer.

proceeds, the phase accumulations of these two states are different due to the g-u energy splitting, so that the final state will
still be a linear combination of $1s\sigma_g$ and $1s\sigma_u$ states but with a
relative phase given by exp [i $\int$ ($\omega_u - \omega_g$)dt]. If this factor is
-1, the capture probability is 100%. If it is 1, the probability
is zero. For a fixed trajectory, the probability will clearly
oscillate regularly with (1/v) with a frequency determined by
the phase integral $\int(\omega_u - \omega_g)$ dz. Physically, the electron
cloud oscillates back and forth between the charge centers.
Such oscillations will generally be washed out in a total
cross section measurement by integration over b.

       While this process has been studied in some detail for
light systems by Everhart et al.,[51] only recently have analogous
oscillations been seen in the transfer between inner shell of
higher charged systems. Some evidence for such oscillations
was reported by Schuch et al.[52] in S - Ar collisions. More
recent data of Hagmann et al.[53] for the F - Ne system are shown
in Fig. 29. The oscillation is seen here as a function of b at
fixed v, as opposed to the Everhart experiment, but the qualitative

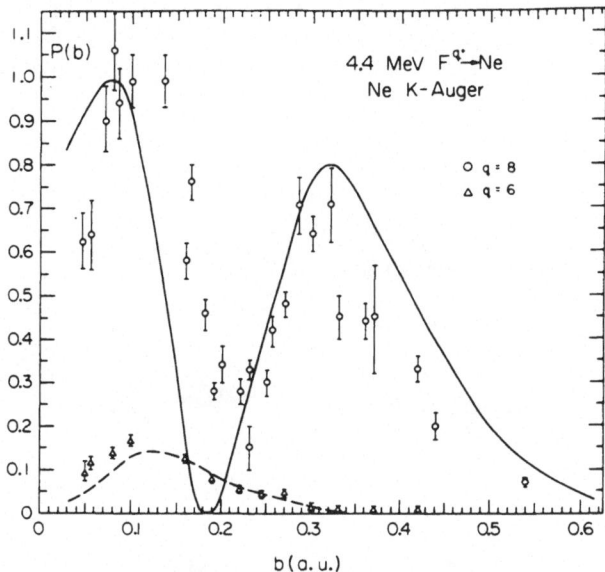

Fig. 29. Ne K-Auger production probabilities for two charge states
of theprojectile fluorine. For q = 8, K-K transfer
dominates the process. The data are from Hagmann et al.,
(ref. 54); the solid lines are a TSAE calculation by Lin
et al.

reason for the dip in P(b) is the same.

As in the total cross section case, a MO or AE treatment
seems appropriate for such a system. The solid lines in Fig. 29
are the TSAE results of Lin et al.[54] and show very good agreement
with the data. It should be pointed out that this process takes
place at very large b, about 4 x r_K. In order to show graphically
what is occurring physically to the electron cloud during the
collision, Day et al.,[55] have used the AE time dependent wave
functions to plot the cloud density as a function of time and
impact parameter, as shown in Figs. 30 and 31. The oscillation
of the cloud between the two centers is evident and helps us to
gain insight into inner-shell transfer in the symmetric case.

Additional reviews are cited in refs. 57, 58.

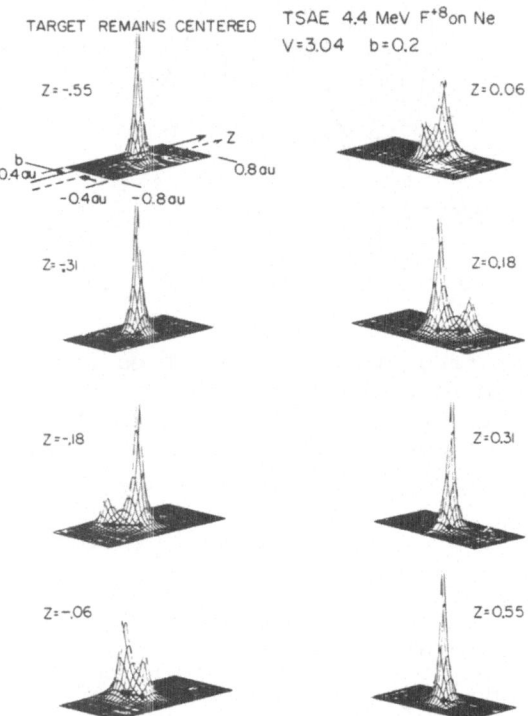

Fig. 30.   The absolute square of the electronic wave function,
           calculated at different times (corresponding to
           different values of Z) in the TSAE calculation (ref. 56).
           Parameters correspond to a minimum in the transfer
           probability.   The target is centered in each picture.

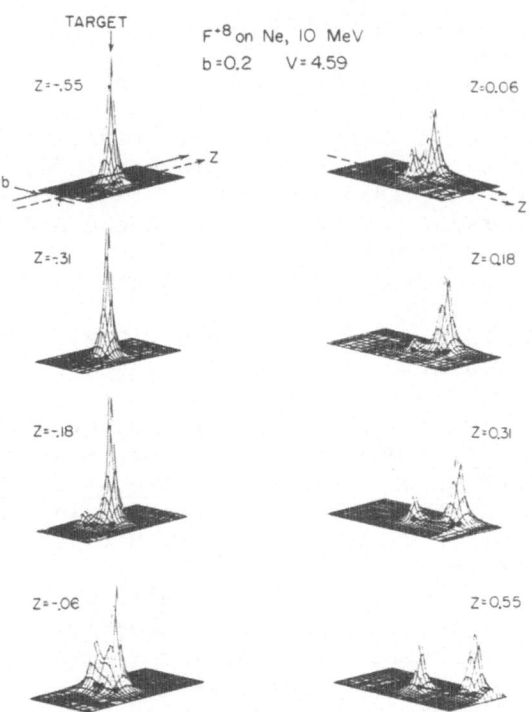

Fig. 31.   As Fig. 30, but for collision parameters corresponding
to a maximum in the transfer probability.

Acknowledgments

The author acknowledges helpful discussions with J. H.
McGuire and C. D. Lin.   This work was supported in part by the
Chemical Sciences Division of the U.S. Department of Energy.

References

1. H. Tawara and A. Russek, Rev. Mod. Phys. 45, 178 (1973).
2. R. A. Mapleton, The Theory of Charge Exchange, (Wiley-Interscience, New York, 1972).
3. D. H. Madison and E. Merzbacher in Atomic Innter Shell Processes, ed. B. Crasemann, p. 1 (Academic Press, New York, 1975).
4. J. R. Macdonald, C. L. Cocke and W. Edson, Phys. Rev. Lett. 32, 648 (1974).
5. J. R. Macdonald, L. M. Winters, M. D. Brown, L. D. Ellsworth, T. Chiao and E. Pettus, Phys. Rev. Lett. 30, 251 (1973); and Phys. Rev. A8, 1835 (1973).
6. M. Rødbro, E. Horsdal Pedersen, C. L. Cocke and J. R. Macdonald, Phys. Rev. A19, 1936 (1979).
7. C. Woods, Ph.D. dissertation, Kansas State University, Manhattan, Kansas 66506 (1975).
8. J. R. Oppenheimer, Phys. Rev. 31, 349 (1928).
9. H. C. Brinkman and H. A. Kramers, Proc. Acad. Sci. Amsterdam 33, 973 (1930).
10. M.R.C. McDowell and J. P. Coleman, Introduction to the Theory of Ion-Atom Collisions (North-Holland, Amsterdam, 1970).
11. E. Merzbacher and H. W. Lewis, Encyclopedia of Physics, ed. Flügge, Vol. XXXIV, p. 166 (Springer Verlag, Berlin, 1958); G. G. Khandewahl, B. A. Choi and E. Merzbacher, Atomic Data 1, 103 (1969).
12. C. D. Lin, S. C. Soong and L. N. Tunnell, Phys. Rev. A17, 1646 (1978); C. D. Lin, J. Phys. B11, L185 (1978).
13. C. D. Lin, J. Phys. B11, L185 (1978).
14. D. R. Bates, Proc. R. Soc. A247, 294 (1958).
15. J. Hall, Ph.D. dissertation, Kansas State University, 1979.
16. V. S. Nikolaev, Zh. Eksp. Teor. Fig. 51, 1263, (1966) [Sov. Phys. JETP 24, 847 (1967)].
17. C. L. Cocke, R. K. Gardner, B. Curnutte, T. Bratton and T. K. Saylor, Phys. Rev. A16, 2248 (1977).
18. J. D. Garcia, Phys. Rev. A1, 280 and 1402 (1970).
19. L. H. Thomas, Proc. R. Soc. London 114, 561 (1927).
20. R. M. Drisko, Ph.D. Thesis, Carnegie Inst. Tech., 1955.
21. R. Shakeshaft and L. Spruch, Rev. Mod. Phys. 51, 369.
22. J. Macek and S. Alston, Phys. Rev. A26, 250, 1982.
23. J. Briggs, J. Phys. B10, 3075 (1977).
24. D. Jakubassa-Amundsen and P. Amundsen, Z. Physik A 297, 203, (1980); -- J. Phys. B14, L705, (1981).
25. J. Macek, R. Shakeshaft, Phys. Rev. A22, 1441 (1980).
26. J. Macek and K. Taulbjerg, Phys. Rev. Lett. 46, 170 (1981).
27. L. Kocbach, J. Phys. B13, L665 (1980).
28. J. McGuire and P. Simony, J. Phys. B14, L737 (1981).
29. J. Eichler and T. Chan, Phys. Rev. A20, 104 (1978).
30. A. L. Ford, J. R. Reading and R. L. Becker, Phys. Rev. A23, 510 (1981).

31. J. Briggs, J. Macek and K. Taulbjerg, Comments on At. Mol. Sci., to be published (1982).

32. R. Shakeshaft, Physics of Atomic and Electronic Collisions, ed. Datz, p. 123 (North-Holland, Amsterdam, 1982.

33. Dž. Belkić, R. Gayet and A. Salin, Phys. Rev. 56, 279 (1979).

34. S. Adriamonje, J. F. Chemin, J. Routurier, H. Laurent and J. O. Schapila, ICPEAC XII, p. 657 (1980).

35. E. Horsdal-Pederson, private communication (1982).

36. T. A. Green, H. E. Stanley and Y. C. Chiang, Helv. Phys. Acta 38, 109 (1965); B. A. Bransden and L. T. Sin Fai Lam, Proc. Phys. Soc. London 87, 653 (1966).

37. C. D. Lin and L. N. Tunnell, Phys. Rev. A22, 76 (1980); and private communication (1982).

38. N. Bohr, Mat.-Fys. Medd.-K. Dan. Vidensk. Selsk. 18, No. 8 (1948).

39. R. McCarroll and A. Salin, J. Phys. B1, 163 (1968).

40. Dž. Belkić and A. Salin, J. Phys. B11, 3905 (1978).

41. C. L. Cocke, J. R. Macdonald, B. Curnutte, S. L. Varghese and R. Randall, Phys. Rev. Lett. 36, 782 (1976).

42. T. R. Bratton, C. L. Cocke and J. R. Macdonald, J. Phys. B, 10, L517 (1977).

43. P. J. Martin et al., Phys. Rev. A23, 2858 (1981).

44. R. D. Rivarola, R. D. Piacentini, A. Salin and Dž. Belkić J. Phys. B 13, 2601 (1980).

45. C. D. Lin, Phys. Rev. A18, 499 (1978).

46. J. H. McGuire, private communication (1976).

47. E. Horsdal Pedersen, F. Folkman and N. H. Pedersen, J. Phys. B 15, 739 (1982).

48. J. H. McGuire and C. L. Cocke, private communication (1974).

49. E. Horsdal Pedersen and N. H. Pedersen, J. Phys. B, to be published (1982).

50. A. L. Ford, J. R. Reading and R. L. Becker, to be published (1982).

51. G. T. Lockwood and E. Everhart, Phys. Rev. 125, 567 (1962); H. Helbig and E. Everhart, Phys. Rev. A136, 674 (1964).

52. R. Schuch, G. Nolte, H. Schmidt-Böcking and W. Lichtenberg, Phys. Rev. Lett. 43, 1104 (1979).

53. S. Hagmann, et al., Phys. Rev. A25, 1918 (1982).

54. C. D. Lin and M. Day, private communication (1981).

55. M. Day and C. D. Lin, private communication (1982).

56. C. D. Lin and P. Richard, Advances in Atomic and Molecular Physics, Vol. 17, 275 (Academic Press, 1981).

57. E. H. Pedersen, Physics of At. El. Coll., ed. S. Datz, p. 139 (North-Holland, Amsterdam, 1982.

CHARGE EXCHANGE PROCESSES INVOLVING MUTICHARGED IONS:

THE QUASI-MOLECULAR APPROACH

Michel Barat

Laboratoire des Collisions Atomiques et Moléculaires
Université Paris-Sud, Bât. 351
91405 Orsay Cedex, France

INTRODUCTION

The physics of multicharged ions has considerably developed during the last years[1]. Aside from purely academic and fundamental interest, the initial motivation arose from fusion research. Indeed from simple theoretical arguments it could be predicted that cross sections for charge exchange between highly charged impurities present in tokomak devices, and hydrogen atom would be very large (10^{-15}-10^{-14} cm^2). Knowledge of the plasma properties would therefore require measurement of the corresponding cross sections as well as a better understanding of the collision process. A large number of experiments were therefore undertaken to determine the *total* (all processes included) electron capture cross section for a variety of multicharged ions. In contrast, very little work was devoted to the identification of the states which are effectively populated by the collision process. A similar effort was undertaken in theory, for the collision energy range of interest for nuclear fusion (say few keV/Nucleon). At these energies the electron velocity v is much larger than the collision velocity V so that the electron motion adapts itself *adiabatically* (or quasi-adiabatically) to the nuclear motion as in a molecule. In this quasi-molecular description of atomic collisions, inelastic processes are induced by *non-adiabatic* (dynamic) couplings between molecular states represented by potential energy curves. Let's take the example of charge exchange processes involving multicharged ions (fig. 1). As will be shown, in section II, the incident channel $A^{q+} + B$ is represented by a potential energy curve which is essentially flat at large internuclear distance in contrast with the energy curves of exit channels,

$A^{(q-1)+} + B^+$, which, due to the Coulomb interaction, displays a repulsive shape. Transitions from the incident channel towards exit channels are enhanced at the crossing between the corresponding energy curves. Since the crossing distance R_c is usually of the order of 10 a.u. (atomic units), the geometrical cross-section would, correspondingly, be very large.

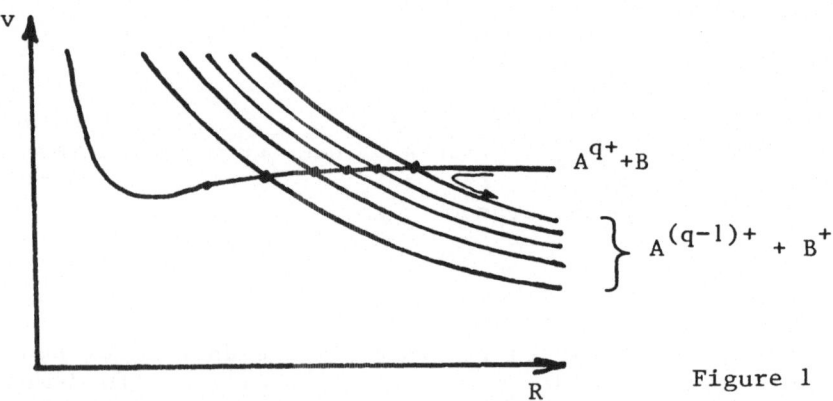

Figure 1

These lectures will be divided into three sections, section I describes the general formalism of the quasi-molecular model of atomic collisions ; section II focuses on the charge exchange process and on various simple models which can be applied to this problem. In the last section, application of the molecular model to collisions involving multicharged ions will be discussed.

SECTION I

THE QUASI-MOLECULAR MODEL

I.1 - GENERAL THEORY. TIME INDEPENDENT FORMALISM

Following the ideas introduced by Born and Oppenheimer[2] The *total* hamiltonian of the colliding system is divided into two parts : (Fig. 2)

(1) $\mathcal{H} = -\dfrac{1}{2M} \vec{\nabla}_R^2 + H_{el} = T_R + H_{el}$

($\hbar = 1$, atomic units will be used throughout these lectures) where the motion of the nuclei (T_R) is separated from the electronic hamiltonian (H_{el}).
The total wave function $\Psi(\vec{R}, \vec{r})$ solution of the Schrödinger equation.

(2) $\mathcal{H} \Psi = E \Psi$

is expanded onto an orthonormal basis set $|\phi_i> - |i>$ which depends parametrically on R :

(3) $\Psi = \sum_i F(\vec{R}) \; \phi_i(R,\vec{r})$, $< \phi_i \; |\phi_j > = < i| \; j > = \delta_{ij}$

Figure 2

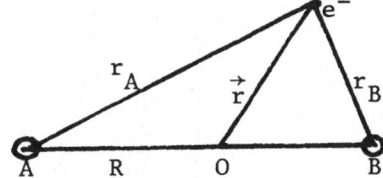

Replacing (3) and (1) in the Schrodinger equation (2), multiplying at left, by $<\phi_j|$ and integrating on the electronic coordinates, one gets the set of coupled equations :

(4) $\left(T_R + < j| H_{el} |j > + \frac{1}{M} < j|\vec{\nabla}_R| \; j >\vec{\nabla}_R - E \right) F_j(\vec{R}) =$

$- \sum_{i \neq j} \left(< j| H_{el} |i > + \frac{1}{M} < j| \vec{\nabla}_R| \; i > \vec{\nabla}_R \right) F_i(\vec{R})$

(the second order terms in $<|\nabla^2|>$ has been neglected)

I.2 - ADIABATIC REPRESENTATION

In traditional molecular physics, the basis functions $|\phi_i >$ are chosen to be eigenfunctions of H_{el}.

(5) $H_{el}| \phi_i >= \varepsilon_i (R) |\phi_i >$, $< i|H_{el}| \; i > = \varepsilon_i(R)$

$< i|H_{el}| \; j > = \varepsilon_{ij} = 0$

(4) can then be written :

(6) $\{ -T_R + (E - \varepsilon_j(R)) \} F_j(R) = \sum_{i \neq j} \frac{1}{M} < j|\nabla_R| \; i> \nabla_R F_i(R)$

Potential scattering (one channel)

In the absence of channel coupling ($< j|\nabla_R| \; i> = 0$)

(6) reduces to :

(7) $(-T_R + E - \varepsilon_j(R)) F_j(R) = 0$

describing the scattering by a potential $\varepsilon_j(R)$ (Fig. 3a) in channel j (elastic collision)

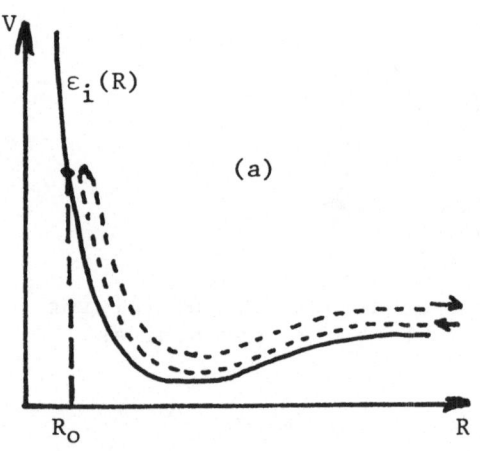
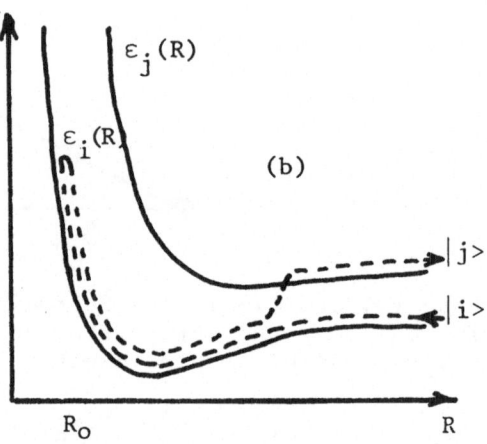

In a (semi-) classical description R_o is the distance of closest approach
--- classical path describing the elastic collision

Representation of a transition between channel $|i>$ and $|j>$ described respectively by potential $\varepsilon_i(R)$ and $\varepsilon_j(R)$

Figure 3

Non adiabatic transitions (Fig. 2b). Transitions between channels $|i>$ and $|j>$ is induced by the *dynamical coupling term* $\frac{1}{M} <j|\nabla_R|i> \nabla_R$ in eq. (6).

Actually the physical meaning of this dynamical coupling is best felt when a semi-classical description is used.

I.3 - TIME DEPENDENT TREATMENT. THE VARIOUS TYPES OF COUPLING

Using the time dependent version of the semi-classical formalism, the expansion (3) is now written as[3] :

(8) $\quad \Psi(t) = \sum a_i(t)\, \phi_i\,(t,\vec{r})\; e^{-i\int^t \varepsilon_i\,(\vec{R}(t'))dt'}$

where $|a_i(t)|^2$ represents the probability that the colliding system be in the state $|i>$ at time t. The set of equations (4) then takes the form[3] :

(9) $i\dfrac{da_j}{dt} = \sum_{j\neq i} \left(+ \varepsilon_{ij}(R) - i <j|\dfrac{\partial}{\partial t}|i> \right)\, a_i\, e^{-i\int^t(\varepsilon_j - \varepsilon_i)dt'}$

which reduces in the adiabatic representation (see eq.5) to :

(10) $\dfrac{da_j}{dt} = -\sum\limits_{j\neq i} a_i < i \mid \dfrac{\partial}{\partial t} \mid j > e^{-i\int^{t}(\varepsilon_j-\varepsilon_i)dt'}$

the coupling terms $< i \mid \dfrac{\partial}{\partial t} \mid j >$ can be rewritten using the "classical trajectory" relation R(t). One has (Fig. 4)

(11) $\dfrac{\partial}{\partial t} = \dfrac{\partial}{\partial R} \cdot \dfrac{dR}{dt} + \dfrac{\partial}{\partial \theta} \dfrac{d\theta}{dt} = v_R \dfrac{\partial}{\partial R} + \dfrac{V_o b}{R^2} \dfrac{\partial}{\partial \theta}$

b is the impact parameter and V_o the incident collision velocity, v_R the "radial" velocity

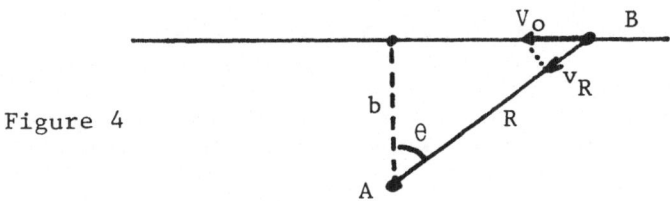

Figure 4

the coupling term is now written as the sum of two terms :

(12) $< i \mid \dfrac{\partial}{\partial t} \mid j > = v_R < i \mid \dfrac{\partial}{\partial R} \mid j > + \dfrac{V_o b}{R^2} < i \mid iL_y \mid j >$

(with $\dfrac{\partial}{\partial \theta} = iL_y$)[4]

The first term $v_R < i \mid \dfrac{\partial}{\partial R} \mid j >$ called *radial coupling* couples molecular states having the same symmetry ($\Sigma\leftrightarrow\Sigma$ $\Pi\leftrightarrow\Pi$...)

The second term called *rotational coupling* couples molecular states, obeying the selection rule $\Delta \Lambda = \pm 1$, (Λ being the projection of the angular momentum along the internuclear axis).

I.4- DIABATIC REPRESENTATION

        As an alternative to the adiabatic representation, one can look for states $\mid \phi_i >$ which do *not* diagonalize H_{el} $< \phi_i \mid H_{el} \mid \phi_j > \neq 0$ but *minimize* the dynamical coupling terms. Such states are called "diabatic states"[5]. In the semi-classical

approach, we can define, in principle, two types of diabatic
states, depending whether the coupling is of the radial or rotational types, since states which are coupled, are different in
these two cases. However diabatic states are usually defined with
respect to radial coupling.

    In fact, there exists no rigorous definition of diabatic
states. However, diabatic states should follow the "prescription":

$$(13) \quad v_R < i|\frac{\partial}{\partial R}|j > \quad \ll \varepsilon_{ij}(R)$$

Remarks : (i)   The formal definition proposed by F.T. Smith[5]

$< i|\frac{\partial}{\partial R}|j > = 0$ leads, if the $\{|\phi_i>\}$ basis set is complete, to

$\frac{\partial}{\partial R}|\phi_i> = 0 \quad \phi_i(R) = c^{te} = \phi (R \to \infty)$. This is the case for a
common *atomic* basis which then constitutes the crudest diabatic
basis.

        (ii) through the condition (13.1), it is readily seen
that the choice of diabatic states may depend on the collision
velocity range.

        For each category of collision problem, the underlying
physics should be the guide to search for the most appropriate
diabatic representation. In this spirit, one can look for diabatic
states within the perturbation theory. ε_{ij} can be considered as
a small perturbation, inducing the inelastic transitions. Quasi-
molecular diabatic states are calculated in diagonalizing *all*
interactions except the one responsible for the transition. For
example, in a one-electron capture process, all interactions
(coulombic, polarization etc..) are diagonalized *except the
exchange interaction*.

I.5 - AN IMPORTANT EXAMPLE : THE MOLECULAR CURVE CROSSING

        The Wigner-Von Neuman[6] non crossing rule tells us that
two adiabatic potential curves of same symmetry do *not* cross (Fig.
5). At the vicinity of an avoided crossing ($R \simeq R_C$), the wave
functions suffer a sudden change of character, which is reflected
in the strongly peaked behaviour of the $<|\partial/\partial R|>$ matrix element.
In such a case, diabatic states are those which cross at R_C and
merge with the adiabatic states outside the crossing. The *smooth*
behaviour of these states results in a very weak $<|\partial/\partial R| >$.
The coupling is given by $\varepsilon_{12}(R)$ which also displays a smooth
behaviour (an interesting feature for practical calculations).

Adiabatic representation 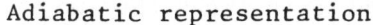   Diabatic representation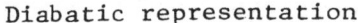

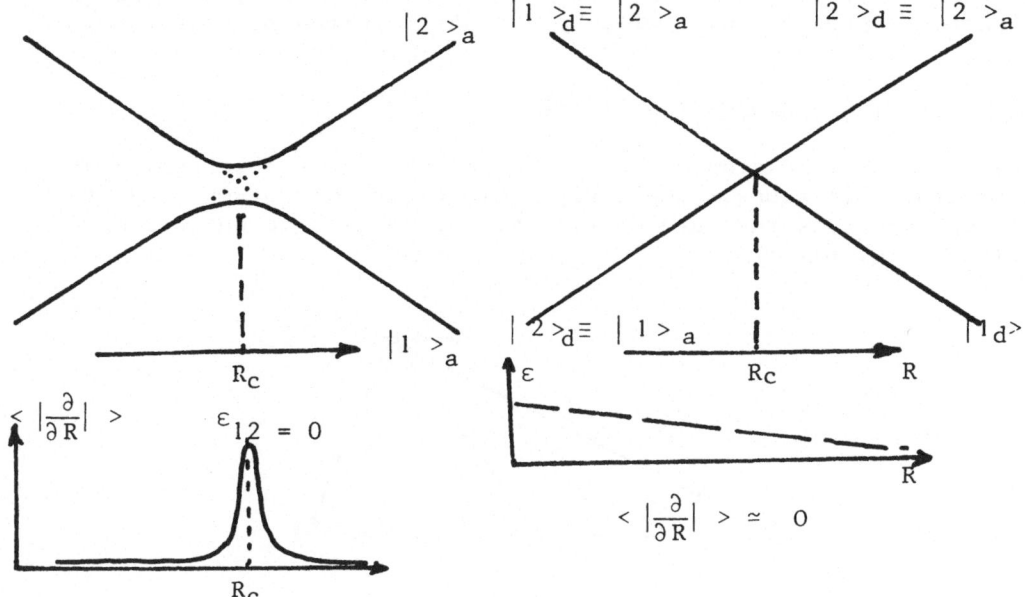

Figure 5

II.6 - COUPLED EQUATIONS WITH DIABATIC STATES

The set of eq.(9) reduces to

$$(14)\ i\frac{da_j}{dt} = \sum_{i \neq j} a_i(t)\ \varepsilon_{ij}(R)\ e^{-i\int^t (\varepsilon_i - \varepsilon_j)dt'}$$

Fortunately, in many cases of practical interest, the coupling between many states can be seen as a successive set of coupling between two states (see section II). The problem is then reduced to solving the two coupled equations

$$(15)\ i\frac{da_1}{dt} = a_2\ \varepsilon_{12}(R)\ e^{i\int^t \Delta\varepsilon\ dt'}$$

$$i\frac{da_2}{dt} = a_1\ \varepsilon_{12}(R)\ e^{-i\int^t \Delta\varepsilon dt'}$$

where $\Delta\varepsilon = \varepsilon_1(R) - \varepsilon_2(R)$ is the energy separation between the two potential curves. The behaviour of the interaction $\varepsilon_{12}(R)$ for the electron capture problem will be discussed in the next section.

I.7 - ELECTRON TRANSLATION FACTOR

As formulated above, the quasi-molecular description of atom-atom collisions, completely neglects the momentum $m\vec{v}$ of the electrons as they are carried by the nucleus to which they belong (Fig. 6).

Figure 6

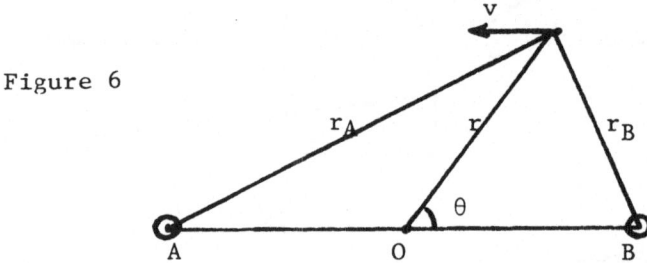

This problem was first pointed out by Bates and McCarroll[7], 25 years ago. Describing the electron capture in H^+H collisions with an atomic expansion, they demonstrated that the atomic wave functions $\chi(\vec{r}_A)$, $\chi'(\vec{r}_B)$ behave incorrectly in the asymptotic region $R \to \infty$. They showed that this drawback is removed if the atomic wave function is multiplied by an electron translation factor (ETF) describing the "traveling" of the electron cloud with the corresponding nucleus. In the semi-classical time dependent description the wave function takes the form (see Fig. 6) :

$$(16) \quad \phi_i(\vec{r}_A, t) = \chi_i(\vec{r}_A) \, e^{-im \frac{\vec{v}}{2} \cdot \vec{r}} \, e^{-i\left(\varepsilon_n + \frac{1}{2} m(\frac{v}{2})^2 \, t\right)}$$

Inclusion of ETF in the framework of the quasi-molecular model is a much more difficult task and the problem can still be considered open[8]. Surprisingly during a long time, this aspect has been neglected or considered as purely academic. It has been shown that, in practice, neglect of ETF leads to spurious long range coupling[9]. Moreover it is found that the result of cross section calculations may depend on the choice of the electron coordinate origin[10]. Since a decade, this problem has been attacked by many investigators and several practical "recipes" has been proposed. A detailed discussion of this problem is outside the scope of this lecture and only the basic ideas and some of these recipes will be mentionned. Although this problem has also been formulated in a fully quantal treatment of the collision[11] most of the works have been carried out within the semi-classical treatment. All attempts to solve this problem start with a "molecular" basis set of the form :

(17) $\quad \Psi = \Sigma \; a_n(t) \; \phi_n(\vec{R},\vec{r})$

$\quad \phi_n(\vec{R},\vec{r}) = \chi_n(\vec{R},\vec{r}).e^{i \; S_m(\vec{R},\vec{r})} \; e^{-i \int \epsilon_n dt'}$

with $\epsilon_n = \; < \phi_n | \; H_{el} | \phi_n >$

The collision problem consists then of solving a set of coupled equations similar to eq.(9) with the new set of wave function ϕ_n. The choice of the best $S_n(\vec{R},\vec{r})$ is, of course, the preoccupation of the theoretical approaches. In any case, $S_n(\vec{R},\vec{r})$ should asymptotically match the atomic ETF of Bates and McCarrol :

$$S(\vec{R},\vec{r}) \rightarrow \alpha \; .\vec{V}.\vec{r}$$

where the proportionality factor depends on the coordinate origin. One simple choice is to neglect the ETF and select the most appropriate coordinate origin (O). In particular Piacentini and Salin[12] show that, if the origin is taken at the nucleus on which the active electron is initially attached (incident channel), the total (all channels included) electron capture cross section is correctly reproduced but the individual charge exchange cross sections cannot be estimated correctly.

Guided by the asymptotic behaviour of the ETF, several authors look for ETF of the form

(18) $\quad S'_n = m \; f(\vec{R},\vec{r}) \; \vec{V}.\vec{r}$

with f being a "switching function" obeing the asymptotic condition

$$f \rightarrow \beta \qquad \text{for } r_A \gg r_B$$
$$f \rightarrow -1+\beta \quad \text{for } r_B \gg r_A$$

where β depends on the coordinate origin. For example, in their study of the $H^+ + H$ resonant charge exchange, Schneiderman and Russek[14] proposed the arbitrary form :

(19) $\quad f = \dfrac{\cos\theta}{2} \Big/ (1 + (\dfrac{R}{a})^2)$

θ (see Fig. 6) is defined with the origin at the center of mass of the nuclei and a is an arbitrary parameter : f is common to all wave functions and is seen to easily fullfil the asymptotic conditions

$$f \rightarrow \frac{1}{2} \quad \text{for } r_A \gg r_B$$

$$f \rightarrow -\frac{1}{2} \quad \text{for } r_B \gg r_A$$

Furthermore f vanishes when $R \rightarrow 0$ to account for the fact that, near the "united atom limit", the electrons belong to both atoms A and B. The parameter $\underline{a}$ is characteristic of the radius of the united atom.

A more elaborate choice of the switching function has been proposed by Vaaben and Taulbjerg[15] in the case of one electron systems. The semi-classical scattering equations takes a form similar to eq.(14) if f is chosen as :

$$(20) \qquad f = \frac{1}{2}(\frac{Z_B - \alpha Z_A}{Z_B + \alpha Z_A} + \frac{Z_A - Z_B}{Z_A + Z_B}) \quad \text{with} \quad \alpha = (\frac{r_B}{r_A})^3$$

Z_A and Z_B are the charges of nuclei A and B respectively.

An interesting feature of both approaches described above, is provided by the fact that a *common* ETF is used for all wave functions. This property automatically insures the orthogonality of the wave functions and therefore greatly simplifies the calculations: the coupled equations are those discussed in § 1. with additional dynamical coupling terms of the form :

$$(21) \qquad < \phi_i \, |\vec{v}.\vec{r}.\vec{\nabla}_r f.\vec{\nabla}_r + \vec{v}.\vec{r} \, \frac{1}{2} \, \vec{\nabla}_r^2 f + \vec{v} \, \vec{\nabla}_r f| \, \phi_j >$$

More rigorous attemps[16] to define ETF specifics to each wave function ϕ_i leads to much heavier calculation, since the orthogonality of the wave functions is lost

SECTION II

THE ELECTRON CAPTURE PROCESS. TWO STATE-MODELS

A. POTENTIAL ENERGY CURVES AND INTERACTION

II.1 Typical cases

In the previous section we have shown, that the two important parameters that govern the exchange processes are the interaction $\varepsilon_{12}(R)$ and the difference of the potential energy curves corresponding to the active channels : $\Delta\varepsilon(R) = \varepsilon_2(R) - \varepsilon_1(R)$. Since we will focus on processes having large cross section we will

discuss here transitions occuring at large internuclear distances.
Whereas the interaction has a similar behaviour whatever the
collision system is, very different situations occur depending on
$\Delta E(R)$

(i) Singly charged ion-atom collisions (Fig. 7)

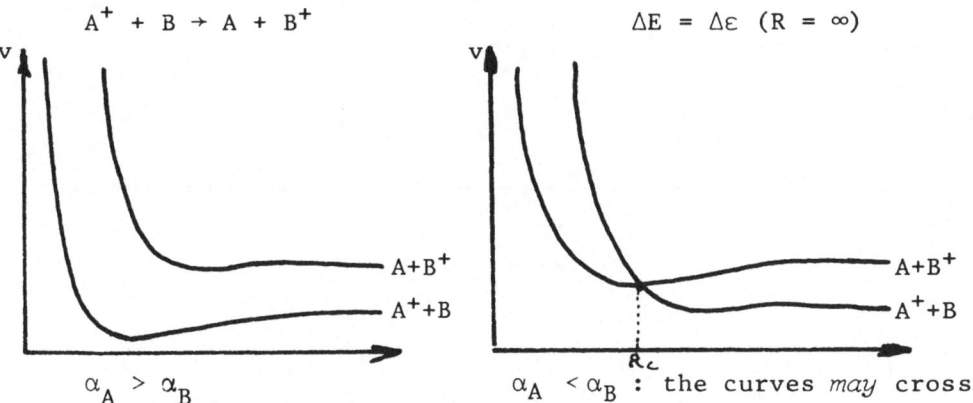

Figure 7

the behaviour of the *diabatic* potential curves is essentially
governed by the polarisability of the neutral partner.

(22) $\qquad \varepsilon_1(R) = - \dfrac{\alpha_B}{R^4} \qquad \varepsilon_2(R) = - \dfrac{\alpha_A}{R^4} + \Delta E \quad \Delta\varepsilon \simeq \dfrac{\alpha_B - \alpha_A}{R^4} + \Delta E$

(ii) Neutralisation $\quad A^+ + B^- \rightarrow A + B$

The attractive Coulomb interaction in the incoming channel leads
often to   curve crossings (Fig. 8a)

$$\varepsilon_1 \simeq \ldots \frac{A}{R^6} \qquad \varepsilon_2 \simeq - \frac{1}{R} + \Delta E \qquad R_c \simeq \frac{1}{\Delta E}$$

this is the well known problem of ionic-covalent interaction in
chemistry.

(iii) Multicharged ion-atom collisions $\quad A^+ + B \rightarrow A^{(q-1)+} + B^+$

The exit channel is dominated by the Coulomb repulsion (Fig. 8b)

(23) $\qquad \varepsilon_1 = - \dfrac{\alpha}{R^4} \qquad \varepsilon_2 = \dfrac{q-1}{R} - \Delta E \qquad R_c \simeq \dfrac{q-1}{\Delta E}$

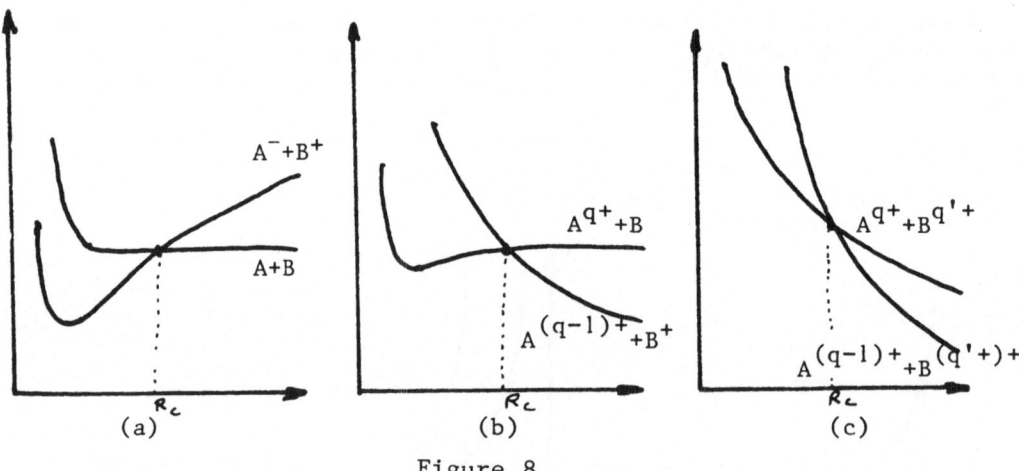

Figure 8

(iv) Ion-Ion Collisions. $A^{q+} + B^{q'+} \rightarrow A^{(q-1)^+} + B^{(q'+1)+}$
(see § III.) both channels have a repulsive character (Fig. 8c)

$$(24) \qquad \varepsilon_1 = \frac{qq'}{R}, \qquad \varepsilon_2 = \frac{(q-1)(q'+1)}{R} + \Delta E \qquad R_c = \frac{q'-q+1}{\Delta E}$$

II.2 The exchange interaction. A simple example:

$$H_A^+ + H_B(1S) \rightarrow H_A(1S) + H_B^+$$

For simplicity we suppose that the atomic wave functions are not perturbed, and that $< 1S_A | 1S_B > \approx 0 \quad \forall R$

$$|\phi_1> = |1S_B>, \qquad |\phi_2> = |1S_A>$$

$$\varepsilon_{12} = <\phi_1|H_{e1}|\phi_2> = <1S_A|-\frac{1}{2}\nabla_r^2 - \frac{1}{r_A} - \frac{1}{r_B}|1S_B>$$

$$(25)$$

$$\varepsilon_{12} = <1S_A|-\frac{1}{2}\nabla^2 - \frac{1}{r_B}|1S_B> + <1S_A|\frac{1}{r_A}|1S_B>$$

$$= E_{1S}<1S_A|1S_B> + <1S_A|\frac{1}{r_A}|1S_B> \approx <1S_A|\frac{1}{r_A}|1S_B>$$

$$|1S_{A,B}> = \frac{2}{\sqrt{4\pi}} e^{-r_{A,B}}$$

$$\varepsilon_{12} = \frac{4}{4\pi} \int e^{-r_A} \frac{1}{r_A} e^{-r_B} d^3r_A$$

Let's introduce the prolate spheroidal coordinates

$$\xi = \frac{r_A + r_B}{2R} \ , \qquad \eta = \frac{r_A - r_B}{2R} \qquad \varepsilon_{12} \text{ is now expressed}$$

(26)
$$\varepsilon_{12} = \frac{1}{\pi} \ \frac{2}{R} \int e^{-(\xi+\eta)\frac{R}{2}} \ e^{-(\xi-\eta)\frac{R}{2}} \frac{i}{\xi+\eta} \ \frac{R^3}{8} \ (\xi^2-\eta^2) d\xi . d\eta . d\phi$$

after integration :

(27) $\varepsilon_{12} = R \ e^{-(R-1)}$, for large R $\varepsilon_{12} \sim R \ e^{-R}$ (28)

As an important consequence, it is seen that the exchange interaction is *exponentially* decreasing at large R.

II.3 Exchange interaction. Asymptotic formula

Very often, the exchange process takes place at relatively large internuclear distance, where the wave-functions have essentially kept their atomic characters. Therefore one can estimate the interaction using asymptotic formula. This approach has mainly been developped by the Soviet School[17]. All attempts start from the Landau-Herring formula[18], which states that the exchange interaction is given by the difference of "flux" of electrons through a surface S perpendicular to the internuclear axis centered at a convenient location R_S. For example, for symmetric charge exchange the choice is obviously $R_S = R_{/2}$. For asymmetric systems R_S can be placed at the top of the potential barrier (see note below and Fig. 9). The Landau-Herring formula is expressed as :

(29) $\varepsilon_{12}(R) = \frac{1}{2} \int_S (\phi_1 \nabla \phi_2 - \phi_2 \nabla \phi_1) \ dS$

ϕ_1 and ϕ_2 are the wave function of the active electron when it is attached to nuclei A and B respectively.

The various results depend on the choice of the wave functions ϕ : hydrogenic[19], hydrogenic perturbed by the other nucleus[20], or quasi-classical[21] . However, all asymptotic formula behave as
$$\varepsilon_{12}(R) = A \ R^\alpha \ e^{-\beta R}$$

Note : The quasi-static description of charge exchange involving multicharged ions[21]. Let's consider for example the reaction
$$A^{q+} + H(1S) \rightarrow A^{(q-1)+} + H^+$$

Fig. 9   represents the static potential created by the two nuclei

*for a given internuclear distance R. It is seen that, due to a
quasi-resonance between $1S_H$ and high Rydberg levels of $A^{(q-1)+}$, an
electron transition can be seen as a tunneling through the barrier
towards the quasi-continuum of $A^{(q-1)+}$ excited states[22].*

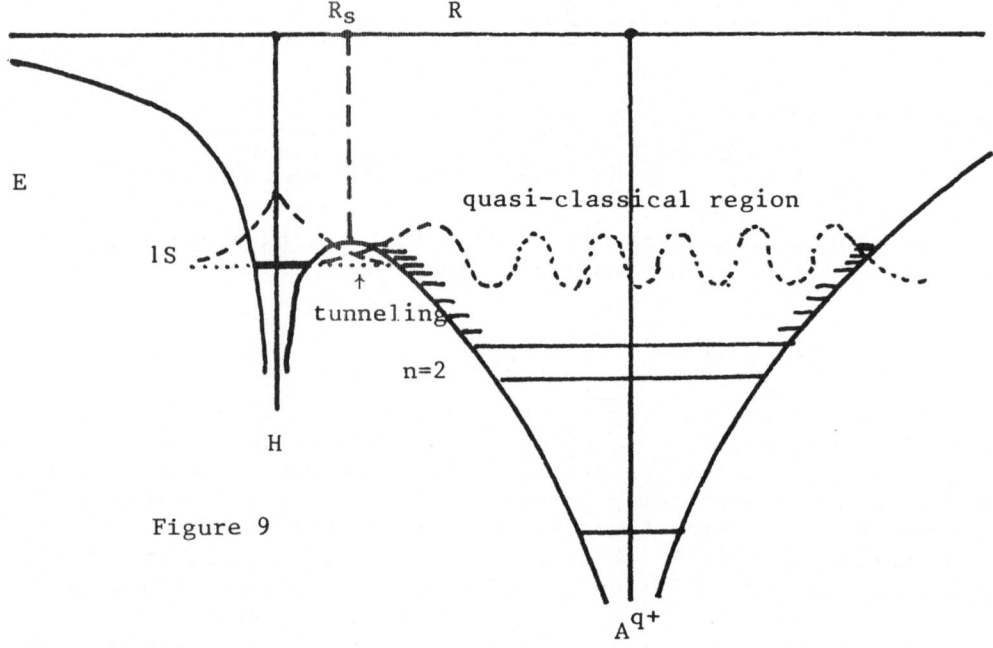

Figure 9

The Landau-Herring method has been applied to various cases,
including neutralisation[23] singly charged ion-neutral[24] and
multicharged ion-neutral charge exchange processes[21,22].

II.4 The exchange interactions : the empirical formula of Olson

Let's consider the charge exchange reaction
$A^{q+} + H \rightarrow A^{(q-1)+} + H^+$. From a compilation of a large number of
exact calculations, Olson[25] shows that the exchange interactions
at the curve crossing are nicely accounted for by a simple
exponential formula

$$(30) \qquad \varepsilon_{12}(R_c) = \frac{9.13}{q^{1/2}}\, e^{-\frac{1.324}{q^{1/2}}R_c}$$

Such a formula provides an easy means to estimate one of the necessary inputs to the Landau-Zener formula (§ II.6)

B. CHARGE EXCHANGE PROBABILITY AND CROSS SECTIONS

II.5 The $H^+ + H(1S) \rightarrow H(1S) + H^+$ resonant charge exchange

Using the atomic treatment (§ II.2) one has $\Delta\varepsilon = \Delta E_\infty = 0$ so that the coupled equations eq.(15) reduce to :

(31) $\quad \dfrac{da_1}{dt} = - ia_2 \, \varepsilon_{12} \qquad \dfrac{da_2}{dt} = -ia_1 \, \varepsilon_{12}$

adding and subtracting these two equations one easily obtains :

(32) $\quad a_1 = A \, e^{-i\int^+ \varepsilon_{12} \, dt} + B \, e^{i\int^t \varepsilon_{12} dt}$

$\quad\quad a_2 = A \, e^{-i\int^+ \varepsilon_{12} \, dt} - B \, e^{i\int^t \varepsilon_{12} dt}$

A and B are determined by the initial conditions :

$\quad\quad a_1(t = -\infty) = 1, \quad a_2(t = -\infty) = 0$

Then the resonant charge exchange probability is given by :

(33) $\quad P(b) = |a_2 (+\infty)|^2 = \sin^2 \int_{-\infty}^{+\infty} \varepsilon_{12}(R) \, dt$

using the impact parameter approximation (straight line trajectories, Fig. 10)

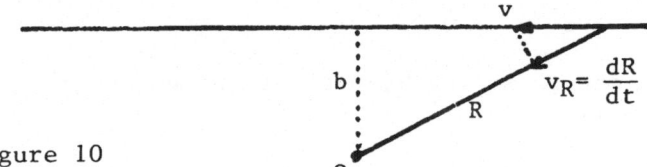

Figure 10

one get $R^2 = b^2 + v^2 \, t^2, \quad \dfrac{dt}{dR} = \dfrac{1}{v} \dfrac{R}{\sqrt{R^2-b^2}}$

(33) becomes

(34) $\quad\quad P(b) = (\tfrac{1}{2}) \, \sin^2 \int_b^\infty \dfrac{\varepsilon_{12}(R) dR}{\sqrt{R^2-b^2}}$

using the $\varepsilon_{12}(R)$ value calculated above (eq. 28)

$\quad\quad P(b) = \sin^2 \dfrac{1}{v} \int_b^\infty \dfrac{R \, e^{-R} \, dR}{\sqrt{R^2-b^2}}$

(35) $\quad\quad = \sin^2 \dfrac{b^2}{v} K_0(b) \simeq \sin^2 \{ \dfrac{2}{v} \sqrt{\dfrac{\pi b}{2}} \, b.e^{-b} \}$

where K_0 is the McDonald's function. The P(b) function is represented on Fig. 11.

The total cross section

$\sigma = 2\pi \int P(b) \, b.db$ can be calculated by splitting the integral into two parts (see Fig. 11)

(36) $\sigma = \sigma_1 + \sigma_2 = 2\pi \int_0^{b_0} P(b) \, db + 2\pi \int_{b_0}^{\infty} P(b) \, db$

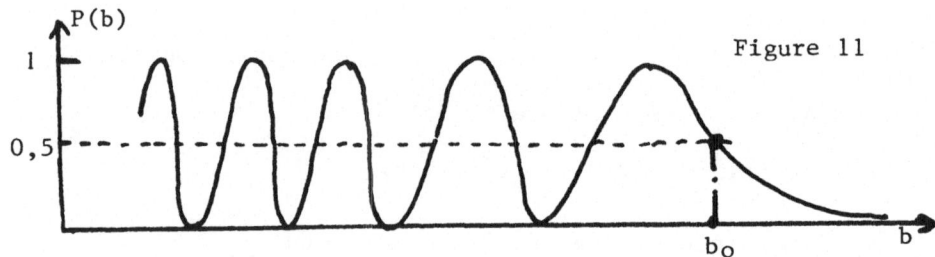

P(b)

Figure 11

b_0

(37)   for σ_1 , $\int \sin^2 \simeq 1/2 \rightarrow \sigma_1 = \dfrac{\pi b_0^2}{2}$

In a first approximation σ_2 can be neglected, then $\sigma \simeq \dfrac{\pi b_0^2}{2}$

Firsov[26] shows that b_0 is such that the argument of the sin function (35) is close to 0.28 :

(38) $\dfrac{2}{v} \sqrt{\dfrac{\pi b_0}{2}} \, b_0 \, e^{-b_0} = 0.28$

b is usually large enough so log b($\ll$ b) can be neglected and we can then extract b_0 from (38)

$b_0 \sim \log \dfrac{2}{0.28} \sqrt{\dfrac{\pi}{2}} - \log v$

the total cross section takes the approximate form :

(39) $\sigma = \dfrac{\pi b_0^2}{2} = \dfrac{\pi}{2} \left(\log \dfrac{2}{0.28} \sqrt{\dfrac{\pi}{2}} - \log v \right)^2$

II.6 The Landau Zener model[27] (LZ)

Approximate solutions of the two coupled equations (15) can be derived in some cases, leading to analytical expressions of the transition probabilities. In particular in the case of a curve crossing Landau and Zener have derived a simple formula which is very useful to estimate the transition probability. In the vicinity of the crossing, the potential energies $\varepsilon_1(R)$ and $\varepsilon_2(R)$ are approximated by their tangents at R_c (Fig. 12). The first important parameter is then the difference of slopes :

(40) $\alpha = \left| \dfrac{d\varepsilon_1}{dR} - \dfrac{d\varepsilon_2}{dR} \right|_{R=R_c}$

Furthermore, the interaction is supposed to be constant and equal to its value at R_c

(41) $\varepsilon_{12}(R) = \varepsilon_{12}(R_c)$

With (40) and (41) equation (15) can be solved approximately and one gets the transition probabilities (Fig. 12a)

$$P_{11} = |a_1(+\infty)|^2 = e^{-\gamma}$$

(42)

$$P_{12} = |a_2(+\infty)|^2 = 1-e^{-\gamma}$$

with $\gamma = \dfrac{2\pi}{v_{R_c}} \dfrac{\varepsilon_{12}^2(R_c)}{\alpha}$

where $v_R = \dfrac{dR}{dt} = v_0(1- \dfrac{V(R)}{E} - \dfrac{b^2}{R^2})^{1/2}$ is the radial velocity

P_{11} and P_{12} correspond to transitions $1\to1$ and $1\to2$, respectively.

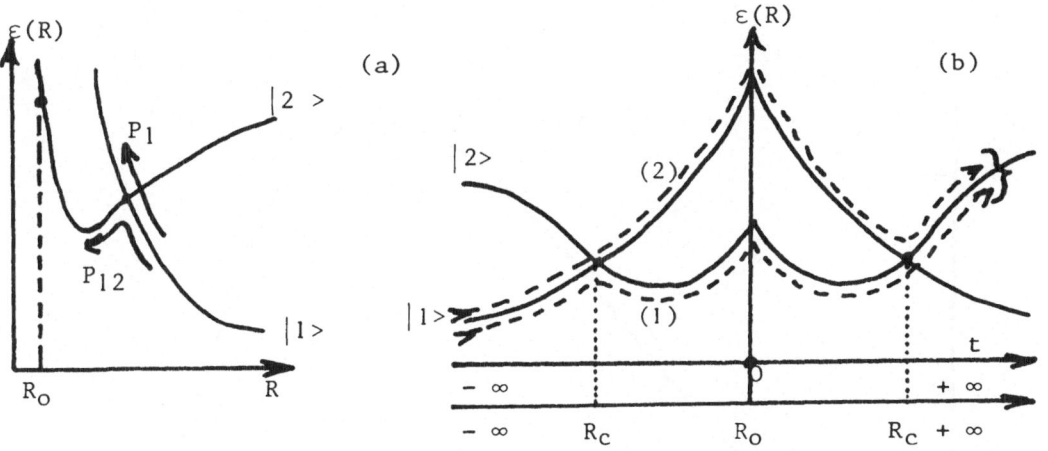

Figure 12

For a *real* collision, the system passes twice through the crossing point (Fig. 12), before and after the turning point R_0 has been reached. The total probability corresponds to the squared modulus of the sum of the amplitudes developed along "paths" 1 and 2 :

$$P = |(a_1(1-a_1))_{(1)} + (a_1(1-a_1))_{(2)}$$

(43)

$$P = 4p_{11}(1 - p_{11}) \sin^2(\Delta\phi + \Delta\omega_{LZ})$$

$\Delta\phi$ is the difference of semi-classical phases developed along paths 1 and 2. $\Delta\omega_{LZ}$ is an additional phase introduced at the crossing[28]. This interference effect, introduced by Stukelberg[24] gives rise to an oscillatory behaviour of the differential cross-section. However when P(b) is integrated along the impact parameter this effect is washed out. So, if only the total cross-section is desired the phase in (43) is averaged:

$$(44) \qquad \tilde{P} = 2p_{11}(1-p_{11})$$

and the total cross section is :

$$(45) \qquad \sigma_{LZ} = 2\pi \int \tilde{P}(b).b \; db = 2\pi \int_o^\infty 2 \; e^{-\gamma}(1 - e^{-\gamma}) \; b \; db$$

which can be written with $v_{R_c} \simeq v_o(1 - \dfrac{V(R_c)}{E})^{1/2}$

$$(46) \qquad \sigma_{LZ} = 4\pi \; R_c^2(1 - \dfrac{V(R_c)}{E}) \; G \; (\gamma)$$

where $G(\gamma)$ is a universal function shown in Fig. 13.

$$G(\gamma) = \int_1^\infty e^{-\gamma x} \; (1 - e^{-\gamma x}) \; \dfrac{dx}{x}$$

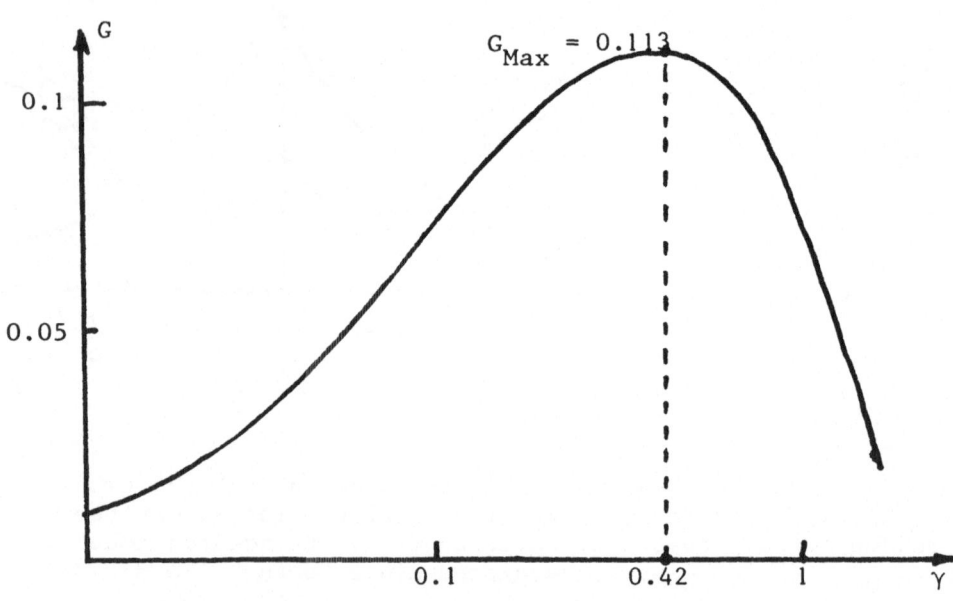

Figure 13

It is worth noting that $\sigma_{Max} = 0.452R_c^2$ and roughly corresponds to half of the geometrical cross-section for $R=R_c$.

II.7   The Demkov model[30]

Let us consider the quasi-resonant charge exchange process

$$A^+ + B \rightarrow A + B^+$$

Neglecting the effect of polarisation forces (§ II.1) one has $\Delta\varepsilon \simeq \Delta\varepsilon(\infty) = Cte$. The Demkov model is particularly well adapted to this problem since this model assumes that :

(47)
$$\Delta\varepsilon = Cte$$
$$\varepsilon_{12} = A\,e^{-\alpha R}$$

Solving the coupled equations (1.15), one gets the transition probability (Fig. 14a) :

(48)
$$P_{12} = |a_2(+\infty)|^2 = \frac{e^{-\gamma}}{1+e^{-\gamma}}$$

with $\gamma = \dfrac{\pi\Delta\varepsilon}{v_R\alpha}$

For an estimation of the transition probability, α can be expressed in terms of the ionization potentials I_A and I_B of atoms A and B, respectively :

(49)
$$\alpha = \frac{1}{\sqrt{2}}\ (\sqrt{I_A} + \sqrt{I_B})$$

and $\Delta\varepsilon = I_A - I_B$

(50)
$$\gamma \simeq \frac{\pi}{v}\sqrt{2}\ (\sqrt{I_A} - \sqrt{I_B})$$

For a real collision, as in the Landau-Zener case, one has to take into account that the system passes twice accross the transition region R_C. R_C being determined by : $\Delta\varepsilon = 2\varepsilon_{12}(R_C)$ [30]. Averaging the oscillations caused by Stueckelberg like interferences (§ II.6, eq (44)) one gets the following formula for the transition probability :

(51)
$$P = 2P_{12}(1-P_{12}) = \frac{2}{(e^{\gamma/2} + e^{-\gamma/2})^2} = \frac{1}{2}\,Sech^2\,\frac{\gamma}{2}$$

The total cross-section is given by

(52)
$$\sigma = 2\pi \int_0^\infty P(b)\ db = \pi \int_0^\infty Sech^2\,\frac{\gamma}{2}\ b\ db$$

which can be expressed in term of a universal function :

(53)
$$\sigma = \frac{\pi R_C^2}{2}\ \sigma^*(\delta^{-1})\qquad \sigma^* = \int_1^\infty \frac{8e^{-\delta x}}{(1+e^{-\delta x})^2}\,\frac{dx}{x^3}\qquad \delta = \gamma(v_R = v_0)$$

σ^* is represented in Fig. 14b. Resolving numerically the coupled equation(15) Olson[31] found a better σ^* function also represented on Fig. 14b.

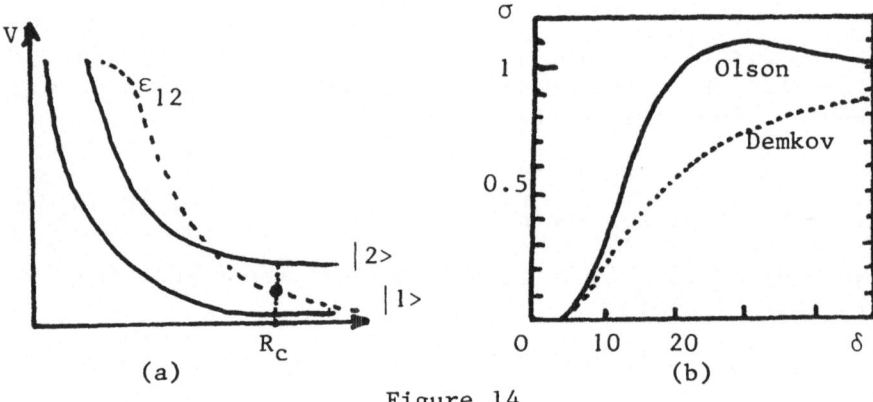

(a)   (b)

Figure 14

II.8 More elaborate models

(i) The exponential model of Nikitin[32]

In this model $\Delta\varepsilon$ and ε_{12} depend on two parameters θ and α :

$$(54) \qquad \Delta\varepsilon(R) = \Delta\varepsilon_\infty \left(1 - \cos\theta \ \exp(-\alpha(R-R_c))\right)$$

$$\varepsilon_{12}(R) = -\frac{\Delta\varepsilon_\infty}{2} \ \sin\theta \ \exp(-\alpha(R-R_c))$$

The limitation of this model arises from the fact that the exponential dependence on R is the same for $\Delta\varepsilon$ and ε_{12}. Typical cases are represented on Fig. 15

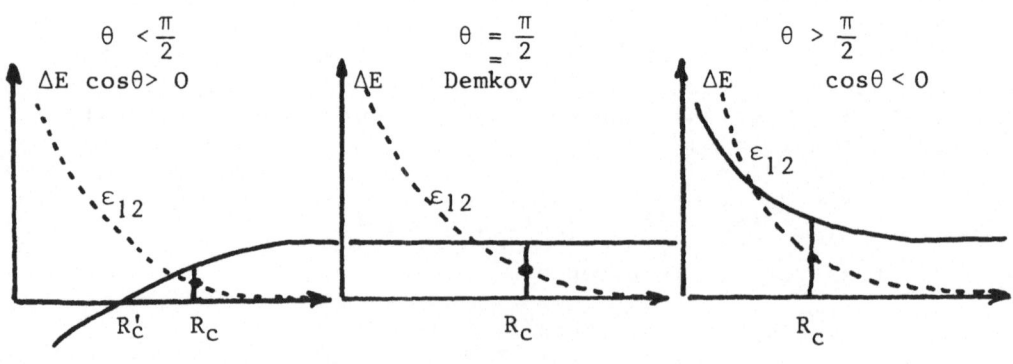

Figure 15

The transition region R_c is such that $\Delta\varepsilon(R_c) = 2\varepsilon_{12}(R_c)(\frac{1-\cos\theta}{\sin\theta})$
The system of coupled equations (15) can again be integrated

analytically and the probabilities for one passage (p_{12}) and for two passages (P) accross R_C are respectively given by

(55) $p_{12} = \exp\left(\frac{\pi\gamma}{2}(1 + \cos\theta)\right) \dfrac{Sh\left(\frac{\pi\gamma}{2}(1-\cos\theta)\right)}{Sh\gamma\pi}$

$P = 2p_{12}(1-p_{12}) = 2e^{\gamma\pi\cos\theta} \cdot \dfrac{Sh\left(\frac{\gamma\pi}{2}(1-\cos\theta)\right)Sh\left(\frac{\gamma\pi}{2}(1+\cos\theta)\right)}{Sh^2\gamma\pi}$

with $\gamma = \dfrac{\Delta\varepsilon_\infty}{\alpha v_R}$

(ii) The exponential linear model of Nikitin[33]

The Landau-Zener model is very useful to determine the transition probability at a curve crossing. However its application to charge exchange processes can be questioned since the actual interaction behaves exponentially. Particularly well adapted to charge exchange involving multiply charged ions, could be the exponential-linear model of Nikitin where

$$\Delta\varepsilon = \alpha\,(R-R_c) \qquad \alpha = \left|\frac{d}{dR}(\varepsilon_1 - \varepsilon_2)\right|_{R=R_c}$$

and $\varepsilon_{12} = A\,e^{-\gamma(R-R_0)}$

Unfortunately no analytical solutions of eq(15) exist in this case. However numerical solutions are given for well selected values of the parameters in ref. 33.

II.9 The Multicrossing problem : Generalized Landau-Zener model.

Let us consider the crossing of a potential curve V_0 with a set of n potential curves v_1 v_2 v_i v_n (Fig. 16). All crossings are supposed to be independent.

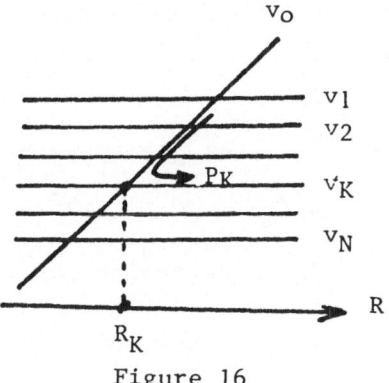

Figure 16

If p_i is the probability that no transition takes place at the v_o-v_i crossing ($p_i = p_{11}$ in § II.6), the final probability $\mathcal{P}_k$ that the system be in channel k is given, for *one* passage by :

(56) $\qquad \mathcal{P}_k = p_1 p_2 \cdots p_{k-1}(1-p_k) p_k = P_k(1-p_k)$

with $P_\ell = p_1 p_2 \cdots p_\ell$

for a real collision (two passages), neglecting interferences, one finds.

$$(57) \qquad \mathcal{P}_k^{(2)} = P_K(1-p_k) + P_{K+1}(1-p_{k+1})^2(1-p_k) + \cdots$$
$$+ P_\ell(1-p_\ell)^2 P_{\ell-1} P_{\ell-2} \cdots P_{k+1}(1-p_k) + \cdots$$

$$\mathcal{P}_N = P_{N-1}(1-p_N)$$

Let us consider now the probability of transitions to *all* states of the series :

$$(58) \qquad \mathcal{P} = \sum_{k=1}^{n} \mathcal{P}_K = \sum_{k=1}^{n} P_K(1-p_k) = 1 - P_n$$

Using the Landau-Zener probability, one obtains

$$(59) \qquad \mathcal{P} = 1 - \exp\left(-2\pi \sum_{k=1}^{n} \frac{\varepsilon_{ok}^2}{\alpha v}\right)$$

The absorbing sphere model [34]

The underlying idea of this model is that after each v_o-v_i crossing part of the flux will be lost from the incident channel v_o. If the number of crossings is very large the incident flux would be almost completely absorbed after a characteristic distance R_k. Then, the exchange probability P, which is equal to the absorption probability, can be approximated by

(60) $\qquad$ P = 1 $\qquad$ for R < R_K
$\qquad\qquad$ P = 0 $\qquad$ for R > R_K

The exchange cross section is then given by the geometrical cross section

$$(61) \qquad \sigma_{exch} = \pi R_K^2$$

Olson[34] has proposed that R_K is determined by :

$$(62) \qquad R_K^2 \, e^{-2.648(I_B/q)^{1/2}R_K} = 2.864 \cdot 10^{-4} \, q(q-1)v_o$$

where I_B is the ionisation potential of the target, q the charge state of the incident ion and v_o the collision velocity.

SECTION III

CHARGE EXCHANGE PROCESSES INVOLVING MULTICHARGED IONS

It is not the aim of this section to give a detailed review of this new field ; this has been done recently by several authors[1]. Our purpose is to discuss typical examples, characterized by increasing complexity.

III.1 - ONE ELECTRON SYSTEMS : FULLY STRIPPED IONS-HYDROGEN ATOM COLLISIONS $A^{q+} + H(1s) \rightarrow A^{(q-1)+}$ (nℓ) + H^+

For the *hydrogenic* $A^{(q-1)+}$(nℓ) excited state the energy defect of the reaction is given by :

(63) $\Delta E = \dfrac{1}{2} (\dfrac{q^2}{n^2} - 1)$

the crossing distance can then be evaluated using (23) :
Let's take the example of Ar^{18+} ions : it can be seen (Fig. 17) that charge exchange towards n = 8-11 will take place for R_c=10-20a.u.

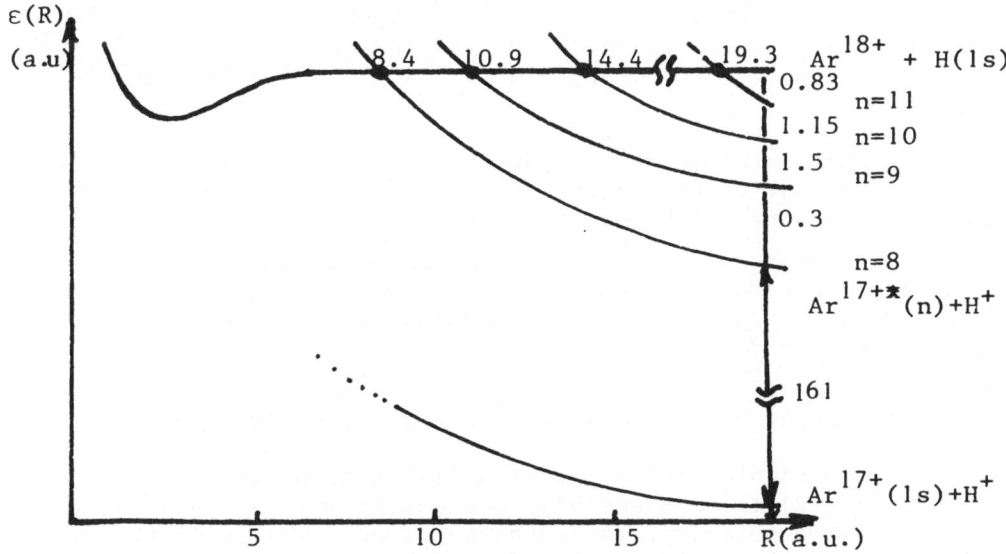

Figure 17

The maximum of the exchange cross section can then be estimated with the L.Z formula (II.6) to be of the order of $0.5\pi R_c^2 = 150-600$ (a.u)^2 $\simeq 40-150$ (Å^2).

Consequently the charge exchange will take place in high Rydberg states and the corresponding cross sections are expected to be large (few $10^{-16}-10^{-14}$ cm2)

In the framework of the quasi-molecule approach discussed in Section I, the *one* electron system has the interesting property that electronic energies ε_i can be calculated *exactly*. The calculations are based on the separation of the Schrodinger equation for HeI in prolate spheroïdal coordinates[35] (see II.2 eq. 26).

The molecular states $|i>$, as eigenfunction of HeI are *adiabatic* . Therefore, in this representation, the transition responsible for charge exchange processes will be given by the $v_R <|\partial/\partial R|>$ dynamical coupling. However diabatic treatment is often carried out in the LZ approximation with an interaction $\varepsilon_{12} = 1/2\ \Delta E(R_c)$ where ΔE is the energy separation of the potential curves at the avoided crossing. Let's consider one electron exchange channel : $A^{(q-1)+}(n\ell) + H^+$. At large distance, the "stark" effect induced by H^+ on the $A^{(q-1)+}(n\ell)$ states lifts the hydrogenic degeneracy. The corresponding assymptotic "stark states" (expressed in parabolic coordinates) becomes in the molecular region the eigenstates $|i> = |n_1, n_2, m>$ of the quasi molecule. (At large R, the elliptical coordinates transform into parabolic coordinates). However, due to an additional symmetry characterizing the one electron system, the Wigner-Von Neumann non crossing rule does not hold and states of same symmetry $|n_1, n_2, m > |n_1', n_2', m >$ do cross except if $n_1 = n_1'$ or $n_2 = n_2'$. This situation is sketched on fig. 18.

Among all the stark states $(n_1', n_2', 0)$, only the $(0, n_2', 0)$ state, called the "effective state", can be populated at the "pseudo crossing" with the incident $(0, n_2, 0)$ state.

If only *the total charge exchange* cross section is desired, the states other than the effective state can be disregarded in the collision problem, reducing to a two state problem for each *n value*.

a) Total charge exchange cross section and n distribution

It has been shown (section II) that the exchange interaction $\varepsilon_{oj}(R)$ is exponentially decreasing with R. Thus for a given collision velocity, the exchange process will selectively populate molecular states for which the interaction is not too large or not too small (Fig. 19) resulting in a *selective excitation of few n principal quantum numbers*.

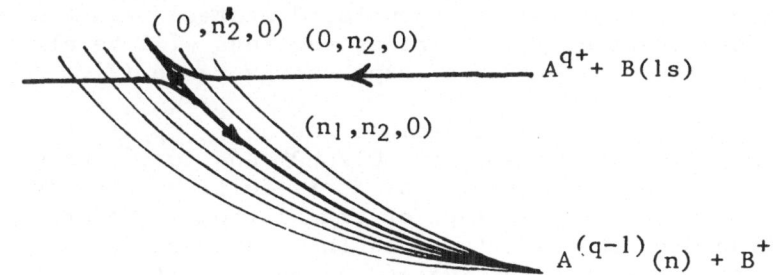

Figure 18

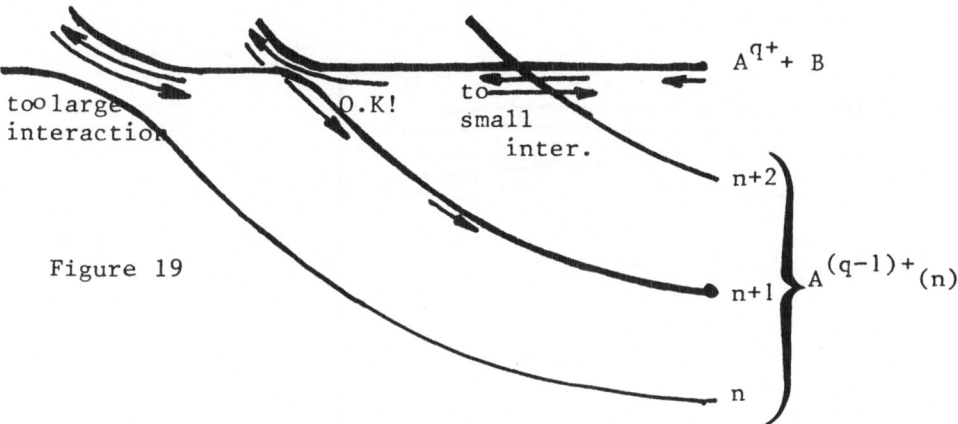

Figure 19

Since $P \simeq e^{-\frac{2\pi}{v}\frac{\varepsilon_{oj}^2}{\alpha}}$, the states selected by the collision process
will depend upon the collision velocity. Using the extension of the
LZ model for several crossings (II.9) we get for each n value, a
LZ shaped cross section centered at different velocities (Fig. 20).

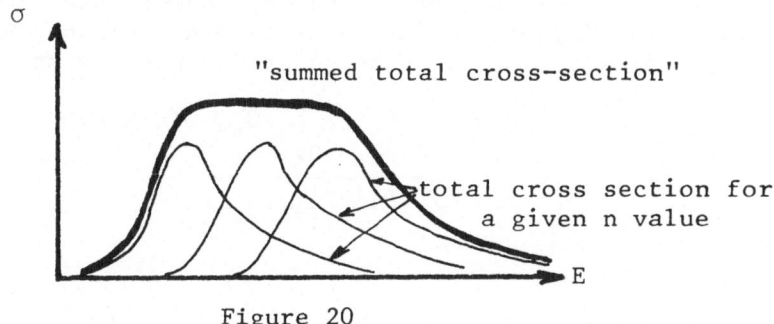

Figure 20

The *total* cross section (for all populated states) resulting from the sum of these various partial cross sections will be almost *independent of the energy* for a large energy range (Fig. 20)

Such theoretical treatment has been applied to a variety of collisional systems by Salop and Olson[36] and *total* charge exchange cross-sections have been determined. It is felt, as already discussed in I.6 that the neglect of translational factors should not affect too much the results since the distribution in the exit channels is not presently considered. Agreement with experiment is satisfactory. As seen in Fig. 21 the total cross section is almost energy independent except for the lightest system where only very few crossings are active in this energy range.

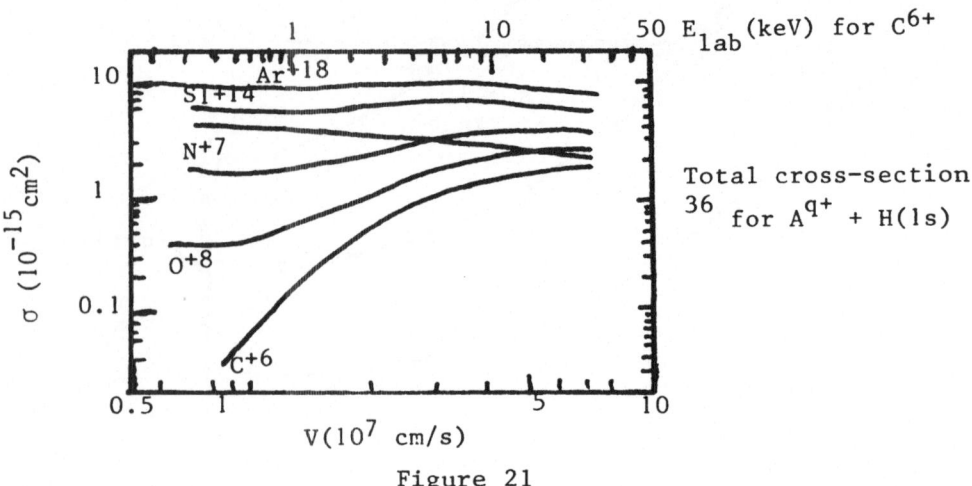

Figure 21

b) Distribution of the population among the ℓ subshell

Referring again to Fig. 18, it is seen that only the "effective" state ψ (0, $n\frac{1}{2}$, 0) is populated at the crossing. At large internuclear distance this state is correlated to ψ_{at}(0,n-1, 0) a "parabolic" eigen-state of the hydrogenic $A(q-1)+$*(n) ion. Then the population of the various ℓ levels could be determined by the projection of the parabolic state onto the spherical states

(37) $$\psi(0, n-1, 0) = \sum_{n,\ell} C_{n\ell} \, \psi(n,\ell,0)$$

with $C_{n\ell} = (-)^{\ell} \sqrt{\dfrac{2\ell+1}{n}} \; \sqrt{\dfrac{n!(n-1)!}{(n-\ell)!(n-\ell-1)!}}$

for example for n=4

$$\psi(0,3,0) = \frac{1}{2} \left(\psi_{4s} - \frac{3}{\sqrt{5}} \, \psi_{4p} + \psi_{4d} - \frac{1}{\sqrt{5}} \, \psi_{4f} \right)$$

However this approach is too crude since the *intrashell* dynamical couplings (radial and rotational) in the exit channels are neglected. The inclusion of these secondary couplings requires the solution of a large set of coupled equations (10 to 33) for $C^{6+}+H$

Furthermore the introduction of appropriate ETF is necessary since population of specific channels is desired. This leads to heavy calculations which have been carried out only in few cases[38]. As an example the n and ℓ-subshell distribution for $C^{6+} + H$ collision shown on Fig. 22 is compared to the statistical distribution ($P \sim 2\ell+1$) and results from Monte-Carlo calculations[39]. Experiments would be welcome ; but for degenerate states, a lengthy and tedious analysis of cascade effects necessitates measurements in various wave length regions (Lyman, Balmer, Pashen..)

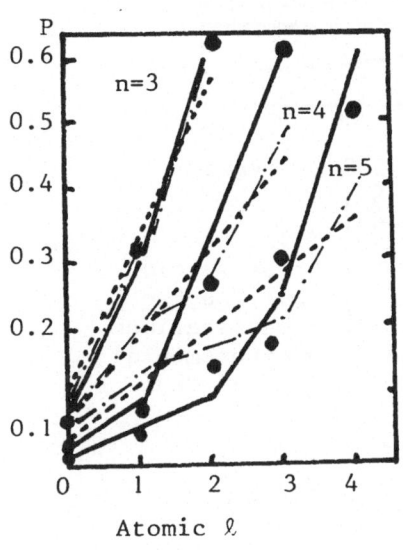

Atomic ℓ

$C^{6+} + H \to C^{5+}(n) + H^+$
Subshell occupation probabilities :
Full line : couple state calculation
(27 keV)
Dashed line : "Statistical" ($\sim 2\ell+1$)
dots : Monte Carlo calculations[39]
(25 keV)
Dash dotted line : Abramov et al.[44]
(high velocity limit)

Fig. 22
(from ref. 38)

III.2 - QUASI ONE ELECTRON SYSTEM. THE C^{4+} + H SYSTEM AS AN EXAMPLE

The *core* electrons carried by a multicharged ion are tightly bound to the nucleus and do not actively participate in the collision. Only the $1s_H$ electron is active in this problem. Then *"model potential"* techniques are indeed appropriate to the treatment of the *active* electron in the coulombic field of the H^+ and the screened coulombic field created by the multicharged ion . These methods have been used by several authors. As an example, potential and cross section for charge transfer[40] in C^{4+}-H are shown in Fig. 23. The ℓ subshell degeneracy is lifted but only the $C^{3+}(n=3)$ states have to be considered. In addition to their LZ shape, the individual cross-sections (σ_{3s}, $\sigma 3p$, $\sigma 3d$)

show a neat structure unambiguously attributable to *subshell couplings in the exit channels* (III.Ib). Particularly noteworthy is the increase of σ_{3d} at high energy.

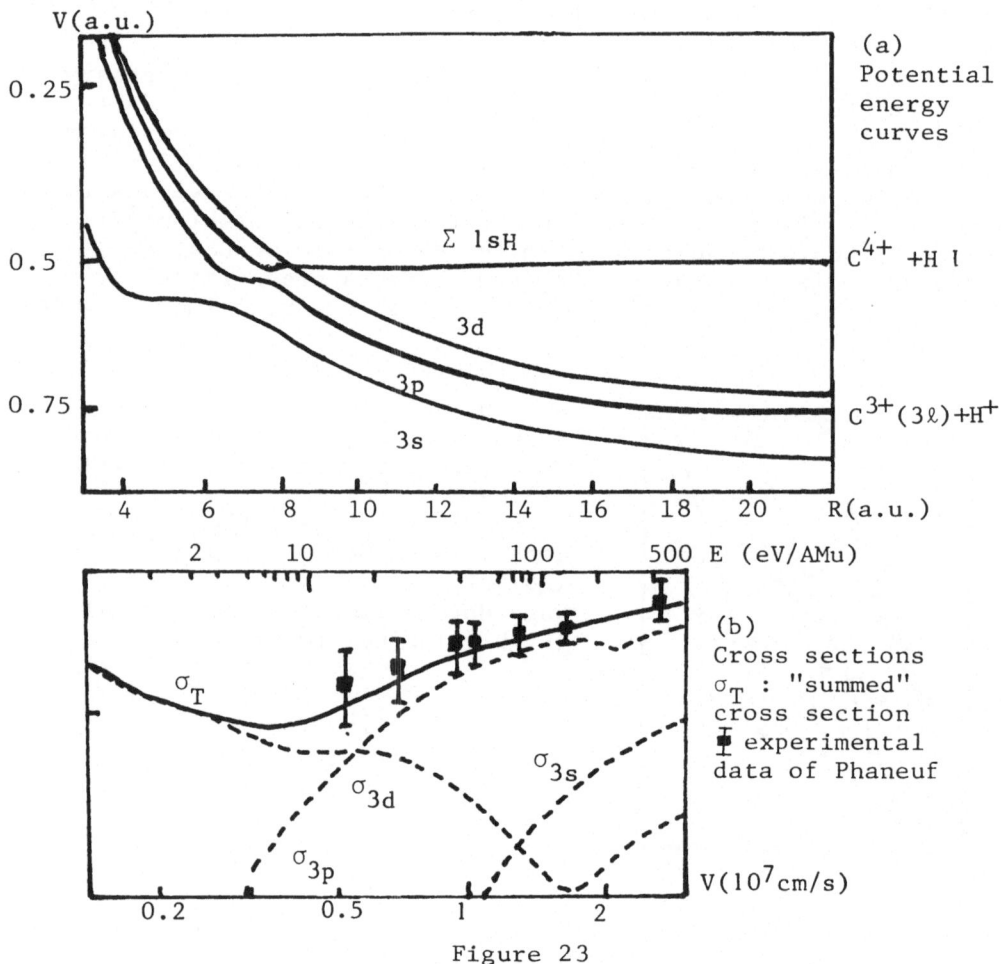

Figure 23

III.3 - TWO ELECTRONS SYSTEMS O^{8+} + He AS AN EXAMPLE

In the case where the target has *two equivalent* electrons (He(1s^2) H$_2$(1sσ_g^2)), application of the one electron model should give a similar role to both electrons. Electron transfer of one and *two electrons on Rydberg states* of the ion should then be very probable. For two electron transfer, this would lead to the production of *doubly excited states* $A^{(q-2)+}(n\ell n'\ell)$. The potential energy curves can be roughly estimated as in the one electron case. The potential energy curves for the incident channel is essentially flat at large internuclear distances whereas the electron transfer channels are dominated by the coulomb repulsion $V_1 = q-1/R$ and

$V_2 = 2(q-2)/R$ for one and two electron transfer respectively. The
strongest repulsivity of V_2 leads to multiple crossings between
the incident channel and the two series of exit channels as well
as crossings between the two series. The situation is illustrated
in Fig. 24 for the O^{8+}+He collision. In this simplified model,
the various $O^{6+}(n\ell, n\ell')$ levels are represented as degenerated.

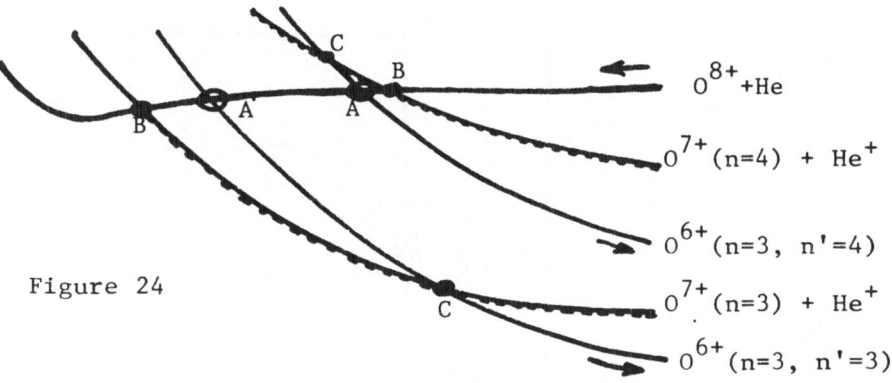

Figure 24

The scheme of the potential curves clearly shows that the two-
electron exchange can be induced by two different mechanisms :

(i) A *direct transition* at the A crossings between V_0 and V_2
potential energy curves. The corresponding interaction, responsible for the transition is given by the *correlation* between the
electrons : $\varepsilon_{02} = <1s^2_{He} |\frac{1}{r_{12}}| n\ell n'\ell'_{O^{6+}}>$ (r_{12} is the distance
between the electrons)

(ii) A *two-step transition* : A first electron is transfered at the
B crossing, then the second electron is transfered at the C crossing.
At each crossing, the transition probability is determined by the
one electron exchange interaction discussed in section II.

        The relative importance of the "simultaneous"
(mechanism (i)) and "successive" (mechanism (ii)) two-electron
transitions is not known. Indeed, very little theoretical work has
been devoted to this problem. To my knowledge, the only molecular
calculations for highly charged ions have been carried out by
Harel and Salin[41] and concerns the O^{8+} + He system. The results
are given in Fig. 25. They demonstrate that, as expected from the
qualitative discussion given above, both electrons are captured
in high Rydberg states. In this process, very *exotic* species
are created, such as ions with electrons in high Rydberg states.
For example, in Ar^{18+} + He collisions $Ar^{18+}(n \simeq 9, n' \simeq 9,10)$ is
expected to be produced , of which the spectroscopy and the study

of decay processes (light emission versus autoionisation) would be of great interest. It also should be pointed out that the autoionisation process complicates the comparison between theory and experiment, since part of the measured O^{7+} flux results from the autoionisation of the O^{6+} ions.

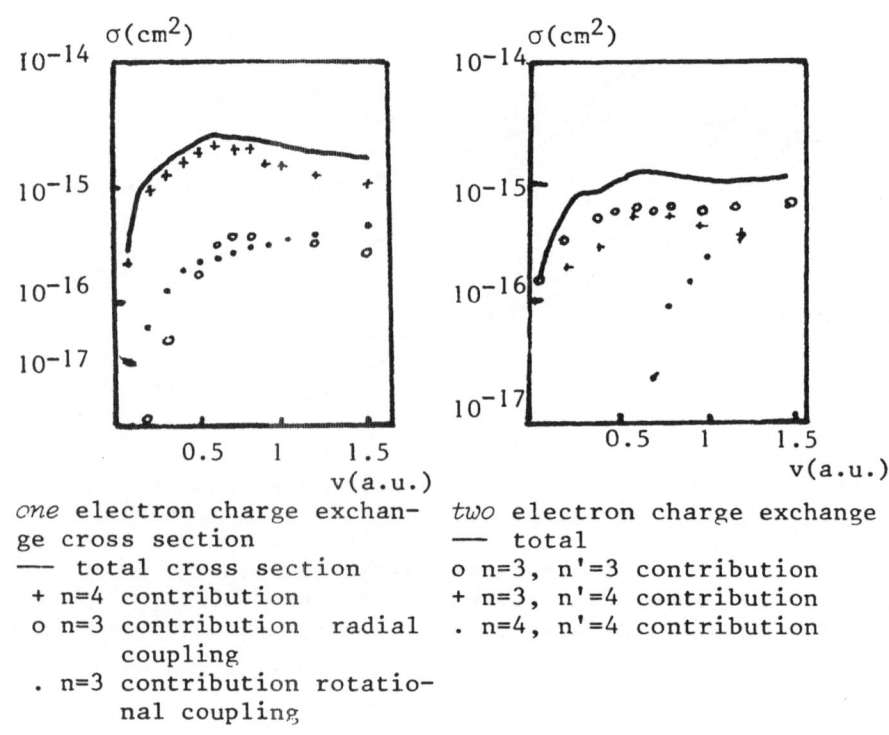

one electron charge exchange cross section
— total cross section
+ n=4 contribution
o n=3 contribution   radial
    coupling
. n=3 contribution rotatio-
    nal coupling

two electron charge exchange
— total
o n=3, n'=3 contribution
+ n=3, n'=4 contribution
. n=4, n'=4 contribution

Figure 25 (from 41)

Similar types of mechanisms would be involved to explain the transfer of more than two electrons observed with heavier targets [42] (Ne, Ar). However, only multi step processes (ii) have to be considered since no interaction can couple molecular states differing by more than two spin-orbitals.

III. 4 - ION-ION CHARGE EXCHANGE PROCESSES

        Little work has been devoted to charge exchange between singly charged ions.[43] But to our knowledge, neither experimental nor theoretical work has been devoted to ion-ion charge exchange involving multicharged ions. However, some predictions can be attempted on the basis of a simple model. Since the incident channel is characterized by a repulsive coulombic curve (Fig. 8c), the crossings responsible for charge transfer would be reached only for collision energies $E > E_{th}$. The existence of such a *threshold* for collision energy is typical of ion-ion charge exchange. Let's consider the collision

$$A^{q_A+} + B^{(q_B-1)+}(1s) \rightarrow A^{(q_A-1)+*}(n_A) + B^{q_B+}$$

The crossing distance will be approximately given by

$$R_c = \frac{2(q_A-q_B)}{(\frac{q_A}{n_A})^2 - q_B^2}$$

It can easily be shown that sizeable charge exchange cross section corresponding to large values of R_c can only be expected if $q_A \gg q_B$ i.e. if *the electron is initially bound to the low charged ion*. Using the LZ model, total cross sections as well as preferentially populated levels can be estimated. The exchange interaction can approximately be derived by an extrapolation of the empirical formula of Olson and Salop[25]

$$\varepsilon_{12}(R_c) = \frac{9.13}{\sqrt{q_A}} \exp \left(- \frac{1.324 \ R_c \cdot q_B}{\sqrt{q_A}}\right)$$

Such estimates for the $He^+(1s) + A^{q+}$ collisions are given in the table below. These results show that *large cross sections can be expected with very asymmetric systems.*

$$He^+(1s) \rightarrow A^{q+} \rightarrow He^{++} + A^{(q-1)+*}(n)$$

Table 1

q	n	R_c	Threshold Energy eV	E_{LZ}^{Max}(eV)	σ_{Max} $10^{-16}cm^2$
Fe^{26+}	6	3.2	217	48.000	4.5
	7	4.9	144	8.300	10
	8	7.3	96	435	23
	9	11	64	70	50
Mo^{42+}	8	3.4	336	27.000	5
	9	4.5	250	14.000	9
	11	7.5	150	1.000	11
	13	12.4	91	106	68

CONCLUSION

        The agreement between experiment and theory for total
charge exchange is now quite good. Thus the main features of the
collision mechanisms are certainly well understood. However
measurements with higher charge states would be desirable to better
establish scaling rules. It is time now for more elaborate
experiments to determine the total cross-sections for *individual*
channels, in order to have some experimental information on the
n and ℓ distribution. Several techniques can be used, including
photon and electron spectroscopy as well as energy loss measurements. Hope that, with the new generation of ion-sources such
experiments will be feasible.

REFERENCES

1. R.E. Olson, XI ICPEAC, book of invited lectures and progress
        reports. N. Oda and K. Takayanagi eds. N. Holland pub.
        p. 391 (1979)

   E. Salzborn and A. Müller, ibid p. 391 (1979)

   R.K. Janev and T.P. Grozdanov, Invited papers and progress
        reports, SPIG-80, Ed. Matic, Belgrad, 181, 1980

2. M. Born and J. Oppenheimer, Ann. Physik. 84, 457 (1927)

3. Mott and Massey, Theory of atomic collisions, Oxford Press
        (1965)

4. See e.g. A. Russek, Phys. Rev. A 4, 1918 (1971)

5. F.T. Smith, Phys. Rev., 179, 111 (1969)

6. J. Von Neumann and E.P. Wigner, Phys. Z. 30, 467 (1929)

7. D.R. Bates and R. McCarroll, Proc. Roy. Soc. A 245, 175 (1958)

8. A. Salin, Com. on At. Phys. (1981)

9. H. Rosenthal, P.R.L., 27, 635-8 (1971)

10. A. Salop and R.E. Olson, P.R. A 19, 1921 (1979)

11. W.R. Thorson and J.B. Delos, P.R. A 18, 117, A18, 135 (1978)

12. R.D. Piacentini and A. Salin, J.Phys.B 7, 1666 (1974)
        9, 563 (1976), 10, 1515 (1977)

13. G.J. Hatton, N.E. Lane and T.G. Winter, J.Phys.B 12, 571 (1979).

14. S.B. Schneiderman and A. Russek, P.R. 181, 311 (1969)

15. J. Vaaben and K. Taulberg, J.Phys.B 14, 1815 (1981)

16. V.H. Ponce, J.Phys.B, 12, 3731 (1979)

G. Bruno Schmid, J.Phys.B 12, 3909 (1979)

17. O.B. Smirnov, Asymptotic Methods in the Theory of atomic
       collisions, Atomizdat, Moscou (1973)

18. C. Herring, Rev. Mod. Phys. 34, 631 (1962)
    L.D. Landau and E. Lifshits, Quantum Mecanics, Moscou (1963)

19. B.M. Smirnov, Sov. Phys. JETP, 19, 692 (1964)

20. E.L. Duman and L.I. Menshikov, Zh. Eks. Teor. Fiz. 40, 130
       (1979)

21. E.L. Duman, L.I. Menshikov and B.M. Smirnov, Sov. Phys. JETP
       49, 260 (1979)

22. T.P. Grozdanov and R.K. Janev, Phys. Rev. A 17, 880 (1978)
    M.I. Chibisov, JETP Lett. 24, 46 (1976)

23. R.K. Janev and A. Salin, J. Phys.B 5, 177 (1972)

24. See e.g. B.M. Smirnov, Sov. Phys. JETP, 20, 345 (1965)

25. R.E. Olson and A. Salop, Phys. Rev. A 14, 579 (1976)

26. O.B. Firsov, JETP 21, 1001 (1951)

27. L. Landau, Phys. Z. Sov. Union. 2, 46 (1932)
    C. Zener, Proc. Roy. Soc. A197, 696 (1932)

28. E.E. Nikitin and A.I. Reznikov, Phys. Rev. A6, 522 (1972)
    M.S. Child, Molec. Phys. 20, 171 (1971)

29. E.C.G. Stuekelberg, Helv. Phys. Acta 5, 369 (1932)

30. Yu.N. Demkov, Sov. Phys., JETP, 18, 138 (1964)

31. R.E. Olson, Phys. Rev. A 6, 1822 (1972)

32. E.E. Nikitin, Opt. Spectrosk. 13, 431 (1962)

33. E.E. Nikitin, Advances in Quantum Chemistry, V5, Academic
       Press. N.Y.p. 135 (1970)

34. R.E. Olson and A. Salop, Phys. Rev. A 14, 579 (1976)

35. Ø. Burrau, Kgl. Danske Vid. Sels. 7, n°14 (1927)
    Gershtein and Krivchenkov, Sov. Phys. JETP, 13, 1044 (1961)

36. A. Salop and R.E. Olson, Phys. Rev. A 13, 1312 (1976)

37. K. Omivdar, Phys. Rev. 153, 121 (1967)

38. T.A. Green, E.J. Shipsey and J.C. Browne, Phys. Rev. A25, 1364
       (1982)

39. R.E. Olson, unpublished

40. M. Gargaud, I. Hanssen, R. McCarroll and P. Valiron, J.Phys.B
       14, 1359 (1981)

41. C. Harel and A. Salin, Contrib. papers at the XII ICPEAC
       Gatlinburg p. 575 (1981)

42. See e.g. S. Bliman, S. Dousson, R. Geller, B. Jacquot
        and D. Van Houtte, J. Physique (Paris) 42, 705 (1981)

43. H.B. Gilbody, XII ICPEAC, book of invited lectures and progress
        reports. S. Datz eds. N. Holland pub. p. 223 (1981)

44. V.A. Abramov, F.F. Baryshnikov and V.S. Lisitsa, JETP Letters
        27, 464 (1978)

ELECTRON-ION COLLISIONS

D. H. Crandall

Oak Ridge National Laboratory
Oak Ridge, Tennessee 37830

1. INTRODUCTION

This discussion will concentrate on basic physics aspects of inelastic processes of excitation, ionization, and recombination that occur during electron-ion collisions. Except for cases of illustration along isoelectronic sequences, only multicharged (at least +2) ions will be specifically discussed with some emphasis of unique physics aspects associated with ionic charge. The material presented will be discussed from a primarily experimental viewpoint with most attention to electron-ion interacting beams experiments.

In the spirit of a NATO Advanced Study, this presentation is intended for persons who are not already experts in the subject but who want to gain some basic understanding of the field and a flavor of current research activities in the field. Two previous NATO Advanced Studies[1,2] contain a number of papers which relate directly to the present topic and will be specifically referenced. In addition there are a number of recent reviews or overviews[3-11] of specific parts of the present topic which can be relied on for some details.

Within the present context, there is little need for a detailed statement of motivation for study of electron-ion collisions. Indeed other discussions within this volume will demonstrate applications of the basic physics in fusion and astrophysical plasmas and in ion source development. The principal motivation of the present paper is taken to be basic atomic physics interest in collision dynamics.

Since the collision system to be discussed will always begin in the initial state of just one free electron plus one multicharged ion, the class of collisions is among the simplest, and detailed correspondence of theoretical predictions and experimental studies can be examined. In fact, progress in the subject has been retarded by a lack of experimental results which can test the theoretical predictions that are rather more advanced than experiments within this subject, especially for multicharged ions.

One general observation which may currently be emerging is that the three processes of excitation, ionization, and recombination are of comparable or nearly equal importance. Even within a single collision, at collision energies for which all three processes are possible, the amplitude for each of the three processes can be roughly comparable. Thus the influence (or possibly even interference) of each of these three processes on the others should be included.

2.   CONCEPTS

2.1 A Specific Sample Case

The inelastic collision processes for a sample case of $e^- + Fe^{15+}$ are illustrated in Fig. 1. The figure is rather complex, but it is useful to conceptually visualize all of the processes in a unified picture to emphasize how they influence each other.

Consider the initial state of the sample $e^- + Fe^{15+}$ system to be that of a free electron of arbitrary energy, k, plus a ground-state ion with electron configuration $1s^2 2s^2 2p^6 3s$ (Fe^{15+} is a sodium-like ion with 11 electrons). This initial state is at zero energy on the figure, but the solid horizontal line represents the state of the ion and the slanted lines attached to the solid horizontal line represent the continuum of energy of the free electron, k (k will be a variable to which we can ascribe a particular value by assigning a subscript such as k_o). What is being represented by the lines with arrows (labeled (1) through (5)) all emanating from the initial state, are examples of inelastic transitions of the $e^- + Fe^{15+}$ system which are driven by specific electron energies, k_i. The energy available as kinetic energy of the free electron must be equal to the energy at the end of each of the arrows (1) through (5) in order to cause that particular illustrated transition.

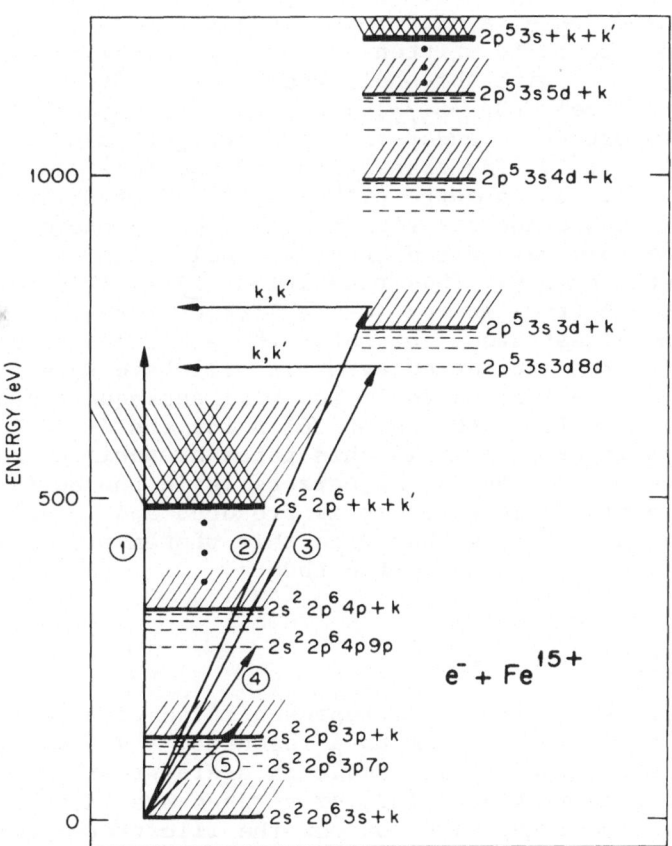

Fig. 1. Schematic representation of the structure and inelastic transitions within the $e^- + Fe^{15+}$ system (see text).

2.2 Excitation

The simplest transition illustrated in Fig. 1 is ⑤ which is direct excitation of the lowest excited levels, $e^- + Fe^{15+}(3s) \rightarrow Fe^{15+}(3p) + e^-$. The actual atomic structure associated with this transition is a little more complicated than the illustration. Since the excited configuration $1s^2 2s^2 2p^6 3p$ has two fine-structure levels, $^2P_{1/2}$ and $^2P_{3/2}$, a different representation of these excitation transitions could be written $e^- + Fe^{15+}(3s\ ^2S_{1/2}) \rightarrow Fe^{15+}(3p\ ^2P_{1/2}) + e^-$ with threshold energy of 34.36 eV and $e^- + Fe^{15+}(3s\ ^2S_{1/2}) \rightarrow Fe^{15+}(3p\ ^2P_{3/2}) + e^-$ with threshold energy 36.96 eV.[12] It is important to understand the energetics of the collisions illustrated on Fig. 1. For transition ⑤ the Fe^{15+} ion is driven to an excited state and absorbs either 34.36 eV or 36.96 eV, but the electron which causes the transition can have any energy greater than these threshold values and after the collision will retain the remaining kinetic energy. Thus transition ⑤ is illustrated for a particular but arbitrary energy just a little above the excited configuration. As illustrated, initially the electron has energy $k_5 = 50$ eV, and after the excitation event it will have $k_f = 50-34.36$ or $50-36.96$ eV. The excited Fe^{15+} ion will subsequently decay by emitting a characteristic photon of 361 Å for the $^2P_{1/2}$ level or 336 Å for the $^2P_{3/2}$ state. Later in this paper a specific example will be given of the use of the 361 Å characteristic photon in studying the nature of a plasma in which it is produced and in using the known plasma parameters together with this photon production to crudely test electron-ion collision theory.

2.3 Recombination Resonance

Transition ④ of Fig. 1 is representative of a specific transition of the $e^- + Fe^{15+}$ system to a resonance state of the Fe^{14+} system. Such a transition can result in dielectronic recombination, a resonance in elastic scattering, or a resonance in the excitation event just described, depending on how the illustrated resonance state decays. As illustrated for ④ the initial system energy of $Fe^{15+}(3s) + k_4$ happens to be a precise match to the energy of a level of Fe^{14+} with configuration $1s^2 2s^2 2p^6 4p9p$. That is, the energy k_4 is a little less than required to promote the 3s outer electron of Fe^{15+} to the 4p level; however, the initially free electron could be in a bound orbital (9p) of Fe^{14+}, and the system energy would then be just right for the initial 3s bound electron to be found in the 4p orbital of Fe^{14+}. To first order the transition involves only two electrons, the initially free electron and the initially outer 3s electron. This two-electron interaction provides the name dielectronic to describe the process. There are, in general, many more of these "resonance" or recombination states of

the Fe^{14+} ion than are illustrated on Fig. 1. However, they will
only occur during $e^- + Fe^{15+}$ transitions at very specific energies.
Thus the distinction in energetics between direct excitations like
transition ⑤ and recombination resonances like transition ④ is
that ⑤ occurs at any electron energy above threshold while ④
only occurs at the energy of a quantum eigenstate of Fe^{14+}. In
terms of measurable quantities like excitation cross sections for
$e^- + Fe^{15+}(3s\ ^2S_{1/2}) \rightarrow Fe^{15+}(3p\ ^2P_{1/2}) + e^-$ the transitions like
④ can have an "indirect" influence if they happen to decay from
$Fe^{14+}(1s^22s^22p^64p9p) \rightarrow Fe^{15+}(1s^22s^22p^63p) + e^-$. Such resonances
in excitation cross sections are particularly important for multicharged ions where the high net charge makes recombination resonances a relatively stronger phenomena than for neutral atoms or
slightly charged ions.

Dielectronic recombination only occurs if the resonance (which
will always involve at least two excited electrons) decays by emission of light rather than by autoionization. After transition ④ of
Fig. 1 a possible radiative stabilization sequence is:

$$Fe^{14+}(2p^64p9p) \rightarrow (2p^63s9p) + h\upsilon(\sim 244\ eV) \quad \text{(stabilized but excited } Fe^{14+})$$
$$\longrightarrow (2p^63s^2) + h\upsilon(\sim 400\ eV) \quad \text{(ground-state } Fe^{14+}),$$

where the first transition carried the system to a state just below
the zero level of Fig. 1, and the photon emitted was shifted
slightly lower in energy than the 4p-3s transition in Fe^{15+}. Note
that total radiative decay of the initial resonance would always
result in at least two photons emitted and a total release of energy
greater than that brought into the collision as kinetic energy of
the incident electron.

The autoionization and radiative relaxation of the recombination resonances are in competition. The autoionization is qualitatively an interaction between two excited electrons and is only
weakly dependent on ionic charge, q. However, the radiative decay
is qualitatively an interaction between an excited electron and the
ionic field of charge, q, and this radiative decay scales roughly as
q^4. Thus for highly charged ions, radiative decay and hence
recombination-stabilization become relatively more likely than autoionization which is usually much faster than radiative decay for
neutral atoms or slightly charged ions.

2.4 Ionization

Finally consider transitions ①, ②, and ③ of Fig. 1 which
can all result in ionization of the form $e + Fe^{15+}(2p^63s) \rightarrow$
$Fe^{16+}(2p^6) + 2e^-$. The threshold energy for removal of the 3s

electron is 489 eV[12] and is illustrated on the figure with two sets
of slanted lines (free electron continua) attached to a horizontal
line representing $Fe^{16+}(1s^2 2s^2 2p^6)$ at 489 eV. The initially free
electron is designated by k and the removed electron by k´ (even
though they are indistinguishable). Transition ① is the direct
ionization transition in which the incident electron has a binary
collision with the bound 3s electron and removes it. Energetically
$k_1 = k_f + k_f' + 489$ eV. Transition ② is qualitatively like transition ⑤ but is the promotion of an inner-shell electron of the
form $e^- + Fe^{15+}(2p^6 3s) \rightarrow Fe^{15+}(2p^5 3s 3d) + e^-$ so that a binary collision caused a 2p → 3d promotion. Transition ② is a dipole-allowed
excitation transition with 801 eV threshold[13] which one might guess
could be fairly strong. Transition ③ is qualitatively like ④
but is the promotion of an inner-shell electron in a binary collision during which the incident electron lost just the right energy
to fall into a bound orbital so that $e^- + Fe^{15+}(2p^6 3s) \rightarrow$
$Fe^{14+}(2p^5 3s 3d 8d)$ which occurs only at the resonance energy of
751 eV.[14] The reason for illustrating transitions ② and ③ is
that they can both decay to the same final state of the system as
transition ①, that is both ② and ③ can contribute to ionization $e + Fe^{15+} \rightarrow Fe^{16+} + 2e^-$.

Figure 2 may be helpful by illustrating transitions ①, ②,
and ③ in a "cartoon." Transition ① is "direct" ionization by
knocking off the outermost (3s) electron. Transition ② is excitation of an inner-shell electron 2p→3d followed by autoionization in
which the excited electron (3d) and the original outer electron (3s)
had an interaction resulting in one falling to the vacant 2p orbital
and the other being ejected. Transition ③ is a resonance recombination event in which an inner electron (2p) was promoted to a (3d)
orbital while the initial electron fell into an (8d) orbital; this
highly excited resonance state then decayed by an autoionization
event (or events) resulting in emission of two of the three electrons (3s3d8d) while the other fell back to the (2p) vacancy.

The point of this involved discussion of transitions ①, ②,
and ③ is that for Fe^{15+} at energies near 800 eV, all three types
of transitions are predicted to be of roughly equal magnitude in
determining the total ionization cross section. The Fe^{15+} case was
selected to illustrate the detailed concepts because the theoretical
predictions of Cowan and Mann[13] and of LaGattuta and Hahn[14] provide
the specific details. Figure 3 shows the ionization cross section
predicted for Fe^{15+}. These predictions were accomplished by independent quantum calculation of processes like transition ② and
transition ③ and by simple addition of the independently calculated cross sections. It remains to be seen whether or not such
predictions are valid but combined experimental and theoretical work
on other Na-like ions indicates that such predictions are roughly
correct. There have not been any calculations which treat all of

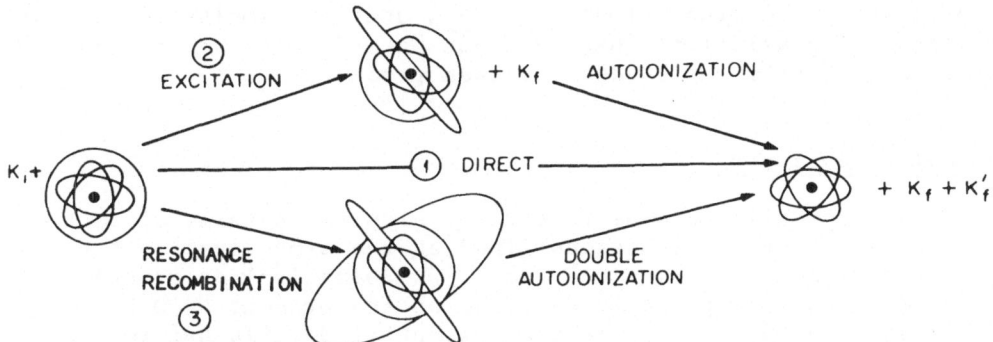

Fig. 2.   A cartoon representation of the electronic transitions
which occur for transitions ①, ②, and ③ of Fig. 1 if
they all proceed to a final state in which ionization has
occurred.

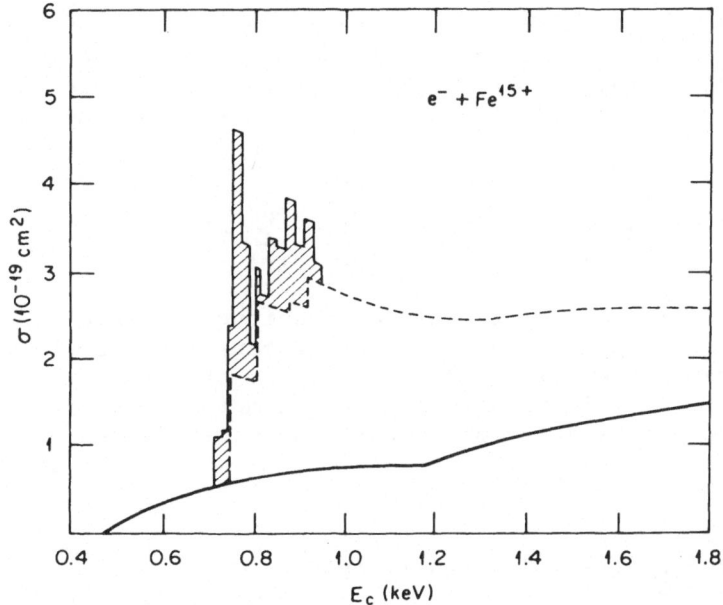

Fig. 3.   The predicted electron impact ionization cross section for
Fe^{15+} (from Ref. 14). The solid curve is direct ionization (represented by transitions like ① on Fig. 1); the
dashed curve is excitation-autoionization (represented by
② on Fig. 1) added to the direct part; the cross-hatched
area is resonance recombination–double autoionization
(represented by ③ on Fig. 1) and added to the other two
contributions.

the processes like transitions ①, ②, and ③ simultaneously and
allow for the possibility that coherent interference could occur
between the quantum amplitudes for each process.

2.5 Collision Theory

No attempt will be made here to develop the details of collision theory. Useful discussions aimed at the same level as this
paper were given by Y.-K. Kim[15] and C. J. Joachain[16] at previous
NATO studies and are given by J. Dubau at the present NATO school.
References 1, 2, and 3 contain most essential details and further
references. However, some of the general expressions are given here
to allow insight and comparisons.

The accepted nonrelativistic theoretical approaches all begin
with the Schroedinger equation for the system of N + 1 electrons:

$$H(Z,N+1)\psi = E\psi \tag{1}$$

where N and Z are the number of electrons and nuclear charge of the
ion, ψ is the total wave function of the N + 1 electron system, and
E is the total energy, $E = E_i (Z,N) + k_i^2$ (which defines the kinetic
energy, k_i^2, of the incident electron in Rydbergs). The Hamiltonian
is

$$H(Z,N+1) = -\sum_{i=1}^{N+1} \left(\nabla_i^2 + \frac{2Z}{r_i} \right) + \sum_{j=i+1}^{N+1} \sum_{i=1}^{N} \frac{2}{r_{ij}} \; . \tag{2}$$

Having written these standard expressions we immediately
encounter difficulty because Eq. (1) is a many-body problem, never
less than three bodies except for collisions of a bare nucleus and a
free electron. In principle there is no compromise in expanding the
many-body wave function in terms of products of single-particle wave
functions:

$$\psi(N+1) = A \sum_{i}^{\infty} \theta_i(N + 1) \; \chi_i(\chi_1 \ldots \chi_N) \tag{3}$$

where A antisymmetrizes the total function and χ_N denotes coordinates of the nth electron. Substitution of Eq. (3) into Eq. (1)
results in a number of coupled differential equations. Generally,
expansion (3) is chosen so that the functions χ_i are the solutions
of the next simpler problem (i.e., they are the eigenvectors of the
N electron ion which is the target), and the $\theta_i(N + 1)$ are coefficients as well as the wave function of the free or incident electron. The wave functions for the bound state system, χ_i, are

usually antisymmetrized products of one-electron functions determined from the solution of a central field model such as the
Hartree-Fock method. It should be further noted that the index, i,
stands for all the system quantum numbers including parity, angular
momentum, and spin. Thus, ψ is a sum over coupled, antisymmetrized,
product wave functions which leads to messy algebra. One additional
modification to Eq. (3) is usually specified:

$$\psi(N + 1) = A \sum_i \theta_i \, \chi_i + \sum_s C_s \, \phi_s(\chi_1 \ldots \chi_{N+1}) \tag{4}$$

where the C_s are numerical coefficients and ϕ_s are bound-state wave
functions of the ion or atom of next lower charge with N + 1 bound
electrons. The addition of the second term in Eq. (4) is needed to
provide a more complete set mathematically, but note that the ϕ_s
states are important physical quantities in this problem as well.
Some of ϕ_s are the bound state resonances shown as dashed lines on
Fig. 1 where they are highly excited but, nevertheless, eigenstates
of the Fe^{14+} ion.

Qualitatively, if expansion (3) is substituted into Eq. (1) and
projected onto the angular momentum components of a particular channel, i, we obtain the close-coupling equations which are of the
form:

$$-\left(\nabla^2 + k_i{}^2\right) \theta_i(X) + \sum_j V_{ij}(X)\theta_j(X) = 0 \tag{5}$$

where X represents the space and spin coordinates and

$$V_{ij}(\chi_1) = \frac{-2Z}{r_1} \delta_{ij} + \int \psi_i(\chi_2 \ldots \chi_{N+1}) \sum_{n=2}^{N+1} \frac{2}{r_{1,n}}$$

$$\cdot \, \psi_j(\chi_2 \ldots \chi_{N+1}) \, d\chi_2 \ldots d\chi_{N+1} \, . \tag{6}$$

The asymptotic form of this potential is

$$V_{ij}(X) \underset{r \to \infty}{\simeq} \frac{-2q}{r} \delta_{ij} \tag{6a}$$

where $q = Z - N$ is the ionic charge and δ is the usual "delta
function" operator.

The ψ_i and ψ_j are any two wave functions of the system. These
expressions are still general and can describe any of the inelastic
collisions, but expressions (5) and (6) cannot be solved without
approximations. This is because the sums in expansions (3) and (5)

are in principle infinite and the potential V_{ij} connects every term
in those expansions to every other term. Thus in solving expressions (5) and (6) various approximations are made and suitable
boundary conditions must be imposed.

2.6 Approximations in Collisional Calculations

For the simplest inelastic collisions, the excitation transitions, direct coupled-state numerical solutions of (5) and (6) are
accomplished by truncating the number of terms in the expansions.
Those recombination resonances which can be represented within the
truncated expansions are then explicitly contained in the solutions
which give the excitation cross sections. Actual numerical (computer) solutions involve computation of a set of one electron basis
functions, χ_i, considerable vector-coupling algebra, and many matrix
inversions. Solutions for simple ions and a few states (expansion
terms) are tractable, but complex ions or many states are not
generally attempted.

The cause of numerical difficulty with Eq. (5) is the coupled
nature of the V_{ij} (i $\neq$ j) part of the potential in Eq. (6). A
substantial simplification results from uncoupling the equation
which can be acomplished by setting

$$V_{ij} = \begin{cases} 0 & \text{if } i \neq j \\ V_{ii} & \text{if } i = j \end{cases}$$

where V_{ii} is given by Eq. (6) with j = i. The differential Eqs. (5)
now have only one term in what was the summation, and each differential equation contains only one index, i, and thus stands alone.
Calculations of a transition from any initial state i to a final
state f becomes a two-state problem. Qualitatively, this is the
distorted-wave approximation. Coupling of states by the incident
electron has been lost but the technique is sufficiently simple to
apply to quite complex ions and to excitation or ionization processes. Of course the actual approximation of Eqs. (5) and (6) can
be done many ways. Henry[3] distinguishes about ten variations of
distorted wave in published work on excitation, and Younger[17] has
applied distorted wave theory to a number of ionization calculations
which include only direct processes like transition ① of Fig. 1.

The Coulomb-Born approximation is a modification of the potential (6) which provides additional simplification. The asymptotic
form of Eq. (6), $V_{if} = \frac{-2q}{r} \delta_{if}$, is used for the problem at all r.
Note that the first term of Eq. (6) is not the same as this in that
the nuclear charge Z has been replaced by the ionic charge, q. The

screening of the bound electrons which was determined by the integral part of Eq. (6) in a flexible (r-dependent) manner even in distorted wave is now fixed to be simply full screening at any r. The plane-wave Born approximation takes $V_{if} = 0$ as $r \to \infty$ which would correspond to $q = 0$ in the Coulomb-Born approach and is not generally appropriate for ions but can provide useful insight at high energies and can be solved analytically. It is characteristic for highly charged ions that distorted-wave or Coulomb-Born theory is believed to be adequate.

Other approximations can be applied to any of the three methods mentioned. Physically, unitarity corresponds to conservation of flux in that the probabilities of all allowed scattering must sum to the initial flux. Naively it would seem that unitarity should always be satisfied but, if the matrix elements obtained are all small, it may not be essential to obtaining a good approximation. Exchange has not been specifically mentioned. In principle, if all the wave functions in the expansion (3) and the potential (6) are fully antisymmetrized, exchange is explicitly included throughout the problem but with a great addition in the real work. Thus, numerous different approximations are in fact made in the way and extent to which exchange is included.

The formulas mentioned thus far are generally applied to excitation and ionization calculations. Transitions involving resonances like those of ③ and ④ in Fig. 1 seem at first glance to present some simplification in that there is no dynamical coupling since transitions occur only at one energy. Thus in expansion (3), θ_i become simply constants, not functions — that is the expansion (3) becomes just the second term of expansion (4). In solving Eq. (1) only noncoupled, well-defined wave functions appear and the problem unambiguously simplifies to something like the Coulomb-Born approximation of the dynamical cases.

Another view of the problem is to recognize that dielectronic recombination is the inverse of autoionization.[18] The autoionization transition occurs at a specified energy between stationary states, and the transition probability is well described by first-order perturbation theory. Microscopic reversibility, along with branching ratios for radiative decay relative to autoionization can be applied to determine dielectronic recombination.

Unfortunately there are thousands of transitions to consider even for simple cases. Dielectronic recombination for a few specific ionic cases has been calculated in detail (Refs. 18-22 for examples) always with some approximations. In fact the general character of the many transitions involved is usually parameterized as a function of ionic charge and transition type ($\Delta n=0 \sim \Delta n=1$ of the initially bound electron). Burgess[23] first provided such a

parameterized general formula and Merts et al.[18] provided a modification for $\Delta n \neq 0$ transitions. The resulting Burgess–Merts formula is
widely used but does not provide physical insight into the nature of
the process.

In addition the highly excited states involved in dielectronic
recombination are easily affected by additional collisions before
decay (density effects) or by external fields. Thus, in many of the
relevant physical environments where dielectronic recombination
occurs, the field and density corrections to the basic description
can dominate.

2.7 General Character of Cross Sections

The excitation cross sections have some particular features
worth noting. First is that the cross sections are finite at
threshold, that is there is a step function from zero to some finite
(often maximum) value as the incident electron energy increases
through the threshold for excitation of the state of interest
(illustrated in Fig. 4a). This step function character results from
the long-range nature of the Coulomb potential [Eqs. (5) and (6)].
In solving (5) a summation is carried out over all possible angular
momentum states of the colliding system. In the usual partial wave
summation it is seen that for ions (the Coulomb-potential) all partial waves contribute at the threshold energy, which is not true for
neutral systems where $V(r) \to 0$ as $r \to \infty$, resulting in the step function for
the ions but not for neutrals.

Solutions of (5) [and even determination of the potential in the
second part of (6)] result in matrix elements of the form

$$\int \psi_i \, V_{if} \, \psi_f \, d(\text{space}) \ . \tag{7}$$

This expression is usually represented or calculated via a multipole
expansion of the potential V_{if}. For the dipole (allowed) transitions (e.g., s → p orbital changes) the matrix elements and cross
sections fall off with collision energy — at least at high energy —
as

$$\sigma \ (\text{dipole allowed}) \propto \frac{\ell nE}{E} \tag{8}$$

while nondipole terms fall off faster. Monopole transitions
(e.g., s → s or p → p excitations) often fall off as

$$\sigma \ (\text{dipole forbidden}) \propto 1/E \ , \tag{9}$$

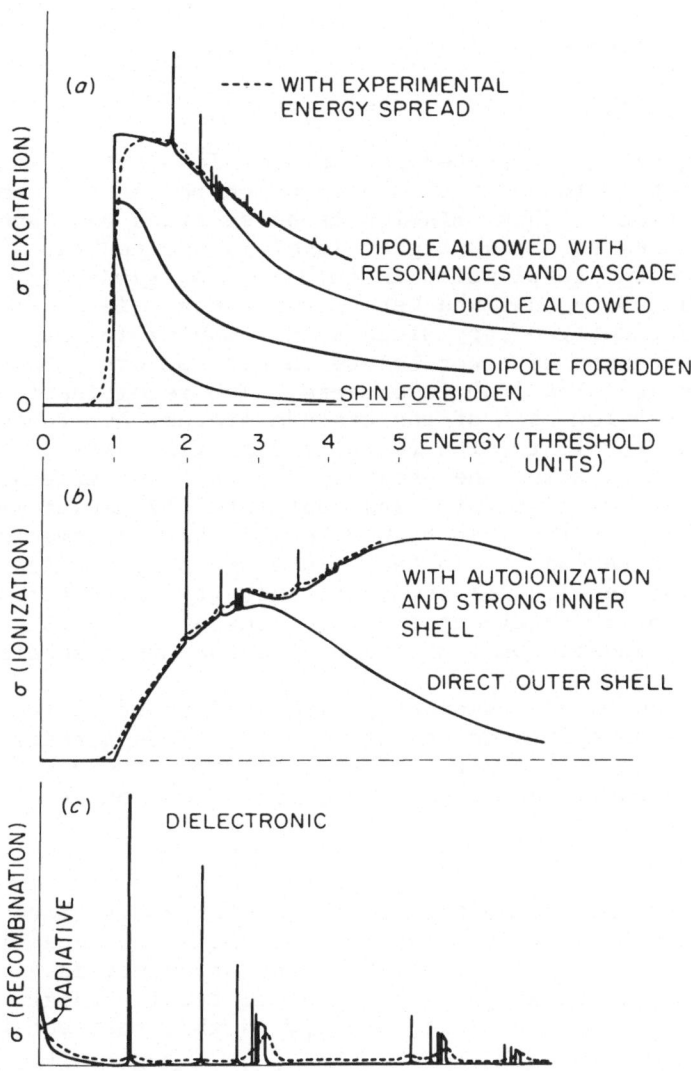

Fig. 4. General character of the cross sections for various inelastic electron-ion collisions. Approximate averaging effects of interacting-beams-type energy resolution are represented by dashed curves.

the 1s → 2s transition in hydrogenic ions being an interesting and still controversial example.[4] For spin forbidden transitions a steep fall off with collision energy is characteristic, such as

$$\sigma \text{ (spin forbidden)} \propto 1/E^3 \quad . \tag{10}$$

See Fig. 4a for illustration of these features.

What is generally observed is the emission (rather than excitation) cross section for a given photon which can be influenced by cascade transitions. Thus emission cross sections can have new step function increases as a next higher level is excited but then radiatively decay to the level which in turn decays, giving the photon being observed. Note that whether or not cascade is observed can depend on experiment geometry since a fast moving ion might pass out of view of the photon detector before cascade occurs. Cascade should be distinguished from resonances. The recombination resonances lie just below each of the cascade states, but these resonances decay by autoionization (usually very fast, say 10^{-14} sec) to form the level from which the observed photons are emitted. Resonances cannot be identified and subtracted by measurement of emission photons for the next higher levels (as has sometimes been accomplished for cascade). In fact since resonances which result in formation of the level under study are only (finally) accompanied by loss of energy by the incident electron they are generally considered to be a generic part of the excitation cross section.

It is common in all theoretical predictions of excitation to present results as collision strengths, Ω_{if}. This quantity is the particle collision equivalent of the oscillator strength. The connection between collision strength and cross section is always

$$\Omega_{if} = w_i E \sigma_{if} \tag{11}$$

where E is again the incident electron kinetic energy here in Rydbergs, w_i is the statistical weight of the lower level [$w_i = (2S_i + 1) \cdot (2L_i + 1)$ or $(2J_i + 1)$ if fine structure is taken into account, where S_i, L_i, and J_i are the usual total angular momentum quantum numbers of the state i], and σ_{if} is the cross section in units of πa_0^2.

Direct ionization cross sections can be thought of as a summation of many excitation cross sections (following Bely and Van Regemorter[24]). Qualitatively this idea suggests that ionization is just excitation of continuum rather than discrete states. According to this qualitative idea, it is easy to understand that ionization cross sections ought to rise, roughly linearly, near threshold since as the collision energy increases more and more continuum states are excited. Neglecting the components due to excitation-autoionization and recombination-double autoionization, the direct ionization cross sections usually peak at 2.5 to 3 times the threshold energy and subsequently fall off roughly like dipole-allowed excitation cross sections. The high-energy behavior is often a useful test of experiment or theory and is well discussed in Ref. 15.

$$\sigma \text{ (direct ionization)} \to A \frac{\ell n E}{E} + B/E \ . \tag{12}$$

However, it is often true that direct, inner-shell ionization is significant which results in a new rising, peaking, and falling component of the total ionization cross section. These features and the autoionization complications are apparent in Fig. 3 and are schematically illustrated in Fig. 4b.

Dielectronic recombination cross sections have not been measured and are not generally calculated. Since so many resonances are involved and since the interest has generally been for plasma conditions where the incident electrons have a broad Boltzman energy distribution, dielectronic recombination rates are usually calculated. Ion-electron beams experiments are in progress which at least will have small energy spreads compared to plasmas and hope to observe some "averaged" or near-cross-section-like dielectronic processes. Figure 4c gives a schematic estimate of the behavior of such dielectronic recombination cross sections with even the small energy spread of interacting beams experiments averaging out much of the structure. The actual resonances are not likely to be so nicely uniform as in Fig. 4c, that is they will vary in width, amplitude, and spacing and at many energies there will be overlapping series of resonances. However, it probably is true that the cross-section space for dielectronic recombination is mostly empty and what little is occupied contains many narrow resonances.

2.8 General Formulas

Provided below are formulas for cross sections and rates which are currently in general use for prediction of excitation, ionization, and recombination. Only for the ionization case are these results generally reliable and then only for the direct part of the ionization cross sections. The most reliable ionization results are those of Younger but these require specific constants from Younger's papers.

The oscillator strengths and energies have to be obtained from tables such as those prepared by NBS in the United States. The subshell ionization energies can be obtained for many ions from Ref. 25.

Excitation. Seaton[26]–Van Regemorter[27]–Gaunt Factor[28]

$$\sigma_{if} = \frac{8\pi}{\sqrt{3}} \frac{f_{if} \, \bar{g}}{E \, \Delta E_{if}} \, \pi a_o{}^2 \tag{13}$$

where E = electron incident energy in Rydbergs
         ΔE_{if} = transition energy in Rydbergs
         f_{if} = oscillator strength (from tables)
         $\bar{g}$ = Gaunt factor (about 1 for $\Delta n = 0$ multicharged ions)

Ionization. Lotz[29]:

$$\sigma_{q,q+1} = 4.5 \times 10^{-14} \sum_j \frac{r_j}{I_j E} \ln E/I_j \quad (\text{cm}^2) \tag{14}$$

(semi-empirical based in part on hydrogenic Coulomb-Born)

Scaled Coulomb-Born:[30]

$$\sigma_{q,q+1} = \pi a_o{}^2 \sum_j \left(\frac{n_j}{Z_{eff}(j)}\right)^2 \frac{I_H}{I_j} \frac{r_j}{X_j} \left[A_j \ln X_j + D_j \left(1 - \frac{1}{X_j}\right)^2 \right.$$
$$\left. + \left(\frac{c_j}{X_j} + \frac{d_j}{X_j{}^2}\right) \left(1 - \frac{1}{X_j}\right) \right] \quad (\text{cm}^2) \tag{15}$$

Younger:[17]

$$\sigma_{q,q+1} = \sum_j \frac{1}{X_j I_j{}^2} \left[A \left(1 - \frac{1}{X_j}\right) + B \left(1 - \frac{1}{X_j}\right)^2 \right.$$
$$\left. + C \ln X_j + \frac{D}{X_j} \ln X_j \right] \quad (\text{cm}^2) \tag{16}$$

where E = electron incident energy in eV
         I_j = ionization energy of subshell j in eV
         r_j = number of electrons in subshell j
         I_H = 13.6 eV
         n_j = principal quantum number
         X_j = E/I_j
         $Z_{eff}(j)$, A_j, D_j, C_j, D_j parameters from Ref. 30.
         A, B, C, D parameters from Younger,[17] and summarized in
Volume 7, Number 6, p.190-201 of the newsletter "Atomic Data for
Fusion" issued by ORNL and NBS from CFADC, Bldg.6003, Oak Ridge
National Laboratory, Oak Ridge, Tenn. 37830.

Dielectronic Recombination. Burgess[23]–Merts[18] Rate:

$$\alpha_{q,q-1} = 3.0 \times 10^{-12} \ T_e^{-3/2} \ B(q) \sum_{m=i+1} f_{im} A(s) \ e^{-\bar{E}/kT_e} \left(\frac{cm^3}{sec}\right) \quad (17)$$

where T_e = electron temperature in units of 10^6 K
   q = initial charge of the recombining ion
   i = initial state of the recombining ion
   m = the excited core configuration j_0 of the recombined ion before radiative stabilization
   f_{im} = oscillator strength of transition $i-m$
   $\bar{E}/kT$ = 0.158 $(q + 1)^2 (n_i^{-2} - n_m^{-2})/a \ T_e$

   [n_i and n_m are principle quantum numbers and
   $a = 1.0 + 0.015 \ q^3/(q + 1)^2$]
   $B(q) = q^{1/2} (q + 1)^{5/2}/(q^2 + 13.4)^{1/2}$
   $A(s) = s^{1/2}/(1 + 0.105 \ s + 0.015 \ s^2)$ for $\Delta n=0$

and

   $A(s) = 0.5 \ s^{1/2}/(1 + 0.21 \ s + 0.03 \ s^2)$ for $\Delta n=1$
   and $s = (q + 1)(n_i^{-2} - n_m^{-2})$

3. EXPERIMENTAL APPROACHES

3.1 Plasmas

Observations and modeling of well-characterized plasmas have been used to determine rates for all three of the inelastic collision processes. The 1972 review of Kunze[31] remains the best source of information on the basic techniques employed for ionization rates and together with the paper of Gabriel and Jordan[32] also provides discussion of excitation rate measurements. All of the plasma-based measurements have relied on observation of emitted radiation. Ionization and (more rarely) recombination rates can be determined from time evolution of line radiation from plasma without absolute calibration of photon detectors while excitation rates require absolute radiometry. All of these measurements require the plasma to be stable and well characterized as to electron temperature and density and all resort to a model of the plasma behavior to extract rates. Problems associated with ion transport within the plasma, nonuniform temperature and density, multiple ionization in a single collision, high electron density and associated secondary collisions, metastable ions, and competing collision processes such as electron capture are difficult to assess in the plasma rate measurements. These rate determinations face difficult assumptions and measurements, all

to obtain a few data values in which much of the physically meaningful details, like structures in cross section vs energy, have been
averaged and lost. Nevertheless two important advantages apply to
plasma rate measurements — the information obtained is directly
applicable to plasma and astrophysical environments and the only
data for ionization of ions of charges greater than +6, or for any
multicharged ions on dielectronic recombination, have come from
plasmas (and EBIS traps).

The most recent paper on ionization rate measurements[33] contains
references to earlier work. There are no recent reports on excitation rate measurements but current work is in progress.[34] There are
a number of observations of dielectronic recombination in plasmas
and astronomy but only a few rate measurements which will be in the
results section.

3.2 Ion Traps

Measurements within ion traps have produced some useful data on
collisions of electrons and ions. Hasted[35] reviewed these at the
previous NATO study and Dunn[36] has given some interesting comparison
of traps vs beams experimental capabilities. The work of Donets and
Ovsyannikov[37] using an EBIS trap has provided the most data of
interest to the present considerations. In addition the Donets ion
trap is one of the exciting new sources of highly charged ions[38]
which holds promise for future atomic collision studies, so it will
be discussed here in modest detail.

A general Electron Beam Ion Source-type (EBIS) trap is represented in Fig. 5. Such devices have been developed to a fairly
sophisticated level of operation. The electron source and magnetic
field components are constructed to quite high tolerances to assure
rectilinear propagation of the electron beam in the solenoid. Magnetic field and electron source axes must coincide to within 0.05
mm.[38] The potential distributions, PD 1, 2, and 3 on Fig. 5, are
applied sequentially in order to create ions, trap and heat the ions
to provide high-charge states, and to eject the ions in a pulsed
beam. The potentials shown on Fig. 5 are specifically like
CRYEBIS-I (ORSAY)[38] but those of KRION-II[37] are qualitatively similar. For the Donets source,[37,38] "KRION-II," the tubes T_4-T_n are at
liquid helium temperature and at ambient pressure about 10^{-12} torr.
With the electron source activated and PD 1 potential distribution a
small amount of cold gas is puffed in only at T_3 position to create
low-charge ions. Then PD 2 is applied for a time selected according
to desired ionic charge state (up to about 1 sec). During operation
of the source as a trap, PD 2, the pressure is still as low as 10^{-9}
torr — a rather thin plasma.

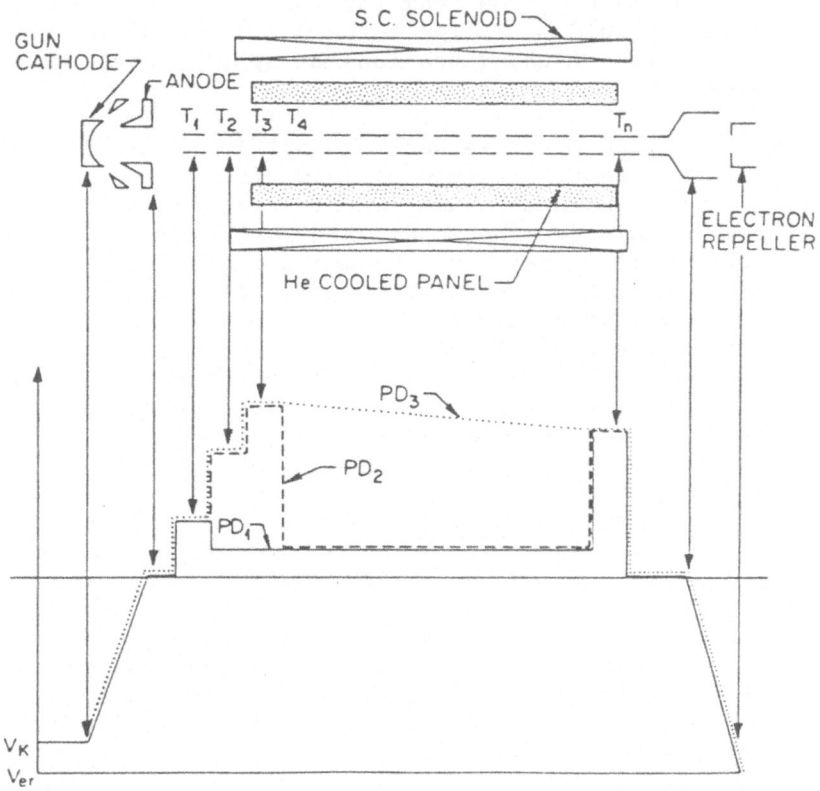

Fig. 5. Schematic arrangement of an Electron Beam Ion Source (EBIS) from Ref. 38. The potential distributions PD 1, 2, and 3 are applied sequentially (see text).

Donets and Ovsyannikov[37b,c] used the KRION-II as a trap to
study ionization collisions. The operating cycle of the ion source
was as follows: (1) activation of the electron beam; (2) introduction of low-charge ions of the investigated element into the electron beam (PD 1 for 0.5-1 msec); (3) confinement of ions in the beam
and their ionization (PD 2 for 0.1-500 msec); (4) extraction of ions
from the electron beam in the longitudinal direction (PD 3 for 10-50
μsec); (5) analysis of the charge spectrum of ions according to
transit time; and (6) electron beam shut-off. The electron beam
energy during PD 2 could be varied between 0.5 and 11 keV which
allowed determination of cross sections at approximately those collision energies.

The determination of cross sections within KRION-II relies on
modeling the time rate of creation of ions of a given charge within
the electron beam in much the same manner as analysis of θ-pinch
ionization rate studies. The input data for the model are the
assumed ionization cross sections and measured intensities of each
ion charge state as a function of time after PD 2 is switched on.
Thus data are collected by operating PD 2 for various times from 0.1
to 500 msec and using the time-of-flight analysis after ion ejection
to obtain the ion intensities. The ion-time-intensity distribution
is modeled by iteratively varying the ionization cross sections to
obtain the best fit to the measured data.

3.3 Electron-Ion Beams in General

Modern beams experiments with electrons and ions began about
1960 and the technique along with the first 15 years of results were
reviewed in 1976 by Dolder and Peart.[7] Together with more recent
but more specialized reviews[4-9] the technique and results are well
covered. The basic approach and apparatus applied to studies of
multicharged ions are presented here for reference and discussion,
but without extensive detail.

A critical part of experiments with multicharged ions is the
ion source. Results published to date have relied on continuous
beams from a PIG-type (Penning Ion Gauge configuration of electrodes) source[39] at Oak Ridge National Laboratory and from an EBIS[40]
at the Justus-Liebig Universität in Giessen. The ORNL-PIG source
employs a magnetic field of about 4 kilogauss to confine a small
plasma, and ions are extracted transverse to the magnetic field.
The Giessen-EBIS is qualitatively like Fig. 5 but without ultrahigh
cryogenic vacuum or superconducting magnets and further it is operated quite differently in order to obtain continuous ion beams.
Ions are allowed to diffuse through the EBIS with a constant potential distribution, roughly like PD 3 of Fig. 5 so that a continuous
ion beam is achieved but at much lower charge states than for the

pulsed operation. The new generation of ion sources[38] will be applied to some electron-ion beams experiments in the future, but the ion source development progress is beyond the scope of the present paper.

A schematic representation of crossed beams is shown in Fig. 6. The beams may cross at any angle, though some special considerations apply for merged beams ($\theta=0$). Cross sections are determined by measurement of the parameters in Eq. (18)

$$\sigma = \left(\frac{S}{K}\right) \frac{qe^2}{I_i I_e} \, VF \tag{18}$$

where S is the rate (counts/sec) of signal detected for the particular experiment, K is a geometry or sensitivity factor which may be different for different types of experiments such as excitation or ionization, I_i is total ion current, I_e is total electron current, q is the ionic charge in electronic units (e), V contains all the angle and velocity terms, and F is the vertical (z axis) beams overlap factor.

$$V = \frac{v_i \, v_e \, \sin \theta}{(v_i{}^2 + v_e{}^2 - 2 \, v_i \, v_e \, \cos \theta)^{1/2}} \tag{19}$$

$$F = \frac{\int i_e(z)dz \, \int i_i(z)dz}{\int i_e(z)i_i(z)dz} \tag{20}$$

In thinking about the nature of interacting beams experiments it is useful to recognize the physical significance of terms in (18) and to consider the limiting cases of $\theta = 0°$ and $\theta = 90°$. Expression (18) is variously derived and expressed. It is given here in the form used by Peart and Dolder[7] and is derived and discussed in some detail by Brouillard and Claeys[41] (however they define the form factor as the inverse of F here and they use y rather than z for the beams overlap/probe direction). Note that $I_i/v_i qe$ and $I_e/v_e e$ are linear particle densities and $(v_i{}^2 + v_e{}^2 - 2v_i v_e \cos \theta)$ is the relative speed of particles. Note also that F is not dimensionless but has units of length as expressed.

The simplest geometry is for $\theta = 90°$. In that case the term (19) reduces to $V = v_i$ if $v_e \gg v_i$ as is often the case. For crossed (90°) beams it is desirable to have v_i as slow as possible, since, for a given cross section, the signal, S, will be inversely proportional to v_i.

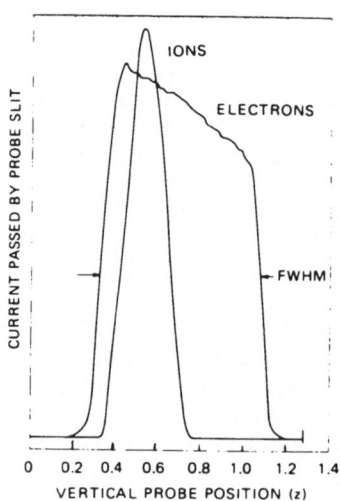

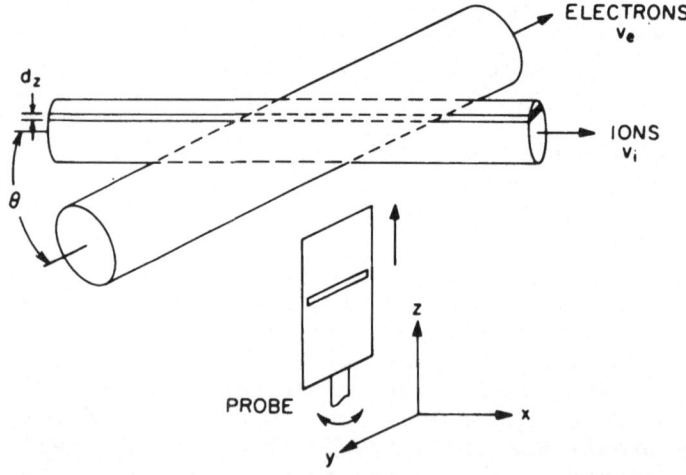

Fig. 6.   A schematic representation of crossed beams. The ion beam
          is along the x axis and the electron beam is at an
          arbitrary angle θ in the x–y plane. The rotatable probe
          can translate vertically through either beam. The upper
          inset shows some actual electron and ion vertical distri-
          bution functions measured with a rotatable probe slit.

For parallel rectilinear, merged beams ($\theta = 0$) propagating in the x direction expression (19) can be shown to be $V = \dfrac{v_i \cdot v_e}{v_i - v_e}$ if expression (20) is written as

$$F = \frac{1}{L} \frac{\iint i_e (z,y) \; dz \; dy \iint i_i(z,y) \; dz \; dy}{\iint i_e (z,y) \; i_i(z,y) \; dz \; dy}$$

which ignores any variation of F along the length of the merge path, L (such variation only occurs if $\theta \neq 0$). The merged beams approach has advantages in allowing an extensive interaction length L for beams of arbitrary velocities. One beam "chases" the other and as the relative energy is compressed, so is the relative spread in energy. This means that high energy resolution can be achieved with merged beams and in fact for crossed or inclined beams the energy resolution also improves as θ approaches 0°. The practical limitation on achieving the ideal energy resolution of merged beams is usually the extent to which $\theta \equiv 0°$. A difficulty of merged beams is measuring F.

Historically it has been considered important to measure F with some form of beam probe as illustrated in Fig. 6. Some imaginative new methods of determining F are now in use,[41] and for some crossed (90°) beams experiments it has been considered sufficient to calculate F. For relatively uniform beams, with one beam contained completely within the other as illustrated in Fig. 6, the value of F is close to the height of the larger beam. The particular beam profiles shown in Fig. 6 are typical for crossed beams experiments at Oak Ridge, and the value of F was always within 10% of the FWHM height of the electron beam. Determination of particle velocities v_i and v_e and beams overlap F is equivalent to measuring density and path length in a static gas target of the type commonly used in beam-static target experiments. However, the determination of velocities and F in interacting beams experiments <u>can be</u> more accurately accomplished than the measurement of equivalent parameters in static-gas-type experiments. In a number of experiments it has proven valuable to be able to measure and to maintain stability of F to approximately ±1% in order to observe detailed structure in cross sections.

3.4 Particular Studies with Electron-Ion Beams

It is useful to consider some of the features of particular interacting beams experiments. Figure 7 is a schematic of the approach used at Oak Ridge (with collaboration of G. H. Dunn and others of the Joint Institute for Laboratory Astrophysics, Boulder). The parallel plate analyzers are employed as "charge sorters" and to

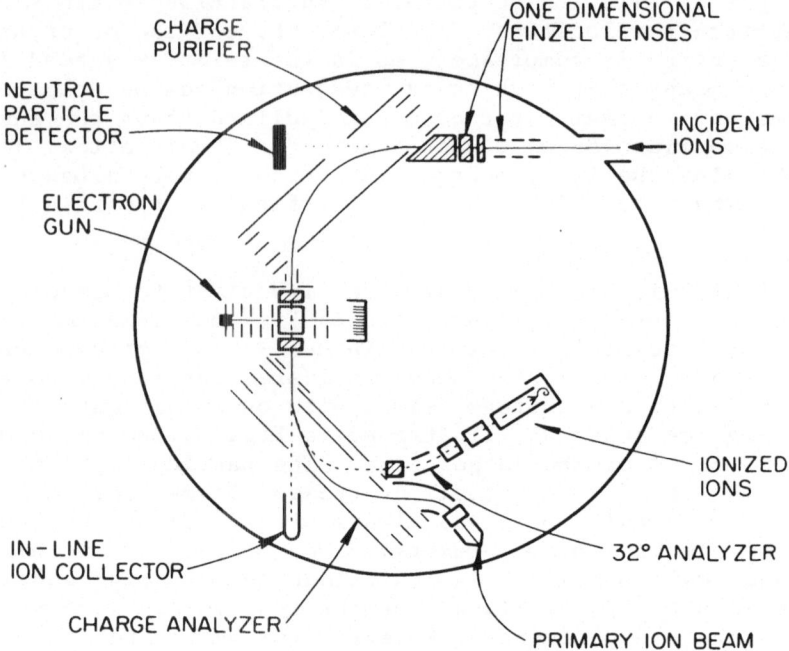

Fig. 7.   A particular crossed beams arrangement used for multi-
          charged ions at Oak Ridge (see text). For some experiments
          a photodetection system, viewing the beams cross region
          from above, detected excitation.

minimize effective path length for interaction of the ion beam with background gas. These analyzers have strong focusing properties in the horizontal or dispersion plane but do not effect vertical-beam trajectories. Thus the flat extended plates called "one-dimensional einzel lenses" were employed to focus and position the beam independently in horizontal and vertical directions. For ionization experiments the principal difficulty has been to adequately separate the "ionized" and incident beam components while maintaining 100% transmission of the beam components from the electron beam crossing point to the ion detectors.

The electron beam of Fig. 7 is confined by a fairly uniform magnetic field of about 200 gauss directed parallel to the electron beam and maintained by permanent magnets.[42] The electron collector consists of razor blades stacked with edges toward the beam to minimize specular reflection of electrons which cannot be compensated by biasing the collector. The ion beam must penetrate the 200-gauss magnetic field transversely so that the ions are deflected vertically (usually a few degrees at most). The cross-hatched electrostatic element in Fig. 7 can provide vertical positioning of the ion beam, in part to counteract the deflection by the magnetic field.

Figure 8 shows some of the features of the electron-ion crossed beams approach used at Giessen, West Germany[43] (with collaborators from Universität Frankfurt and Technishe Universität Wien). The electron beam is shaped and intensified by electrostatic focusing electrodes. In addition all of the electron gun parts can be raised to a potential of several hundred volts. Thus ionizing events within the electron beam result in ions of increased charge at a somewhat different energy than those ionized outside the electron beam. This "energy tagging" of the experiment region, together with high resolution ion beam analysis after the interaction region, allows separation of unwanted signal due to stripping of ions on background gas outside the electron beam.

It is common in crossed beams experiments to employ beam switching schemes which allow detection of background signal from either beam alone which can then be subtracted from the total signal with both beams on. Usually detectors are gated so that signal is collected only when beams are completely on or completely off. If it can be proved that all the background signal arises from one beam, then only the other beam need be switched on and off to determine the background. Caution should be exercised since the evolution of gas by a beam, or the focusing or deflection of one beam by the other, can change the background. Thus beam modulation techniques do not guarantee correct identification of backgrounds. One of the common diagnostics is to change the frequency at which beams are switched on and off. For gas evolution, beam switching times which are much faster than characteristic vacuum pumping times can

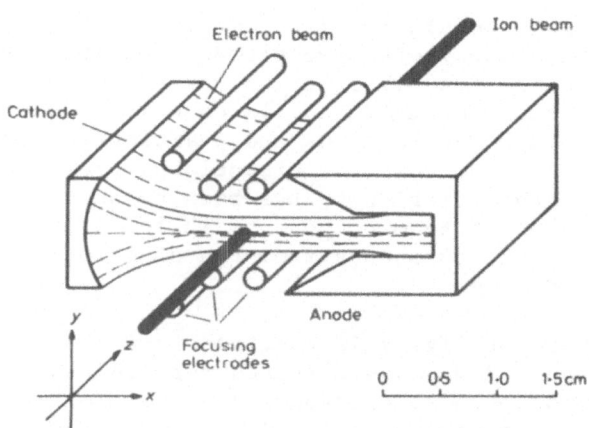

Fig. 8.   A particular crossed beams arrangement used for
           multicharged ions at Giessen (see text) from Ref. 43.   The
           electrostatic focusing electrodes confine and intensify the
           electron beam.

suppress evolved gas problems.   Switching times from near 1 sec to
near 1 μsec have been employed in beams experiments, and signal to
noise as poor as 1 to 100 has been properly accounted via beams
switching techniques.

     The devices illustrated in Figs. 7 and 8 have been used to
measure electron impact ionization of multicharged ions.   The apparatus of Fig. 7 has also been used to measure excitation cross sections (see Ref. 4 for review).   The excitation cross-section
measurements relied on detection of radiation at the collision
center from the excited ions by a photomultiplier viewing the collision region from above.   These experiments are quite tedious since
both beams create significant photon background, and the photon
detector only counts about 10^{-4} of the true beams interaction signal
due to solid angle and detector sensitivity limitations.

     Table 1 gives parameters of two excitation experiments[44,45]
which employed photon detection with crossed beams.   As discussed in
Ref. 4, for the N^{4+} excitation case the acquisition of a single
cross-section value with ±6% statistical uncertainty required about
10^5 sec (28 hr) of data collection time for the parameters shown and
the both-beams switching employed.   In addition, systematic variation of beams-switching frequency and both electron and ion current
intensities not only required more time but revealed spurious
signals probably due to creation and trapping of background ions by
the electron beam and subsequent interaction of the trapped ions
with the incident ions.   It seems unlikely that more than a few
absolute experiments on excitation cross sections for multicharged
ions can ever be performed using crossed beams and photon detection.

Table 1. Parameters from two electron impact
excitation experiments.

Parameters	Ba^+ (Ref. 44)	N^{+4} (Ref. 45)
Electron energy	4 eV	15.5 eV
Electron current	10 μA	90 μA
Ion energy	750 eV	40 keV
Ion current	0.1 μA	1.0 μA
Pressure (both beams on)	1×10^{-9} torr	1.2×10^{-9} torr
Wavelength	455 nm	124 nm
Band pass	10 nm	30 nm
$D(z_0, \lambda)$	7.4×10^{-4}	4×10^{-4}
Signal (S) (both beams)	10 Hz	2 Hz
Background (B) (both beams)	3 Hz	90 Hz
Emission cross section	17.4×10^{-16} cm^2	2.7×10^{-16} cm^2

Severe difficulties are also encountered in attempting interacting beams measurements of dielectronic recombination. Figure 9
illustrates an experiment pursued at the Center for Astrophysics at
Harvard[46] with C^{3+} ions. The approach is to measure the cross section associated with a transition

$$e^- + C^{3+} (1s^2 2s) \rightarrow C^{2+} (1s^2 2p \, n\ell)$$
$$\phantom{e^- + C^{3+} (1s^2 2s) \rightarrow C^{2+}} \downarrow C^{2+} (1s^2 2s \, n\ell) + h\upsilon$$

which is qualitatively similar to transition ④ in Fig. 1. The
photons produced are near in wavelength to the 2s-2p, 1550 Å transition which will also be directly excited. The relatively few
dielectronic stabilizing photons might be detected by coincidence
between the total photons near 1550 Å and the C^{2+} ions. A principal
anticipated problem could be field ionization of the C^{2+} ($1s^2 2s \, n\ell$)
in the analyzing field, so electrons produced by field ionization
are accelerated into the channel plate detector and can also be
counted in coincidence with the photons. Electron capture by the

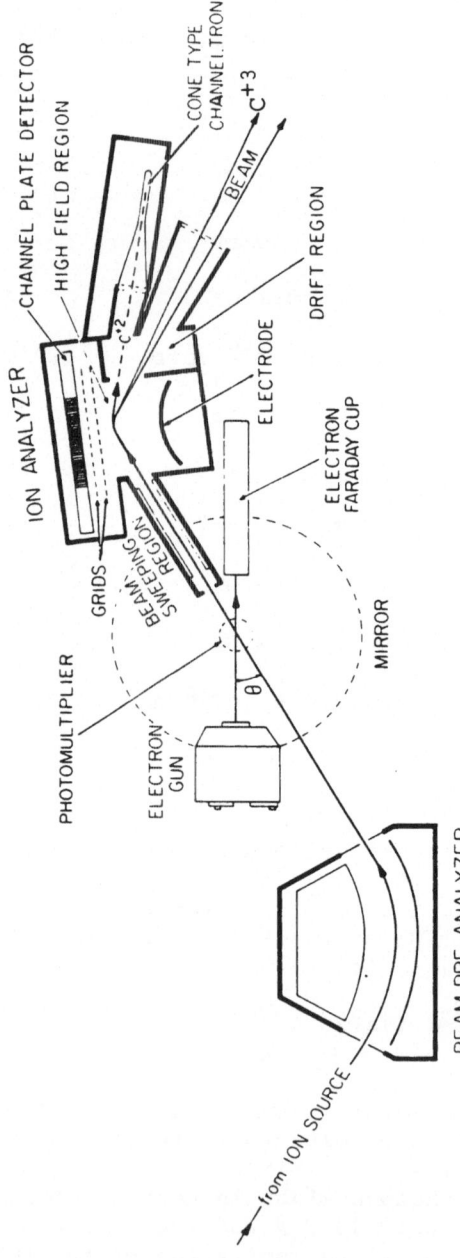

Fig. 9. Inclined beams arrangement used at Harvard–Smithsonian (from Ref. 46) to study collisions of electrons and C^{3+}. The mirror helps concentrate photons from excitation and dielectronic recombination onto the photomultiplier viewing only the beams-crossing region from above.

C^{3+} from background gas at about 10 keV has a cross section near 5 x 10^{-15} cm^2 and provides a formidable C^{2+} signal in the experiment even at 10^{-10} torr pressure. Nevertheless the experiment might provide the first definitive averaged "cross-section" measurement for the dielectronic recombination resonances associated with a specific excitation transition in a multicharged ion.

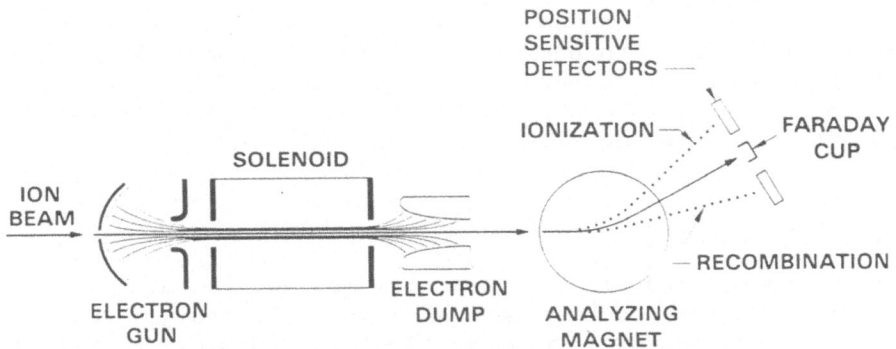

Fig. 10. Apparatus in use for merged electron-ion studies of recombination (at Oak Ridge from Ref. 47). The ions are from a tandem Van de Graff accelerator.

One approach to reducing the electron capture background in recombination experiments is to use faster ion beams. At MeV energies the capture from background gas is of the order 10^5 times smaller than at keV energies. At Oak Ridge an approach is being developed to use electron-ion merged beams with a fast ion beam from a tandem accelerator.[47] Figure 10 illustrates the approach which relies on injection of an ion beam through the cathode of an electron gun, with subsequent collapse of the electron beam onto the ion beam employing a strong solenoid magnet. Signal associated with recombination has been obtained but additional analysis and systematic diagnostics are required.

Interacting beams experiments have significant advantages over other experimental approaches in that nearly all experiment parameters can be controlled and systematically varied for diagnostic purposes. However, the achievable beam particle interaction densities are quite low and backgrounds are high so that experiments are tedious. Systematic variation of parameters has proven essential to detection and elmination of problems in most electron-multicharged ion experiments. The single most significant test, when applicable, is to measure the cross section for the process under study <u>below its threshold energy</u> and <u>obtain a zero value</u> for the cross section.

4. RESULTS AND DISCUSSION

In spite of the difficulties, interacting beams studies with multicharged ions have been quite rewarding. Enough detail in cross section vs energy has been obtained to illuminate physical mechanisms which provide unique aspects to electron-ion collisions for highly charged ions. Some of the results obtained to date which illustrate the physics are presented here.

4.1 Excitation Studies

Excitation of ions has been recently reviewed[3,4] and only a few experimental measurements have been made for multicharged ions. Figure 11 presents measured cross sections and theoretical calculations for excitation of 2s-2p in Li-like ions,[45,48] N^{4+} and Be^+. The Be^+ case nicely illustrates the convergence of theoretical approximations. The theoretical approximations[49,50] are: CBI – Coulomb-Born, CBXI – Coulomb-Born with exchange, CBII – unitarized Coulomb-Born, CBXII – unitarized Coulomb-Born with exchange, UDWPOII – unitarized distorted-wave with polarized orbitals and exchange, 2cc – 2-state close coupling, 5cc – 5-state close coupling. These theories progressively improve (that is converge toward experiment) in exactly the order anticipated from the quality of the approximations (Sects. 2.5 and 2.6). However, the convergence to experiment is never complete suggesting inadequacy of even the best theoretical result. The $\bar{g}$ result [Eq. (13)] with $\bar{g} \simeq 0.2$ at threshold is a simple scaled prediction which came closest to the experiment. By contrast the N^{4+} experiment[45] shows no discrepancy with the 2cc theory[51] and CBXII-type theory agrees with 2cc to within 3% (see Ref. 3). Additional consolation for theory is that use of the same $\bar{g}$ value for N^{4+} as for Be^+ misses the N^{4+} experimental values by a factor of about 4. It has been understood for some time that $\bar{g}$ should be near 1.0 for $\Delta n=0$ transitions of highly charged ions. Apparently the increase in ionic charge, and consequently more tightly bound electrons, has led to less distortion and coupling via the potential V_{ij} in Eqs. (5) and (6) as had been

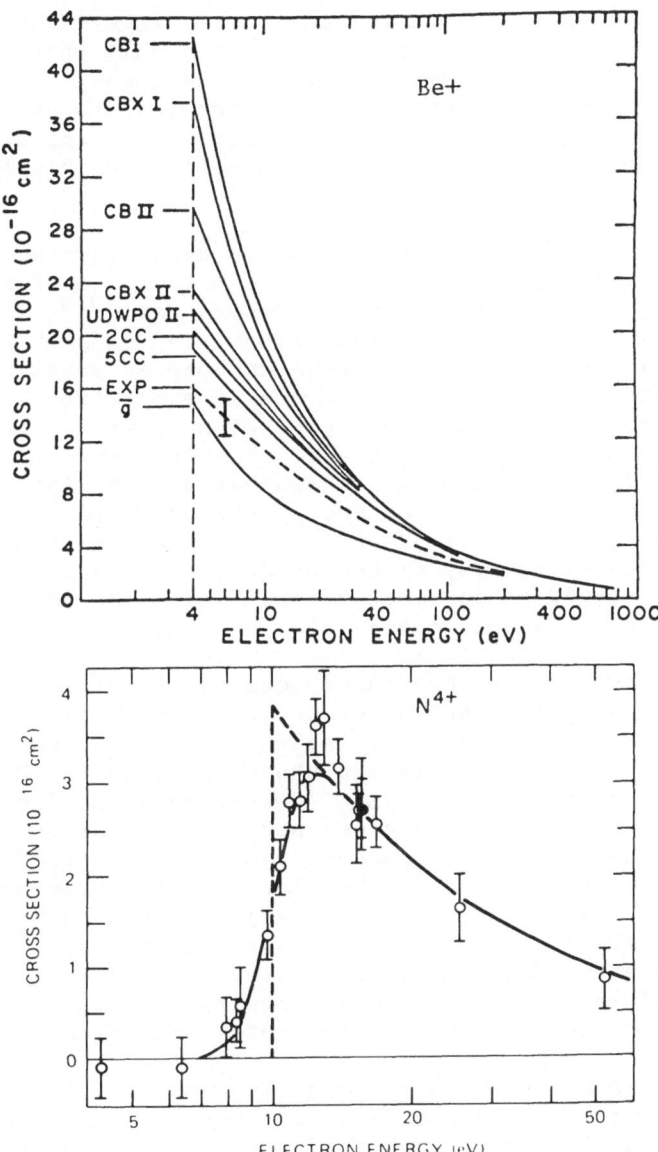

Fig. 11. Electron impact excitation of 2s-2p transition in Li-like Be+ (Refs. 48-50) and N⁴⁺ (Refs. 45 and 51). The error bar shown for Be+ experiment -EXP- is 98% confidence level total uncertainty while for N⁴⁺ the bars are 90% C. L. relative uncertainty except for outer bar on solid point at 15.5 eV which is total uncertainty at 90% C. L. The various theories are identified in the text. For Be+ none of the theories are satisfactory while for N⁴⁺ the solid curve represents either CBXII or 2cc folded with ±2 eV Gaussian energy spread.

hoped. The result is that good quantum theory with almost any realistic potential is accurate in the more highly charged case. Whether or not this improvement in excitation theory with ionic charge extends to other types of transitions besides strong dipole-allowed cases like 2s-2p remains questionable, but the unique aspect here is the improved reliability of theory for more highly charged ions.

Resonances do occur in the excitation cross section for the Li-like 2s-2p. Figure 12a shows part of the calculated cross section for C^{3+} (2s-2p) excitation, including resonances.[52,3] These are the recombination resonances of configuration $(1s^2 3\ell 3\ell')$ which lie just below the $1s^2 3\ell$ excited states. These resonances autoionize to $(1s^2 2p) + \varepsilon\ell$ giving the structure in the 2p excitation. Figure 12a demonstrates that such resonances occur with a variety of shapes, amplitude, and spacing, even in the relatively simple Li-like structure. However for this case the resonances are of little consequence. An experiment with 1-eV energy spread would have to be able to detect structure of about 3% of the direct excitation cross section in order to observe any of the predicted resonances in this C^{3+} case. The single cross-section value which was measured[53] in this energy range is shown on Fig. 12a.

The resonances are rather inconsequential for the strong dipole-allowed 2s-2p excitation but for He-like ions; for forbidden transitions such as $(1s^2)$ $^1S \rightarrow (1s2s)$ ^3S the resonances are predicted to dominate.[54a,3] Figure 12b shows the case for Be^{2+} for resonances of configuration $(1s3\ell3\ell')$ which decay to $(1s2s)$ $^3S + \varepsilon\ell$ giving the structure in the ^3S cross section. These same resonances for Fe^{24+} were calculated to enhance even plasma excitation rates by up to an order of magnitude.[54b] However the radiative decay of the resonances to final recombination (Fe^{23+}) rather than by autoionization to the excited states of Fe^{24+} was not included.[54c] The effects of radiative decay of these states will be shown later as observations of recombination and it appears that the resonances will not enhance the Fe^{24+} excitation rates significantly.

Such resonance effects have been observed in interacting beams experiments for singly charged ions (see Ref. 4). Recently studies of the Na-like 3s-3p excitation (structure as shown on Fig. 1) have observed resonances for Mg^+ and Al^{2+} with crossed beams.[55] Even for this strong dipole transition the resonances are not negligible. Resonance structures in excitation cross sections add severe complexity. Since most structures are so narrow in energy compared to observable energy resolution, it is useful to approximate their effect. For highly charged ions, theoretical effort is currently devoted to understanding and approximating resonances in electron-ion excitation cross sections, but the theory must proceed without experimental tests at present.

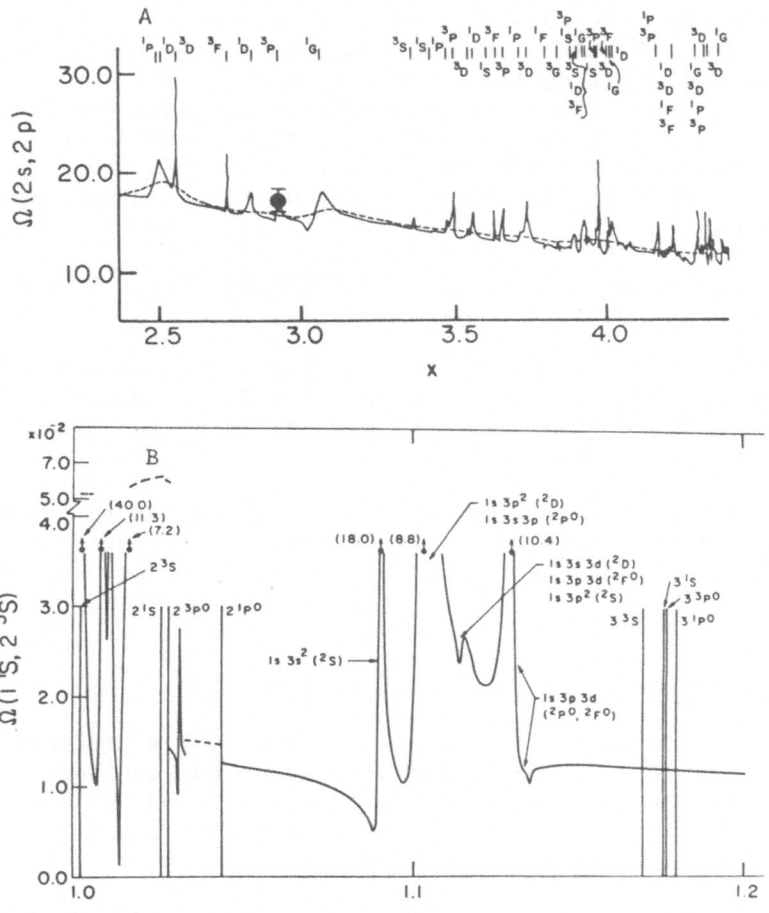

Fig. 12. (a) The calculated C^{3+} (2s-2p) collision strength vs energy
in threshold units including resonances within 5cc approximation (Ref. 52). Dashed line is the 5cc averaged with a
Gaussian energy width of 1 eV. Within this energy range
only one cross-section value was measured (Ref. 53), and it
is plotted at X=2.84 with 1 s. d. relative uncertainty
error bar.

(b) The calculated collision strength for
$(1s^2)$ ^1S → $(1s2s)$ ^3S in Be^{2+} near threshold (Ref. 54 — see
also Ref. 3). Resonances of the form $(1s3\ell3\ell')$ are
included and dashed curve represents averaged, resonance
enhanced values where many resonances are occurring.

Recently another question was raised about calculated excitation cross sections for heavy, highly-charged ions. Again taking
the Fe^{24+} case but the spin forbidden transition
$(1s2s)^3S \rightarrow (1s2p)^1P$, Clark et al.[86] calculated the excitation rate
in the usual L–S coupling and in intermediate coupling with relativistic corrections. The intermediate coupling case, because of the
spin orbit coupling of the forbidden and allowed components, gives a
rate about one order greater than with L–S couping, raising questions about the reliability of other calculations ignoring intermediate coupling.

To summarize current insights on electron impact excitation of
ions:

1.   Distorted–wave and Coulomb–Born theories (unitarized and
     including exchange) appear to be reliable for excitation of
     low–lying bound states and are more reliable for highly charged
     ions than for singly charged ions.

2.   Resonances in the excitation cross sections can be quite
     strong, especially (but not exclusively) for nondipole–allowed
     transitions, and require further detailed studies for assess-
     ment.

4.2   Ionization Studies: H–Like and He–Like Cases

Discussion in Sects. 2.4 and 2.7 and Figs. 1–4 provides the
impression that both recombination resonances and excitation can
have significant effect on ionization cross sections. Thus the
study of ionization, in detail, can give insight into all of the
physics of inelastic collisions of electrons and ions.

However, the simplest ions, H–like and He–like (ground state),
cannot exhibit excitation and resonances in ionization which rely on
inner–shell excitation since there is only one shell populated. The
work of Donets and Ovsyannikov, based on modeling charge state evolution in their EBIS provides the only cross–section data for ions
of truly high charge, including hydrogenic ions up through Ar^{17+}.
Figure 13a shows their data[37c] for hydrogenic ions compared to
Coulomb–Born calculations.[56] The calculations are for nuclear
charge Z = 128, and all results are scaled by Z^4 as suggested by the
Coulomb–Born theory for these hydrogenic ions. The agreement is
good although Donets and Ovsyannikov note that their results are on
average about 1 standard deviation greater than the predictions.
Other theories such as distorved wave (including exchange)[17a] are
nearly indistinguishable from the one chosen for comparison.

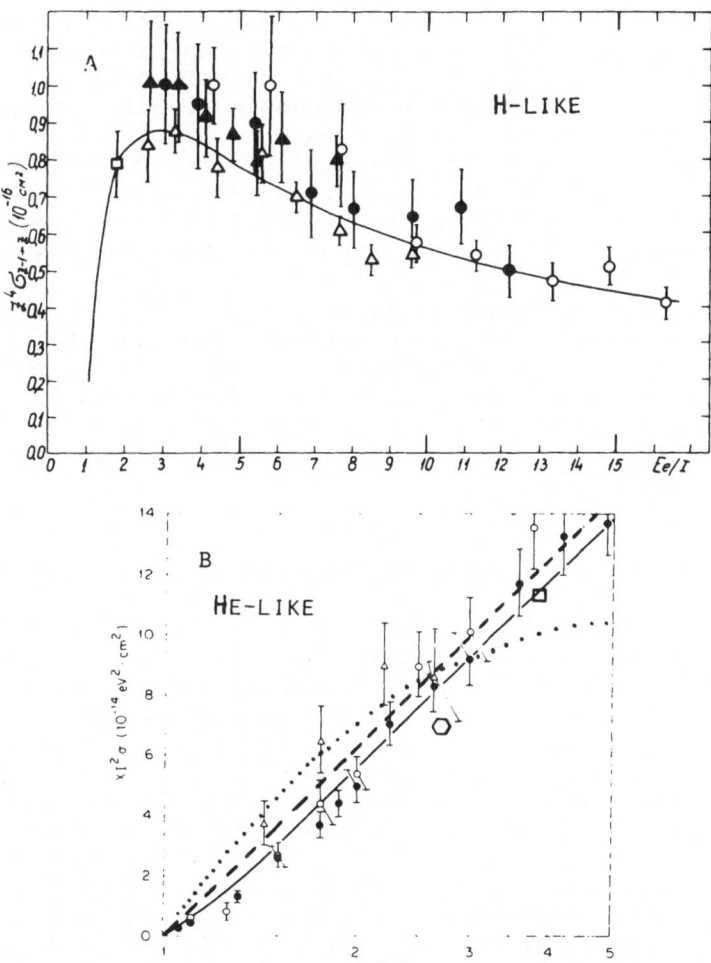

Fig. 13. Scaled cross sections for electron impact ionization of
         H-like and He-like ions. (a) Measured values are: 0 $-C^{5+}$,
         ● - N^{6+}, Δ - O^{7+}, ▲ - Ne^{9+}, ■ - Ar^{17+} (from Ref. 37c).
         Cross sections are scaled by nuclear charge, Z^4, and com-
         pared with calculations in Coulomb-Born approximation —
         solid line.

         (b) Measured values are ● - B^{3+}, 0 - C^{4+}, Δ - N^{5+} (from
         Ref. 57), also ■ - Ne^{8+} and O - Ar^{16+} (from Ref. 37c).
         Cross sections are scaled by ionization potential squared,
         I^2, and collision energy in threshold units, X, as
         suggested by Younger (Ref. 17). Solid curve is distorted
         wave theory by Younger; dashed curve is the Lotz formula,
         Eq. (14); dotted curve is classical theory (Ref. 58).

Results for He-like ions are shown in Fig. 13b. The scaling of these data is as chosen by Younger for these specific data and sometimes referred to as a Fano plot.[17] The experimental data are nearly all from crossed beams[57] but a couple of points from Donets and Ovsyannikov[37c] are included for comparison. The Lotz semiempirical formula[29] was chosen to agree with early Coulomb-Born calculations so it is not surprising that it agrees so well with the recent distorted wave.[17b] Even the original classical result of Thomson[58] is not grossly different from the other results, however.

In spite of the fact that ionization theory is never fundamentally satisfying (because the free motion of three particles within their mutual potential is not properly represented) the theory of these H and He-like cases seems to be well established. In fact it is probably more reasonable to assert that the theory for these cases proves that the experiments are basically correct rather than the other way around.

4.3 Ionization Studies: Cases with Electronic Shell Structures

With the addition of one more electron, electronic shell structure is established and the complexity of the ionization process begins to emerge. Until the first interacting beams experiments on Li-like ions[59] were reported it was not generally expected that excitation-autoionization would play an important role in ionization of such simple systems. Figure 14a shows the N^{4+} case which has been studied in the greatest detail. The excitation-autoionization transition $(1s^22s) \rightarrow (1s2s2\ell) \rightarrow (1s^2) + \epsilon\ell$ is clearly established near the appropriate 420 eV threshold. Subsequent addition of the excitation cross sections[60-62] to Coulomb-Born ionization theory[63,30] is seen to reasonably represent, but slightly underestimate, the measurements in the N^{4+} case. The long-dashed curve beginning at 500 eV is a Bethe approximation[64] which is supposed to include all of the inelastic processes that could lead to ionization in the high energy limit. The long-dashed curve is then supposed to have the correct slope and in fact is the only theory showing agreement in slope with the high energy trend of the crossed beams experiment.

However, the agreement between theory and experiment for the excitation-autoionization component is not satisfying as is illustrated by Fig. 14b. Simply plotting the ratio of the second peak (at excitation-autoionization onset) to the first peak (due to direct ionization) for the Li-like ions gives different trends between experiment and theory. The disagreement depends completely on the O^{5+} experimental result, which is the least definitive data. Table 2 gives more definitive comparison of theory and experiment for the inner-shell excitation assuming that interference and resonances are unimportant in the experiment. The basic question

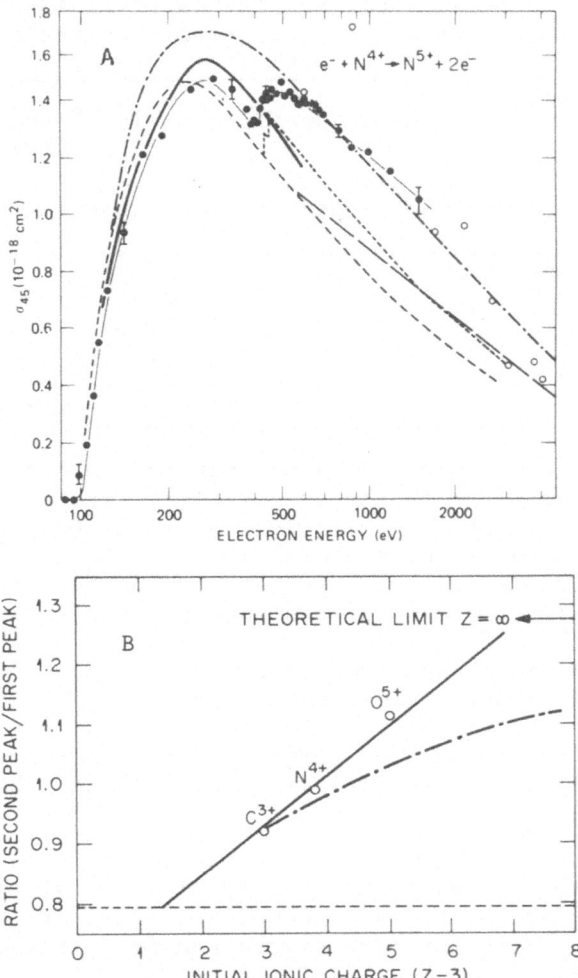

Fig. 14. (a) Electron impact ionization of N^{4+}. Solid circles are
          crossed beams data (Ref. 59); open circles are measurements
          in an EBIS ion source (Ref. 37b). Solid curve is Coulomb-
          Born by Moores (63); dot-dashed curve is Lotz, Eq. (14);
          short-dashed curve is scaled Coulomb-Born, Eq. (15); dotted
          curve is excitation result of Henry (Ref. 60) added to the
          scaled Coulomb-Born; long-dashed curve is Bethe result of
          Kim and Cheng (Ref. 64).

          (b) Increasing relative importance of excitation autoioni-
          zation for Li-like ions along the isoelectronic sequence.
          Data points and inferred straight line are experiment (Ref.
          59) and dot-dashed curve is theory (Ref. 61).

Table 2. Comparison of theoretical and experimental cross
sections for inner-shell excitation of Li-like ions,
$1s^22s \rightarrow 1s2s2\ell$ at energies about 1.1 times excitation threshold.

		Cross sections in 10^{-19} cm^2		
Ion	Energy (eV)	Experiment	Scaled Coulomb-Born	Six-state close-coupling	Ratio exp/6cc theory
Be$^+$	130	17	23	12.2	1.4
C^{+3}	340	3.2	3.7	2.15	1.5
N^{+4}	460	1.8	2.0	1.27	1.4
O^{+5}	612	2.8 (1.4)	1.1	0.74	3.8 (1.9)

Cross sections determined by estimating the increase in total ionization cross sections at energy 1.1 times excitation threshold. The Be$^+$ case is from Ref. 65 while C^{3+}, N^{4+}, and O^{5+} are from Ref. 59. For O^{5+} the uncertainties are large, and the value in parentheses is the smallest excitation cross section derived from the ionization data allowing for the 90% confidence level error bars.

From Ref. 61 which cautions that the technique is not appropriate to low ionic-charge cases such as Be$^+$.

From Ref. 60 with Be$^+$ case extrapolated according to $(Z - 1.4)^3 \sigma = $ a constant for a given energy in threshold units.

remains whether the O^{5+} experiment is spurious or if simple addition of direct excitation theory is inadequate to explain the ionization process. Detailed data for more highly charged Li-like ions would be quite valuable.

For Be-like ions a different, more traditional question arises — what is the role of metastable ions? The ground state of these ions is $(1s^22s^2)$ ^{1}S and the metastables $(1s^22s2p)$ ^{3}P are quite long-lived and are higher in energy by roughly 10% of the ground-state ionization potential. The statistical weight of the levels favors the metastable 9 to 1. Beams experiments for B$^+$, C^{2+}, N^{3+}, and O^{4+} reveal that the metastable population is substantial in ion beams from different sources and that in some cases the population may even be changed by change of ion source parameters.[66] It seems likely that the metastable content in almost any collection of Be-like ions is likely to be high and may influence behavior of the ions via differences in inelastic collision processes for metastable

vs ground-state ions. However, for highly charged ions the metastable lifetime decreases significantly and metastable populations
will be lower. The role of excitation-autoionization has been
addressed for these ions[61,62] but in the experiments this contribution has only been observed for the O^{4+} case.[57,66] Thus, the
scaling of the indirect ionization contribution with charge state
along the isoelectronic sequence for Be-like ions is at least as
uncertain as for the Li-like ions.

The Na-like ions are obviously an interesting class for study
of ionization (see Sect. 2.4). The original crossed beams experiment[67] for Mg^+ did not reveal the expected excitation-autoionization
contribution.[68,69] Recent beams experiments for Mg^+, Al^{2+}, and Si^{3+}
demonstrate that the indirect contributions to ionization do occur
in these cases.[70] While distorted wave theory, which included
excitation-autoionization,[71] predicted some specific features of the
observed ionization cross sections, it failed to predict the correct
magnitude for the indirect contributions and did not give the specific detailed shape of the cross sections either.

Recent experiment[70] and theory[71] for ionization of Al^{2+} are
shown in Fig. 15. The positions and orbital promotions associated
with inner-shell excitation are shown and the distorted wave calculations of excitation-autoionization are shown to be about a factor
of two greater than observed. In particular the dominant 2p-3p
transition, which proceeds primarily via the monopole term in the
expansion of expressions (5) and (6), appears to be missing in the
experiment.

Henry and Msezane[72] have carried out coupled-state calculations
to further investigate the observed features of indirect ionization
in Na-like ions. Figure 16 shows the coupled-state results and
introduces the suggestion that the shape of the measured Al^{2+} cross
section near 75 eV can be attributed to recombination resonances
similar to those discussed in Sect. 2. A set of model resonances of
configuration $(2p^53s3\ell n\ell)$ which could decay by double-autoionization
are shown in Fig. 16a. Figure 16b shows the coupled-state calculation of indirect contributions associated with 2p-3ℓ promotions
only. The coupled-state theory still overestimates the 2p-3p excitation component somewhat, but the estimated effect of resonances
between 73 and 80 eV, just below this excitation threshold, matches
the observed cross section in shape.

These studies on Al^{2+} illustrate the value of the details which
can be obtained in interacting beams studies. In addition they provide some credence to the suggestion of significant excitation and
resonance contributions to the ionization of Fe^{15+} (Fig. 3 and
Ref. 14). However, the trend of the indirect effects with increasing charge along the isoelectronic is in question and the magnitude

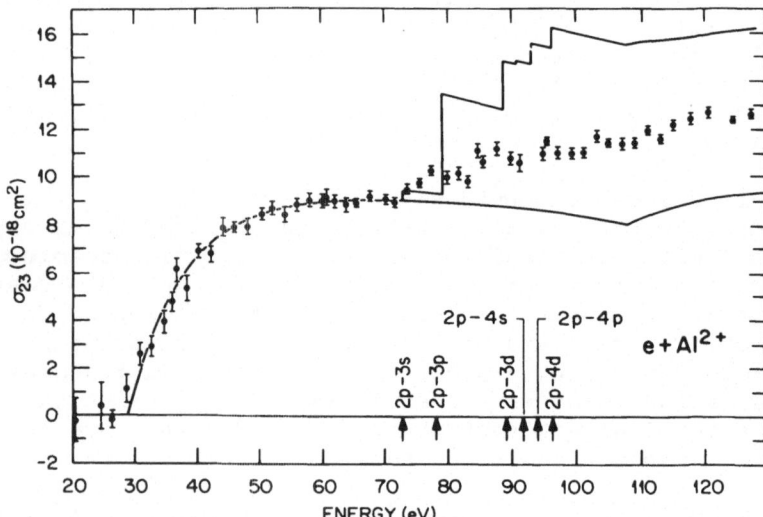

Fig. 15. Electron impact ionization of Al^{2+} (from Ref. 70). The
solid curve is distorted wave direct ionization (Ref. 17d)
normalized to experiment at 70 eV and with distorted wave
excitation-autoionization (Ref. 71) for each orbital promotion added at the arrows.

has not yet been correctly predicted (compared to experiment) for
any of the Na-like ions.

A number of experiments with even more complex ions demonstrate
that the indirect effects on ionization can completely dominate the
cross section. Experiments on singly charged alkali-like ions Mg^+,
Ca^+, Sr^+, and Ba^+ (see Ref. 7) demonstrated indirect contributions
of: roughly negligible, factor of 2, factor of 2.5, factor of 4.5
respectively. However for isoelectronic 3+ ions of Si^{3+}, Ti^{3+},
Zr^{3+}, Hf^{3+}, the indirect contributions have been found to be 65%, a
factor of 10, a factor of 20, and a factor of 10 respectively.[73]
(Note that Hf^{3+} is only similar to Ba^+, not isoelectronic.) The
experiment and distorted wave excitation calculations[73,74] are shown

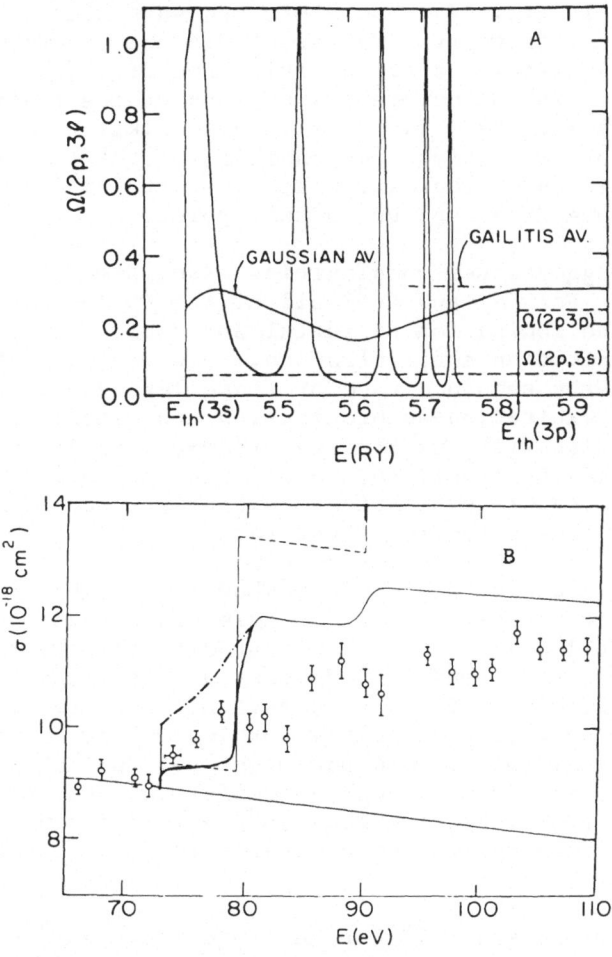

Fig. 16. (a) Model calculation of resonances ($2p^53s3\ell n\ell'$) for electron + Na-like ions in a 3-state close coupling approxima-
tion (from Ref. 72). These resonances decay dominantly by
double-autoionization resulting in contribution to ionization cross sections for Na-like ions. The Gaussian average
is with electron energy spread of 2 eV. There will be
other resonance series in the same energy range.

(b) Comparison of calculated and measured cross sections
for Al^{2+} ionization (from Ref. 72). Dashed curve is distorted wave excitation-autoionization (Ref. 71); solid
curve is 2-state close coupling for excitation-autoionization including only 2p-3ℓ excitations and folded with a
2-eV Gaussian energy spread; dot-dashed curve between 73
and 80 eV is estimated close-coupling result including
excitation-autoionization and resonances like those of the
upper figure (a).

for the Ti^{3+} case in Fig. 17. The distorted wave theory overestimates the indirect effects by a factor of 2.5 but predicts the shape
faithfully. The excitation-autoionization contributions are attributed to 3 pairs of fine structure transitions of the configurations
$(3p^63d) \rightarrow (3p^53d^2) \rightarrow (3p^6) + e$. In all these alkali-like cases,
except Na-like, the transitions responsible for the very large
indirect contributions to ionization are $\Delta n=0$ transitions rather
than $\Delta n=1$ which were discussed up to this point.

Figure 17b suggests new complexities associated with these
indirect effects. For Ca^+ and Sc^{2+} all of the 45 levels of $3p^53d^2$
are higher than the ionization potential and could contribute to
ionization via excitation-autoionization. However, for V^{4+} and
higher members of the sequence none of these levels can autoionize.
Presumably at V^{4+} excitation-autoionization contributions near
threshold will vanish with the binding of $\Delta n=0$ transitions while
$\Delta n=1$ transitions may (probably do) contribute at higher energies.
The possible effect of recombination resonances — double autoionization has not been considered for these cases.

Alkali-like ions were expected to show the largest indirect
(inner shell) effects because of the single outer electron and
numerous inner-shell electrons. Recent studies with 3+ rare gas
ions,[75] Ne^{3+}, Ar^{3+}, Kr^{3+}, and Xe^{3+} demonstrate that indirect effects
are likely to be important in many cases. All of these ions have
p^3 outer orbitals so that opportunity for low energy excitation-
autoionization transitions occurs but might not dominate. The
indirect effects observed are roughly negligible, 20%, 60%, and over
a factor of 2 for the respective ions $Ne^{3+} \rightarrow Xe^{3+}$ if the Lotz formula [Eq. (14)] is taken as representative of the direct ionization
cross sections.

Very recent measurements[76] for Xe^{5+} are shown in Fig. 18. The
enhancement, at threshold, due to indirect processes is again an
order of magnitude. It now appears that such large indirect effects
will not be unusual for complex but highly charged ions.

Beams experiments by the German group at Giessen have convincingly demonstrated a related enhancement of ionization cross
sections. The measurements of Müller and Frodl[77] on multiple ionization of Ar ions in a single collision are shown in Fig. 19. In
all cases the onset of L-shell ionization is accompanied by a significant increase in multiple ionization attributed to inner-shell
ionization followed by autoionization. This effect also appears to
increase in relative importance with increasing initial ionic charge
(much larger for σ_{35} than for σ_{24}) and again the magnitude of the
effect is up to about a factor of 10 already for 3+ ions.

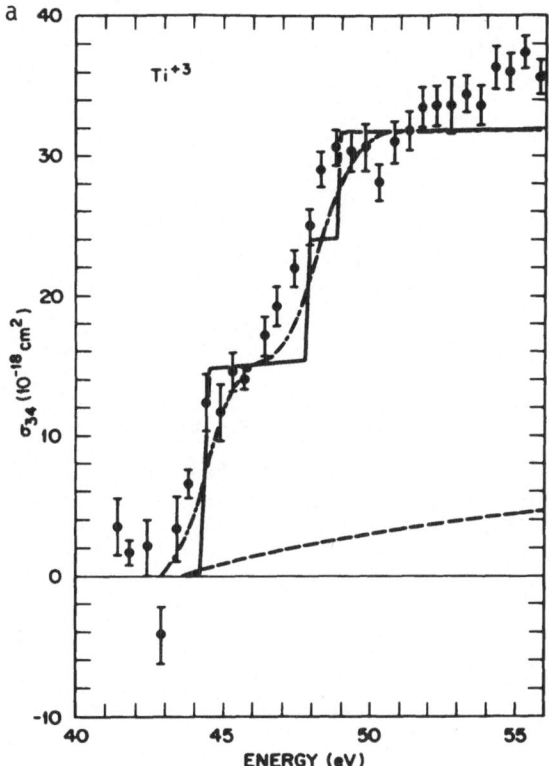

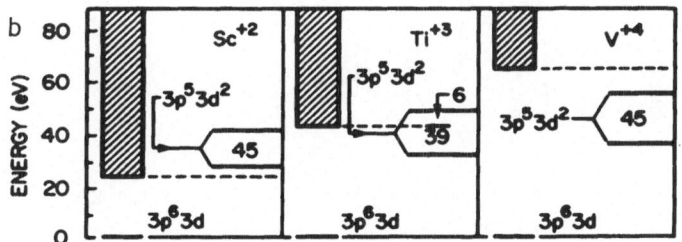

Fig. 17. (a) Electron impact ionization of Ti^{3+} (from Ref. 73).
Dashed curve is for direct ionization [Lotz formula, Eq.
(14)]; solid curve is excitation-autoionization via
(3p^{6}3d) → (3p^{5}3d^2) → (3p^6) + e divided by 2.5 and added to
Lotz; dot-dashed curve is the solid curve averaged with a
2-eV Gaussian energy spread.

(b) The energy positions of the 45 fine structure levels of
3p^{5}3d^2 K-like ions relative to the ionization potential.

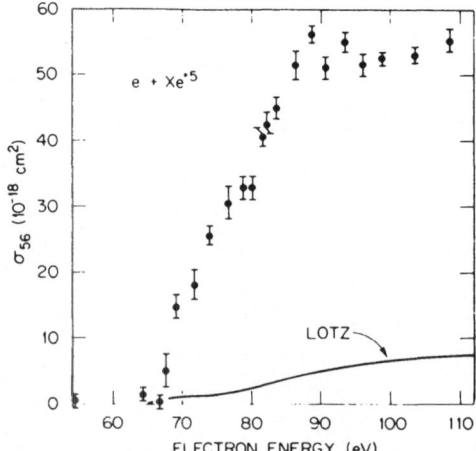

Fig. 18. Electron impact ionization of Xe^{5+} showing significant
         indirect enhancement of the cross section beginning at
         threshold (from Ref. 76).

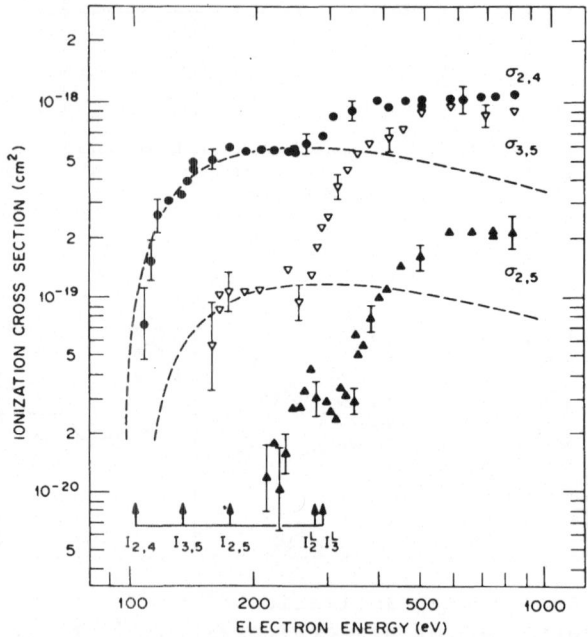

Fig. 19. Multiple ionization of Ar^{2+} and Ar^{3+} by electron impact
         ionization measured by Müller and Frodl.[77] Data points are
         labeled by initial and final charge states. Ionization
         potentials for initial and final charge states are indi-
         cated as I_{if} and direct ionization of an L shell electron
         by I_i^L. Dashed curves are normalized Bethe calculation for
         direct double electron ionization.

To summarize current insights on electron impact ionization of highly charged ions:

1. Direct ionization is well represented by theory.

2. Indirect processes involving inner-shell electrons are impor-
    tant, ranging from a few percent to in excess of an order of
    magnitude over direct ionization, and are not as well repre-
    sented by present theory.

3. It is likely that recombination resonances, as well as
    excitation-autoionization, influence ionization cross sections.

4. The indirect effects on ionization generally increase with
    ionic charge (possibly except case where a class of transitions
    such as $\Delta n=0$ become bound).

5. The indirect effects also increase with the complexity of the
    ion for a given ionic charge.

4.4 Dielectronic Recombination

Some of the current beams experiments on dielectronic recombination with multicharged ions have been described (Sect. 3.4), but
we have no definitive cross-section measurements. However, there is
convincing evidence about dielectronic recombination and some estimates of rates from plasma-based spectroscopic measurements.

High resolution x-ray spectra from the PLT tokamak[78] reveal a
rich spectrum near 1.85 Å including satellite lines attributed to
dielectronic recombination. The observed and calculated spectra are
shown in Fig. 20. The accompanying table (Table 3) assigns spectroscopic transitions. Transitions w, x, y, and z are He-like Fe^{24+}
transitions most likely produced by electron impact excitation in
the hottest part of the plasma. The other lines above 1.855 Å are
satellite lines of the Li-like iron but of the form $(1s2\ell2\ell') \rightarrow$
$(1s^2 2\ell')$ which can be produced by radiative stabilization following
formation of dielectronic resonances. The lines A, B, and d_{13} are
attributed to dielectronic recombination via n=3 resonances of the
form $(1s2\ell3\ell') \rightarrow (1s^2 3\ell)$ and the dielectronic recombination to
$n \geqslant 4$ presumably broadens the primary optically allowed He-like
line, I, on the long wavelength side. The calculated spectra of
Bely-Dubau et al.[79] are remarkably consistent with the observed
spectra and convincingly attribute the observations to dielectronic
recombination. These calculations allow nearly all of the recombination resonances of Fe^{24+} to decay via radiation to Fe^{23+} (some of
the lines observed in Fig. 20) rather than by autoionization which
would enhance the excitation of Fe^{24+} states (particularly lines x,
y, and z of Fig. 20).

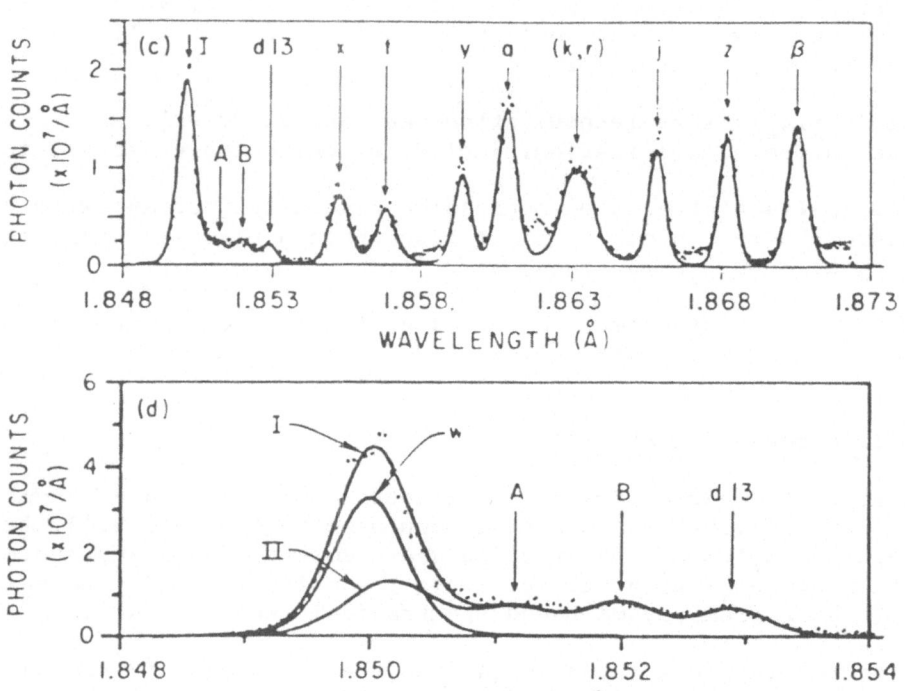

Fig. 20. High resolution spectra from the PLT tokamak at wavelengths
near the Fe^{24+} $(1s^2)$ → $(1s2p)$ transition (from Ref. 78).
The satellite lines are identified in the table and are
ascribed principally to dielectronic recombination transitions in Fe^{25+}. The solid curves are calculated spectra
(Ref. 79). Curve II is the calculated contributions for
dielectronic satellites in which the recombining electron
is in n=3 levels for features A, B, and d_{13} and is in
n=4-11 levels for the broad feature near 1.85 Å. Curve II
plus the principal line transition, w, give curve I.
Reprinted by permission of the author.

Table 3.　Identification of Spectral Features

Key	Transition	λ (Å)	Key	Transition	λexpt (Å)	λtheor (Å)
A	$1s^23d(^2D_{3/2})-1s2p3d(^2D_{5/2})$	1.8512[a]	w	$1s^2(^1S_0)-1s2p(^1P_1^o)$	1.8500	1.8500
h9	$1s^23d(^2D_{5/2})-1s2p3d(^2F_{7/2})$	1.8509[b]				1.84992
h15	$1s^23d(^2D_{5/2})-1s2p3d(^2D_{5/2})$	1.8509[b]	x	$1s^2(^1S_0)-1s2p(^3P_2^o)$	1.8552	1.8551
9	$1s^24p(^2P_{3/2})-1s2p4p(^2D_{5/2})$	1.8509[b]				1.85519
B	$1s^23s(^2S_{1/2})-1s2p3s(^2P_{1/2})$	1.8520[a]	t	$1s^22s(^2S_{1/2})-1s2p2s(^2P_{1/2}^o)$	1.8567	1.8570
a2	$1s^23d(^2D_{5/2})-1s2p3d(^2D_{5/2})$	1.8513[b]	y	$1s^2(^1S_0)-1s2p(^3P_1^o)$	1.8592	1.8591
h7	$1s^23s(^2S_{1/2})-1s2p3s(^2P_{3/2})$	1.8514[b]				1.85947
a1	$1s^23s(^2S_{1/2})-1s2p3s(^2P_{3/2})$	1.8515[b]	q	$1s^22s(^2S_{1/2})-1s2p2s(^2P_{3/2}^o)$	1.8608	1.8604
d5	$1s^23p(^2P_{3/2})-1s2p3p(^2P_{3/2})$	1.8516[b]	a	$1s^22p(^2P_{3/2}^o)-1s2p^2(^2P_{3/2})$	1.8618	1.8618
d15	$1s^23p(^2P_{1/2})-1s2p3p(^2D_{3/2})$	1.8518[b]	k	$1s^22p(^2P_{1/2}^o)-1s2p^2(^2D_{3/2})$	1.8632	1.8631
			r	$1s^22s(^2S_{1/2})-1s2p2s(^2P_{1/2}^o)$		1.8635
			j	$1s^22p(^2P_{3/2}^o)-1s2p^2(^2D_{5/2})$	1.8657	1.8657
d13	$1s^23p(^2P_{3/2})-1s2p3p(^2D_{5/2})$	1.8529[a] / 1.8526[b]	z	$1s^2(^1S_0)-1s2s(^3S_1)$	1.8681	1.8677
						1.86801
			β	$1s^22s^2(^1S_0)-1s2s^22p(^1P_1)$	1.8705	1.8710

[a]Experimental value.
[b]Theoretical prediction (Ref. 10).

Similar measurements for He-like Cl^{15+} on the Alcator C tokamak show much smaller contributions by dielectronic satellites.[80] This decrease in the relative intensity of dielectronic satellites may be due to the lower charge, q, and the scaling of radiative stabilization of dielectronic resonances as q^4. Also, A. Gabriel discussed similar spectra from the Solar Maximum Mission in his lectures at the NATO school.

There have been attempts to obtain recombination rates from θ-pinch and tokamak plasmas. The θ-pinch measurements assume coronal equilibrium and adjust both ionization and recombination rates to match the time evolution of spectral lines observed from the plasma. Rates for Fe^{8+} through Fe^{10+} recombination obtained from a θ-pinch[81] are somewhat lower than predicted by the Burgess-Merts formula, Eq. (17). However, it is likely that density effects at $n_e \approx 3 \times 10^{16}$ cm^{-3} might modify the recombination rates.[82]

Tokamak measurements[83] on Mo ions assumed that "sawtooth"-type variation of soft x rays were caused by MHD instability which rapidly changed electron temperature but not other plasma parameters such as density. The recombination rates inferred from time behavior of radiation during "sawtoothing" gave recombination rates for Mo^{30+} and Mo^{31+} a little lower than the Burgess formula.

The radiation from Fe^{15+}, Fe^{16+}, Fe^{17+}, and Fe^{18+} from the ISX-B tokamak have been studied during neutral beam injection and recombination rates inferred.[84] The time behavior of the Fe^{15+}, 361 Å transition (discussed in Sect. 2.2 with Fig. 1) is shown in Fig. 21. During the neutral beam injection (NBI) counter to the plasma current direction, the iron impurities collected in the interior of the plasma and remained for relatively long times in regions of almost constant electron temperature. Dielectronic recombination coefficients are deduced from these assumptions, the time history of the radiations (like Fig. 21), and a model of the ionization equilibrium which incorporates ionization, electron capture (between injected H^0 and Fe ions as a recombination mechanism) as well as radiative and dielectronic recombination rates. Assuming the other rates are correct (the ionization rates included excitation-autoionization as in Fig. 3), the dielectronic rates are deduced. The rates deduced are shown in Table 4 and compared with the Burgess-Merts formula and other direct calculations.[19a,85] The comparisons are good for Fe^{18+} but poor for Fe^{15+}. The authors point out that even at their low density of 6×10^{13} cm^{-3} electron collisions with excited (recombining) ions could reduce dielectronic rates and that such effects may not be uniform between ions.

To summarize current insights on dielectronic recombination:

1. This resonance process typically involves summation over many
    narrow transitions to electronic configurations with more than
    one excited electron which must stabilize radiatively.

2. Calculated rates always involve approximations or truncations
    and typically provide only modest agreement with rates deduced
    from modeled plasma observations.

3. Interacting beams "cross-section" measurements could observe
    the averaged recombination associated with recombination into a
    given n level and <u>might</u> provide more definitive tests of theory
    in some specific cases.

4.5 Directions

The theoretical tools for prediction of cross sections for the
three inelastic processes of electron ion collisions have been
developed. These tools are being refined and put into forms which
could allow predictions of detailed and specific data for a large
number of individual highly charged ions.

However, there are only a few measurements against which these
theoretical results can be tested particularly for high-charge
states. Experiments on electron impact ionization are currently
providing the most detailed test cases. Effects due to resonances
are important and are the most elusive of experimental investigation. The new generation of ion sources should allow more interacting beams experiments and with more highly charged ions. These
experiments will all be technically challenging and tedious, but
should provide significant tests of current understanding.

In all collisions of highly charged ions it is the understanding
of the formation and decay of fairly highly excited electronic configurations that provide the most challenge and are of the greatest
importance in physical environments where these ions occur.

ACKNOWLEDGMENTS

The author was supported through the Office of Fusion Energy
Applied Plasma Physics Branch of the U.S. Department of Energy under
contract W-7405-eng-26 with the Union Carbide Corporation during
preparation of this manuscript. He is particularly indebted to all
of his coauthors on research papers and to many of the other
researchers cited in this paper for the insights and information on
which this paper is based.

Table 4. Recombination rates for Fe ions from Ref. 84. The num-
          bers 14, 15, 16, and 17 are the charge states of the ions
          <u>before</u> recombination.

Rates in 10^{-11} cm^3/s	exp	BM^a	JDKB^b	Hahn^c
α^d_{17} (680 eV)	1.50 ± 0.45	1.80	1.35	2.04
α^d_{17} (480 eV)	1.82 ± 0.54	1.76	1.69	2.40
α^d_{14} (480 eV)	0.74 ± 0.22	3.30	1.15	2.50
α^d_{14} (400 eV)	0.70 ± 0.21	3.50	1.49	
α^d_{16} $^d_{15}$ (480 eV)	2.17 ± 1.10	5.45	1.04	4.83
α^d_{16} (480 eV)	1.75^d	2.35	1.20	2.30
α^d_{15} (480 eV)	1.24^d	2.32	0.61	2.10

^{a}Burgess-Merts [Eq. (18) and Ref. 18].
^{b}Jacobs, Davis, Kepple, and Blaha (Ref. 19a).
^{c}Y. Hahn (Ref. 85).
^{d}Individual values assumed in computer codes.

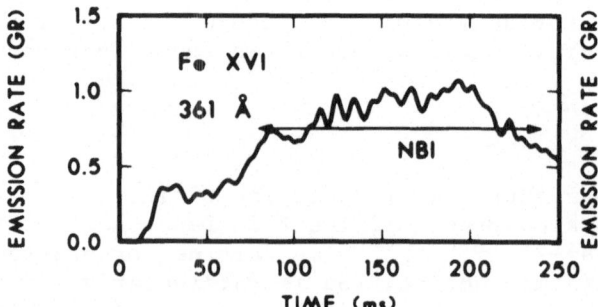

Fig. 21. Radiation from Fe^{15+} (3p-3s) during "counter" neutral
          beam injection in ISX-B tokamak (from Ref. 84). Together
          with other line radiation and modeling these data are
          employed to estimate dielectronic recombination.

REFERENCES

1. The Physics of Ion-Ion and Electron-Ion Collisions, organized
    by F. Brouillard and J. Wm. McGowan at Baddeck, Nova Scotia,
    September 1981 to be published by Plenum Press, New York and
    London, 1982.
2. Atomic and Molecular Processes in Controlled Thermonuclear
    Fusion, organized by M.R.C. McDowell and A. M. Ferendeci at
    Chateau de Bonas in France, August 1979, published by Plenum
    Press, New York and London, 1980.
3. R.J.W. Henry, "Excitation of Atomic Positive Ions by Electron
    Impact," Phys. Rep. 68, 1-91 (1981).
4. D. H. Crandall, "Electron Impact Excitation of Ions," within
    Ref. 1.
5. E. Salzborn, "Electron Impact Ionization of Ions," within
    Ref. 1.
6. D. H. Crandall, "Electron Impact Ionization of Multicharged
    Ions," Phys. Scr. 23, 153-162 (1981).
7. K. T. Dolder and B. Peart, "Collisions Between Electrons and
    Ions," Rep. Prog. Phys. 39, 693-749 (1976).
8. D. H. Crandall, "Experiments on Collisions of Electrons and
    Multicharged Ions," in Physics of Electronic and Atomic
    Collisions. Invited papers of XII ICPEAC edited by S. Datz,
    North-Holland Publishing, Amsterdam, New York, and Oxford,
    pp. 595-608 (1982).
9. J.B.A. Mitchell and J. Wm. McGowan, "Experimental Studies of
    Electron-Ion Recombination," within Ref. 1.
10. J. Dubau and S. Volonté, "Dielectron Recombination and Its
    Applications in Astronomy," Rep. Prog. Phys. 43, 199-251
    (1980).
11. G. H. Dunn, "Electron-Ion Collisions," in The Physics of
    Ionized Gases, invited lectures of SPIG-80, edited by
    M. Matić, Boris Kidrič Institute of Nuclear Sciences,
    Beograd, pp. 49-95 (1981).
12. J. Reader and J. Sugar, J. Phys. Chem. Ref. Data 4, 353 (1975).
13. R. D. Cowan and J. B. Mann, Astrophys. J. 232, 940 (1979).
14. K. J. LaGattuta and Y. Hahn, Phys. Rev. A 24, 2273 (1981).
15. Y.-K. Kim, "Theory of Electron-Atom Collisions," within Ref. 1.
16. C. J. Joachain, "Theoretical Methods for Atomic Collisions: A
    General Survey," pp. 147-183 within Ref. 2.
17. S. M. Younger, (a) Phys. Rev. 22, 111 (1980); (b) Phys. Rev. A
    22, 1425 (1980); (c) Phys. Rev. 23, 1138 (1981); (d) Phys.
    Rev. A 24, 1272 (1981); (e) Phys. Rev. A 24, 1278 (1981);
    (f) J. Quant. Spectrosc. Radiat. Transfer 26, 329 (1981);
    (g) J. Quant. Spectrosc. Radiat. Transfer 27, 541 (1982).
18. A. L. Merts, R. D. Cowan, and N. H. Magee, Jr., "The Calculated
    Power Output from a Thin Iron-Seeded Plasma," Report
    LA 6220-MS, Los Alamos Scientific Laboratory, Los Alamos,
    New Mexico (March 1976).

19. V. L. Jacobs, J. Davis, P. C. Kepple, and M. Blaha,
    (a) Astrophys. J. 211, 605 (1977); and (b) Astrophys. J.
    215, 690 (1977); also (c) V. L. Jacobs, J. Davis, J. E.
    Rogerson, M. Blaha, M. Cain, and M. Davis, Astrophys. J.
    239, 1119 (1980).
20. J. N. Gau, Y. Hahn, and J. A. Retter, J. Quant. Spectrosc.
    Radiat. Trans. 23, 131 and 147 (1980).
21. L. J. Roszman, Phys. Rev. A 20, 673 (1979).
22. F. Bely-Dubau, A. H. Gabriel, and S. Volonté, Mon. Not. R.
    Astron. Soc. 189, 801 (1979).
23. A. Burgess, Astrophys. J. 141, 1588 (1965).
24. O. Bely and H. Van Regemorter, Ann. Rev. Astron. Astrophys. 8,
    329 (1970).
25. E. Clementi and C. Roetti, At. Data Nucl. Data Tables 14,
    177-478 (1974).
26. M. J. Seaton in Atomic and Molecular Processes, edited by D. R.
    Bates (Academic Press, New York, 1962), p. 374.
27. H. Van Regemorter, Astrophys. J. 136, 906 (1962).
28. S. M. Younger and W. Wiese, J. Quant. Spectrosc. Radiat.
    Transfer, 22, 161 (1979).
29. W. Lotz, Z. Phys. 216, 241 (1968) and Z. Phys. 220, 466 (1969).
30. L. B. Golden and D. H. Sampson, (a) J. Phys. B 10, 2229 (1977);
    (b) D. H. Sampson and L. B. Golden, J. Phys. B 11, 541
    (1978); (c) L. B. Golden, D. H. Sampson, and K. Omidvar, J.
    Phys. B 11, 3235 (1978); (d) D. L. Moores, L. B. Golden, and
    D. H. Sampson, J. Phys. B 13, 385 (1980); and (e) L. B.
    Golden and D. H. Sampson, J. Phys. B 13, 2645 (1980).
31. H. J. Kunze, Space Sci. Rev. 13, 565 (1972).
32. A. H. Gabriel and Carole Jordan in Case Studies in Atomic
    Collision Physics II, edited by E. W. McDaniel and M.R.C.
    McDowell, p. 209, North-Holland Publishing Co., Amsterdam
    (1972).
33. P. Greve, M. Kato, H.-J. Kunze, and R. S. Hornady, Phys. Rev. A
    24, 429 (1981).
34. H. Griem, Univ. of Maryland, private communication (1981).
35. J. B. Hasted, "Confinement of Charged Particles for Collision
    Studies," within Ref. 1.
36. G. H. Dunn, IEEE Trans. Nucl. Sci., NS-23, #2, 929 (1976).
37. E. D. Donets, (a) IEEE Trans. Nucl. Sci., NS-23, #2, 897
    (1976); (b) E. D. Donets and V. P. Ovsyannikov, "The
    Ionization of Positive Ions of N, O, Ne, and Ar by Electron
    Impact," report P7-10780 of Joint Institute for Nuclear
    Research, Dubna, U.S.S.R. (1977), [translation
    ORNL-tr.4616]; (c) E. D. Donets and V. P. Ovsyannikov, "A
    Study of the Ionization of C, N, O, Ne, Ar, Kr, and Xe
    Positive Ions by Electron Impact," report P7-80-404, Joint
    Institute for Nuclear Research, Dubna (1980), [translation
    ORNL-tr.4702]; and (d) E. D. Donets and V. P. Ovsyannikov,
    "The Cryogenic Electron Beam Ionizer 'KRION-II'," report

P7-80-515 of the Joint Institute of Nuclear Research, Dubna (1980), [translation ORNL-tr.4703, available through Technical Information Center, P.O. Box 62, Oak Ridge, TN 37830, U.S.A.].

38. Joël Arianer and Richard Geller, "The Advanced Positive Heavy Ion Sources," Ann. Rev. Nucl. Part. Sci., $\underline{31}$, 19 (1981).

39. M. L. Mallory and D. H. Crandall, IEEE Trans. Nucl. Sci. NS-23, #2, 1069 (1976) and M. L. Mallory and E. D. Hudson, IEEE Trans. Nucl. Sci. NS-22, #3, 1669 (1975).

40. G. Clausnitzer, H. Klinger, A. Müller, and E. Salzborn, IEEE Trans. Nucl. Sci. NS-23, #2, 1027 (1976) and Nucl. Instrum. Methods $\underline{128}$, 1 (1975).

41. F. Brouillard and W. Claeys, "On the Measurement of Ion-Ion Charge Exchange," within Ref. 1.

42. P. O. Taylor, K. T. Dolder, and W. E. Kaupilla, Rev. Sci. Instrum. $\underline{45}$, 538 (1974).

43. A. Müller, E. Salzborn, R. Frodl, R. Becker, H. Klein, and H. Winter, J. Phys. B $\underline{13}$, 1877 (1980).

44. D. H. Crandall, P. O. Taylor, and G. H. Dunn, Phys. Rev. A $\underline{10}$, 141 (1974).

45. D. Gregory, G. H. Dunn, R. A. Phaneuf, and D. H. Crandall, Phys. Rev. A $\underline{20}$, 410 (1979).

46. J. L. Kohl and G. P. Lafyatis, Center for Astrophysics, Cambridge, Mass., private communication.

47. P. F. Dittner, G. D. Alton, W. B. Dress, C. D. Moak, P. D. Miller, and S. Datz, Oak Ridge National Laboratory, Oak Ridge, Tenn., private communication.

48. P. O. Taylor, R. A. Phaneuf, and G. H. Dunn, Phys. Rev. A $\underline{22}$, 435 (1980).

49. M. A. Hayes, D. W. Norcross, J. B. Mann, and W. D. Robb, J. Phys. B $\underline{10}$, L429 (1977).

50. R.J.W. Henry and W. L. van Wyngaarden, Phys. Rev. A $\underline{17}$, 798 (1978).

51. W. L. van Wyngaarden and R.J.W. Henry, J. Phys. B $\underline{9}$, 146 (1976).

52. J. Callaway, J. N. Gau, R.J.W. Henry, D. H. Ozu, Voky Lan, and M. Le Dourneuf, Phys. Rev. A $\underline{16}$, 2288 (1977).

53. P. O. Taylor, D. C. Gregory, G. H. Dunn, R. A. Phaneuf, and D. H. Crandall, Phys. Rev. Lett. $\underline{39}$, 1256 (1977).

54. A. K. Pradhan, D. W. Norcross, and D. G. Hummer, (a) Phys. Rev. A $\underline{23}$, 619 (1981); (b) Astrophys. J. $\underline{246}$, 1031 (1981); (c) A. K. Pradhan, Phys. Rev. Lett. $\underline{47}$, 79 (1981).

55. G. H. Dunn, R. A. Falk, D. S. Belić, D. H. Crandall, D. C. Gregory, and C. Cisneros, experiments at ORNL and JILA, in preparation for publication (1982).

56. M. R. Rudge and S. B. Schwartz, Proc. Phys. Soc. London, $\underline{88}$, 563 (1966).

57. D. H. Crandall, R. A. Phaneuf, and D. C. Gregory, Report ORNL/TM-7020, Oak Ridge National Laboratory (1979).

58. J. J. Thomson, Philos. Mag. 23, 449 (1912).
59. D. H. Crandall, R. A. Phaneuf, B. E. Hasselquist, and D. C. Gregory, J. Phys. B 12, L249 (1979).
60. R.J.W. Henry, J. Phys. B 12, L785 (1979).
61. D. H. Sampson and L. B. Golden, J. Phys. B 12, L785 (1979) and J. Phys. B 14, 903 (1981).
62. H. Jakubowicz and D. L. Moores, J. Phys. B 14, 3733 (1981).
63. D. L. Moores, J. Phys. B 11, L403 (1978).
64. Y.-K. Kim and K.-T. Cheng, Abstracts of Papers at XI International Conference on the Physics of Electronic and Atomic Collisions, edited by K. Takayanagi and N. Oda, p. 218, North-Holland Publishing Co., Amsterdam (1979).
65. R. A. Falk and G. H. Dunn, "Electron Impact Ionization of Be^+," submitted to Physical Review A.
66. R. A. Falk, G. Stefani, R. Camilloni, G. H. Dunn, R. A. Phaneuf, D. C. Gregory, and D. H. Crandall, "Experimental Measurements of Electron Impact Ionization: Cross Sections for Be-Like Ions, B^+, C^{2+}, N^{3+}, and O^{4+}," in preparation for publication.
67. S. O. Martin, B. Peart, and K. T. Dolder, J. Phys. B 1, 537 (1968).
68. O. Bely, J. Phys. B 1, 23 (1968).
69. D. L. Moores and H. Nussbaumer, J. Phys. B 3, 161 (1970).
70. D. H. Crandall, R. A. Phaneuf, R. A. Falk, D. S. Belić, and G. H. Dunn, Phys. Rev. A 25, 143 (1982).
71. D. C. Griffin, C. Bottcher, and M. S. Pindzola, Phys. Rev. A 25, 154 (1982).
72. R.J.W. Henry and A. Z. Msezane, "Cross Sections for Inner-Shell Excitation of Na-Like Ions," submitted for publication.
73. R. A. Falk, G. H. Dunn, D. C. Griffin, C. Bottcher, D. C. Gregory, D. H. Crandall, and M. S. Pindzola, Phys. Rev. Lett. 47, 494 (1981).
74. D. C. Griffin, C. Bottcher, and M. S. Pindzola, Phys. Rev. A 25, 1374 (1982).
75. D. C. Gregory, P. F. Dittner, and D. H. Crandall, "Electron Impact Ionication of Triply Charged Rare Gas Ions" (in preparation for publication).
76. D. C. Gregory, D. H. Crandall, C. Bottcher, M. S. Pindzola, and D. C. Griffin, "The Effect of Inner-Shell Excitation by Electron Impact on Ionization of Multicharged Ions — Xe^{+5} and Xe^{+6}" (to appear in Abstracts of International Conference on X-Ray and Atomic Inner-Shell Physics, Eugene, Oregon, August 1982, edited by B. Craseman).
77. A. Müller and R. Frodl, Phys. Rev. Lett. 44, 29 (1980).
78. M. Bitter, K. W. Hill, N. R. Sauthoff, P. C. Efthimion, E. Meservey, W. Roney, S. von Goeler, R. Horton, M. Goldman, and W. Stodiek, Phys. Rev. Lett. 43, 129 (1979); and M. Bitter, S. von Goeler, K. W. Hill, R. Horton, D. Johnson, W. Roney, N. Sauthoff, E. Silver, and W. Stodiek, Phys. Rev. Lett. 47, 921 (1981).

79. F. Bely-Dubau, A. H. Gabriel, and S. Volonte, Mon. Not. R.
     Astron. Soc. 186, 405 (1979) and 189, 801 (1979).
80. E. Källne, J. Källne, and J. E. Rice, "Observation of H- and
     He-like X-Ray Line Emission in High Density Tokamak
     Plasmas," submitted for publication.
81. R. L. Brooks, R. U. Datla, and H. R. Griem, Phys. Rev. Lett.
     41, 107 (1978).
82. V. L. Jacobs and J. Davis, Phys. Rev. A 18, 697 (1978).
83. C. Breton, C. DeMichelis, M. Finkenthal, and M. Mattioli, Phys.
     Rev. Lett. 41, 110 (1978).
84. R. C. Isler, E. C. Crume, and D. E. Arnurius, "Ionization and
     Recombination Coefficients for Fe XV-Fe XIX," to be in Phys.
     Rev. A 26(4) (October 1982).
85. Y. Hahn, Phys. Rev. A 22, 2896 (1980).
86. R.E.H. Clark, N. H. Magee, Jr., J. B. Mann, and A. L. Merts,
     Astrophys. J. 254, 412 (1982).

PRODUCTION OF MULTIPLY CHARGED IONS

FOR EXPERIMENTS IN ATOMIC PHYSICS

Hannspeter Winter

Institut für Allgemeine Physik

Technische Universität Wien

Karlsplatz 13, A-1040 Wien, Österreich

SURVEY

Techniques for production of intense beams of multiply charged ions are reviewed and discussed. Emphasis is given to low ion velocity which is of special interest for experiments in atomic physics. Properties of various ion sources as their attainable ion charge states, ion beam qualities, ion yields and technical peculiarities are described. Additionally, techniques for measuring metastable ion beam fractions are covered. Where possible, it is tried to assess the expectable further developments.

CONTENTS

1. INTRODUCTION

    During the last decade interest in atomic physics
with multiply charged ions has increased considerably.
This is caused by several reasons, e.g.:
- Demand for accurate experimental data on various atomic
  collision processes involving multiply charged ions
  (energy balance and spectroscopy of thermonuclear
   fusion plasmas and astrophysical objects, development
  of hot plasma-based X-ray lasers etc.)
- Test of fundamental atomic behaviour as predicted by
  quantum electrodynamics, which becomes more marked at
  high ion charge states
- The status of multiply charged ion source development
  has improved considerably, thus offering higher ion
  charge states and more intense ion beams than available
  still some years ago.

In a high quality experiment, the successful application
of multiply charged ions demands their delivery under
well defined conditions. This includes their kinetic
energy, mass number (isotopic species), charge state,
geometrical trajectory and internal energy (electronic
configuration). Additionally, there are also some practical requirements as long-time availability, temporal
constancy, sufficient intensity and ion beam quality.
Fullfillment of all these requirements poses stringent
conditions on useful multiply charged ion sources.

In contrary to widespread belief the construction and
successful operation of multiply charged ion sources is
not a simple task. It requires both great experience
and technical skill. Also a certain amount of artistic
gift will be helpful.
It is well known that in hot plasmas (eg. Tokamak discharges, sparks, pinch discharges) multiply charged ions
are produced quite efficiently. For useful plasma ion
sources, however, low ion temperatures and stable discharge conditions are mandatory. The quite limited size
of an ion source leads also to a complete domination
of the respective discharge plasmas by the processes of
plasma wall interaction with all their consequences.
The necessarily high power density in the ionisation
region causes extensive wear of all exposed ion source
components. Therefore, the construction and operation
of powerful multiply charged ion sources is always a compromise between the demands of ion beam users and various
unavoidable technical restrictions.

Systematical development of ion sources became necessary
as soon as they were identified as the critical parts
in rather expensive experimental setups as eg. heavy ion
accelerators or neutral beam injectors. The first powerful
multiply charged ion sources were used for heavy ion
cyclotrons in the early fifties. Further development was
caused by the possibility to improve the final energy of
heavy ion accelerators by using multiply charged ion
sources. In 1969, the first International Ion Source Conference in Saclay (GR 1) contained a session on multiply
charged ion sources in which the concept of electron beam
ion sources was dealt with. In the following years, several
other international meetings were devoted either completely or in part to multiply charged ion sources (GR 2 -
5). Also at the annual US particle accelerator conferences
this topic was taken care of, and beginning in 1977, in
Japan annual meetings on Ion Sources and Application
Technology (GR 6) were organized.

The development of multiply charged ion sources until 1976 has been treated in a review by Winter (1978).

We describe the present status of multiply charged ion sources and put special emphasis on the production of relatively slow highly charged ions, which are of primary interest in experiments in atomic physics.

For convenience, we will use the following abbreviations and symbols:

CSD   charge state distribution
c.w.   continuously working
EBIS   electron beam ion source(s)
ECRIS   electron cyclotron resonance-heated plasma-MCIS
MCI   multiply charged ion(s)
MCIS   multiply charged ion source(s)
PIG   Penning ionization gauge-type ion source

I_e, j_e electron current, electron current density

n_e, T_e electron number density, electron temperature

n_q number density of MCI in charge state q

Q   general symbol for cross section

q   ion charge state/ "multiply charged ions": $q \geqslant 2$

           "highly charged ions": $q \geqslant 10$ or $q \geqslant Z/2$

           "fully stripped ions": $q \equiv Z$

t_c characteristic ion confinement time

t_d characteristic ion diffusion time

v_e electron velocity

Z   nuclear charge of atom or ion/ "low-Z"-ion:

                                        $Z \leqslant 10$

Other, less frequently used abbreviations or symbols are defined in the text at their first appearance.

2. MULTIPLY CHARGED IONS - PRODUCTION AND LOSS PROCESSES

Production of MCI can result from various atomic collision processes, but in most MCIS electron impact ionization is the dominant contribution. This belongs to all plasma MCIS (cf. chapter 4.), but also to the EBIS configuration (cf. chapter 5.) and the stripping processes (cf. chapter 6.1.).
In recent years, the production of slow MCI in collisions of highly charged fast ions with gas atoms (recoil ions) has become important, too (cf. chapter 6.2.).

As far as electron impact ionization is concerned, it may proceed either via single collisions or in a step-by-step manner, depending on the respective ion dwell time in the MCIS configuration. The latter ionization scheme is more interesting because the single ionization cross sections decrease relatively slowly with increasing q, whereas cross sections for removal of many electrons in one single collision are both small and decrease rapidly with q. General information on electron impact ionization cross sections can be found in a recent review of Crandall (1982).

The loss of ions from a MCIS configuration, in first approximation, is due to diffusion of MCI out of the ionzation region. Only for relatively high q volume recombination processes (electron capture, electronic recombination) will become of comparable importance. Therefore, MCIS ionization media are far off from thermodynamic equilibrium and the expectable CSD

$$f(q) = n_q / \sum_{i=1}^{Z} n_i$$

will result primarily from equilibrium between electron impact ionization and diffusion losses. The latter can be characterized by mean diffusion times $t_d(q)$, which in first approximation can be assumed as independent of q.

2.1. Ionization Balance for Plasma-MCIS

In discharge plasmas a steady net influx F_o of neutral particles into the ionization region (volume V_i) is given. Consequently, the density of neutral species, $\bar{n}_o$, shows a time dependence given by

$$\frac{dn_o}{dt} = \frac{F_o}{V_i} - n_e \sum_{q=1}^{Z} n_o S_{o,q} \tag{1}$$

$$S_{p,q} = <Q_{p,q} \, v_e> \text{ .. rate coefficient describing}$$
$$\text{the ionization step } p \to q$$
$$Q_{p,q} \text{ .. respective ionization cross section}$$

For the ionized atoms the respective number densities n_q follow the time-dependences

$$\frac{dn_q}{dt} = n_e \left(\sum_{p=0}^{q-1} n_p \, S_{p,q} - \sum_{r=q+1}^{Z} n_q \, S_{q,r} \right) - \frac{n_q}{t_d} \qquad (2)$$

Additionally, plasma quasineutrality demands

$$n_e = \sum_{q=1}^{Z} q \, n_q \qquad (3)$$

In steady state the CSD follows from setting $\frac{dn_q}{dt} = 0$ and solving the system of coupled differential equations (1), (2) and (3).
The fundamental parameter is $n_e \, t_d$ which defines the extent of ionization by multiple collisions. This product can be increased either by increasing the electron density n_e or by reducing ion losses due to diffusion.

If $\quad n_e \, t_d \ll \left(\sum_{r=q+1}^{Z} S_{q,r} \right)^{-1}$ for all q, the MCI production follows pure single collision conditions, i.e.

$$n_q \sim S_{o,q} \qquad (4)$$

In deriving (4), we have approximated $\quad \sum_{r=q+1}^{Z} S_{q,r} \simeq S_{q,q+1}$

For low q and typical plasma electron temperatures,

$$S_{q,q+1} \leqslant 10^{-7} \text{ cm}^3 \text{ s}^{-1}$$

In figs. 1 a, b, c we show results of calculations according to equs. (1), (2) and (3) for various $n_e \, t_d$ and different energies of ionizing electrons.
These results are from Winter (1978). It is clearly seen how with increasing values for $n_e \, t_d$ the influence of multiple collisions leads to larger fractions of multiply charged ions.

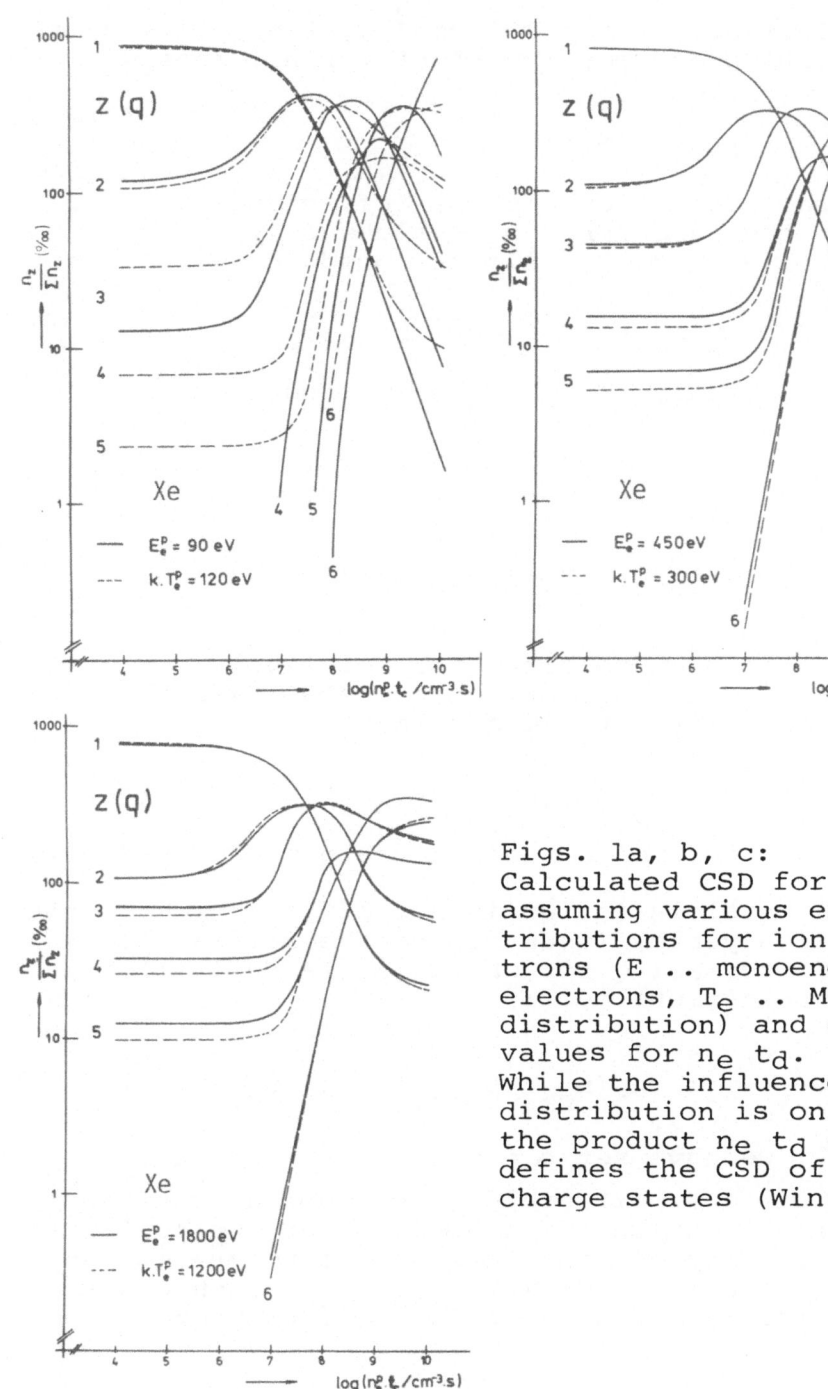

Figs. 1a, b, c:
Calculated CSD for Xe ions, assuming various energy distributions for ionizing electrons (E .. monoenergetic electrons, T_e .. Maxwellian distribution) and different values for n_e t_d.
While the influence of energy distribution is only minor, the product n_e t_d crucially defines the CSD of higher ion charge states (Winter 1978)

If $n_e t_d$ increases above 10^7 cm^{-3} s, the influence of multiple ionization becomes important for all ions with

$$\sum_{r=q+1}^{Z} S_{q,r} > (n_e t_d)^{-1} \qquad (5)$$

In the classical plasma ion sources (cf. chapters 4.1., 4.2.), this is the case for low q-values only. According to Geller (1977) fully stripped low-Z ions can be produced if

$$n_e t_d \simeq 10^{10} \text{ cm}^{-3} \text{ s, } k T_e \simeq 5 \text{ keV}$$

and production of fully stripped heavy ions demands

$$n_e t_d \simeq 10^{13} \text{ cm}^{-3} \text{ s, } k T_e \geqslant 30 \text{ keV}$$

2.2. Ionization Balance for Confined Ions

If the ions can be confined for relatively long times t_c within the ionization region, step-by-step ionization is dominant for all q with

$$t_c > t_q \equiv (n_e \sum_{r=1}^{q} S_{r-1,r})^{-1} \qquad (6)$$

t_q .. "ionization time" for charge state q, Redhead (1967)

If the neutral particle influx is cut off at the beginning of the ionization period (thus fixing the initial condition for n_o), for all q for which (6) is fullfilled the following conditions hold:

$$\frac{dn_q(t)}{dt} \simeq n_e(n_{q-1} S_{q-1,q} - n_q S_{q,q+1}) \qquad (7)$$

In deriving (7) we have set $p = q-1$ and $r = q+1$ only.

Solution of equs. (7) yields the temporal evolution of the CSD for all ion charge states. After the confinement time t_c has elapsed, the ions are found predominantly in those charge states q for which $t_c \simeq t_q$. Such conditions are given for the EBIS configuration, cf. chapter 5.

3. FORMATION AND QUALITY OF BEAMS OF MULTIPLY CHARGED IONS

3.1. Extraction from Multiply Charged Ion Sources

Ion extraction is achieved by imposing an electric field between the area of MCI production and an adjoining extraction electrode. Successively, the ion beam must be shaped and guided to the experimental region such that the highest possible flux of wanted ion species becomes available. In the extraction field region no space charge compensation (cf. chapter 3.3.) of the ion beam is possible. Therefore, high extraction field strengths are useful.

3.1.1. Extraction from Plasmas

Generally the ion extraction process from a plasma follows Bohm's sheath criterion (Bohm 1949, Septier 1967):

$$j_i \simeq e \, n_i \, \left(\frac{k \, T_e}{m_i}\right)^{1/2} \qquad (8)$$

j_i ion current density at plasma boundary

n_i ion number density within the plasma

T_e plasma electron temperature

Generalization of (8) for extraction of MCI (Winter and Wolf 1974/1) leads to

$$j_q \simeq q^{3/2} \, e \, n_q \, \left(\frac{k \, T_e}{m_i}\right)^{1/2} \qquad (9)$$

and therefore to the relation

$$I_q \sim q^{3/2} \, n_q \qquad (10)$$

j_q, I_q current density and total extracted current of MCI with charge state q

From relation (10) we obtain for the respective CSD:

$$f_q \simeq \frac{I_q/q^{3/2}}{\sum\limits_{r=1}^{z} I_r/r^{3/2}} \qquad (11)$$

According to (11) higher ion charge states are preferred
by the extraction process, which will cause depletion
in the plasma adjoining to the extraction region of such
higher charged ions. Thus, a gradual decrease of the
respective f_q-values within a short time toward the final
CSD will occur as given by

$$f_q \sim \frac{n_q}{\sum_r n_r} \qquad (12)$$

This effect is possibly encountered when measuring the
CSD in short-pulsed operation of MCIS-plasmas which
gives higher CSD for high q-values than in c.w. operation.
Another reason for this, off course, is the higher
achievable discharge power density in pulsed operation.
Correct measurement of CSD-values necessitates complete
collection of ion currents for all charge states produced.
Without magnetic field in the ion beam formation region
beam blow-up due to space charge effects does not cause
discrimination for different q, but magnetic fields due
to e.g. lenses may lead to some discriminative effects.
Usually, beams extracted from plasma-MCIS along their
magnetic field axis yield better CSD for high q-values
if only the central ion beam part is inspected. This can
be explained by the higher electron density in the MCIS-
axis. Because of the same reasons, enlargement of extraction apertures with otherwise unchanged parameters will
cause smaller f_q-values for higher q.
With high plasma density and small extraction apertures
expansion cups will facilitate the formation of a good
ion beam. Under unstable plasma conditions (which are often
found due to preference of smallest possible neutral
gas influx) ion beam formation will be greatly inhibited.
Additionally, increased ion energy spread will occur.

3.1.2. Extraction from Ion Traps

After the ion confinement time has elapsed, all ions
are expelled from the ionization region. Therefore, the
resulting CSD will correspond to the respective total
ion charges and will therefore be given by relation (12):

$$f_q \simeq \frac{I_q/q}{\sum_{r=1}^{z} I_r/r} \qquad (13)$$

3.2. Characterization of Ion Beam Quality

The quality of an ion beam can be described quite simply by its respective "perveance":

$$P \equiv I_i / U_b^{3/2} \qquad (14)$$

I_i/A Ion beam current

U_b/V Ion beam acceleration voltage

The kinetic energy spread of ions is important in all experiments where a well defined ion energy is needed. For ions extracted from plasmas the typical energy spread (FWHM-values) is of the order of k T_e but may increase considerably for unstable plasma conditions.

These concepts, however, do not include any informations about the ion beam structure which is of interest for all ion beam formation and handling processes. A more useful characterization is found if one considers the flow of beam particle representation points in the phase-space, which leads to the concepts of ion beam emittance and brightness (Septier 1967, Green 1974), cf. fig. 2:

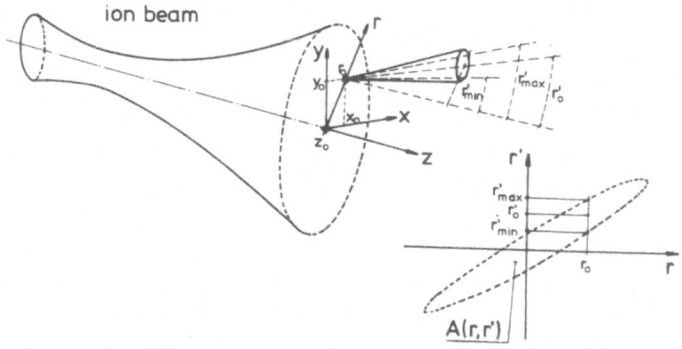

Fig. 2: Principal configuration for
        ion beam emittance measurements

The ion beam quality depends on how accurately both the kinetic energy and the trajectories of the beam particles are defined. In other words, the quality of a beam will be the higher, the smaller the phase-space hypervolume

containing the beam particle representation points becomes.
If mutual particle interaction can be neglected, one may
apply Liouville's theorem to the flow of representation
points in the phase space, which means that the respective
hypervolume stays constant with time. Although for real
ion beams this relation will never hold exactly, it can
well serve for a concept of ion beam quality description
and for comparison of different MCIS-types.

Emittance

At a certain position on the ion beam axis (z-axis, cf.
fig. 2) we consider the trajectories and momenta of all
ions passing through a point (x,y), or, if radial symmetry
can be assumed, through an annular element at radius r.
We denote momenta by

$$x' = \frac{p_x}{p_z} \simeq \frac{p_x}{p} \; , \quad y' \simeq \frac{p_y}{p} \; , \quad r' \simeq \frac{p_r}{p}$$

p_x, p_y, p_r respective components of ion momentum p

With A(x,x'), A(y,y') and A(r,r') we denote the areas
obtained if for all x the corresponding possible x'-
values, for all y the corresponding y'-values or for
all r the corresponding r'-values are plotted (cf. fig. 2).
These areas represent two-dimensional phase-space volumes
and therefore should remain constant for all z along a
drifting ion beam. The same should hold for the corresponding emittances defined as

$$E_x \equiv \frac{1}{\pi} A(x,x') \; , \quad E_y \equiv \frac{1}{\pi} A(y,y') \; ,$$

$$E_r \equiv \frac{1}{\pi} A(r,r') \tag{15}$$

Imaging systems as e.g. lenses will shape the ion beams
such that their corresponding emittances do not change,
but that they can be fitted into the acceptances of
the subsequent beam lines in the best possible way.
For ion beams under acceleration the emittances will
decrease. A constant value for ion beam quality characterization can be obtained by multiplication of emittance
with the respective ion momentum, thus obtaining
so-called normalized emittances:

$$E_n = \frac{p}{m_i c} \quad E = \frac{v}{c} \frac{E}{(1-\frac{v^2}{c^2})^{1/2}} = \frac{\beta \, \gamma}{\pi} A \qquad (16)$$

Brightness

By definition, the brightness of an ion beam gives the beam particle density in phase-space:

$$B \equiv \frac{2 \, I_i}{\pi^2 E_x \, E_y} \, , \qquad B_n \equiv \frac{2 \, I_i}{\pi^2 E_{x,n} \, E_{y,n}}$$

$$(17)$$

$$\text{or} \quad B_r \equiv \frac{2 \, I_i}{\pi^2 \, E_r^2} \, , \quad B_{r,n} \equiv \frac{2 \, I_i}{\pi^2 \, E_{r,n}^2}$$

During acceleration the ion beam current remains constant and thus also the normalized brightness.

Emittance and brightness can be measured by scanning the ion beam with a slotted or pierced plate behind which the resulting beamlets are investigated by either photographic techniques or with a moving Faraday cup assembly. Emittance areas are given usually for 50 % or 5 % full widths of investigated beamlets.

3.3. Influence of Ion Beam Space Charge

Interaction of the ion beam with the background gas leads to production of slow electrons which can compensate the ion space charge. In case of rapidly changing ion beam intensities (either very short pulsed operation or unstable ion source conditions) this compensation mechanism cannot work properly and therefore deterioration of ion beam quality will result.
For completely space charge-compensated ion beams a difference in emittances for various q-values can only be caused by the MCIS itself. If the space charge compensation is disturbed, however, this may act differently according to the respective ion current densities and may therefore lead to a dependence of emittance on ion charge state.

4. PLASMA-TYPE MULTIPLY CHARGED ION SOURCES

Electron impact ionization in plasmas is the most common way of MCI-production. This belongs to all discharge type ion sources, e.g. the Penning ionization gauge (PIG)-type (see chapter 4.1.) and the Duoplasmatron-type (see chapter 4.2.). Despite a great variety of well known low-pressure arc ion sources only the above mentioned species are of greater interest for MCI-production.
Electron cyclotron resonance-heating has become a very successful method to produce hot, dense plasmas for MCI-production. MCIS based on this method (ECRIS) are now able to deliver intense beams of fully stripped low-Z ions (see chapter 4.3.).
Finally, various approaches have been made to produce MCI by means of hot and very dense plasmas (e.g. inter-action of focussed laser radiation with solids, spark discharges, pinch discharges etc.). Such methods are shortly described in chapter 4.4.

4.1. Plasmas of PIG-Type Discharges

The discharge configuration named after F. Penning has found many different applications. It covers an astonishing variety of discharge modes, its operating parameters span neutral pressures from some Torr down to the UHV and discharge currents from 10^{-6} A to many A. The typical discharge geometry is shown in fig. 3.

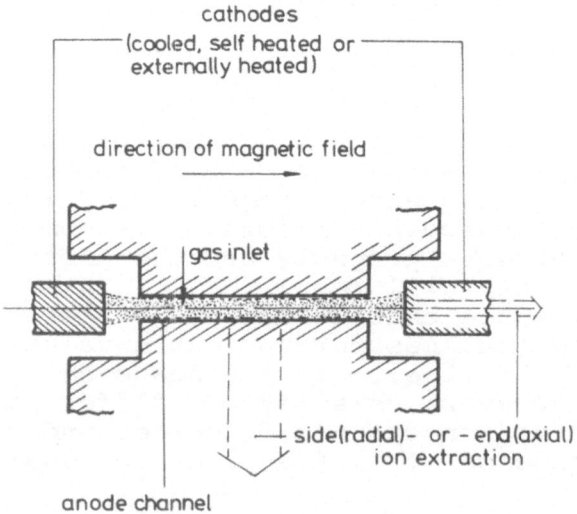

Fig. 3: PIG-discharge geometry, schematically

For MCI-production, PIG-discharges are usually operated
in homogeneous magnetic fields of some kG and at pressures
between 10^{-4} and 10^{-1} Torr.
Primary electrons can be released from the cathodes either
due to ion impact (cold cathode-PIG) or thermionically
(heated cathode-PIG). These electrons cross the very
narrow cathode fall (extension of mm) and are rapidly
thermalized within the plasma. Ions and electrons are
transported toward the anode by diffusion across the magnetic field. Usually, the discharge voltage is consumed
almost exclusively by the cathode falls and the anode fall
is quite small. Therefore, the velocity of primary electrons when entering the plasma corresponds to the discharge voltage. Without external cathode heating, two
discharge modes can be found:
      At relatively low discharge currents the latter can
only be increased by rising the discharge voltage (positive
discharge characteristic). As soon as the ion bombardment
heats up the cathodes sufficiently, onset of thermionic
emission changes the discharge characteristic into a negative one which is similar to that being found with externally heated cathodes. Cathode heating can be achieved
either by external electron bombardment or direct heating
("filament" cathodes). With hot cathodes, rather high
discharge currents can be drawn at discharge voltages
of some 100 V only.
      Ions can be extracted from the plasma either through
one of the cathodes (axial extraction) or through the
anode (radial extraction), cf. fig. 3. In the first case
beam formation is easier and higher ion current density
can be achieved, while side-extracted ion currents are
usually more stable. Moreover, immediate separation of
the different ion charge states takes place. Early PIG-
MCIS were predominantly used in cyclotrons and therefore
essentially side-extraction was used, while axial extraction
is still very less common.
An extensive review of PIG-MCIS and their historical development has been given by Bennett (1972).
Development of cold cathode-PIG-MCIS is due to Anderson
and Ehlers (1956) with further fundamental investigations
by Wolf (1973) as well as Gavin (1976), see e.g. figs. 4
and 5 a,b. Such ion sources can also produce ions from
solid materials by applying either sputtering (cf. table 1)
or vaporization into the discharge with an oven. In both
cases an auxiliary gas (mostly noble gases) has to be used.
      When enhancing the discharge current sufficiently,
the cold cathode-PIG-MCIS can be smoothly converted into
a heated cathode-type.

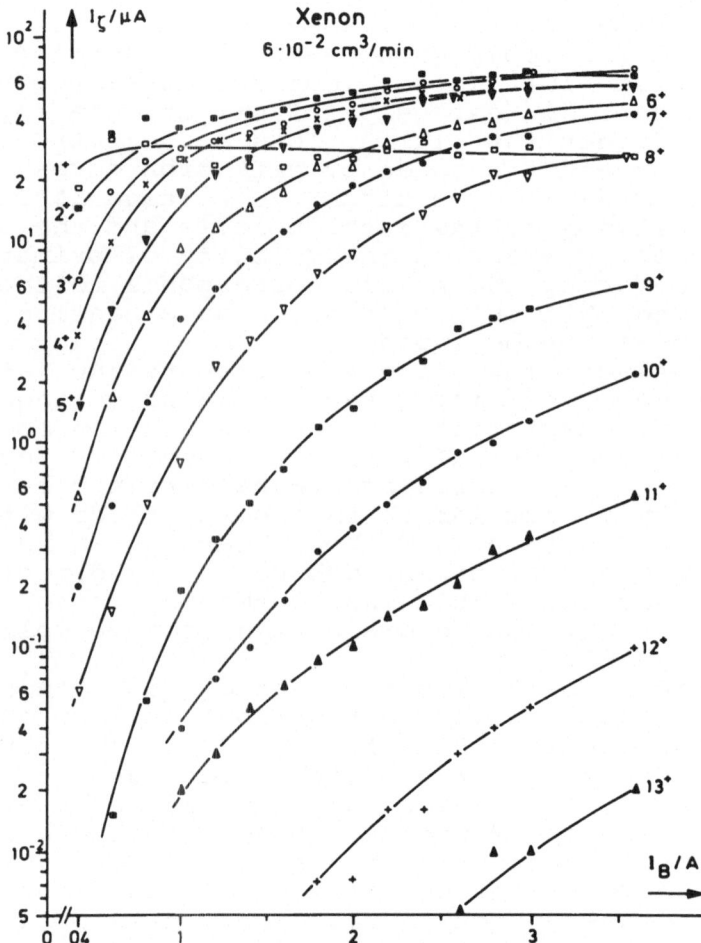

Fig. 4: Different charge state-currents for
a cold-cathode PIG-MCIS (Wolf 1973).
The ion charge state is denoted by ζ.
For a constant discharge voltage of
ca. 2 keV (pulse duration 1 ms with
50 Hz repetition frequency) various
discharge currents have been used.
The gas influx was held constant.
Above ca. 4 A discharge current
thermionic emission from the cathodes
led to heated cathode-characteristic.

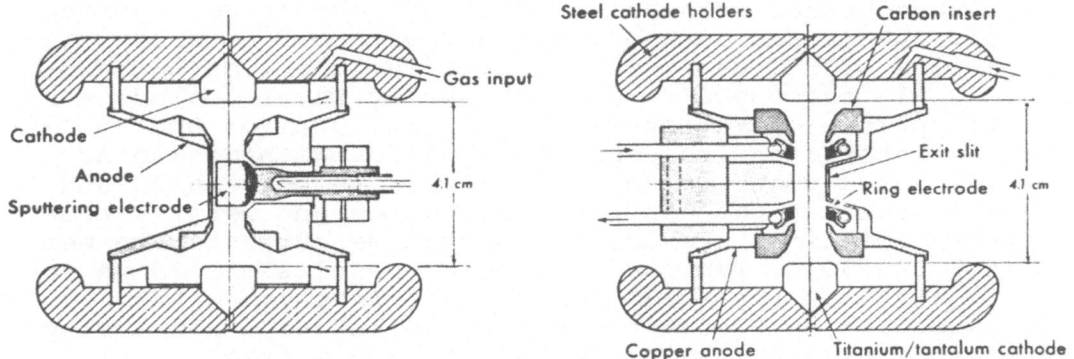

Fig. 5 a,b: Cold cathode-PIG-MCIS with different sputtering electrodes (Gavin 1976).

Table 1: Some results for a cold cathode-PIG-MCIS with annular sputtering electrode (Jacoby and Müller 1975, Schulte et al. 1976). Pulsed operation (10 % duty factor) with Ti cathodes was applied (Ar-operation).

Ion species	Ta^{3+}	Ta^{4+}	Ta^{5+}	Ta^{6+}	Ta^{7+}	Ta^{8+}
$I_q/10^{-6}$ A	8.7	11.7	24.3	22.5	14.4	10.5
Ion species	Ta^{10+}		Ti^{+}	Ti^{2+}	Ti^{3+}	Ti^{4+}
$I_q/10^{-6}$ A	4.5		15.0	30.0	14.0	14.0

Emittance measurements for a cold cathode-PIG-MCIS by Wolf)1973) gave for Ar^{2+} at 18 keV typical values of 65 mm mrad (5 %). Strong dependence from the alignment of the extraction electrode was found.

PIG-MCIS with externally heated cathodes by applying electron bombardment were first used by Makov around 1950 (see Makov 1966), and by Jones and Zucker (1954) which also applied directly heated filament cathodes. Subsequent work at JINR Dubna/USSR led to the development of powerful PIG-MCIS which nowadays deliver by far the most intense beams of all PIG-MCIS (Morozov et al 1957, Pigarov and Morozov 1961, Pasyuk and Tret'yakov 1972/1).

The Dubna-heated cathode-PIG-MCIS has a cathode separation
of ca. 240 mm and an anode chimney with 8x8 mm cross section. One of the cathodes is heated by electron bombardment while the opposite one is usually water-cooled. Normally pulsed operation with 1 - 2 ms duration and 150 -
200 Hz repetition frequency is applied. With Ne and Ar
as discharge gases, usual life times are between 20 and
40 hrs, but for the heavier noble gases and metal vapours
shorter lifetimes are common. Discharge currents are between 5 and 50 A. Typical test bench results are 15 µA for
Kr^{11+} and 5 µA for Xe^{13+} (Pasyuk and Tret'yakov 1972/2).
For application of the sputtering technique a sputtering
electrode is arranged oppositely to the extraction slit.
Sputtering voltage is switched on during the discharge
pulse only. Some typical results are given in table 2.

Table 2: Results for the Dubna-PIG-MCIS with
         sputtering operation (Pasyuk and Tret'yakov
         1972/2, Tret'yakov et al. 1973).

ion species	discharge parameters current/A	voltage/V	$I_q/\mu A$	I_{tot}/mA
Mg^{7+}	7.5	660	5	140
Al^{6+}	8.2	300	40	105
Si^{5+}	11.0	620	1600	104
Ca^{8+}	9.5	600	35	66
Ti^{8+}	15.2	400	40	95
Cu^{7+}	10.0	540	1900	150
Zn^{9+}	7.5	400	24	170
Ge^{5+}	9.0	700	3400	–
Mo^{8+}	9.5	380	400	100
Ta^{8+}	9.8	470	3000	54
W^{9+}	9.0	360	120	61
Re^{9+}	20.0	580	600	86

The Dubna-group has also attached directly heated ovens
made from stainless steel onto their PIG-MCIS. This enabled
processing of solid charge materials up to ca. 1000 $^{\circ}$C.
The anode was always kept hotter than the oven. Respective
results have been described by Pasyuk and Tret'yakov
(1972/2).

Following the above described Dubna-design, a somewhat smaller heated cathode-PIG-MCIS was developed for the UNILAC accelerator in Darmstadt/FGR, see fig. 6.

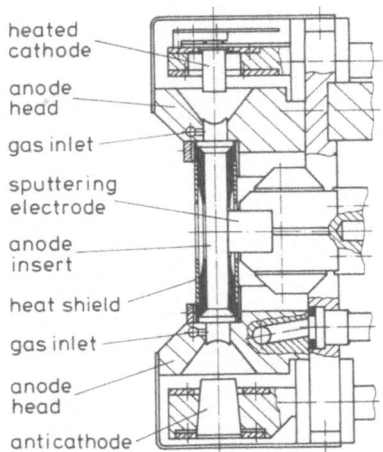

heated
cathode

anode
head

gas inlet

sputtering
electrode

anode
insert

heat shield

gas inlet

anode
head

anticathode

Fig. 6: Heated cathode PIG-MCIS with sputtering assembly as used by GSI Darmstadt/FGR (Schulte et al. 1976).

Some typical results obtained with this source are shown in figs. 7 and 8. A similar source was also recently used by Gavin et al. (1981).

Generally, PIG with radial extraction yield much higher currents of MCI than with axial extraction (see also Bennett 1972). On the other hand, axially extracted PIG sources need much smaller magnets and, at comparable discharge power, they deliver much higher total current densities for extracted ions. Therefore, they are of special interest if only small discharge power is available or lower yields of MCI are sufficient.

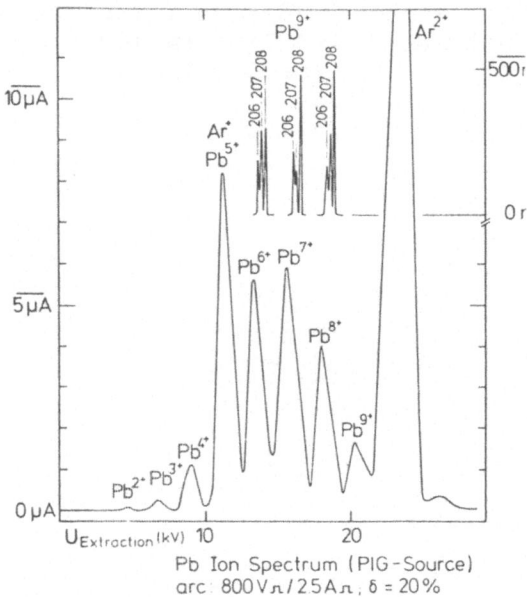

Fig. 7: GSI heated cathode PIG-MCIS, charge spectrum
for operation with Pb (Müller and Jacoby 1977)

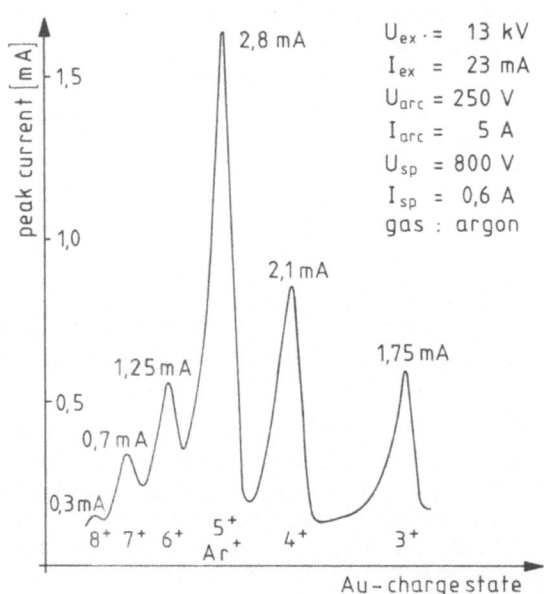

Fig. 8: GSI heated cathode PIG-MCIS, charge spectrum
for operation with Au (Wolf and Keller 1981)

4.2. Plasmas of DUOPLASMATRON-Type Discharges

The DUPLASMATRON-discharge is a low pressure arc with
plasma compression by a "Zwischenelectrode" (ZE) and an inhomogeneous magnetic field (Von Ardenne 1956, Fröhlich 1959,
Demirkhanov et al. 1962). In fig. 9 the principal discharge
structure is shown. The magnetic field between ZE-tip and
anode together with a focusing action of one or more
double-layers along the anodic discharge plasma cause
greatly enhanced plasma density in front of the anode
aperture. Typically, $10^{13} - 10^{14}$ cm^{-3} can be achieved
with discharge currents of a few A only.

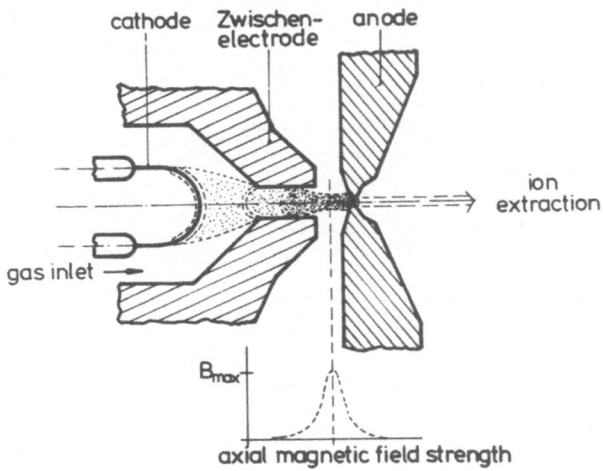

Fig. 9: DUOPLASMATRON discharge configuration,
        schematically

Investigations on MCI-production with the DUOPLASMATRON
were made by Von Ardenne (1956) and Braams et al. (1965).
The most important contributions in this field have been
made by studies at GSI Darmstadt/FGR. Krupp (1968) used
pulsed (0.1 ms/10 Hz) discharges with rather high discharge currents up to 25 A, to obtain quite appreciable
currents of multiply charged noble gas ions. Furtheron,
Illgen (1968), Illgen et al. (1972) and Schulte in den
Bäumen et al. (1972) investigated MCI-production from
both noble gases and solid feeding materials. Systematical
studies were carried out by Winter and Wolf (1974/1, 1975)
and by Krupp and Winter (1975), using in all cases c.w.-
discharge conditions with currents up to 20 A.
Pulsed operation was studied by Keller and Müller (1976,
1977), which also introduced important technical improvements, cf. figs. 10, 11 and 12.

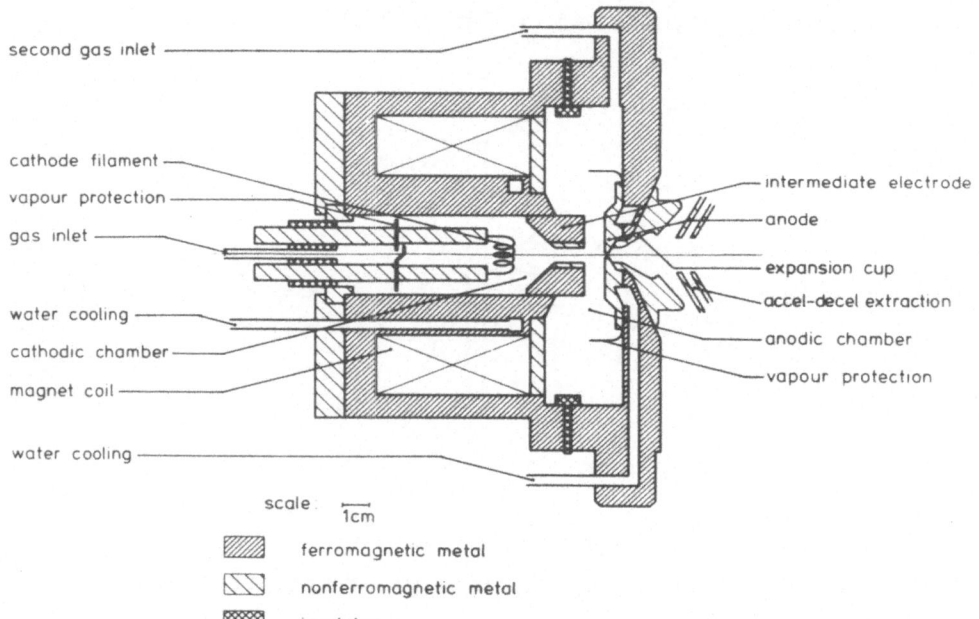

Fig. 10: DUOPLASMATRON-MCIS with directly cooled
        anode aperture region, after Keller and
        Müller (1976, 1977)

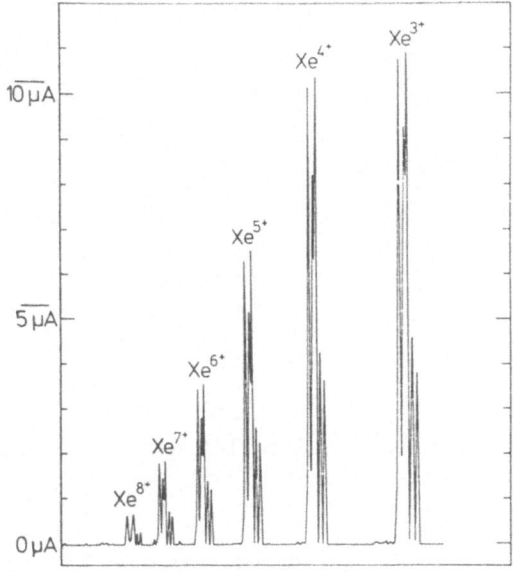

Xe Ion Spectrum (Duoplasmatron)
arc : 200 V$_n$ / 5 A$_n$; U$_{Ex}$ 35 kV, δ =20%

Fig. 11: Ion charge state spectra obtained from a pulsed DUOPLASMATRON-MCIS for Xe, from Keller and Müller (1977).

The achievable CSD for MCI is greatly influenced by discharge stability. Generally in pulsed operation higher
yields of MCI are possible. Keller and Müller (1976) describe optimized operation conditions and obtained up
to 0.1 µA for Xe^{11+} (peak value for pulsed operation).

Ions from solid feeding materials have been produced by
the sputtering technique, cf. fig. 12.

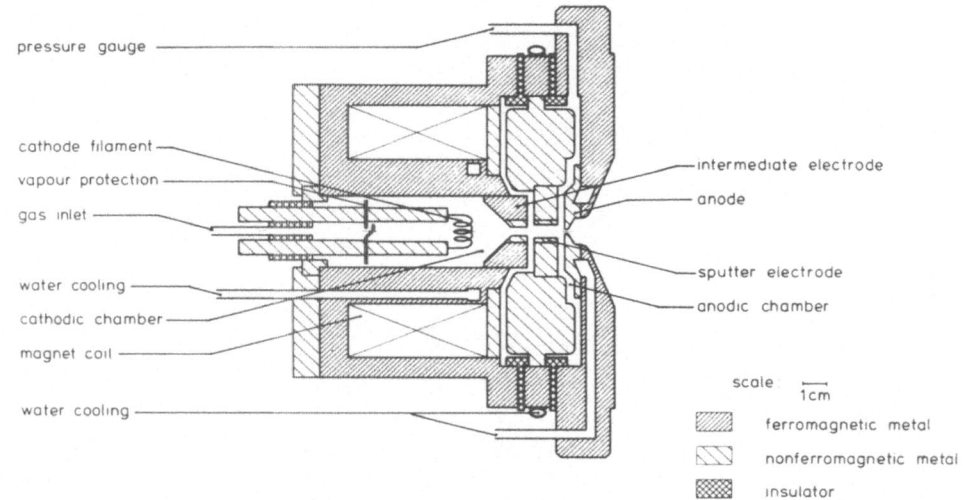

Fig. 12: DUOPLASMATRON-MCIS with sputtering electrode
         (Keller and Müller 1978)

The sputtering electrode is kept typically at - 500 V
against the anode potential. The feeding material is
attached as an annullar insert of the sputtering electrode.
By using appropriate auxiliary gases typically 10 % of
the total extracted ion current belonged to the respective
solid ion species. Some typical data are 6.5 µA for $^{48}Ti^{3+}$
and 3.5 µA for $^{58}Ni^{3+}$. The sputtering technique can be
applied to obtain multiply charged ions of all refractory
materials.

        A combination between PIG- and DUOPLASMATRON-MCIS
has been developed for efficient singly charged ion production by Demirkhanov et al. (1962), Morgan et al. (1967)
and Davis et al. (1975). This configuration was named
DUOPIGATRON and has been investigated in respect to MCI-
production by Winter (1974) and Keller and Winter (1976).
In fig. 13 the respective discharge geometry is shown,
which has been obtained by a simple modification of the
GSI-type DUOPLASMATRON-MCIS.

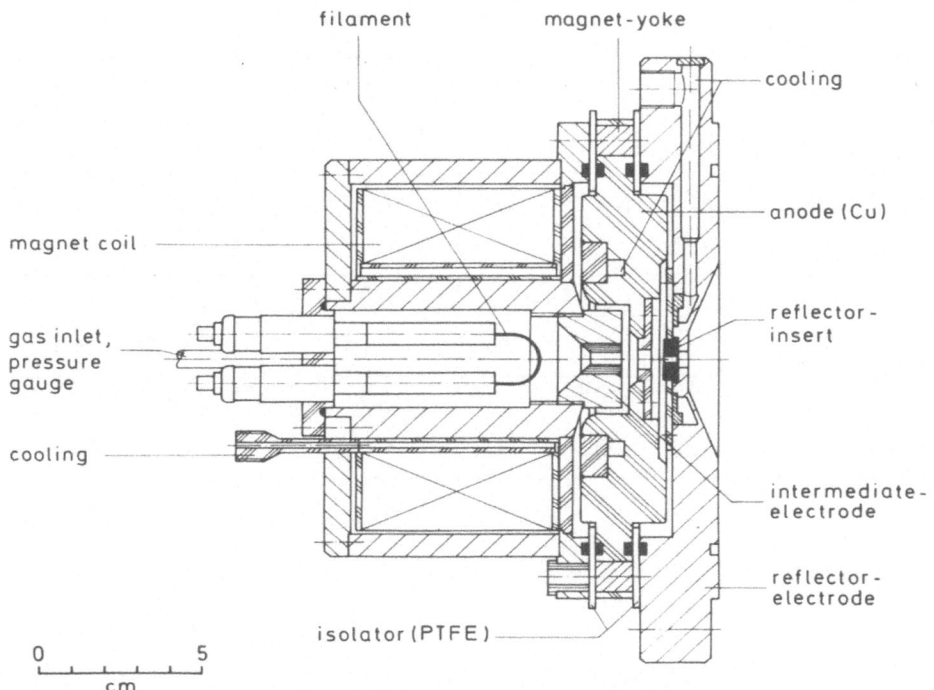

Fig. 13: DUOPIGATRON-MCIS (Wolf and Keller 1981)

The typical discharge parameters for the DUOPIGATRON are
some A discharge current and ca. 100 V discharge voltage.
At lower gas consumption, however, also a high discharge
voltage mode has been found (Keller and Winter 1976) which
offers similar CSD as an axially extracted PIG-MCIS.
The DUOPIGATRON-MCIS has been proven to deliver multiply
charged ions from almost all feeding materials with
quite reliable operating conditions. For solid feeding
materials, inserts in the reflector electrode are used
(cf. fig. 13). There were also investigations with
an expansion cup extraction geometry, see Wolf and Keller
(1981). Results for various ion species have been collected
during some 5 years of operation and some typical data
have been summarized in table 3.

Table 3: Some typical results of DUOPIGATRON-MCIS
c.w. operation (from Wolf and Keller 1981)

Ion species	N^{2+}	Ne^{3+}	Al^{3+}	Si^{2+}	S^{2+}	Ar^{3+}	Ca^{2+}
$I_q/\mu A$	10	1	1	2	5	50	3
Ion species	Cr^{3+}	Ni^{3+}	Cu^{3+}	Zn^{3+}	Ge^{3+}	Kr^{4+}	Zr^{3+}
$I_q/\mu A$	0.5	0.1	0.5	0.2	2	0.1	0.1
Ion species	Nb^{3+}	Mo^{3+}	Ag^{3+}	In^{4+}	La^{4+}	Er^{4+}	W^{4+}
$I_q/\mu A$	1.5	1	0.5	0.1	0.1	0.5	0.1
Ion species	Ir^{4+}	Pt^{4+}	Au^{4+}	Hg^{4+}	Pb^{4+}	Bi^{4+}	U^{4+}
$I_q/\mu A$	0.5	0.5	0.5	0.5	0.6	0.3	1.5

Both the DUOPLASMATRON and the DUOPIGATRON ion sources
are known to provide exceptionally high brightness when
operated with hydrogen feeding gas. There are no systematic studies on ion beam quality for MCI, so far.

The balance models shortly described in chapter 2.
have been applied to compare calculated CSD with results
from measurements, see Winter and Wolf 1974/2.
Typical parameters for the DUOPLASMATRON-MCIS are shown
in fig. 1 a, for the heated cathode-PIG-MCIS in fig. 1 b
and for cold cathode-PIG-MCIS in fig. 1 c.

In conclusion, one can obtain abundant multiply charged
low-Z ions and heavy ions up to q = 13 with PIG-MCIS.
Typical ion source lifetimes depend very much on the
mode of operation but can reach easily more than 10 hrs.
The DUOPLASMATRON-MCIS offers lower q-values than the
PIG-MCIS, but its operation is easier and less costly.
Finally, the DUOPIGATRON-MCIS is a very versatile ion
source for production of up to triply and quadruply
charged ions. Both the DUOPLASMATRON- and the DUOPIGATRON-
MCIS under careful operation reach lifetimes of 50 hrs.
and more.

4.3. Electron Cyclotron Resonance-Heated Plasmas

4.3.1. General Principles

RF-discharges are commonly used for ion production with the main advantage of long ion source lifetime and quite clean ion beams because no heavily loaded electrodes are needed. However, in the common RF-ion sources both plasma density and electron temperature are small and therefore MCI-production is not very efficient, cf. Valyi (1970). At given plasma density electromagnetic waves can only penetrate into the plasma if the frequency is above the respective cut off-value f_{co}

$$f_{co}/kHz \sim f_{ep} = (\frac{e^2}{\pi m_e} n_p)^{1/2} \approx 9(n_p/cm^{-3})^{1/2}$$

f_{ep} electron plasma frequency   (18)

For a given frequency the respective plasma cut off-density follows from (18):

$$n_p/cm^{-3} \approx 10^4 x (f/MHz)^2 \qquad (19)$$

Following chapter 2., for efficient MCI-production in a plasma frequencies at and above 10 GHz (microwave range) are needed to assure sufficiently dense and hot plasmas. Moreover, the ion confinement time has to be sufficiently long which can be achieved by embedding the plasma into a "min B" geometry to avoid excessive diffusion losses and MHD instabilities.
Electron cyclotron resonance heating forces the electrons of a plasma within a magnetic field to oscillate with frequencies of multiples of the respective electron cyclotron frequency

$$f_{ec} = \frac{e B}{2\pi m_e c} \approx 2.8 \ B/kG \quad GHz \qquad (20)$$

For efficient electron heating electron oscillations should take place transversally to the magnetic field lines. This is easily achieved within magnetic mirrors. Electrons moving along the field lines assume kinetic energy transversally to the field line direction by so-called stochastic heating processes. The heating occurs in a zone where equ. (20) is fullfilled, and electrons passing this resonance zone many times can assume kinetic energies of up to many keV.

Since a simple mirror field is subject to MHD-plasma instability, superposition of magnetic multipole fields
(creation of min-B configuration) will improve greatly
the ion confinement time. The magnetic hexapole geometry
is the best compromise between desirable high number
of poles and engineering restrictions.
There are cold electrons present in the plasma, too, which
can give rise to MCI-recombination. It is therefore ad-
visable to pose the ECR-heating zone next to one of the
magnetic mirror throaths where the produced MCI can be
extracted.
Under the above described conditions highly charged ions
will be produced by step-to-step ionization processes and
will therefore appear after their respective ion confinement time only (cf. chapter 2.). Because of this the ECR-
heated MCIS (ECRIS) are most suited for c.w. applications
or pulsed operation with relatively long pulse duration.
Most conveniently, the heating electrons are not produced
by ion impact on material electrodes as e.g. in the PIG-
MCIS, but within the plasma which therefore permits very
long ion source lifetimes.

4.3.2. Survey on ECRIS Development

The ECRIS concept was introduced by R. Geller and
coworkers at C.E.N. Grenoble out from experiments on
plasma heating for thermonuclear research. In fig. 14
an early setup is shown where a simple magnetic mirror
assembly served to study ECR heating. By means
of a similar configuration ("MAFIA") Apard et al. (1972)
demonstrated production of fully stripped ions of nitrogen.
It was then realized that higher performance necessitated
the application of min B-geometry which was added as a
superimposed magnetic hexapole field. Moreover, the region
of plasma production was kept at relatively high pressure
(ca. 10^{-3} Torr) in a first stage, while plasma heating
was achieved at ca. 10^{-5} Torr background pressure in a
second stage - "SUPERMAFIOS", cf. Geller (1976). A still
more complicated apparatus used a third stage to improve
ion beam formation - "TRIPLEMAFIOS", cf. Briand et al.
(1977). Both "SUPERMAFIOS" (cf. fig. 15) and "TRIPLEMAFIOS"
showed very encouraging results, but the use of normal-
conducting magnetic coils caused an unacceptable high
electric power consumption. Therefore, the magnetic hexapole field coils were replaced by an assembly of permanent
magnets made from $SmCo_5$ - "MICROMAFIOS", cf. Bechtold et
al.(1980 a). This permitted considerable reduction both
in physical size (ca. 50 cm length vs. 2 m for "SUPER-
MAFIOS") and power consumption (ca. 100 kW vs. up to 3 MW).

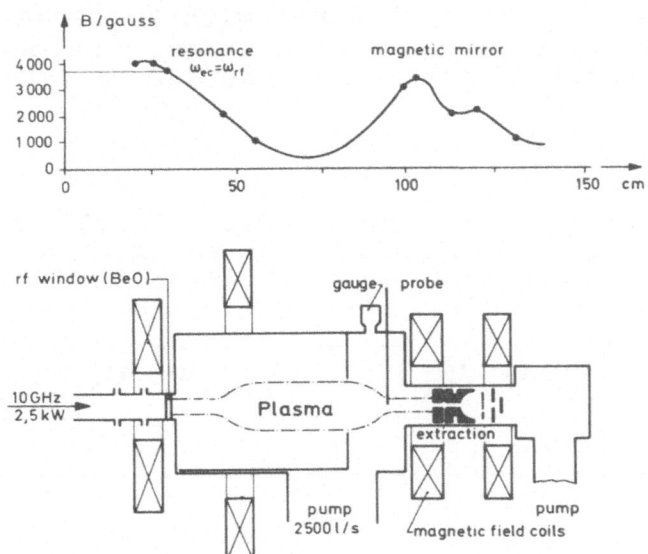

Fig. 14: Early ECR-heated plasma ion source
(Consolino et al. 1969). With a similar
setup ("MAFIA") the production of fully
stripped ions was demonstrated .

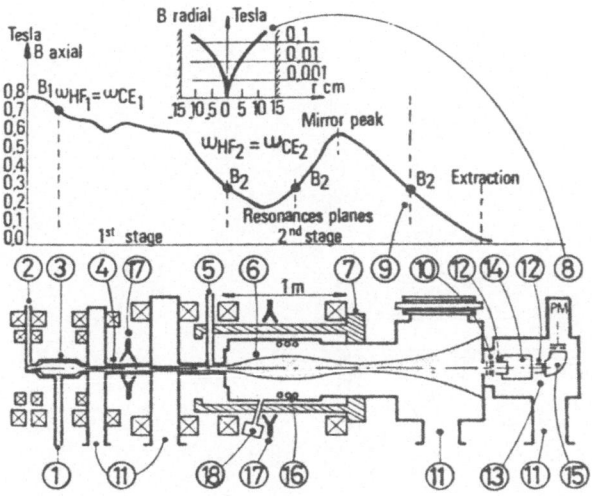

Fig. 15: "SUPERMAFIOS"-ECRIS (Geller 1976)

In table 4 some results obtained with "MICROMAFIOS" are given. Fractions of lower charge states than noted are much higher and follow approximately the CSD of PIG-MCIS.

Table 4: Electrical currents (in 10^{-6} A) obtained with "MICROMAFIOS" at an extraction voltage of 8 kV, cf. Arianer and Geller (1981).

Ion species	$^{12}C^{6+}$	$^{15}N^{7+}$	$^{18}O^{8+}$	$^{20}Ne^{9+}$	$^{40}Ar^{13+}$	$^{84}Kr^{17+}$
Ion current	0.2	0.1	0.1	< 0.1	< 0.1	0.05

Typical ion energy spreads of 20 q eV and emittances of ca. 150 mm mrad at 7 kV extraction voltage have been obtained.
Whereas in the "MICROMAFIOS"-setup both ionization stages were fed separately by microwave lines, the latest development concerned a still simpler configuration where only one microwave power connection is necessary - "MINIMAFIOS", see fig. 16 from Bourg et al. (1982).

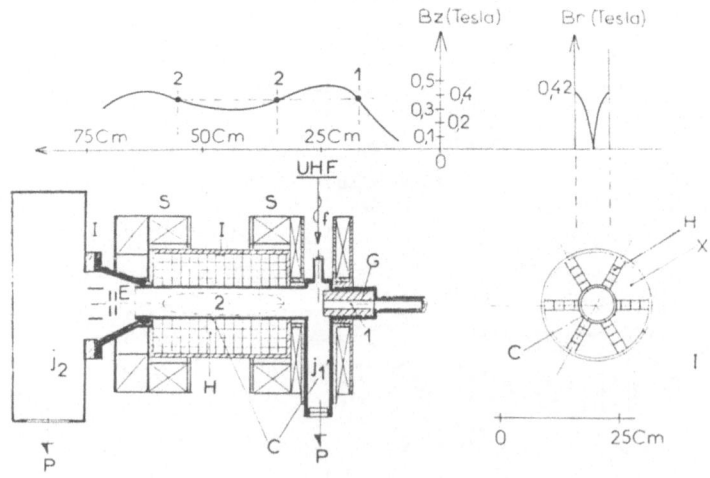

Fig. 16: "MINIMAFIOS"-ECRIS (Bourg et al. 1982)

In one version of "MINIMAFIOS" the first stage of
the earlier design was replaced by a dielectrique tube
within which a first resonance zone led to plasma production (cf. fig. 16. The plasma diffused into the second
stage where heating was performed. It was found, however,
that operation of this configuration was somewhat critical
and therefore the design was slightly altered. Now the
first stage is again a box-shaped cavity as in the case
of the earlier "MICROMAFIOS", cf. Bourg et al. (1982).
In table 5 some results are given, which had been obtained by A. Drentje from KVI, University of Groningen,
The Netherlands, when testing a "MINIMAFIOS"-ECRIS to be
delivered for a cyclotron in the above mentioned laboratory.

Table 5: Some results obtained with a "MINIMAFIOS"-
         ECRIS (version 2), A. Drentje, March 1982
         (private communication)
         a) extraction voltage 12 kV
         b) extraction voltage 20 kV

         The source was optimized for high yield
         of highest charge state by varying both
         gasflow and microwave power (ca. 500 W)

Ion species	$^{15}N^{7+}$	$^{20}Ne^{9+}$	$^{40}Ar^{12+}$
Ion current / μA	0.1 (a) 0.09 (b)	0.01 (a) 0.035 (b)	0.1 (a) 0.07 (b)

At the tests summarized in table 5, emittance measurements
yielded typical values of 200 mm mrad.

ECRIS development was also carried out by several other
groups: Jongen et al. (1980, 1981, 1982) constructed a
fully superconducting ECRIS ("ECREVIS") for a cyclotron.
A small-scale version "ECREVETTE" has been put recently
into operation. We also mention work proceeding at the
Kernforschungsanlage Karlsruhe/FGR, where an ECRIS with
superconducting mirror coils and $SmCo_5$-hexapole magnets
is under construction ("HISKA", Bechtold et al. 1980 b, 1982).
Also this group built a small-scale version which is
already operating.

A very important development concerns the use of solid
feeding materials in ECRIS: Bliman et al. (1978) applied
a high temperature-environment arc discharge within the
first stage of "TRIPLEMAFIOS" to obtain MCI from Ta and U.
More recently Geller and Jacquot (1981) introduced a new
method: At first, the solid feeding material is put into
the resonance zone where it quickly vaporizes and becomes
deposited onto a liner within the plasma heating cavity.
The material is reemitted into the discharge due to
sputtering from the places on the liner walls, where the
magnetic surface intersects the liner. Such, the solid
material atoms are ionized and converted into MCI, while
unused atoms are redeposited onto the walls and recycled.
By use of this technique in "MINIMAFIOS", Geller and
Jacquot (1981) obtained e.g. W^{33+} and Mo^{21+} with operation
times of several days without a need for reloading.
Further improvement of the ECRIS-performance seems possible
if one realizes that the experiments carried out so far
showed an increase of plasma density with the square of
the microwave frequency, cf. Geller (1982). Moreover,
the achieved electron temperature was found to follow
roughly the applied microwave power. Increasing the
frequency necessitates higher magnetic field strengths
(see equ.(20), which seems not too difficult with presently
available permanent magnets (up to 0.9 T vs. 0.4 T in
"MINIMAFIOS"). Therefore, Geller and coworkers plan to
increase their microwave frequency from now 10 GHz up to
16 GHz and eventually even higher.

The status of ECRIS-development has reached an almost
commercial level of high performance and reliability.
With noble gases as feeding material, source lifetimes
of many thousand hours are now available which is of
great interest for atomic collision experiments with
highly charged and fully stripped low-Z ions. Typical
ion currents for these are now in the microampere range
at beam energies as low as 1 keV x q.
Further work concerns both still higher ion charge states
also for heavy ions and the routine production of MCI
from solids. Moreover, the use of permanent magnets for
the multipole field configuration as well as of superconducting magnet coils is very promising and already
operating devices have been demonstrated to work
according to expectations.
The interest in ECRIS-development and application has
stimulated the organisation of specialized workshops,
which are also concerned with all technical problems
in connection with the building and operation of various
ECRIS, cf. GR 7.

4.4. Dense Plasmas Produced by Laser Irradiation, Spark Discharges and Condensed Discharges

Deposition of sufficiently high power onto a solid surface or into a small volume of dense gas causes production of hot, dense plasmas which can act as an efficient source of multiply charged ions. In the following, we describe some approaches which have used this principle. The common features of such MCIS are on the one hand the very simple configuration; on the other, however, the relatively poor ion beam qualities because of appreciable ion heating taking place. If well defined ion energies are desired, means for kinetic energy selection have to be provided, which usually result in strong decrease of ion beam intensities.

4.4.1. Laser-MCIS

Production of MCI by Laser beam-irradiation of solid targets has been investigated in view to MCI production, as a byproduct of inertial plasma confinement studies for thermonuclear fusion. For sufficiently high ion yields, high power Lasers are needed. To avoid intolerable target degradation, on the other hand, such Lasers must have short pulse characteristics. In this context, the following Lasers are of interest:

Nd glass Laser (wavelength 1060 nm, pulse durations 0.1 - 2 ns, pulse power up to 10^{15} W)

CO_2 Laser (wavelength 10600 nm, pulse duration ca. 1 ns, pulse power up to 10^{12} W)

Iodine Laser (wavelength 1340 nm, short pulse operation not easy, energy/pulse some 100 J)

Focusing of visible and near-IR Laser radiation down to spot sizes of 10^{-4} cm^2 makes irradiation power densities of up to 10^{17} W/cm^2 technically feasible. Vaporisation with subsequent ionization from solid surfaces sets in at typically 10^8 W/cm^2.

For plasma densities up to the respective cut off-values (cf. chapter 4.3.1.) the absorption of radiation can take place and will proceed via inverse bremsstrahlung, thus accelerating the charged plasma particles. Up to limiting values which are not yet fully understood, the plasma electron temperature will be roughly proportional to the irradiation power density. At 10^{13} W/cm^2 typically 1 keV can be achieved. With present day facilities, in such plasmas He-like ions can be produced up to Zn, cf. Peacock (1982). At 10^{10} W/cm^2, electron temperatures will reach ca. 20 eV and fully stripped low-Z ions can be produced already.

MCI-production in Laser-produced plasmas has been studied
since long time, cf. Faure et al. (1971). Such MCIS were
also applied for high energy accelerators, cf. Anan'in
et al. (1974), and for a rewiev on more recent work,
Kutner (1981). More recently, Laser-MCIS were also applied
for low energy atomic collision experiments, especially
in studies of electron capture into highly charged low-Z
ions, see Goldhar et al. 1976, and Phaneuf (1981 a,b).
In fig. 17 we show the experimental setup of the latter
author, and in figs. 18 a, b some CSD obtained in these
experiments are given.

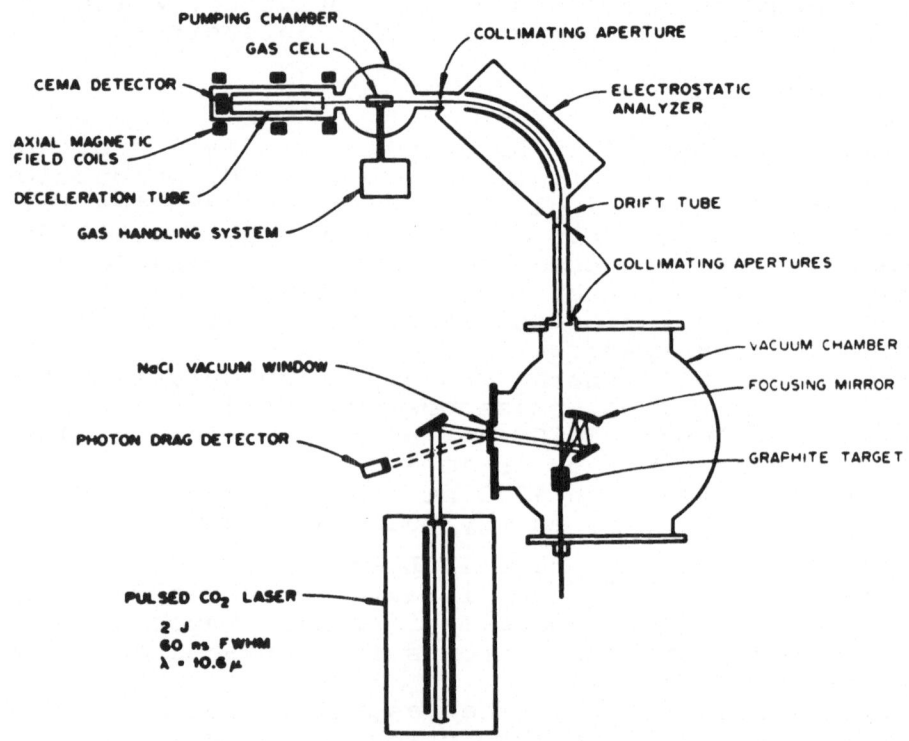

'Fig. 17: Experimental set-up for total electron capture studies with highly charged low-Z ions (Phaneuf 1981)

For production of O^{q+} ions, Phaneuf used a target made
of ThO_2. In his experiments, the considerable low
number of MCI available after energy selection (for
high q typically 1 ion per shot, repetition frequencies
of up to 2 Hz were possible) made many shots necessary
for one cross section measurement.

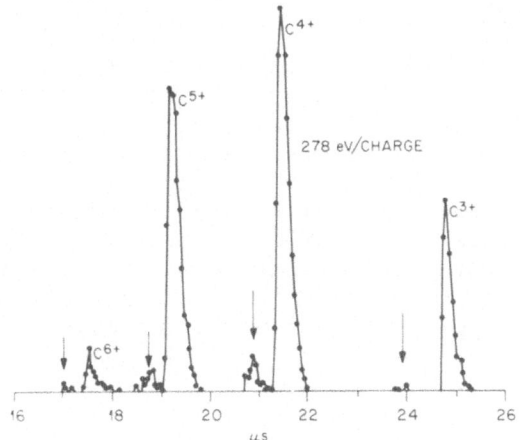

Fig. 18 a: Charge spectrum for C^{q+} ions from
a Laser-produced plasma (Phaneuf 1981 a)

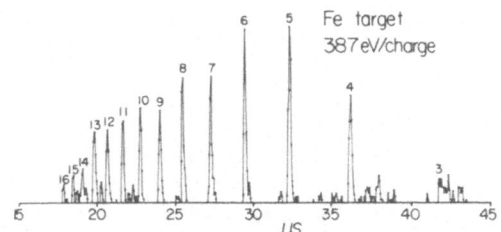

Fig. 18 b: Charge spectrum for Fe^{q+} ions from
a Laser-produced plasma (Phaneuf 1981 a)

Since Laser plasmas expand very rapidly, after a short
time the initially recombination-dominated CSD remains
"frozen". For highly excited states of the MCI thermal
equilibrium can be found, whereas for lower levels
considerable deviation from thermal equilibrium toward
population inversion can be found.

The ions are emitted from the plasma with velocities of
up to 10^7 cm/s with their mean energy increasing linearly
with the charge state q. The energy spread of ions
increases with the Laser irradiation power and reaches
typically some keV at irradiation power of 10^{10} W/cm^2.
Most target surfaces are very quickly deteriorated because of the efficient vaporization. This makes the
achievment of reproducible ion yields very difficult
and necessitates frequent readjustment of the target
position.

4.4.2. <u>Spark Discharge-MCIS</u>

By connecting a high voltage source to a suitably shaped discharge gap, breakdown with very high currents can be initiated. Self-pinching of the discharge channel results in high current densities (up to 10^8 A/cm^2) and thus in production of very dense plasmas. If the spark gap is carefully matched to an oscillating circuit, periodical breakdown can be obtained.
Such spark-MCIS have been widely used for spectroscopical studies up to very high charge states (H-like ions of Fe). There were also applications with accelerators, cf. fig. 19.

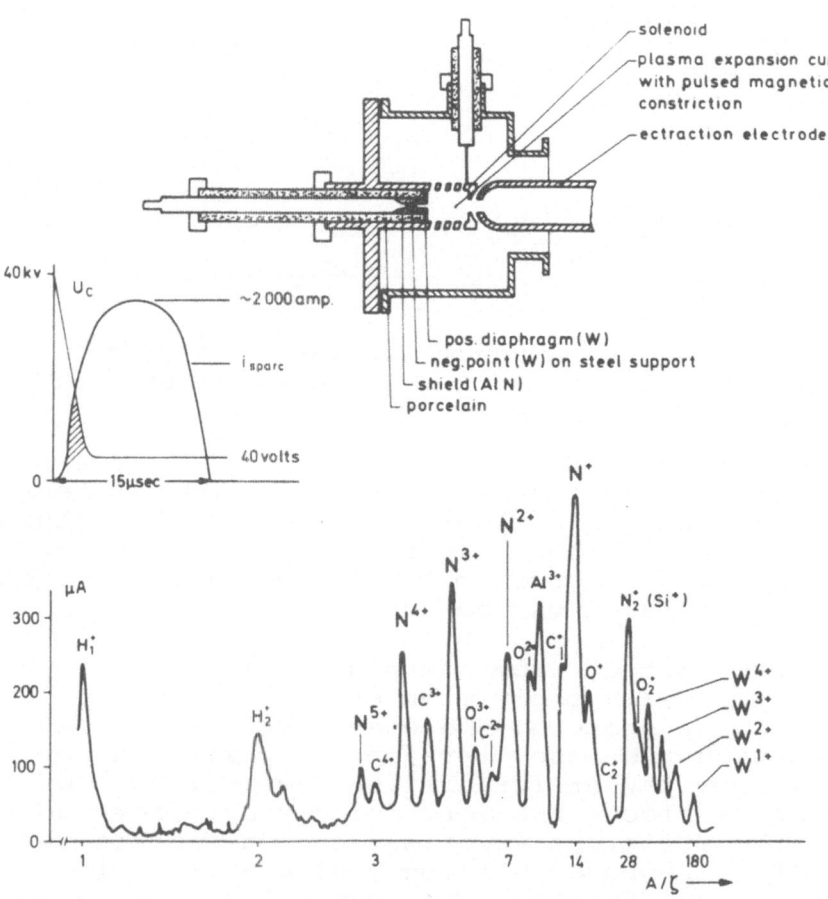

Fig. 19: Spark discharge-MCIS from Bolotin et al. (1961)

Typical break-down durations are of the order of 10^{-6} s,
but the time intervall during which the MCI are efficiently
produced is much shorter (see fig. 19, insert showing
the current-voltage characteristic of the discharge).
Both the very small duty cycles and the large variation
of emitted ion currents are inconvenient for practical
applications in heavy ion accelerators. However, the
very simple construction of spark discharge-MCIS is
an advantage and has led to frequent applications in low
energy atomic collision studies. We mention the work
of Zwally et al. 1969, and more recently, of Takagi
et al. 1982. The latter authors used a YAG Laser beam-
triggered spark gap without ion extraction. The discharge
circuit contained a capacitor of 10 μF charged up to a
voltage of 20 kV with an inductance of 150 nH.
The anode material delivered highly charged ions of
Be and B (beryllium-lanthanum hexaboride), C (graphite)
and Fe. At discharge energies of 125 J up to Fe^{16+} has
been observed. The ions were emitted with kinetic energies
between some 100 eV and several keV. Selection of ion
charge states and -kinetic energies was achieved along
a time-of-flight line with subsequent electrostatic
energy analysis. At 1 % resolution, typically 100 – 1000
highly charged ions/pulse could be obtained.

4.4.3. Condensed Discharge-MCIS

Rhee (1981) described a Marx generator-charged Blumlein
system which drives a plasma focus discharge from which
accelerated fully stripped ions of He, N and Ar have
been obtained. In a similar way also MCI from solid
feeding materials could be produced. The ion energies
were typically 500 keV x q and the total number of
charged particles per shot was ca. 10^{11}.
This technique is still in an early stage and practical
application for atomic collision experiments remains
to be demonstrated.

5. Electron Beam Ion Sources

An electron beam which carries a current I_e at an electron energy E_e produces in its radial cross section a potential through with a depth of

$$\Delta U/eV \approx 480 \times \frac{I_e/A}{(E_e/keV)^{1/2}} \qquad (21)$$

If also care is taken that positive ions within this electron beam cannot leave it in axial direction (by providing appropriate potential barriers), these ions will be confined and further ionized by the beam electrons. This principle of ion trapping was used in ion mass spectrometry by Plumlee (1957) and Baker and Hasted (1966) to increase the ion source yield. Redhead (1967) demonstrated that in such traps multiply charged ions can be produced in step-by-step ionization processes. Donets et al. (1968, 1969) constructed a powerful MCIS by following the same ideas, which is nowadays called electron beam ion source (EBIS), cf. fig. 20.

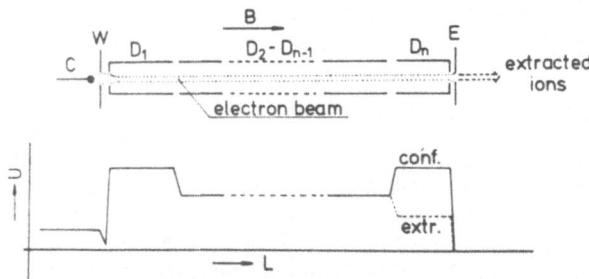

Fig. 20: Principal Configuration of EBIS

A dense electron beam is injected into a homogeneous magnetic field inside a series of drift tubes, where ultrahigh vacuum conditions are provided. Atoms of the desired ion species are leaked into the electron beam and trapped by the electronic space charge after their first ionization. The time to reach a certain charge state q is given by equ. (6) in chapter 2.2.

In equ. (6) we can use the expression

$$S_{r-1,r} \equiv Q_{r-1,r} \, v_e \qquad (22)$$

for the respective ionization rate coefficient, because the ionizing electrons are practically monoenergetic. It is seen that ions of high charge states will appear after a time which is the smaller the higher the electron beam current density is chosen. Moreover, the electron beam energy has to be sufficiently high.
This applies only if the respective ionization time is not larger than the achievable confinement time t_c, which in turn will be reached as soon as the electronic space charge is fully compensated by the ions produced. t_c is limited by the background gas pressure and can be estimated by assuming the background gas to be mainly hydrogen and using empirical ionization rates (Arianer and Geller 1981):

$$t_c/s \simeq 3.5 \times 10^{-10} (E_e/eV)^{1/2}/(p/Torr) \; 3(E_e/eV) \qquad (23)$$

Equ. (23) yields e.g. $t_c = 1$ s for $E_e = 10$ keV at
$$p = 3 \times 10^{-9} \text{ Torr}$$

In table 6 we give estimated values for $n_e \, v_e \, t_z$ which have to be reached to produce some fully stripped ions (q = Z).

Table 6 : Conditions for production
of fully stripped ions
(from Donets 1982)

ion species	$n_e \, v_e \, t_z \, (cm^{-2})$	E_e/keV
Ne^{10+}	2×10^{20}	2.5
Ar^{18+}	2×10^{21}	9
Xe^{54+}	2×10^{23}	80
U^{92+}	2×10^{24}	250

As already mentioned, the necessary vacuum conditions for a given electron beam follow from the respective t_q (t_z) – values and the requirement of $t_q < t_c$

The total number of ions in charge state q (only one q is assumed to be present) produced during one confinement period will be (cf. Arianer and Geller 1981):

$$N_q \simeq 1/q \times 10^{13} \ (I_e/A)/(E_e/eV)^{1/2} \qquad (24)$$

If a number of various ion charge states up to q_{max} are produced, we obtain instead of N_q

$$q \ N_q \rightarrow \sum_{r=1}^{q_{max}} r \ N_r \qquad (25)$$

Further interesting relations for the EBIS principal working conditions were discussed recently by Vella (1981).

Realization of an EBIS necessitates solution of the following problems:

a) Production of a very dense stable electron beam within a homogeneous magnetic field configuration (cf. Amboss 1981 and references therein)
b) Realisation of a sufficiently low background pressure in the ionization region
c) Setup of an appropriate potential distribution along the electron beam to enable for injection of atoms, ionization during trapping and ejection of MCI after elapse of ion confinement time.

The first EBIS of Donets et al. (1968, 1969) is shown in fig. 21. The magnetic field with up to 4 kG on axis was provided by a solenoid of 165 mm length. The electron beam was produced with a fully immersed LaB_6 electron gun having a microperveance of ca. 1 (the microperveance is defined by

$$\mu P \equiv I_e/\mu A \ (U_e/V)^{-3/2} \qquad (26)$$

and values of 1 or higher belong to "highly perveant" electron guns). Five drift tubes with a potential distribution as shown in fig. 21 served for setting up the axial potential distribution which during the trapping period featured potential barriers on each side. For ion extraction one barrier was removed.
Considerably high ion charge states have been detected by means of a time-of-flight spectrometer. The achievable confinement times were limited, however, because the background gas pressure could not be made lower than 2×10^{-8} Torr.

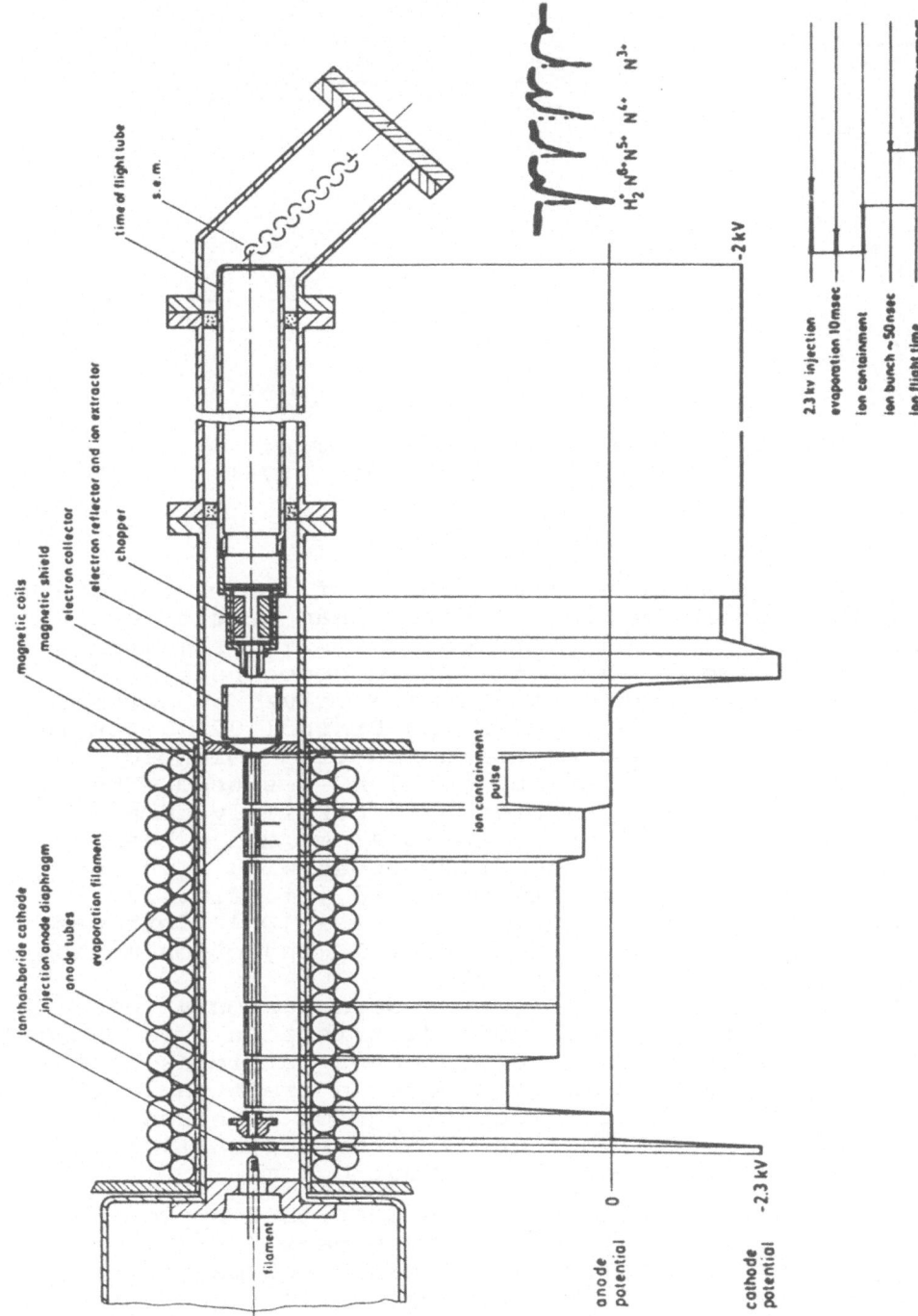

Fig. 21: First EBIS built by Donets et al. (1968, 1969)

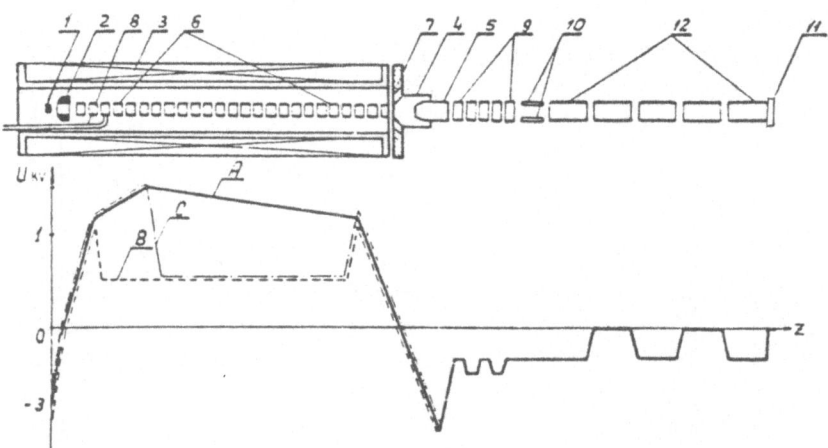

Fig. 22: Schematic view of cryogenic
         EBIS "KRION I" (Donets 1976)

With a new enlarged model an increase of ion charge states
and - production rates with electron beam density and
trap length could be demonstrated. The still remaining
problems of incomplete magnetic field homogeneity and
too high background gas pressure were solved by switching
to cryogenic techniques, Donets and Pikin (1974), Donets
and Ilyushchenko (1974). A superconducting solenoid
with a length of 120 cm and up to 15 kG was used. The
electron beam was now generated by a partially immersed
gun with current densities up to 30 A/cm^2 and a beam
diameter of 3 mm passing the drift tubes with 5 mm inner
diameter. 25 drift tubes were provided (cf. fig. 22).
A low background gas pressure of ca. 2 x 10^{-11} Torr was
achieved by cryopumping. A new particle injection method
was also introduced (cf. fig. 22):
Ions generated within the electron beam can only reach
the trapping region if potential distribution A is changed
into B. Well defined number of injected particles is obtained by switching from B to C after a presettable time.
Ion ejection after the confinement period is caused by
switching back to distribution A, after which all ions
are expelled within a period of about 0.1 ms.

Further progress of their work was described by Donets
and Ovsyannikov (1977, 1980) and by Donets (1982).
The most recent EBIS-parameters include now magnetic
field strengths up to 2.5 T, an achievable background
pressure of ca. 10^{-12} Torr and electron beam current

densities of 150 A/cm^2 with a design goal of 600 A/cm^2
at an energy of 20 keV. Ion confinement times of some
seconds could be reached which recently permitted the
formation of up to 2×10^6 Xe^{52+} ions/per pulse, corresponding to a value of ca. 10^{22} for n_e v_e t_q (cf. table 6).
Other results include production
of fully stripped carbon (ca. 2×10^9/pulse), fully
stripped argon and He-like Krypton.

Other important work on EBIS-development has been due
to Arianer and coworkers at Orsay/France:
Their first EBIS ("SILFEC", Arianer et al. 1975, Arianer
and Goldstein 1976) included an 80 cm long solenoid
for magnetic field strengths up to 8 kG. With an external
electron gun a compressed electron beam with 1 mm diameter
and up to 60 A/cm^2 current density was injected into
a bakeable structure with 15 drift tubes (inner diameter
8 mm). With background pressures of ca. 10^{-9} Torr ion
confinement times of up to 0.2 s could be reached.
Arianer et al. (1979) described a new cryogenic version
("CRYEBIS I") with up to 3 T and an electron beam with
current of up to 2 $A/1000$ A/cm^2 at 10 keV. A technical
description of "CRYEBIS I" was recently given by
Arianer et al. (1982).

With this EBIS, Arianer and coworkers could produce up
to Kr^{34+} and Xe^{44+}, cf. Arianer and Geller (1981). At some
instants unexpectedly low ionization times were observed
and explained by "supercompressed" electron beams: The
ultimate compression of an electron beam launched from
an external gun into a magnetic field results in the
so-called Brillouin flow, see Brewer (1967) and Amboss
(1981). However, partial space charge compensation
might permit to exceed even this limiting value, see
also the discussion by Becker (1981).
At present, the Orsay group works on the completion of
"CRYEBIS II" which has the currently most ambitious
design parameters of B = 5 T, $E_{e.max}$ = 50 keV and
$j_{e,max}$ = 10^4 A/cm^2.
With these parameters one hopes to produce fully stripped
xenon and He-like uranium (Arianer 1981).

Other work concerned with EBIS is carried out presently
at Cornell University/USA, cf. Kostroun et al. (1981),
at Frankfurt University/FGR, cf. Becker et al. (1981)
and at IPP Nagoya/Japan, cf. Kaneko (1979).
The first project mentioned concerns a quite simple
version with a normal conducting solenoid (up to 4 kG),
whereas the other ones involve cryogenic structures.

A comprehensive review on the very interesting possibilities to study atomic collision processes (e.g. electron impact ionization of ions or ion-ion interaction) within an EBIS (the so-called electron beam ionization method) has been given recently by Donets (1982).

Small EBIS-type MCIS were built and applied for atomic collision studies at Gießen/FGR (Clausnitzer et al. 1975, Müller and Salzborn 1979) and at IPP Nagoya, cf. Imamura et al. (1981).

Typical emittance values for EBIS are below 100 mm mrad, and energy spreads being independent on the ion charge state of typically 50 eV have been observed (Arianer and Geller 1981).

The construction and successful operation of EBIS is very intimately connected to a satisfactory understanding of all phenomena related to formation, compression and stability of dense electron beams within magnetic fields. Various effects have been observed but are not fully understood, so far: This includes the behaviour of secondary electrons during the trapping process as well as the question whether the primary electron energy remains well defined until the end of ion confinement time.

At higher electron energy an unsatisfactory high power dissipation at the electron collector can probably be avoided by deceleration. This, in turn, can give rise to primary electron oscillations with some still not understood consequences. Further studies are also needed to find out the relation between ion ejection time and emittance. Finally, there is also interest in production of ions from solid or agressive species which can probably be injected into the EBIS as ion beams.

Despite all these problems, according to present knowledge the EBIS concept is the most advanced way to produce highly charged and even fully stripped heavy ions. Highest ion charge states reached so far were produced with EBIS and further progress in regard to both ion charge states and achievable ion yields will be made within the next years.

The EBIS is especially well suited for short pulsed operation with a duty cycle depending on the desired ion species and charge states. However, also c.w.-operation is possibly, but in this case the CSD for higher ion charge states will decrease accordingly.

For further detailed information on the technical questions connected to EBIS development the reader is referred to the proceedings of specialized workshops, cf. GR 8.

6. Production of Multiply Charged Ions by Energetic Collisions of Ions with Atoms or Solids

6.1. MCI-Production by Stripping

Fast atomic particles passing through thin foils or gas cells will usually loose some of their electrons if their initial charge state is well below the respective "equilibrium charge state" for a thick target. The equilibrium charge state may be estimated from Bohr's criterion (cf. Betz 1972):

$$\bar{q} \simeq z^{1/3} \ (v/a.u.) \qquad (27)$$

v   particle velocity, 1 a.u. = 2.2 x 10^8 cm/s

A review on various aspects of the stripping process as well as many data was given recently by Alton (1981 a). The main field of application of stripping is in heavy ion accelerator technology and beam foil spectroscopy (i.e. the study of light emitted from highly charged ions which are produced and excited in passing a foil).

However, stripping was also applied to produce highly charged low-Z ions by using two tandem accelerators and particle deceleration after stripping (Bayfield et al. 1980, Saylor et al. 1981). Because of rather costly equipment and considerable energy spread resulting from the stripping process, this method will probably not be useful for the low energy region.

6.2. Production of Multiply Charged Recoil Ions

In collisions between very fast multiply charged
projectile atoms with a gas target inner shell vacancies
are produced, which together with the action of the
rapidly changing electric field seen by the atoms will
result in efficient electron removal from the latter.
Such processes have been shown to work well at impact
parameters of several a_o (1 a_o = 0.53 x 10^{-8} cm).
Because of this the energy imparted onto the nucleus
of the collision partner by the projectile remains very
small, i.e. in the order of a few eV only. The resulting
very slow "recoil ions" will be scattered under almost 90°
to the primary ion beam direction.
Such impact of fast MCI on atoms was found to result in
efficient ionization by Richard et al. (1969). Furthermore,
Mowat et al. (1973) found that the ionization efficiency
increased with the projectile charge state.
Cross sections for recoil ion production for impact of
25 - 45 MeV Cl^{q+} (q = 6 - 13) on He, Ne and Ar were
measured by Cocke 1979. For e.g. impact of 34 MeV Cl^{13+}
on Ne, the Ne^{8+} was produced with a cross section of
ca. 10^{-17} cm^2. The total ionization cross sections for
similar processes in many collision systems were measured
by Schlachter et al. (1981). Still more recently, Schlachter
et al. (1982) obtained cross sections for recoil ion
production by 330 MeV U^{44+} ions. Some results of these
studies are given in table 7.

Table 7: Cross sections for recoil ion production
by 1.4 MeV/amu U^{44+} impact on rare gases,
from Schlachter et al. (1982)

recoil ion species	Ne^{8+}	Ar^{10+}	Kr^{12+}	Xe^{18+}
$Q/10^{-16}$ cm^2	1	1	1.2	ca. 1.2

Apart from emission of x-rays or Auger electrons by
excited recoil ions, one can also study inelastic
collisions between already produced recoil ions and
their surrounding gas atoms (cf. Mann et al. 1981,
Beyer et al. 1982).
Going one step further, the recoil ions can be extracted
from their place of birth and made into a beam of highly
charged ions. Using this technique, Cocke et al. (1981)
investigated total electron capture from various noble

gas atoms into slow multiply charged argon ions.
Such recoil ions can also be introduced into suitably
shaped ion traps for studies on atomic interactions at
very low kinetic energies. This was done e.g. by Vane et
al. (1981), who investigated collisions between Ne atoms
and fully stripped neon ions. Experiments of similar type
were reviewed recently by Sellin (1982).

Finally, we mention that recoil ion production by impact
of much slower MCI on gas targets is also quite efficient,
cf. Groh et al. (1981).

Because beams of fast highly charged ions are now available
at various heavy ion accelerator laboratories, recoil ion
production technique is an interesting new MCIS which is
especially suited for atomic collision experiments at
low impact energy.

7. Techniques for Measurement of Metastable Ion Beam Fractions

In experiments with MCI quite often the internal energy of the latter can play a decisive role. This is the case e.g. for electron impact excitation or ionization of MCI, or for electron transfer processes in atomic collisions.

In such cases the metastable fractions of ion beams become of interest. In this context we use the term "metastable" for all excited MCI whose mean lifetime exceeds the time interval between ion production and the collision processes of interest. Such times are usually not longer than 10^{-4} s.

In all plasma-MCIS as well as the EBIS production of long-lived excited MCI cannot be excluded if existing at all for the respective ion species. Therefore, clean experimental conditions require appropriate means for at least rough determination of metastable ion beam fractions.

Various techniques for metastable ion detection have been devised which, however, are either applicable to specific cases only or do not allow quantitative measurements. Moreover, most of these techniques have been used for singly charged ions only. This applies for quenching by electric or magnetic fields, selective electron impact ionization (Latypov et al. 1964), conversion to metastable atoms with subsequent surface detection (Kadota and Kaneko 1975), drift tube studies based on different mobilities (see e.g. Glosik et al. 1978), preferential stripping of metastable ions (Seim et al. 1981) and Laser fluorescence studies (see e.g. Rosner et al. 1976).

The most versatile technique is ion beam attenuation in a gas cell (Turner et al. 1968, Vujovic et al. 1972):

For mutually independent attenuation of ground state- and metastable ions in a suitable target gas (respective target thickness denoted by Π) and if all reaction products are immediately removed, the ion current attenuation (α, defined as ratio of ion currents measured with Π and $\Pi = 0$, respectively) will be given by

$$\alpha = (1-f)\exp - Q_g \Pi + f \exp - Q_m \Pi \qquad (28)$$

f   metastable ion beam fraction (we assume one metastable
            component only)

Q_g, Q_m cross sections denoting all processes causing ion
      beam attenuation, for ground state- and metastable
      ions, respectively.

Q_g and Q_m have to be sufficiently different to permit the evaluation of f from the attenuation characteristics, cf. fig. 23.

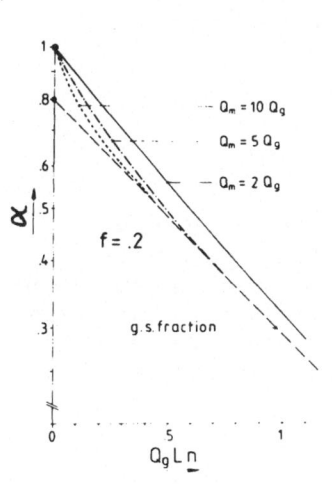

Fig. 23:
Syntesized attenuation curves for a two-component ion beam with f = 0.2 and assumption of various ratios Q_m/Q_g.

With increasing target thickness $\Pi \equiv Q$ L n
(n   target atom number
      density
 L   target cell length)
the less attenuated component remains. If Q_g and Q_m are not very different, large target thicknesses are needed to extrapolate f-values and under such conditions the attenuation method is unreliable

For multiply charged ions the ion beam attenuation is always dominated by single electron capture, thus causing production of ions with charge state q-1. If the attenuation products are not removed (cf. fig. 24), the attenuation ratio is given by:

$$\alpha = (1-f)(1 - \frac{q-1}{q})\exp - Q_g\Pi + f(1 - \frac{q-1}{q})\exp - Q_m\Pi + $$

$$+ \frac{q-1}{q} \qquad (29)$$

Toward higher q, however, single electron capture cross sections for ground state- and metastable ions become more and more similar, cf. fig. 25. In such cases the "Optical Attenuation Method - OAM" recently introduced by Matsumoto et al. (1982) will be more successful: Under the basic assumption, that for the predominant single electron capture channels the primary ion configuration remains unchanged, the secondary ion states formed by the capture process are essentially different.

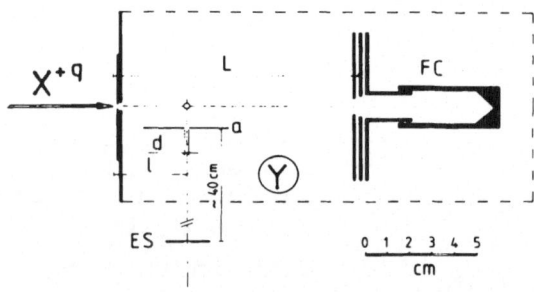

Fig. 24: Simple ion beam attenuation geometry
without removal of secondary products.
FC Farady cup for attenuation measurements
a, ES aperture and spectrometer entrance
slit respectively, for OAM measurements (from Brazuk and Winter 1982)

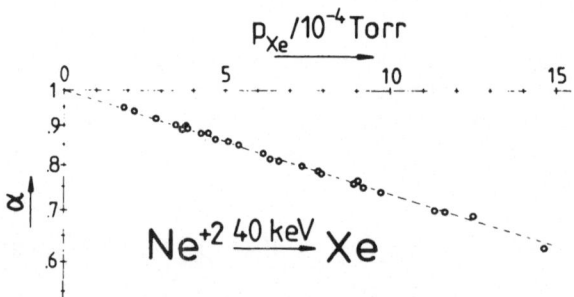

Fig. 25: Ion beam attenuation characteristic obtained with experimental geometry as
shown in fig. 24, for Ne^{2+} - Xe collisions
at 40 keV. The straight line shows that
Q_g and Q_m are not very different, cf.
fig. 23 (from Brazuk and Winter
1982).

Since for single electron capture into MCI mainly excited
final states are populated, the resulting radiation will
be typical for the respective primary ion state. From the
decrease of characteristic emission line intensities
with increasing ion beam attenuation the respective metastable fraction can be deduced. The OAM was recently also
used in the vuv spectral region, cf. Brazuk and Winter
(1982).

Another method for metastable ion detection was devised
by Hagstrum (1956, 1960): Potential emission of electrons
caused by slow ion impact on well defined solid surfaces
depends on the internal energy of the ions. Therefore,
the secondary emission process will show whether ground
state- or metastable ions have been involved, cf. fig. 26.

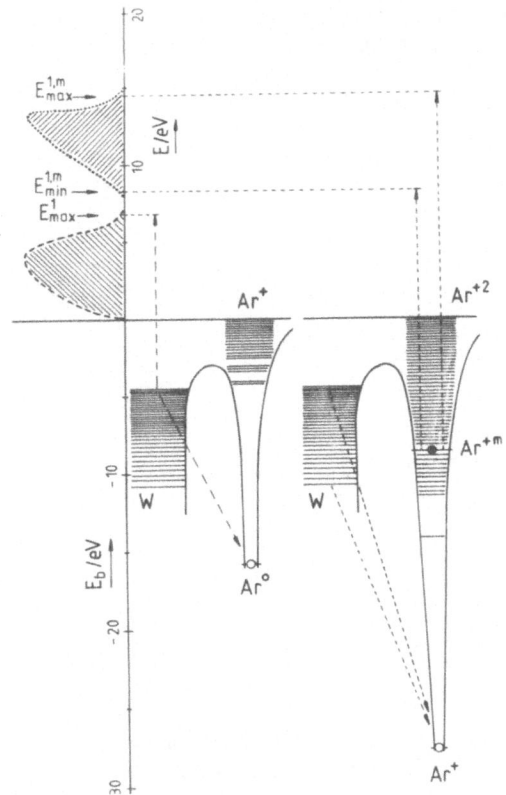

Fig. 26:
Various processes guiding
potential emission due
to impact of slow ions
(Ar$^+$ ground state argon
        ion
 Ar^{+m} metastable argon ion)
on clean tungsten.
Ar$^+$ is directly neutralized by Auger neutrali-
zation, giving rise to
the low electron energy
distribution.
Ar^{+m} at first is deexcited
by Auger deexcitation,
causing the high energy
electrons, and subsequently
is neutralized in the same
way as Ar$^+$

Varga and Winter (1978) and Varga et al. (1981) have
applied Hagstrum's technique for determination of metastable ion beam fractions by measuring the secondary
ion yields with and without metastable admixtures
in beams of singly and doubly charged noble gas ions.

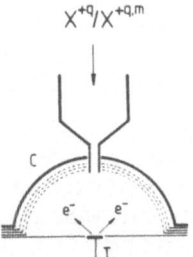

Fig. 27: Experimental setup used
by Hofer et al. (1982)
for INS measurements.
The ejected electron
energy distribution was
obtained by means of
three retarding grids in
front of the electron
collector C.

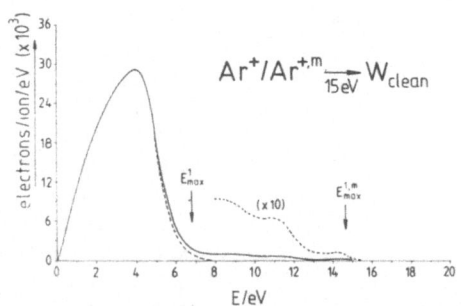

Fig. 28: Typical example for measured ejected electron
energy distribution for impact of ground
state (without high energy tail) and mixed
state Ar^+ on clean tungsten. The example
belongs to a metastable fraction of ca. 4 %
(Hofer et al. 1982).

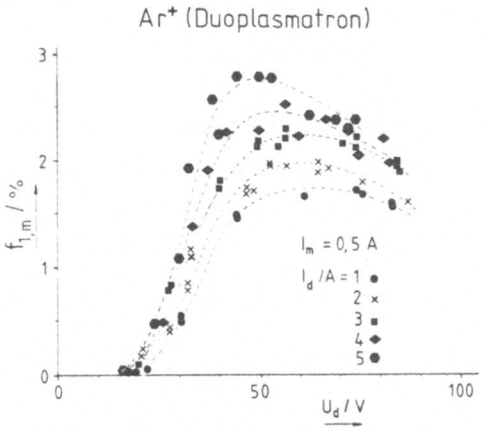

Fig. 29: Metastable ion beam fractions for
         a DUOPLASMATRON ion source run with
         argon, from Hofer et al. (1982).
         U_d, I_d discharge voltage and - current,
                respectively
         I_m magnetic field coil current

More recently, "Ion Neutralization Spectroscopy - INS" was
applied by Hofer et al. (1982) to measure metastable ion
beam fractions for different ion sources, cf. figs. 27,
28 and 29. The measurements consisted in determination
of ejected electron energy distributions for impact
of decelerated ions on clean tungsten.

The INS-technique can be used for all metastable ions
with excitation energies above ca. 10 eV, irrespective
of the ion charge state.

8. Technical Aspects of Multiply Charged Ion Sources

As commonly found in reality, there is a long way from an ingenious concept of MCI-production toward a properly working MCIS. Usually one has to consider many aspects of ion source technology, e.g.:

- Ion source construction materials for optimized design
- Transport of feeding material into the ion source and processing therein
- Type of feeding material for desired ion species
- Ion source conditioning after recharge or repair

We refer to a quite exhaustive treatment on these questions by Freeman and Sidenius (1973), and also to Alton (1981 b), who described apart from ion source technology various types of ion sources and technical problems associated with the ion extraction.

In addition to this, some special points should be mentioned in connection with MCIS-operation:

- The extensive power dissipation on critical parts inside MCIS requires both careful design and special conditioning before full power load is exerted.
- The feeding material should be carried into the MCIS as the clean element whenever possible. Additional compounds will always deteriorate the achievable CSD for high ion charge states.
- Production of high ion charge states is favoured by adjustment to lowest possible feeding material number density in the ionization region. Such conditions, on the other hand, may cause extremely critical operating conditions, specifically unstable behaviour, and thus may affect both the achievable CSD and ion beam quality.
- Finally, the high power dissipation can make pulsed operation of the MCIS indispensable. In many cases, however, this will be no disadvantage.

9. Summary and Outlook

In this review we have tried to demonstrate the present status of multiply charged ion source development, which has greatly improved during the last ten years. The classical plasma-MCIS (PIG-type, DUOPLASMATRON etc.) have now been routinely used to such an extent that their ultimate capability is well known. New ion source concepts have been brought to a status of almost routine operation (ECRIS and EBIS). They are now able to deliver appreciable currents of fully stripped low-Z ions and highly charged heavy ions.

The ECRIS is especially suited for c.w.-operation, can be operated without critical handling and has very long lifetime. The use of permanent magnets for the magnetic hexapole fields and of cryogenic techniques for superconducting magnetic field coils has greatly decreased the physical size and electric power consumption of ECRIS. Operation for MCI from solid feeding material is also becoming possible.

The EBIS is the most advanced method for producing highly charged heavy ions but in such cases can be operated with relatively low duty cycle only. Also here the use of cryogenic techniques for magnetic field coils and ultrahigh vacuum generation was responsible for the progress achieved. There are still some problems in connection with production, formation and deceleration of the fast intense electron beams which are needed in powerful EBIS.

Both for ECRIS and EBIS further improvements are under way and the production of fully stripped heavy ions up to Xe^{54+} will be achieved in the near future.

Another interesting method of MCI-production are the highly charged recoil ions resulting from collisions of fast MCI with atoms. This technique is very promising for the study of MCI-atom collisions at low impact energies.

The present status of MCIS-development has led to a new trend in experimental atomic physics, i.e. the test of quantum-electrodynamics with highly charged ions. In astrophysics and thermonuclear fusion research there is need for various data on collision processes with MCI, which now can be obtained. With the MCIS-configurations which are available at present, many collision experiments can be made more detailed than some years ago. This applies especially to state-selective investigations in atomic collisions with MCI, being of interest also for theory. All these very interesting new possibilities also justify further efforts in developing and improving the known techniques of multiply charged ion production.

ACKNOWLEDGEMENTS

The author is indebted to many of his colleagues for useful comments and informations in context with this work.
In particular, he thanks Drs. E. Donets, R. Geller and
B. Wolf for interesting discussions on work not yet
published.

GENERAL REFERENCES

GR 1 Proc. 1. Int. Conf. on Ion Sources, June 18-20,1969,
      I.N.S.T.N. Saclay/France, ed. A. Septier et al.

GR 2 Proc. Int. Conf. on Multiply Charged Ion Sources and
      Accelerating Systems, October 25-28, 1971, Gatlinburg,
      TN/USA, published in IEEE Trans. Nucl. Sci. NS-19/2 (1972)

GR 3 Proc. 2. Int. Conf. on Ion Sources, September 11-15,
      1972, Wien/Austria, ed. F. Viehböck et al.

GR 4 Proc. 2. Symposium on Ion Sources and Formation of
      Ion Beams, October 22-25,1974, Berkeley,CA/USA,
      ed. C.P. Pezzotti

GR 5 Proc. Int. Conf. on Heavy Ion Sources, October 27-30,
      1975, Gatlinburg,TN/USA, published in IEEE Trans.
      Nucl.Sci. NS-23/2(1976)

GR 6 Proc. Symposium on Ion Sources and Application
      Technology, February 14-15,1977, Kyoto/Japan,
      ed. T. Takagi

GR 7 3. Int. Workshop on ECRIS and Related Topics,
      December 8,1980, GSI Darmstadt/FGR, ed. B.H. Wolf,
      Report GSI 81-1
      4. Int. Workshop on ECRIS and Related Topics,
      January 14-15,1982, C.E.N. Grenoble/France, ed.
      R. Geller (in preparation)

GR 8 Workshop on EBIS and Related Topics, June 15-16,1977,
      GSI Darmstadt/FGR, ed. B.H. Wolf and H. Klein,
      Report GSI-P-3-77
      II EBIS Workshop, May 12-15,1981, Saclay-Orsay/France,
      ed. J. Arianer and M. Olivier

REFERENCES

Alton G D (1981 a) IEEE Trans.Nucl.Sci. NS-28/2,1043
Alton G D (1981 b) Nucl.Instr.Meth. 189,15
Amboss·K (1981) loc.cit. GR 8, p. 59
Anderson C E and Ehlers K W (1956) Rev.Sci.Instr. 27,809
Apard P, Bliman S, Geller R, Jacquot B and Jacquot C (1972)
      C.R.Acad.Sc. Paris 275,621
Anan'in O B, Baldin A M, Beznogikh Yu D, Bykovskii Yu A,
      Zinov'ev L P, Kozyrev Yu P, Monchinskii V A and
      Semenyushkin I N (1974) ZhETF Piz. Red. 19,19
Arianer J, Baron E, Brient M, Cabrespine A, Liebe A,
      Serafini A and Ton That T (1975) Nucl.Instr.Meth.
      124,157
Arianer J and Goldstein C (1976) loc.cit. GR 5, p. 979
Arianer J, Cabrespine A, Goldstein C and Deschamps G
      (1979) IEEE Trans.Nucl.Sci. NS-26/3,3713
Arianer J (1981) IEEE Trans.Nucl.Sci. NS-28/2, 1018
Arianer J and Geller R (1981) Ann. Rev.Nucl.Part.Sci. 31,19
Arianer J, Cabrespine A and Goldstein C (1982) Nucl.Instr.
      Meth. 193, 401

Baker F A and Hasted J B (1966) Phil.Trans.Roy.Soc. 261,33
Bayfield J E, Gardner L D, Gulkok Y Z, Saylor T K and
      Sharma S D (1980) Rev.Sci.Instr. 40,869
Bennett J R J (1972) loc.cit. GR 2, p. 48
Bechtold V, Chan-Tung N, Dousson S, Geller R, Jacquot B
      and Jongen Y (1980 a) Nucl.Instr.Meth. 178,305
Bechtold V, Friedrich L and Schweickert H (1980 b),
      loc.cit. GR 7, p. 23
Bechtold V et al. (1982) loc.cit. GR 7
Becker R (1981) loc.cit. GR 8, p. 164
Becker R, Kleinod M and Klein H (1981) loc.cit. GR 8, p. 48
Betz H D (1972) Rev.Mod.Phys. 44,465
Beyer H F, Mann R and Folkmann F (1982) J.Phys.B:At.Mol.
      Phys. 15,1083
Bohm D (1949) in "The Characteristics of Electrical Dis-
      charges in Magnetic Fields", ed. A. Guthrie and
      R.K. Wakerling, chapter 9, McGraw-Hill, New York
Bliman S, Dousson S, Fremion L and Geller R (1978) Nucl.
      Instr.Meth. 148,213
Bourg F, Geller R, Jacquot B, Lamy D, Pontonnier M and
      Rocco J C (1982) Nucl.Instr.Meth. 196,325
Braams C M, Zieske N and Koefoed L (1965) Rev.Sci.Instr.
      36,1411
Brazuk A and Winter H (1982) J.Phys.B:At.Mol.Phys. 15
      (in print)
Brewer (1967) in "Focusing of Charged Particles", vol. 2,
      ed. A. Septier, chapter 3.3, Academic Press, New York

Briand P, Chan-Tung N, Geller R and Jacquot B (1977) loc.
        cit. GR 8, p. 42
Bolotin L I et al. (1961) Prib.Tekh.Eksp. 6,86

Cocke C L (1979) Phys.Rev. A 20,749
Cocke C L, DuBois R, Gray T J, Justiniano E and Can C
        (1981) Phys.Rev.Letters 46,1671
Crandall D H (1982) these proceedings
Clausnitzer G, Klinger H, Müller A and Salzborn E (1975)
        Nucl.Instr.Meth. 128,1
Consolino J, Geller R and Leroy C (1969) loc.cit. GR 1,
        p. 537

Davis R C, Jernigan T C, Morgan O B, Stewart L D and
        Stirling W L (1975) Rev.Sci.Instr. 46,576
Demirkhanov R A et al. (1962) Proc. 1 Int. Conf. High
        Energy Accelerators, Brookhaven/USA, Report
        BNL 767, p. 224
Donets E D, Ilyushchenko V I and Alpert V A (1968) pre-
        print JINR R7-4124, Dubna
Donets E D, Ilyushchenko V I and Alpert V A (1979) loc.
        cit. GR ;, p. 635
Donets E D and Ilyushchenko V I (1974) preprint JINR-P7-
        8310, Dubna
Donets E D and Pikin A J (1974) preprint JINR-P7-7999,
        Dubna
Donets E D (1976) loc.cit. GR 5, p. 897
Donets E D and Ovsyannikov V P (1977) preprint JINR P 7-
        10438, Dubna
Donets E D and Ovsyannikov V P (1980) preprint JINR P 7
        80-515, Dubna
Donets E D (1982) Proc. Symposium on Production and Physics
        of Highly Charged Ions, Stockholm, June 1-5,1980
        (to be published in Physica Scripta)
Faure C, Perez A, Tonon G, Aveneau B and Parisot D (1971)
        Physics Letters 34 A,313
Freeman J H and Sidenius G (1973) Nucl.Instr.Meth. 107,477
Fröhlich H (1959) Nukleonik 1,183

Gavin B F (1976) loc.cit. GR 5, p. 1008
Geller R (1976) loc.cit. GR 5, p. 904
Geller R (1977) Proc. 13. Int. Conf. on Phenomena in
        Ionized Gases, invited papers, p. 103, September
        12-17, Berlin/FGR
Geller R and Jacquot B (1981) Nucl.Instr.Meth. 184,293
Geller R (1982) loc.cit. Donets (1982)
Glosik J, Rakshit A B, Twiddy N D, Adams N G and Smith D
        (1978) J.Phys.B:At.Mol.Phys. 11,3365
Goldhar J, Mariella R and Javan A (1976) App.Phys.Letters
        29,96

Green T S (1974) Rep.Progr.Phys. 37,1257
Groh W, Müller A, Achenbach C, Schlachter A S and
    Salzborn E (1981) Physics Letters 85 A,77

Hagstrum H D (1956) Phys.Rev. 104,309
Hagstrum H D (1960) J.Appl.Phys. 31,897
Hofer W, Vanek W, Varga P and Winter H (1982) to be
    published

Illgen J (1968) Thesis, University of Heidelberg
Illgen J, Kirchner R and Schulte in den Bäumen J (1972)
    loc.cit. GR 2, p. 35
Imamura H, Kaneko Y, Iwai T, Ohtani S, Okuno K, Kobayashi
    N, Tsurubuchi S, Kimura M and Tawara H (1981)
    Nucl.Instr.Meth. 188,233

Jacoby W and Müller M (1975) report GSI-PB-2-75, p. 12,
    GSI Darmstadt
Jones R J and Zucker A (1954) Rev.Sci.Instr. 25,562
Jongen Y, Pirart C and Ryckewaert G (1980) loc.cit GR 7,
    p. 1
Jongen Y, Pirart C and Ryckewaert G (1981) IEEE Trans.
    Nucl.Sci. NS-28/3,2696
Jongen et al. (1982) loc.cit. GR 7
Kadota K and Kaneko Y (1975) J.Phys.Soc. Japan 38,524
Kaneko Y (1979) Proc. Nagoya Seminar on Atomic Processes
    in Fusion Plasmas, September 5-7,1979, ed. Y. Itikawa
    and T. Kato, p. 28
Keller R and Müller M (1976) loc.cit. GR 5, p. 1049
Keller R and Müller M (1977) loc.cit. GR 8, p. 7
Keller R and Müller M (1978) in "Low Energy Ion Beams",
    ed. K.G. Stephens, I.H. Wilson and J.L. Moruzzi,
    p. 40, The Institute of Physics, Bristol/UK
Keller R and Winter H (1976) Particle Accelerators 7,77
Kostroun V O, Ghanbari E, Beebe E N and Janson S W (1981)
    loc.cit. GR 8
Krupp H (1968) Unilac-Bericht Nr. 1-68, GSI Darmstadt/FGR
Krupp H and Winter H (1975) Nucl.Instr.Meth. 127,459
Kutner V B (1981) preprint JINR P-9-81-139, Dubna

Latypov Z Z, Kupriyanov S E and Tunitskii N N (1964) Zh.
    Eksp.Teor.Fiz. 46,833

Makov B N (1966) report IAE-1051, Kurchatov Institute for
    Atomic Energy, Moscow/USSR
Mann R, Folkmann F and Beyer H F (1981) J.Phys.B:At.Mol.
    Phys. 14,1161
Matsumoto A, Ohtani S and Iwai T (1982) J.Phys.B:At.Mol.
    Phys. 15/12
Morgan O B et al. (1967) Rev.Sci.Instr. 38,467
Morozov P M, Makov B N and Joffe M S (1957) Atomnaya
    Energiya 3,272

Mowat J R, Sellin I A, Pegg D J, Peterson R S, Brown M D
    and Macdonald J R (1973) Phys.Rev.Letters 30,1289
Müller M and Jacoby W (1977) loc.cit. GR 7, p. 1
Müller A and Salzborn E (1979) Nucl.Instr.Meth. 164,607

Pasyuk A S and Tret'yakov Yu P(1972 a) loc.cit. GR 3, p. 512
Pasyuk A S and Tret'yakov Yu P(1972 b) report JINR-P7-6668,
    Dubna
Peacock N J (1982) loc.cit. Donets (1982)
Phaneuf R A (1981) IEEE Trans.Nucl.Sci. NS-28/2,1182
Pigarov Yu D and Morozov P M (1961) Sov.Phys.-Techn.Phys.
    6,336
Plumlee R H (1957) Rev.Sci.Instr. 28,830

Redhead P A (1967) Can.J.Phys. 45,1791
Rhee M J (1981) IEEE Trans.Nucl.Sci. NS-28/3,2663
Richard P, Morgan I L, Furuta T and Burch D (1969) Phys.
    Rev. Letters 23,1009
Rosner S D, Gaily T D and Holt R A (1976) J.Phys.B:At.
    Mol.Phys. 9.L489

Saylor T K, Bayfield J E, Gardner L D, Gulkok Y Z and
    Sharma S D (1981) IEEE Trans.Nucl.Sci. NS-28/2,1024
Schlachter A S, Berkner K H, Graham W G, Pyle R V, Schneider
    P J, Stalder K R, Stearns J W, Tanis J A and Olson R E
    (1981) Phys.Rev. A 23,2331
Schlachter A S, Groh W, Müller A, Beyer H F, Mann R and
    Olson R E (1982) to be published
Schulte H, Jacoby W and Wolf B H (1976) loc.cit. GR 5, p.
    1042
Schulte in den Bäumen J, Illgen J and Wolf B H (1972)
    loc.cit. GR 3, p. 549
Sellin I A (1982) loc.cit. Donets (1982)
Seim W, Müller A and Salzborn E (1981) Z.Phys. A 301,11
Septier A (1967) "Focusing of Charged Particles", vol. 2,
    chapter 3.4, Academic press, New York

Takagi S, Ohtani S, Kadota K and Fujita J (1982) reports
    IPPJ-564 and 565, Institute of Plasma Physics, Nagoya/
    Japan
Tret'yakov Yu P et al. (1973) report JINR-P7-7092, Dubna
Turner B R, Rutherford J A and Compton D M J (1968) J.
    Chem.Phys. 48,1602

Valyi L (1970) Nucl.Instr.Meth. 79,315
Vane C R, Prior M H and Marrus R (1981) Phys.Rev. Letters
    46,107
Varga P and Winter H (1978) Phys.Rev. A 18,2453
Varga P, Hofer W and Winter H (1981) J.Phys.B:At.Mol.Phys.
    14,1341
Vella M C (1981) Nucl.Instr.Meth. 187,313

Von Ardenne M (1956) "Tabellen der Elektronenphysik, Ionen-
    physik und Übermikroskopie", vol 1, part C, VEB Verlag
    der Wissenschaften, Berlin
Vujovic M, Matic M, Cobic B and Gordeev Yu S (1972) J.Phys.
    B:At.Mol.Phys. $\underline{5}$,2085

Winter H (1974) in Proc. Physics of Ionized Gases 1974,
    Sept. 16-21,1974, Rovinj/Yugoslavia, p. 79
Winter H and Wolf B H (1974 a) Plasma Physics $\underline{16}$,791
Winter H and Wolf B H (1974 b) loc.cit. GR 4, p. V-1
Winter H (1978) in "Experimental Methods in Heavy Ion
    Physics", ed. K. Bethge, Lecture Notes in Physics,
    vol. 83, Springer, Berlin
Winter H and Wolf B H (1975) Nucl.Instr.Meth. $\underline{127}$,445
Wolf B H (1973) Thesis, University of Heidelberg/FGR
Wolf B H and Keller R (1981) Proc. U.M.I.S.T., June 22-26,
    1981, Manchester/UK, to be published by Pergamon press

Zwally H J, Koopman D W and Wilkerson T D (1969) Rev.Sci.
    Instr. $\underline{40}$,1492

CHANNELING OF IONS AND ELECTRONS IN CRYSTALS AND

RADIATION FROM CHANNELED ELECTRONS

J.U. Andersen

Institute of Physics, University of Aarhus

DK-8000 Aarhus C, Denmark

INTRODUCTION

Channeling of ions in crystals was discovered in the early six-
ties in measurements and calculations of ion ranges. It was found
that for incidence directions parallel to close-packed rows in the
crystal, the ions moved to much larger depths before coming to rest.
A comprehensive theoretical description of channeling was given in
1965 by Lindhard[1], and most of the basic phenomena had been investigated in detail by 1970[2]. Since then, the main emphasis in channeling studies has been on application in solid-state physics[3], in nuclear physics[4], and in high-energy physics.

At an early stage, also channeling of electrons was studied,
and later the connection to the diffraction phenomena used in electron microscopy was established[5]. Because the electron mass is much
smaller than that of positive ions, the classical description of
ion channeling must be replaced by a quantum theory. In recent years,
radiation originating in transitions between the quantum levels of
channeled electrons has been observed[6]. The motion of the electrons
may be described as free-particle motion (plane waves) parallel to
the rows or planes, and transverse motion in bound states. In a classical picture, the electrons move in spirals around an atomic row
or oscillate around an atomic plane. The channeling radiation may
therefore be described as characteristic radiation from two- or
one-dimensional atom-like systems, moving with a velocity parallel
to a crystal axis or plane, which for electrons of a few MeV is close
to the velocity of light. The lines of characteristic radiation are
then in the keV region and may be observed with a solid-state detector. As for atoms, the study of such line spectra gives very detailed

information about the electron states[7,8], and it is now being developed into a kind of channeling spectroscopy.

ION CHANNELING

Applicability of Classical Mechanics

It was shown by Lindhard in 1965[1] that in the description of
channeling of particles heavy compared to the electron, $M_1 \gg m_0$,
classical mechanics may be applied. This is not obvious since for
single collisions, a classical description breaks down for high
particle velocities. Let us briefly recall the argument of Bohr for
scattering in a Coulomb potential. The scattering angle ϑ is given
by

$$\text{tg}(\vartheta/2) = b_0/2b ,\tag{1}$$

where b is the impact parameter and b_0 is the collision diameter,

$$b_0 = \frac{Z_1 Z_2 e^2}{\frac{1}{2} M_1 v^2} .\tag{2}$$

Here $Z_1 e$ and $Z_2 e$ are the charges of the projectile and the scattering centre, and $\frac{1}{2} M_1 v^2$ is the energy of the projectile. An attempt to
define a wave packet, which follows the classical trajectory with
average impact parameter b and scattering angle ϑ shows that this is
possible provided that

$$\kappa \equiv b_0/\lambdabar = \frac{2Z_1 Z_2 e^2}{\hbar v} > 1 ,\tag{3}$$

where $\lambdabar$ is the de Broglie wavelength, $\lambdabar = \hbar/M_1 v$. Here and above, M_1
should be replaced by the reduced mass if the target is not very
heavy compared to the projectile. The condition results from the
requirement that the spread $\Delta\vartheta$ in direction of motion of the wave
packet after the collision should be small compared to the scattering angle ϑ. For a wave packet of transverse extension $2\Delta b$, there
are two contributions to this spread, one from the uncertainty re-
lation, $\Delta(p\vartheta)\Delta b \gtrsim \hbar/2$, which gives $\Delta\vartheta_1 \simeq \lambdabar/2\Delta b$, and the other from
the scattering law in Eq. (1), which for small angles ϑ leads to a
spread in the scattering angle $\Delta\vartheta_2 \simeq (\Delta b/b)\vartheta$. Optimizing b to give
the minimum value of $\Delta\vartheta^2 \simeq \Delta\vartheta_1^2 + \Delta\vartheta_2^2$, one then finds that $\Delta\vartheta < \vartheta$ implies the inequality (3).

For typical projectiles (protons, α particles), a classical
description of the deflection in single collisions is then not permitted at high velocities. (Note that in the limit $v \to c$, we ob-
tain $\kappa \to 2Z_1 Z_2/137$ since $\alpha = e^2/\hbar c \simeq 1/137$.) The situation is even
worse than it appears from Eq. (3) since the Coulomb field from

nuclear charge Z_2e is screened by atomic electrons, and this leads to a stronger requirement than given by Eq. (3)[1].

How can the use of classical mechanics then be justified in the description of channeling? The basic channeling mechanism is illustrated in Fig.1. A particle is approaching a row of atoms at a small angle ψ (the angle should be much smaller than shown in the figure). Each time the particle passes an atom, it receives a small kick (repulsive), and gradually the combined efforts of the atoms turn the trajectory away from the row. In an analysis of this correlated scattering with a wave packet representing the projectile, the effective strength of the potential, represented by Z_1e in Eq. (3), is increased by the number of atoms contributing to the deflection. Since the condition for the channeling phenomenon to take place is that many atoms contribute to govern the particle trajectory, as shown in Fig. 1, one may then always use a classical orbital picture of the projectile motion.

String Effect

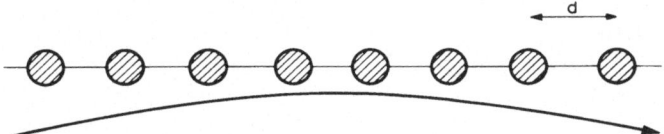

Fig. 1. Governing of ion path by correlated collisions with atoms in the string.

This result is of course of great importance in the theory of channeling since classical mechanics is much simpler than quantum mechanics, but it also contains a general lesson regarding the application of Bohr's criterion (3). The violation of this inequality does not automatically preclude the application of a classical picture to processes involving a collision between a projectile and an atom, as is often assumed. It is only the relation between the impact parameter b and the scattering angle ϑ (in a single collision), which becomes indefinite. Another example of a situation, where a classical impact-parameter description may still be used, is a collision leading to a nuclear excitation, in which the impact parameter must be zero on the atomic scale.

Continuum Potentials and Transverse Energy

Let us return to the governing of trajectories by correlated collisions, shown in Fig. 1. Since the deflection angles are very small, they may be evaluated in the momentum approximation from the momentum transfer,

$$\overline{\Delta p} \simeq -\int_{-\infty}^{\infty} \frac{dz}{v} \overline{\nabla}_r V_a(\sqrt{z^2 + r^2}) \ . \tag{4}$$

Here, $p = M_1 \overline{v}$ is the momentum, and the momentum transfer $\overline{\Delta p}$ is nearly perpendicular to the direction of motion, which almost coincides with the z direction. The projectile-atom interaction potential is $V_a(R)$, and the vector $\overline{R}$ is decomposed into $\overline{R} = (\overline{r}, z)$, where $\overline{r}$ is the projectile distance from the row when it passes the atom. The gradient of this potential with respect to $\overline{r}$ then gives the force perpendicular to the z direction, which is responsible for the deflection.

If $\overline{r}$ varies little over a distance d equal to the spacing of atoms in the row, we may replace the expression (4) by an average over a time interval $\tau = d/v$,

$$\overline{\Delta p} \simeq -\int_{-d/2v}^{d/2v} dt \ \overline{\nabla}_r \ U(\overline{r}(t)) \ , \tag{5}$$

with

$$U(\overline{r}) \equiv \frac{1}{d} \int_{-\infty}^{\infty} dz \ V_a(\sqrt{z^2 + r^2}) \ . \tag{6}$$

Here, $t = z/v$ is a time parameter for the motion, and the integral in Eq. (5) may be interpreted as the total momentum transfer during the time τ due to the force in the two-dimensional potential $U(\overline{r})$. As seen from Eq. (6), this potential is obtained by smearing the atomic changes in the z direction. It is therefore usually called the 'continuum string potential'. Since close to the string we can ignore the interaction with other atoms in the crystal than those sitting on the row, the potential also corresponds to an average of the crystal potential in the z direction.

Equations (5) and (6) represent the basic approximation in channeling theory, which usually is denoted 'the continuum approximation'. The discrete deflections in collisions with atoms are

replaced by the action of a continuous force from the crystal potential averaged in the axial direction. A similar approximation may be
introduced for planar channeling, and the continum potential from
atoms in a single atomic plane is

$$V(x) \quad = \quad Nd_p \int dydz \; V_a(\sqrt{x^2+y^2+z^2}) \; , \tag{7}$$

where Nd_p is the density of atoms in the plane, N being the atomic
density in the crystal and d_p the planar spacing. If the interaction
with crystal atoms in different rows or planes, respectively, cannot be ignored, Eqs. (6) and (7) should be replaced by the corresponding averages of the full crystal potential $W(\bar{R})$,

$$U(\bar{r}) \quad = \quad \frac{1}{d} \int_{-d/2}^{d/2} dz W(\bar{r},z) \tag{6'}$$

and

$$V(x) \quad = \quad \frac{1}{A} \int_A dydz W(x,y,z) \; . \tag{7'}$$

In the continuum approximation, there is a complete separation between the longitudinal z motion, which is that of a free particle, and the transverse motion governed by a transverse Hamiltonian, which for the axial case is

$$H(\bar{p}_\perp,\bar{r}) \quad = \quad \frac{p_\perp^2}{2M_1} + U(\bar{r}) \; . \tag{8}$$

The total energy may be separated into longitudinal and transverse
energy,

$$E \quad = \quad E_z + E_\perp \; , \tag{9}$$

which are separately conserved. Of course, there is a gradual degradation of the energy due to stopping, and also the separate conservation of $E_\perp$ is only an approximation, which gradually breaks
down ('dechanneling'). However, the continuum picture with the associated conservation laws may be a good approximation over a considerable depth of penetration.

Dips in Yield

The two-dimensional potential in Eq. (6') is illustrated in
Fig. 2. The potential has sharp peaks at the position of strings in
the transverse xy plane. A plane is included as a finite value of $E_\perp$
to illustrate the restriction in the transverse plane. Particles
with this transverse energy are allowed to move only in the region
where $U(\bar{r}) < E_\perp$, and there will be small areas around string posi-

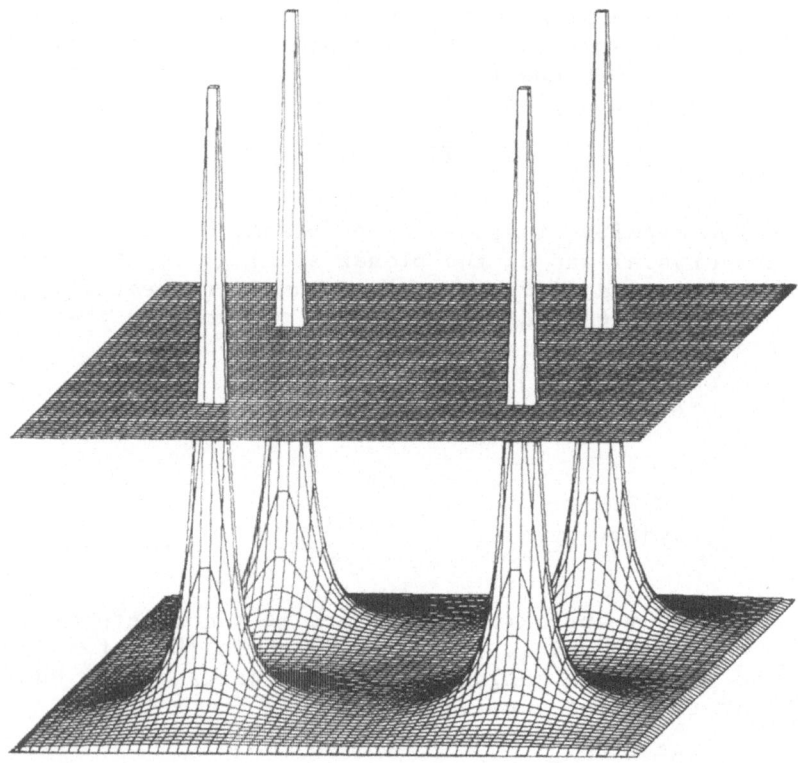

Fig. 2. Axial continuum potential. An additional plane has been in-cluded to show the allowed area in the transverse plane for a par-ticle with fairly high transverse energy.

tions which are forbidden. If the radius of these areas is large compared to vibrational amplitudes ρ of crystal atoms, the particles cannot hit the crystal atoms directly, and hence nuclear reactions or Rutherford backscattering is prevented. This is the most drama-tic consequence of channeling and was by Lindhard, who predicted it, denoted the 'string effect'. A dip in backscattering yield is shown in Fig. 3. For small angles of incidence with the axis, the yield drops by about two orders of magnitude!

From the theory of channeling, one may calculate such dips in detail, but here we shall only give simple estimates of two charac-teristic parameters, the half-width of the dip $\psi_{\frac{1}{2}}$ and the minimum yield χ_{min}, defined as the yield at $\psi = 0$ relative to that for large ψ.

From our discussion, $\psi_{\frac{1}{2}}$ should correspond roughly to the incid-ence angle, at which the transverse energy of the particles is large

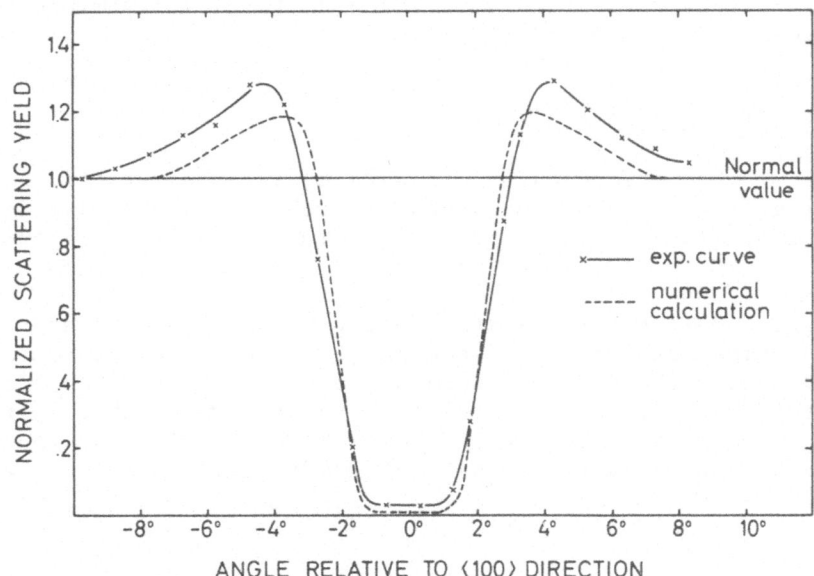

Fig. 3. Dip in backscattering yield for 480-keV protons incident along a <100> direction in W.

enough for the particles to be able to penetrate to a distance $\sim\rho$ from a string. The transverse momentum of particles incident at an angle ψ to the axis is $p_\perp = p\sin\psi \simeq p\psi$, and the transverse energy of such particles will therefore be

$$E_\perp \simeq \frac{(p\psi)^2}{2M_1} = E\psi^2 . \tag{10}$$

If the radius of the forbidden areas around strings is $\sim\rho$, we have the relation

$$E_\perp \simeq U(\rho) . \tag{11}$$

Since ρ is typically much smaller than lattice spacings of atoms ($\rho \simeq 0.1$ Å, $d \simeq 2$-5 Å), we may use the single-string potential in Eq. (6). A very useful Thomas-Fermi-type estimate was given by Lindhard, the so-called 'standard potential',

$$U(r) \simeq \frac{Z_1 Z_2 e^2}{d}\log((Ca/r)^2 + 1) , \tag{12}$$

where Ca is a characteristic screening length, $Ca \simeq 1.5\, a_0 Z_2^{-1/3}$, a_0 being the Bohr radius, $a_0 = 0.53$ Å.

If we insert r = ρ into this formula, the logarithm becomes ∿2, and we obtain from Eqs. (10)-(12)

$$\psi_{\frac{1}{2}} \simeq \psi_1 = (\frac{2Z_1Z_2e^2}{d \cdot E})^{\frac{1}{2}}.$$ (13)

The angle ψ_1 was introduced by Lindhard as a characteristic angle for scaling of channeling phenomena with the parameters Z_1, Z_2, d, and E. A typical magnitude is $\psi_1 \sim 1°$ for α particles of about 1 MeV.

Consider next the yield at zero angle, χ_{min}. According to Eq. (10), the transverse energy of the particles will be (close to) zero, and there should be no yield of backscattering since the particles would be confined to move in the centre of 'channels', i.e., far from the strings (see Fig.2). However, Eq. (10) is an approximation to be applied only for not too small angles. The more correct relation is

$$E_\perp = E\psi^2 + U(\bar{r}) ,$$ (14)

where $\bar{r}$ is the point of entry into the crystal of the particle, which will be uniformly distributed over the transverse plane. For very small angles ψ, it becomes important that a small fraction of the particles hit the surface very close to a string, thereby acquiring a large transverse potential energy. We could also say that the angle ψ relative to the axis is modified for these particles by their first collision with a string, and that the modification is given by the conservation of the quantity in Eq. (14) after entry into the crystal. When the particles move away from the string, the potential energy $U(\bar{r})$ is converted into transverse kinetic energy. If only particles with transverse kinetic energy larger than that given by Eq. (11) contribute to the yield, we may estimate χ_{min} as the fraction of the incident particles, which enter the crystal within a distance ρ from a string,

$$\chi_{min} = Nd\pi\rho^2 ,$$ (15)

where $(Nd)^{-1}$ is the transverse area per string. It should be noted that this formula may be derived in a more precise way within the continuum approximation[1]. On the other hand, corrections to this approximation are important for the magnitude of χ_{min} and lead to an estimated increase of a factor of 2 to 3 (Ref. 2). Still, the reduction of the yield is very large, usually about two orders of magnitude.

As we have seen, thermal vibrations play a very important role in channeling, and we should also mention that the continuum potentials in Eqs. (6) and (7) should be corrected for thermal vibrations of atoms. This introduces a smearing.

$$U_T(\bar{r}) = \int \frac{d^2\vec{r}'}{\pi\rho^2} e^{-r'^2/\rho^2} U(\bar{r}-\bar{r}') ,$$

and   (16)

$$V_T(x) = \int \frac{dx'}{\sqrt{\pi}\rho} e^{-x'^2/\rho^2} V(x-x') ,$$

where we have assumed the distribution of thermal displacements to
be Gaussian. The main result of this smearing is a rounding off at
small distances $r \lesssim \rho$ or $x \lesssim \rho$, and the effect on the parameters
$\psi_{\frac{1}{2}}$ and χ_{min} is small. In fact, in the continuum approximation, the
minimum yield χ_{min} depends only on geometric quantities, as seen
from Eq. (15), and the only feature of the potential used in the
derivation of this formula is that it is repulsive.

Application of Ion Channeling

     The aspect of channeling, which led to its discovery, is the
increase in ion ranges. This also led to the name 'channeling' since
the attention was focused on the open channels in a crystal, bordered
by rows or planes of atoms. For applications, however, the most useful consequence of the governing of ion trajectories by the continuum potentials is the strong reduction of the yield of processes
requiring a close collision with an atom, the 'string effect'[1]. One
example was shown in Fig. 3 for Rutherford backscattering.

     Without doubt, the most important application of the string
effect has been to the analysis of crystals after implantation of
impurities[3]. First, the implantation of ions with a typical energy
of 50 keV introduces damage into the crystal due to recoils of atoms
hit by the incident ions. If the lattice order is destroyed, dips
like that shown in Fig. 3 disappear, and, in general, the increase
of the minimum yield χ_{min} may be used as a measure of the extent of
the disorder. Second, it is important to know where in the lattice
the implanted imputities are located after the implantation and
after subsequent annealing treatments to restore the lattice order.
This can be decided from a measurement of the channeling dips for
interaction between the projectiles and the impurities, which may
be singled out through observation of, e.g., a specific nuclear reaction. If the impurity is replacing a host atom, i.e., is substi-
tutional, the same dip will be observed as for backscattering from
host atoms. If, on the other hand, the impurity atoms occupy inter-
stitial positions, no dip, or at least a different dip, will be observed.

     Another interesting application is the measurement of very
short nuclear lifetimes, which also utilizes the possibility of
distinguishing between lattice sites and positions displaced from

a lattice site by distances $\gtrsim 0.1$ Å. By the crystal-blocking technique, which in principle is a time-of-flight method, one may measure lifetimes down to 10^{-17} s, shorter by orders of magnitude than
may be reached by other direct lifetime measurements. For a discussion of this application, we refer to the review by Gibson[4].

ELECTRON CHANNELING AND RADIATION FROM CHANNELED ELECTRONS

Quantum Theory of Electron Channeling[5]

    The penetration of electrons through crystals is very differ-
ent from that of heavy, positive particles, due partly to their
negative charge, partly to their small mass. Let us first consider
the consequences of the change in sign of the charge. If for a moment we assume that electrons incident at a small angle to a crystal axis are steered by correlated collisions with atoms in atomic
rows, and that the motion then may be pictured as a classical trajectory, then the only difference from channeling of ions will be
the change in sign of the continuum potentials in Eqs. (6)-(7').
We may once again use Fig. 2 as an illustration, but the figure
should be turned upside-down! The allowed region in the transverse
plane for particles with a transverse energy corresponding to the
off-set plane in Fig. 2, now a negative value of $E_\perp$, will then be
the the small circular areas close to strings, which were forbidden
for positive particles. The motion then has a completely different
character, being confined to the vicinity of one atomic string. The
potential there has axial symmetry, and angular momentum with respect to the string will be conserved. The electrons will spiral
around a string. It will be useful in the following to think of the
transverse motion in a frame of reference moving with velocity v_z
in the z direction, i.e., following the uniform z motion of the
electron. In this frame of reference, which we shall denote the
'rest frame', the motion will resemble that of the electron in a
two-dimensional hydrogen atom.

    The question then arises whether classical orbital pictures of
electron motion, which we know are not applicable to a real hydrogen
atom, can be used here. One way of attacking this problem is to
calculate the angular momentum for a spiralling electron. If it is
large compared to h, we would expect a classical description to apply, in analogy to Bohr's correspondence principle for the hydrogen
atom.

    We may for this estimate use the simple potential in Eq. (12)
and assume the electron to spiral around the string with circular
transverse motion at the characteristic screening distance Ca from
the string. The magnitude of the force is then

$$|\overline{\nabla}U(Ca)| = \frac{Z_2 e^2}{Ca \cdot d} \tag{17}$$

Let us for a moment keep the notation M_1 for the particle mass in the Hamiltonian (8) for the transverse motion. If the angular frequency of the motion is ω, and the centripetal force is given by Eq. (17), we obtain

$$M_1 \omega^2 Ca = \frac{Z_2 e^2}{Ca \cdot d} . \tag{18}$$

This relation determines the angular momentum L, which we express in units of $\hbar$,

$$L/\hbar = \frac{M_1 \omega (Ca)^2}{h} = \left[\frac{M_1}{m_0} 2.25 Z_2^{1/3} \frac{a_0}{d}\right]^{\frac{1}{2}} \tag{19}$$

where m_0 is the electron rest mass; the Bohr radius is $a_0 = \hbar^2/m_0 e^2$, and we have inserted the value $Ca = 1.5 a_0 Z_2^{-1/3}$. For typical values of Z_2 and d, the factor multiplying M_1/m_0 is close to unity, and we may simplify Eq. (19) to

$$L/\hbar \simeq (\frac{M_1}{m_0})^{\frac{1}{2}} . \tag{20}$$

We see then that for negative particles very heavy compared to an electron, a classical picture should apply, just as for the channeling of positive ions. (Experiments with negative pions have been performed already!) For electrons, the effective mass in the transverse Hamiltonian in Eq. (8) turns out to be the relativistic mass, $M_1 = \gamma m_0$, where

$$\gamma = (1-\beta^2)^{-\frac{1}{2}} , \quad \beta = v/c . \tag{21}$$

This is straightforward to show from Eq. (5), but we shall not go through the argument. The easiest way to calculate γ is from the total energy,

$$E = E_{acc} + m_0 c^2 = \gamma m_0 c^2 , \tag{22}$$

where E_{acc} is the energy of acceleration. As an example, for $E_{acc} = 4$ MeV, we obtain $\gamma \simeq 9$. Equation (20) then tells us that for electrons of a few MeV, the angular momentum is not very large compared to h, and a quantum description of the motion should be used. On the other hand, in the limit of very large γ, the classical limit is approached.

In a quantum description, we interpret the Hamiltonian in Eq. (5) as an operator with quantized momentum

$$\bar{p}_\perp = -i\hbar\bar{\nabla}_r \ , \tag{23}$$

and the stationary states of channeled motion are given by

$$\psi_n(\bar{R}) = e^{ik_z z} u_n(\bar{r}) \ , \tag{24}$$

where hk_z is the z momentum, and $u_n(\bar{r})$ is the transverse wave function obtained from an equation of type of a stationary Schrödinger equation,

$$\left\{\frac{p_\perp^2}{2m_0\gamma} + U_T(\bar{r})\right\}u_n(\bar{r}) = E_\perp^{(n)} u_n(\bar{r}) \ . \tag{25}$$

For bound states, the states may be classified according to their angular momentum as s, p, d, ... states, just as for the hydrogen atom.

For the planar case, the situation is simpler since the transverse motion is one-dimensional, and the bound transverse states are obtained from

$$\left\{\frac{p_x^2}{2m_0\gamma} + V_T(x)\right\}u_n(x) = E_\perp^{(n)} u_n(x) \ . \tag{26}$$

For free transverse motion, the eigenstates become two- and one-dimensional Bloch waves, respectively, since the potentials in Eqs. (6') and 7') are periodic. Then also a Bloch wave vector $\bar{k}_\perp$ or k_x should be specified in Eqs. (25) and (26), and the eigenvalues form a band structure, $E_\perp^{(n)}(\bar{k}_\perp)$ or $E_\perp^{(n)}(k_x)$.

Channeling Radiation

When we want to consider radiative transitions between bound channeling states, it is useful to transform the description to the 'rest frame', in which only the transverse motion remains. Except for very large values of γ, this is nonrelativistic, and the standard, nonrelativistic theory of emission of dipole radiation may be applied. Since the transverse energies are exceedingly small compared with the total energy of the projectiles, we may use the parameters specified in Eqs. (21) and (22) for the Lorentz transformation.

The first thing to notice is that distances and momenta perpendicular to the velocity of the transformation are invariant under a Lorentz transformation. The transverse wave function must therefore be invariant, and hence the stationary Schrödinger equation determining the eigenstates must be the same. The transverse energy will, however, be very different! To see this, consider first the kinetic-energy term in Eq. (25). The nominator is invariant, but the mass in the denominator is now the rest mass, $\gamma m_0 \rightarrow m_0$ since the motion is nonrelativistic. Hence the kinetic energy is increased by a factor γ. This is also the case for the potential $U(\bar{r})$ since it is defined as the crystal potential averaged in the z direction, and the crystal is contracted by a factor γ as viewed from the 'rest frame' ('Lorentz contraction'). Equation (25) should therefore be multiplied by a factor γ, leaving eigenstates $u_n(\bar{r})$ unchanged but increasing the transverse energies by a factor γ.

In view of these results, we may interpret the transition towards the classical limit for large γ in a different way. In the 'rest frame', the projectile mass is always m_0, but the centripetal force is proportional to γ for a fixed distance. The situation is then analogous to that for highly charged ions. For fixed radius of an outer-electron orbit, the classical limit is approached with increasing ionic charge.

For radiative dipole transitions, one obtains as usual that the intensity is proportional to the square of a dipole matrix element, and the usual selection rules follow. For axial-channeling radiation, the angular momentum must change by one unit, $\Delta\ell = 1$, and for the planar case, the requirement of a change in parity leads to Δn odd for transitions between bound states.

The frequency of the radiation in the 'rest frame' is determined by the Bohr relation,

$$\hbar\omega_R = \gamma(E_\perp^{(n)} - E_\perp^{(m)}) \tag{27}$$

for a transition between levels n and m. Since the radiation must be detected in the laboratory frame, the frequency should be modified according to a Lorentz transformation of the photon four-momentum, which in the forward (z) direction leads to the well known formula for the relativistic Doppler shift,

$$\omega_L = (1+\beta)\gamma\omega_R \simeq 2\gamma^2|\Delta E_\perp|/\hbar . \tag{26}$$

For 4-MeV electrons, the energy gain due to relativistic effects is a factor of ~160! This crucial point, which makes channeling radiation interesting, was first realized by the Russian physicist Kumakhov, who has since then contributed much of the theoretical work on channeling radiation, and this radiation is therefore often called 'Kumakhov radiation'[6].

We end this discussion by quoting a formula for the intensity of the radiation in the forward direction,

$$N = \alpha^4 \gamma^{-2} \left(\frac{\hbar\omega_L}{e^2/a_0}\right)^3 \frac{\left|\langle\bar{r}\rangle_{mn}\right|^2}{a_0^2} \frac{t}{a_0} \frac{\delta\Omega}{4\pi} \quad [\text{photons/electron}] \quad (29)$$

This gives the number of photons detected within the solid angle $\delta\Omega$ in the forward direction, per electron in the channeling state n in a crystal of thickness t. It has been assumed that the electrons are relativistic, $\gamma \gg 1$ and $\beta \simeq 1$, and the opening angle of the detector should be small compared to γ^{-1}, $\delta\Omega \ll \gamma^{-2}$.

Observation of Channeling Radiation

We shall finally give a couple of examples of the observation of channeling radiation[7],[8]. They are both for 4-MeV electrons in silicon, the reason being that this is the maximum energy available at our accelerator in Aarhus, and that thin crystals of silicon are easy to make!

Our experimental setup is shown in Fig. 4 and described in detail in Refs. 7 and 8. The silicon crystal is mounted in a goniometer, which allows us to chose freely the direction of the incident beam relative to a crystal axis or plane. The beam is collimated to a divergence of $\sim 0.02°$, and after collimation, it is bent into the target area. This detail is extremely important since otherwise the photon detector would be killed by the background radiation from electrons hitting the slits. After passing the crystal, the beam is bent into a Faraday cup, shielded by a lot of lead bricks from the forward photon detector.

Two photon spectra for incidence close to a <111> axial direction in silicon (cube diagonal with average atomic spacing d = = 5.43 Å×√3/2) are shown in Fig. 5. The solid lines indicate the

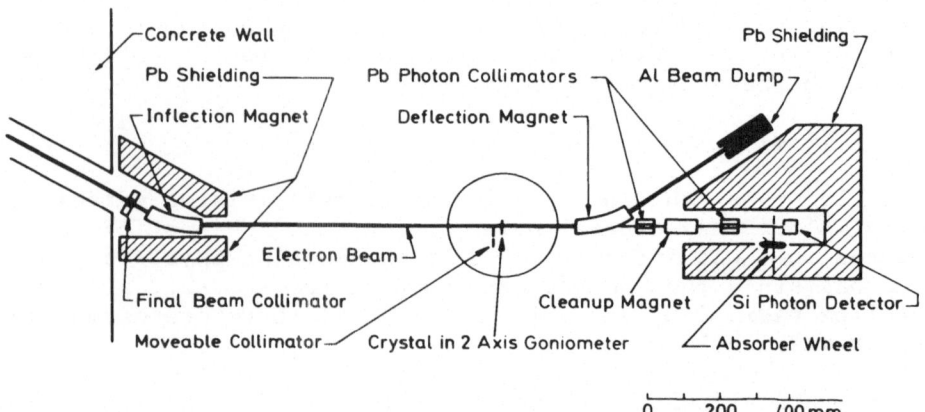

Fig. 4. Experimental setup for measurement of channeling radiation.

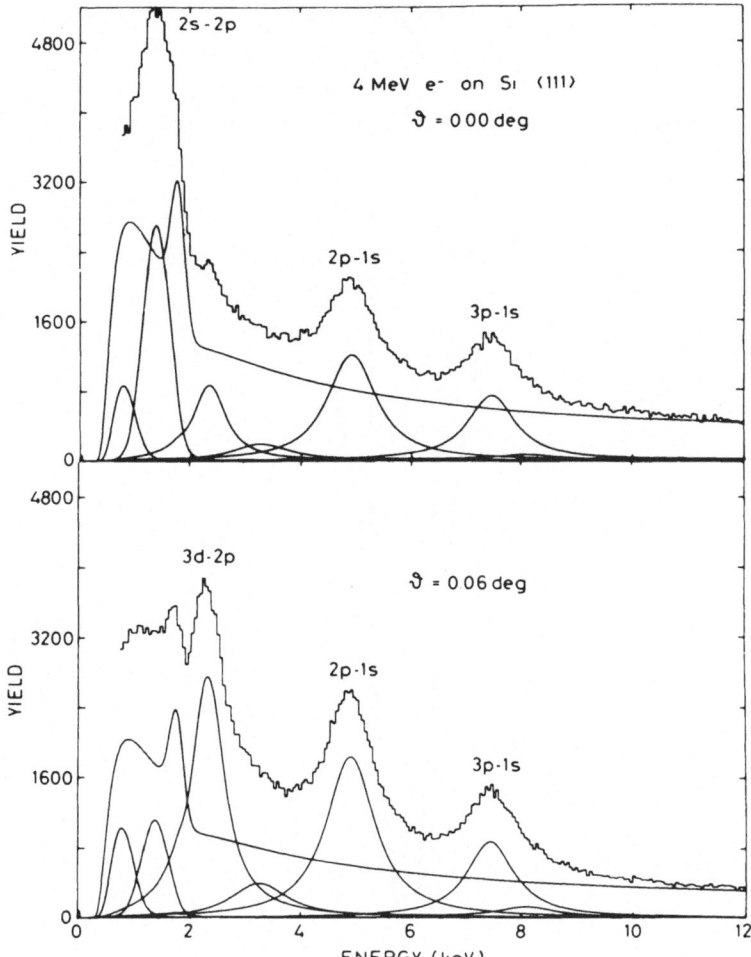

Fig. 5. Radiation spectra for 4-MeV electrons incident at two dif-ferent angles to a <111> direction in Si.

components in fits to the spectra. First, there is a background of
incoherent or 'normal' bremsstrahlung, with a 1/E shape modified
partly by absorption (note the K-absorption edge at ∿1.8 keV),
partly by a contribution from Si-K xrays (the peak just below the
absorption edge). Apart from a scaling in magnitude, the background
components are identical in the two spectra.

Second, there are a number of peaks with shapes determined
partly by an intrinsic Lorentzian shape, partly by the Gaussian
resolution function (∿300 eV FWHM). Four of the lines have been
identified as transitions between bound states in the <111> axial
potential. This is shown in Fig. 6, with the calculated energy

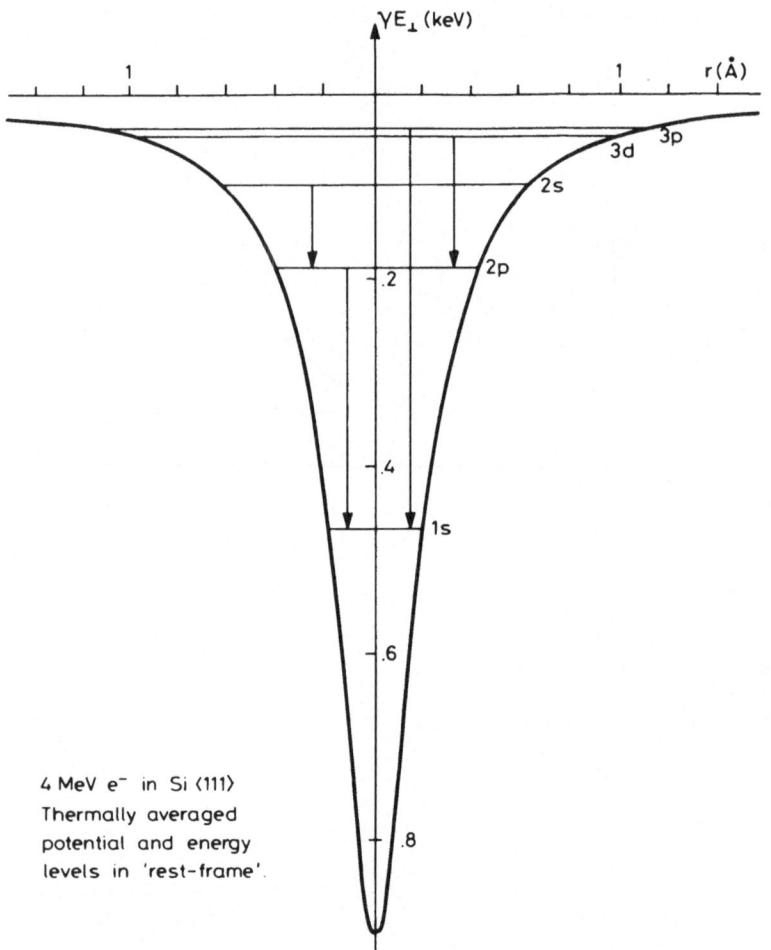

Fig. 6. *Thermally averaged continuum potential for 4-MeV electrons channeled along <111> direction in Si, with transverse-energy levels of bound states.*

levels. Note that the energy scale corresponds to the 'rest frame', i.e., includes the factor $\gamma \sim 9$ for 4-MeV electrons. The measured peak energies agree within a few percent with those determined from Eqs. (27) and (28). It is important here to include the thermal average, Eq. (14), in the definition of the potential.

The intensity of the lines of channeling radiation carries information about the population of the different channeling states, which for a thin crystal are determined by the overlaps at the surface of the crystal between the incident plane wave and the channeling eigenstates,

$$P_n \propto |<u_n(\bar{r})|e^{i\bar{k}_\perp \cdot \bar{r}}>|^2 , \qquad\qquad (30)$$

where $\hbar\bar{k}_\perp$ is the transverse momentum of the incident electrons. For varying incidence angle ψ, the intensity of a line $n \to m$ is proportional to $P_n(\psi)$, and according to Eq. (30), the observation of this variation determines the wave function u_n in a Fourier representation (or momentum representation). Such a measurement is shown in Fig. 7 in a two-dimensional representation. Spectra such as those shown in Fig. 5 have been recorded at angular intervals of 0.02°. The results are seen to be in qualitative agreement with the assignments in Fig. 5. Thus only the line corresponding to a transition from an s state (2s-2p) has a maximum at $\psi = 0$, and the deepest-bound levels with the highest momenta have the broadest Fourier spectrum. A detailed analysis also gives good quantitative agreement with calculations. There are a number of other features in Fig. 7, which we shall not discuss in detail, but we may mention that some of the 'ridges' can be interpreted as transitions from free states of transverse motion to bound states, i.e., they are analogous to the radiative capture of electrons from the continuum observed for ions penetrating solids.

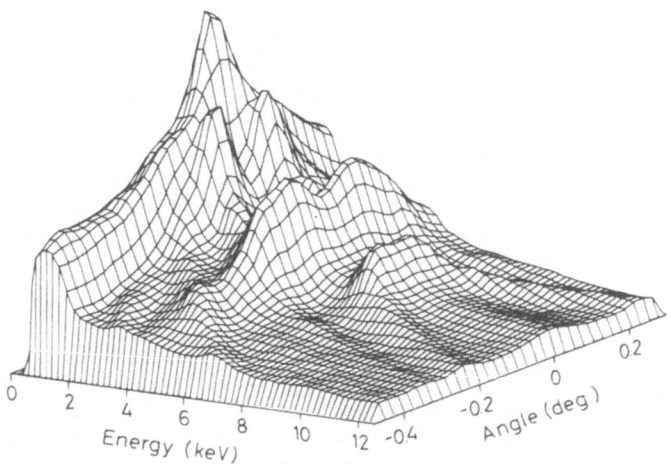

Fig. 7. Photon spectra vs. angle of incidence to a <111> axis for 4-MeV electrons in a 0.5 μm thick Si crystal.

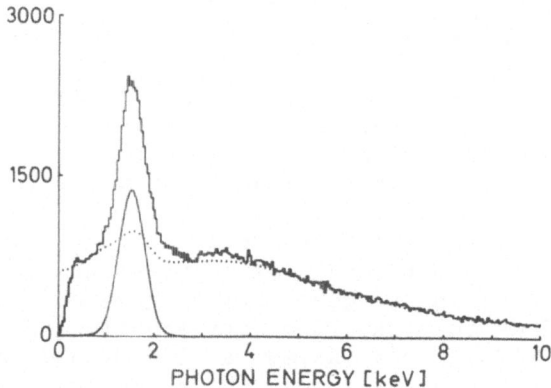

Fig. 8. Radiation spectrum for 4-MeV electrons incident at a small angle to a {110} plane in Si. The line at ∿1.5 keV is from transitions between the two energy levels shown in Fig. 9.

An example of planar-channeling radiation is shown in Fig. 8, once again for 4-MeV electrons in silicon but now for incidence nearly parallel to a {110} plane (a diagonal plane in the elementary cube, which is the closest-packed plane for the diamond structure of silicon). The spectrum is decomposed into a background similar to that in Fig. 5 and a single line at ∿1.5 keV. The interpretation of this line may be seen in Fig. 9, showing one period of the planar potential in Eq. (7') and the two bound states in this potential. Once again, the line energy observed agrees to within a few percent with that calculated for the transition between these levels, from Eqs. (27) and (28). For larger angles of incidence, free states of transverse motion are populated, and from measurements similar to those shown in Fig. 7, the band structure of these states may be mapped out[7]. We shall not here go into these details, but it may be mentioned that the radiation from free states connects the channeling radiation to the earlier-studied phenomena of 'coherent bremsstrahlung'.

CONCLUDING REMARKS

We have given here only a broad general discussion of channeling phenomena, but the interested student should be able to penetrate deeper into the subject through the references given below. We have tried to emphasize the basic ideas and also at places the connection to other fields of atomic physics, where similar concepts are important. The most important example is the analysis of the

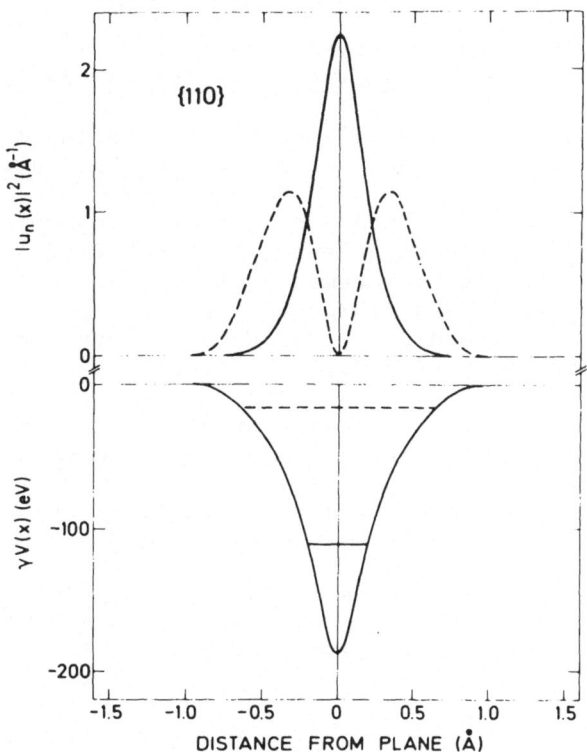

Fig. 9. One period of {110} planar potential for 4-MeV e⁻ in Si and the transverse-energy levels for the two bound states. The squares of the corresponding wave functions are shown above.

applicability of classical mechanics in the description. The channeling phenomena are rich in applications, and we have discussed only few of these. The emphasis given here to channeling of electrons and channeling radiation is out of proportion to the present importance of these phenomena, but partly they represent a new development in the field, partly they demonstrate in a very direct way the existence of well defined channeling states determined by continuum averaged potentials, in the same way as spectroscopy gives the most direct evidence for the existence of discrete bound states in atoms, as was realized already by Bohr in his first quantum analysis of the hydrogen atom. Of course the emphasis also reflects a personal bias since my own work has been concentrated on these subjects for the last few years.

REFERENCES

1. J. Lindhard, Influence of crystal lattice on motion of energetic charged particles, K.Dan.Vidensk.Selsk.Mat.Fys.Med. 34:No 14 (1965)
2. D. Gemmel, Channeling and related effects in the motion of charged particles through crystals, Rev.Mod.Phys. 46:129 (1974)
3. J.W. Mayer, L. Eriksson, and J.A. Davies, *Ion Implantation in Semiconductors*, Academic Press, New York (1970)
4. W.M. Gibson, Blocking measurements of nuclear lifetimes, Ann. Rev.Nucl.Sci., 25:465 (1975)
5. J.U. Andersen, S.K. Andersen, and W.M. Augustyniak, Channeling of electrons and positrons, K.Dan.Vid.Selsk.Mat.Fys.Med. 39:No 10 (1977)
6. W. Wedell, Electromagnetic radiation of relativistic positrons and electrons during axial and planar channeling, Phys.Status Solidi B99:11 (1980)
7. J.U. Andersen, K.R. Eriksen, and E. Lægsgaard, Planar-channeling radiation and coherent bremsstrahlung for MeV electrons, Phys.Scripta 24:588 (1981)
8. J.U. Andersen, E. Bonderup, E. Lægsgaard, B.B. Marsh, and A.H. Sørensen, Axial-channeling radiation from MeV electrons, Nucl.Instrum.Methods 194:209 (1982)

THEORY OF LOW ENERGY ELECTRON-ION PROCESSES FOR MEDIUM TO HIGHLY

CHARGED IONS

J. Dubau and C.J. Zeippen

Observatoire de Paris

92190 Meudon, France

CONTENTS

1. Introduction

In the present lectures, we shall be concerned mainly with the atomic physics involved in the interpretation of spectroscopic obser-vations and in the construction of theoretical models. In low-density plasmas, such as coronal and Tokamak plasmas, the electron impact excitation of positive ions is responsible for the emission of most line spectra. In the X-ray region, an additional type of lines are

observed: they are the unusual satellite lines often produced during
the electron-ion recombination process. A third electron-ion interaction which plays an important role in plasma models is the ionisation by electron impact. However, we have chosen to ignore this process in our talk since the methods used to calculate ionisation cross
sections are less elaborate and accurate than those employed to study
the other two electron-ion processes mentioned above. The analysis of
high-density plasmas, such as laser produced plasmas, is more complicated. In this case, the simple picture of isolated binary processes
does not hold. Moreover, a proper interpretation of spectroscopic
observations requires the resolution of radiative transfer equations.

Depending on the nuclear charge Z, and the number of electrons
N of the ionic target and on the accuracy required, different levels
of approximation are used. For instance, a non-relativistic theoretical formalism is often appropriate for lowly charged ions, but sophisticated close-coupling calculations are necessary to obtain accurate
results. On the contrary, for very highly charged ions ($z = Z - N \gg 30$)
a fully relativistic approach must be used but a simple hydrogenic
approximation is sufficient to calculate reliable data. For medium
range target charges, Breit-Pauli relativistic corrections to the
non-relativistic Hamiltonian implemented in a Distorted Wave formalism give excellent results. Of course, in all cases, the contribution of electron-ion resonances must be taken into account. Also,
great attention must be paid to the radiation field, which can modify
substantially the decay modes of such resonances.

As stated in the title of the talk, we shall restrict ourselves
to the low-energy processes, for which the velocity of the incident
or scattered electron is of the same order as the excitation energy
of the ionic electrons directly involved in the collision. In stationary plasmas, this does not result in any great practical limitation,
since low-energy processes are markedly dominant, as shown by the
amplitude of associated probabilities. Transient plasmas are more of
a problem, because high-energy processes are then important and also
because more complex excitation processes take place. Generally, the
analysis of such cases is very difficult. Fortunately, simplified
methods give atomic results which are often sufficiently accurate,
considering the huge uncertainties in the plasma model itself.

The present lectures attempt to describe in some detail a number of very important approximations implemented in the electron-
ion collision formalism and to illustrate the physical aspects of
the problem of medium to highly-charged targets with a few typical
examples. Also, we mention some major computer programs used in the
field. Of course, to provide full detail about any of the methods
and phenomena considered here would be beyond the scope of the talk!
That is why we shall refer, as we go along, to a number of good
papers presenting original work or reviews.

2. Quantum description of an electron-ion system

For low density plasmas, such as coronal and Tokamak plasmas, it is sensible to assume that a given electron-ion collision is unaffected by any third particle. The electron-ion collision theory attempts to describe such an "ideal" process. When considering real plasmas, this basic hypothesis must be checked case by case. For example, when considering laser plasmas, it is not always justified to include theoretical excitation rates in a plasma model without precautions !

A quantum description of the collision is required for low-energy scattering because it is then impossible to distinguish the colliding electron from the target ionic electrons during the whole collision process: the Pauli exclusion principle must be applied; exchange can take place. Also, at some well defined energies, through capture of the colliding electron by the target ion, autoionising states can be created, whose properties are similar to those of normal excited states.

In sections 2 and 3, we shall neglect relativistic effects. This is a reasonable approximation for low-ionised atoms, but for highly-ionised atoms, relativistic corrections in the electron-ion Hamiltonian become important: they will be examined in section 4.

2.1 Some basic concepts on cross sections and scattering matrices.

Consider an ion A^{+z} with N electrons. In a non-relativistic description of the electron-ion system, the wave function Ψ_j of the total (N+1)-electron system must satisfy the Schrödinger equation

$$H (x_1, x_2, \ldots x_{N+1}) \Psi_j = E \Psi_j \qquad (2.1)$$

where j labels an entrance channel corresponding to a particular asymptotic boundary condition, E is the energy of the total system, x_n denotes the space and spin coordinates of the nth electron : $x_n = (\vec{r}_n, \sigma_n)$.

When the colliding electron and the ionic target are separated by a large distance, they do not interact appreciably. The total wave function Ψ_j can then be expressed as a simple product of the target wave function Θ_i and of the colliding electron wave function for a given exit channel i. Thus

$$\Psi_j(x_1, x_2, \ldots, x_{N+1}) \underset{r_{N+1} \to \infty}{\sim} \frac{1}{\sqrt{N+1}} \sum_i \Theta_i(x_1 \cdots x_N) \varphi_{ij}(x_{N+1}) \qquad (2.2)$$

where the summation runs over all exit channels. When $i \neq j$, the collision is inelastic, while it is elastic when $i = j$. The target wave function Θ_i is usually a configuration-interaction-type expansion, in terms of Slater-type orbital wave functions, for which the radial function is denoted $P_{n\ell}(r)$. In (2.2), i= $(\alpha_i S_i L_i M_{S_i} M_{L_i})$ where α_i is a degeneracy parameter and we use the conventional notation for angular momenta.

Let us first consider a neutral target. Before the collision, due to the absence of Coulomb interaction, the wave associated with the colliding electron behaves as a plane wave. After the collision, the diffusion process changes part of the wave into a spherical wave; the other part, unperturbed, remains plane. This physical situation can be represented by using the following boundary condition for φ_{ij}:

$$\varphi_{ij}(x) \underset{r_{N+1} \to \infty}{\sim} e^{i\vec{k}_i \cdot \vec{r}} \delta(m_{s_i}|\sigma)\delta_{ij} + \frac{e^{ik_i r}}{r} f_{ij}(\hat{r},\sigma) \qquad (2.3)$$

where r and $\hat{r}$ are the radial and angular parts of $\vec{r}$: $\vec{r} = (r,\hat{r})$, $\vec{k}_i$ is the wave vector : $\vec{k}_i = m_e \vec{v}_i/\hbar$. m_s is the spin projection of the colliding electron, $\delta(m_s/\sigma)$ is the spin function which is a Kronecker symbol. For the sake of simplicity, atomic units will be used from now on, i.e. we shall put $m_e = e = \hbar = 1$, where m_e and e are the electron mass and charge, respectively and $\hbar$ is the reduced Planck constant.

The flux relation between the incident and scattered waves provides a direct link between $f_{i,j}(\hat{r},\sigma)$ in (2.3) and the differential cross section $d\sigma_{ij}/d\hat{r}$ for transition (i→j), and also the total cross section $\langle\sigma_{ij}\rangle$ for the same transition, averaged on the solid angle of the incident electron, $\hat{k}$: (indeed, in a plasma, collisions occur in random directions)

$$\langle \sigma_{ij}\rangle = \frac{1}{4\pi}\int d\hat{k}\, \sigma_{ij}$$

$$\qquad (2.4)$$

$$\sigma_{ij} = \frac{1}{2}\sum_\sigma \int d\hat{r}\, \frac{d\sigma_{ij}}{d\hat{r}} = \frac{1}{2}\sum_\sigma \int d\hat{r}\, |f_{ij}(\hat{r},\sigma)|^2$$

Finally, the excitation rate coefficient is obtained by averaging the total cross section over the electron velocity distribution (usually, a Maxwellian distribution of temperature $T_e = f(v_j,T_e))$:

$$C_{ij}(T_e) = \int dv_j\, v_j\, f(v_j,T_e)\langle\sigma_{ij}\rangle \qquad (2.5)$$

Of course, the relatively simple description above is not valid when the target is ionised, because of the presence of the Coulomb asymptotic potential of charge z = Z - N, where Z is the nuclear charge and N the number of ionic electrons. The distortion imposed on the colliding electron wave function $\varphi_{ij}(x)$ must then be taken

into account : the plane and spherical waves in (2.3) become "coulombic plane" and "coulombic spherical" waves, respectively. This transformation is achieved by redefining $\vec{k}_i \cdot \vec{r}$:

$$\vec{k}_i \cdot \vec{r} \;\Rightarrow\; \vec{k}_i \cdot \vec{r} \;+\; \frac{\gamma}{k_i}\, \ln 2 k_i r \;+\; \xi(\hat{r}) \qquad\qquad (2.6)$$

where it is important to note that $\xi(\hat{r})$ is a function of the
sole $\hat{r}$.

   In the case of low energy collisions, it is convenient to use
a partial wave expansion of the wave function because convergence
can be obtained rapidly, as opposed to the case of high-energy collisions.

   Assuming that the asymptotic colliding electron wave function
φ_{ij} can be separated into angular, spin and radial parts, we write

$$\varphi_{ij}(x) \;\underset{r\to\infty}{\sim}\; Y_{\ell_i}^{m_{\ell_i}}(\hat{r})\, \delta(m_{s_i}|\sigma)\, \frac{F_{ij}(r)}{r} \qquad\qquad (2.7)$$

where Y_ℓ^m is a standard spherical harmonic function, ℓ_i is the orbital
momentum and m_{ℓ_i} , m_{s_i} are the orbital momentum and spin components.
An expression like (2.7) is justified because the colliding electron
Hamiltonian has the asymptotic form

$$H_{N+1}(x_{N+1}) \;\underset{r_{N+1}\to\infty}{\sim}\; -\frac{\nabla_{N+1}^2}{2} \;-\; \frac{\gamma}{r_{N+1}} \qquad\qquad (2.8)$$

Inserting (2.7) in the Schrödinger equation for the system, we obtain
that $F_{ij}(r)$ must satisfy, for r large, the coulombic radial equation

$$\left\{ \frac{d^2}{dr^2} \;-\; \frac{\ell_i(\ell_i+1)}{r^2} \;+\; \frac{2\gamma}{r} \;+\; k_i^2 \right\} F_{ij}(r) = 0 \qquad\qquad (2.9)$$

where $\varepsilon_i = \frac{k_i^2}{2}$ is the colliding electron energy (in a.u). The solutions of equation (2.9) are the confluent hypergeometrical functions.
Among the independent solutions of (2.9) which are real, let us
consider functions f (regular solution) and g (irregular solution),
with the asymptotic forms

$$f_{k_i\ell_i}(r) \;\underset{r\to\infty}{\sim}\; \frac{1}{\sqrt{k_i}}\, \sin(\eta_i) \quad,\quad f_{k_i\ell_i}(0) = 0$$

$$\qquad\qquad\qquad\qquad\qquad\qquad\qquad\qquad (2.10)$$

$$g_{k_i\ell_i}(r) \;\underset{r\to\infty}{\sim}\; \frac{1}{\sqrt{k_i}}\, \cos(\eta_i) \quad,\quad |g_{k_i\ell_i}(0)| = \infty$$

where

$$\eta_i = k_i r \;-\; \frac{\ell_i \pi}{2} \;+\; \frac{\gamma}{k_i}\, \ln 2 k_i r \;+\; \arg \Gamma\left(\ell_i+1-\frac{i\gamma}{k_i}\right)$$

A general solution $F_{ij}(r)$ can be expressed asymptotically as a linear combination of f and g :

$$F_{ij}(r) \underset{r \to \infty}{\sim} a_{ij} f_i(r) + b_{ij} g_i(r) \qquad (2.11)$$

where i stands for (k_i, ℓ_i) (see (2.10)).

Inserting (2.7) in (2.2), we build a γ representation, with $\gamma_i = (\alpha_i \, S_i \, L_i \, M_{S_i} \, M_{L_i} \, l_i \, m_{l_i} \, m_{s_i})$, where the symbols are the usual quantum numbers for the target (large letters) and the colliding electron (small letters). Since we consider here the non-relativistic case, the total orbital momentum and spin L and S, as well as their components M_L and M_S, and the total parity π are "good" quantum numbers and can be used in the representation for the total system.

From the energy conservation principle, we know that the total energy of the system is

$$E = E_i + \frac{k_i^2}{2} \qquad (2.12)$$

where E_i is the target ($\alpha_i \, S_i \, L_i$) state energy. In fact, an alternative representation to γ_i, $\Gamma_i = (\alpha_i \, S_i \, L_i \, l_i \, L \, S \, \pi \, M_L \, M_S)$, allows a simplification of the calculations because the number of non-zero transitions to be considered is reduced. The two representations are related by the orthogonal transformation

$$(\gamma_i | \Gamma_i) = C^{L_i \, l_i \, L}_{M_{L_i} \, m_{l_i} \, M_L} \, C^{S_i \, \frac{1}{2} \, S}_{M_{S_i} \, m_{s_i} \, M_S} \qquad (2.13)$$

where $C^{d_1 d_2 d_3}_{m, m_1 m_3}$ is a real Clebsch-Gordan coefficient. Substituting φ_{ij} of (2.7) to φ_{ij} of (2.3) and performing transformation (2.13), (2.2) supplies a set of solutions ψ_j of the form

$$\psi_j (x_1 \dots x_{N+1}) \underset{r_{N+1} \to \infty}{\sim} \frac{1}{\sqrt{N+1}} \sum_i \Theta_i (x_1 \dots x_N ; \hat{r}_{N+1}, \sigma_{N+1}) \frac{F_{ij}(r_{N+1})}{r_{N+1}} \quad (2.14)$$

for each L S π of the total system. Note that ij = $\Gamma_i \Gamma_j$. i takes all integer values from 1 to NCHOP, the number of open channels in the representation. In (2.14), function Θ is defined thus :

$$\Theta_i = \sum_{\gamma_i} (\gamma_i | \Gamma_i) \, \theta_i (x_1 \dots x_N) \, Y^{m_{\ell_i}}_{\ell_i} (\hat{r}_{N+1}) \, \delta(m_{s_i} k_m) \quad (2.15)$$

and function $F_{ij}(r)$ is solution of (2.9) for r large. F_{ij} could be either real or complex but one can choose it to be real. It is always possible to operate a change on the set ψ_j in (2.14) through a matrix transformation. For example, we can choose a real F_{ij} which satisfies

$$F_{ij}(\underset{\sim}{R}|r) \underset{r\to\infty}{\sim} \frac{1}{\sqrt{k_i}} \left(\sin(\eta_i) \delta_{ij} + \cos(\eta_i) R_{ij} \right) \qquad (2.16)$$

where R_{ij} is an element of matrix $\underset{\sim}{R}$, which is called the reactance matrix (sometimes denoted $\underset{\sim}{K}$). It is easy to see, from (2.10) and (2.11), that the following complex asymptotic form is also valid for F_{ij} :

$$F_{ij}(\underset{\sim}{S}|r) \underset{r\to\infty}{\sim} \frac{1}{\sqrt{k_i}} \left(\exp(-i\eta_i) \delta_{ij} - \exp(i\eta_i) S_{ij} \right) \qquad (2.17)$$

where S_{ij} is an element of matrix $\underset{\sim}{S}$, which is called the scattering matrix. The two F_{ij}'s are related by the matrix transformations

$$\underset{\sim}{F}(\underset{\sim}{S}|r) = \underset{\sim}{F}(\underset{\sim}{R}|r) \frac{2}{i} \left(1 - i\underset{\sim}{R} \right)^{-1}$$

$$\underset{\sim}{S} = \left(1 + i\underset{\sim}{R} \right)\left(1 - i\underset{\sim}{R} \right)^{-1} \qquad (2.18)$$

In collision theory, the reactance matrix $\underset{\sim}{R}$ is shown to be symmetric and real. Therefore, from (2.18), this results in $\underset{\sim}{S}$ being unitary and symmetric, properties interpreted as being linked to the electron flux conservation during the collision and to the microreversibility principle. The real $F_{ij}(R/r)$ is convenient for computational work, while the complex $F_{ij}(S/r)$ is convenient for its relation with the particle flux.

To relate the cross section σ and the scattering matrix $\underset{\sim}{S}$, it is necessary to expand the solution defined in (2.2) and (2.3) on the spherical solutions (2.14). In (2.3), part of the expression is already a spherical wave. The expansion of the coulombic plane wave part requires the well-known development :

$$e^{i\vec{k}\cdot\vec{r}} + \dots \text{(see (2.6))} \underset{r\to\infty}{\sim} \frac{4\pi}{r\sqrt{k}} \sum_{l=0}^{\infty} \sum_{m=-l}^{+l} i^l Y_l^{m*}(\hat{k}) Y_l^m(\hat{r}) f_{bkl}(r) \qquad (2.19)$$

where $f_{kl}(r)$ is a function already mentioned in (2.10). The resulting expansion corresponds to the representation χ_i. A shift to representation Γ_i is secured through the transformation (χ_i / Γ_i) defined in (2.13). The final expression giving the scattering cross section for the transition between ionic states j and i turns out to be

$$\sigma_{ij} = \frac{\pi}{(2L_j+1) k_j^2} \Omega_{ij} \qquad (2.20)$$

where

$$\Omega_{ij} = \frac{1}{2} \sum_{LS\pi} \sum_{(ij)} (2L+1)(2S+1) \left| T_{(ij)} \right|^2 \qquad (2.21)$$

Ω_{ij} is called the collision strength. (ij) refers, for a given $LS\pi$, to the set of channels $\Gamma_i \Gamma_j$ which include the ionic states i and j, respectively.

The relation between the transition matrix $\underset{\sim}{T}$ and the scattering matrix $\underset{\sim}{S}$ is

$$\underset{\sim}{T} = \underset{\sim}{S} - 1 \qquad (2.22)$$

where 1 is the unit matrix.

2.2 The Close-Coupling approximation (CC)

Section 2.1 of the present course was concerned mainly with the asymptotic behaviour of the electron-ion system wave function. In particular, separable wave functions were defined (see (2.2) and (2.14)). In the short range, the picture must be modified considerably as account must be taken of the interactions between colliding and ionic electrons, other than the asymptotic Coulomb interaction.

In the close-coupling (CC) approximation, the electron-ion wave function is taken to be of the form (2.14) for all values of r. However, to refine the model, one can include in the expansion a number of ionic states lying higher than the excited states present in (2.14). The "trial" wave function is of the type

$$\psi_j^t = \sum_{i=1}^{NCHF} \mathcal{A}\, \Theta_i \, \frac{F_{ij}^t}{r} \qquad (2.23)$$

where $\mathcal{A}$ is the antisymmetrisation operator and NCHF $\geqslant$ NCHOP. The F_{ij} radial functions associated with the extra ionic states (NCHF–NCHOP) correspond to negative energies. Therefore, they disappear for r large. Those extra channels are called closed channels because they cannot be excited. They can, however, create resonance effects. Of course, as the total energy E increases, the closed channels can become open.

To derive radial functions $F_{ij}(r)$ for any r, a variational method is convenient (see, for example, BURKE and SCHEY (1962) or KISSNER and SEATON (1972)). One can show that

$$\delta\left\{ \langle \psi_i | H-E | \psi_j \rangle - \frac{1}{2} R_{ij} \right\} = \langle \delta\psi_i | H-E | \psi_j \rangle$$

$$+ \langle (H-E)\psi_i | \delta\psi_j \rangle + \langle \delta\psi_i | H-E | \delta\psi_j \rangle \qquad (2.24)$$

where
$$\delta \psi_n = \psi_n^t - \psi_n \tag{2.25}$$

ψ stands for a set of solutions of the Schrödinger equation. They are of the type (2.14). ψ^t stands for a set of trial functions of the type (2.23).

Applying the Kohn-Hulthen variational principle, (2.24) implies that

$$\delta \left\{ < \psi_i | H - E | \psi_j > - \frac{1}{2} R_{ij} \right\} = 0 \tag{2.26}$$

to first order, for small variations around the set of exact solutions ψ. We can now impose on the approximate solution in the class (2.23) to satisfy (2.26) and (2.24), thus applying the "Hartree-Fock" variational principle. This means that

$$< \delta \psi_j^t | H - E | \psi_i^t > = 0 \tag{2.27}$$

where
$$\delta \psi_j^t = \sum_i \mathcal{A} \, \Theta_i \, \frac{\delta F_{ij}^t}{r} \tag{2.28}$$

Equations (2.27) and (2.28) lead to the famous coupled equations

$$\left\{ \frac{d^2}{dr^2} - \frac{\ell_i (\ell_i + 1)}{r^2} + \frac{2\gamma}{r} + k_i^2 \right\} F_{ij}^t (r) = \sum_{i'} W_{ii'} F_{i'j}^t \tag{2.29}$$

$W_{ii'}$ is an integro-operator in two parts :

- a local potential whose asymptotic form is

$$\sum_{\lambda=1}^{\infty} \frac{a_{ii'}^{\lambda}}{r^{\lambda+1}} \tag{2.30}$$

- a non-local exchange potential operator whose terms decrease exponentially in the asymptotic region. When solving (2.29), instabilities due to the non-orthogonality of $F_{ij}(r)$ and $P_{n\ell_i}(r)$ for the same l_i appear. It was found useful to impose orthogonality conditions on the radial functions $F_{ij}(r)$ and the target radial functions $P_{n l_i}(r)$ for the following reasons : (i) this greatly simplifies the algebraic formulation; (ii) when such conditions are not imposed, the functions $F_{ij}(r)$ are not, in general, uniquely defined. EISSNER and SEATON (1972) include Lagrange multipliers in the variational formalism to allow for the orthogonality conditions and they compensate for the restrictions thus imposed by adding extra functions in (2.23) : they are the correlation functions Φ_m:

$$\psi_j^t = \sum_i \mathcal{A} \, \Theta_i \, \frac{F_{ij}^t}{r} + \sum_m c_m \Phi_m \tag{2.31}$$

(N+1)- electron functions Φ_m are built on the target radial functions $P_{n l_i}(r)$. Of course, the set of equations (2.29) is replaced by a

more complex formula which will not be given here. Some extra correlation functions ϕ_m can also be used to limitate the error caused by
the fact that the set of coupled equations must be finite in practice.
The ϕ_m can produce resonance effects (see Section 3).

There exist many computer programs which solve the close-coupling
equations. Two well-known and widely used codes are IMPACT (CREES et
al.,1978), RMATRX(BERRINGTON et al.,1978). The CC approximation gives
its best results for electron-ion collisions involving lowly and
medium-ionised atoms. For highly ionised atoms, it is rather expensive as compared to other methods such as Distorted Wave which produces comparable results if correlation functions are included in
the distorted wave expansion (see Section 3).

2.3 The Quantum Defect Theory (QDT)

In practice, it is impossible to solve the CC equations (2.29)
for a very large number of energy values. In particular, when many
resonances come very close to each other, the number of energy values
required to give a good profile of a cross section "point by point"
would be far too large; also, the intricacies become, in such a case,
a real obstacle, particularly because the numerical accuracy suffers
considerably.

To overcome such difficulties, it is very interesting to extrapolate or interpolate quantities which are not   affected by the
resonance structure. This is exactly the aim of the multichannel
QDT which has proved itself to be a particularly powerful tool.

Considering the CC radial equations (2.29) and the corresponding asymptotic form (2.9), it appears that energy k_i^2 plays a more
important role in the long range than in the short range. For negative values of k_i^2 the asymptotic coulombic potential of charge z
is responsible for the existence of Rydberg series of bound states
and autoionising states (ie., collisional resonances) in the ion-
electron system. The expression for the energy of such states is

$$\mathcal{E}_{i,nl} = E_i - \frac{z^2}{2\nu_{nl}^2} \tag{2.32}$$

where (i, nl) represents a state made of the ionic target state i
plus a bound electron nl. The ionic target energy, E_i, is also an
excitation threshold for the ion system. The Rydberg series converge
towards the corresponding thresholds. Quantity ν_{nl} is defined thus:

$$\nu_{nl} = n - \mu_{nl} \tag{2.33}$$

where μ_{nl} is called the quantum defect. μ_{nl} differs from zero
because the short range potentials are not identical to the coulombic
long range potential. The quantum defect is a quantity which remains
unaffected by the bound state or resonance structure.

Implementing this physical idea, HAM (1955) and SEATON (1955) developed a theory for the one-channel case, relating quantities calculated above the unique ionisation threshold (elastic collision) to the converging bound state series. A relation was found between the reactance matrix in the continuum and the quantum defect in the negative energy region. Later on, GAILITIS (1963) and BELY et al. (1963) generalized the theory to the many-channel case.

Consider, near the origin $r = 0$, a series expansion in terms of r for F_{ij} such that the first term $a\, r^{l+1}$ is independent of k_i^2. This is possible because k_i^2 is negligible compared to the centrifugal term $l_i(l_i + 1) / r^2$ or to the coulombic potential z/r. Calculations show that for a given r in the short range, F_{ij} varies slowly with k_i^2. Now take two independent and analytical functions of k_i^2 which are solutions of (2.9) for all values of r, i.e. "asymptotic coulombic potential" functions which can be built using series expansions in terms of k_i^2. Call them $f_i(r)$ and $g_i(r)$. They also vary slowly in the short range region. The limit of the short range region, r_o, is taken as the radial point where the coulombic potential begins to dominate the ionic short range potential in $1/r^{\lambda+1}$, with $\lambda \geqslant 1$ (see (2.30)).

From point r_o on, function F_{ij} can be expressed as a linear combination of f_i and g_i :

$$F_{ij}(r) = a_{ij}\, f_i(r) + b_{ij}\, g_i(r) \tag{2.34}$$

Coefficients a_{ij} and b_{ij} are slowly-varying functions of k_i^2. To know their evolution in the resonance region is of particular interest, because the convergent or divergent behaviour of f_i and g_i for negative energies k_i^2 is usually felt above point r_o, except sometimes in the case of n low.

Knowing the asymptotic behaviour of f_i and g_i, function F_{ij} in (2.34) can be related to functions $F_{ij}(R/r)$ and $F_{ij}(S/r)$ defined in (2.16) and (2.17). Consequently, a_{ij} and b_{ij} can be related to matrices $\underset{\sim}{R}$ and $\underset{\sim}{S}$ and to the cross section.

For example, SEATON (1969) demonstrated that, for a single and unperturbed Rydberg series converging to a threshold E_p, R_{ij} takes the following form as a function of energy E :

$$R_{ij} = \mathcal{R}_{ij} + \frac{\mathcal{R}_{ip}\mathcal{R}_{pi}}{\mathcal{R}_{pp} + \tan \pi \nu_p} \tag{2.35}$$

where elements of $\underset{\sim}{\mathcal{R}}$ vary slowly with energy, since

$$\frac{k_p^2}{2} = E - E_p = -\frac{1}{2\nu_p^2} < 0 \tag{2.36}$$

Now, above threshold E_p, the order of the reactance matrix increases by one because it contains one extra open channel, labelled p in (2.35). This new matrix is denoted $\mathcal{R}$ to avoid confusion with $\underset{\sim}{R}$, the reactance matrix below threshold E_p; its elements are the continuation, for $k_p^2 \geqslant 0$, of the elements of $\mathcal{R}$ in (2.35), which are calculated for $k_p^2 < 0$. Consequently, matrix $\underset{\sim}{\mathcal{R}}$ is continuous through threshold E_p and it is not affected by the resonance structure.

Just under the threshold E_p, the density of resonances per unit of energy increases to infinity for n large (i.e. for ν_p large), as apparent in (2.36). In that narrow region, the elements of $\underset{\sim}{\mathcal{R}}$ are almost constant, as the energy variation on which they depend is very small. Therefore, ν_p ends up as the only varying quantity in (2.35). Through the relation between $\underset{\sim}{R}$ and $\underset{\sim}{S}$ (2.18), one can show that

$$S_{ij} = \mathcal{S}_{ij} + \frac{\mathcal{S}_{ip}\,\mathcal{S}_{pj}}{e^{-2i\pi\nu} - \mathcal{S}_{pp}} \tag{2.37}$$

and that the average value of the scattering matrix over a resonance under threshold E_p is

$$\int_{\nu_0}^{\nu_0+1} |S_{ij}|^2 \, d\nu = |\mathcal{S}_{ij}|^2 + \frac{|\mathcal{S}_{ip}|^2 |\mathcal{S}_{pj}|^2}{\sum_{i'} |\mathcal{S}_{pi'}|^2} \tag{2.38}$$

To derive (2.38), SEATON (1969) makes use of the unitarity of the $\underset{\sim}{\mathcal{S}}$ matrix, continuation for $k^2 < 0$, of the scattering matrix $\underset{\sim}{\mathcal{S}}$ above threshold E_p. Here again, the difference in notation aims at avoiding confusion between $\underset{\sim}{S}$ and $\underset{\sim}{\mathcal{S}}$. It is worth noticing that relation (2.18) is valid for matrices $\underset{\sim}{\mathcal{S}}$ and $\underset{\sim}{\mathcal{R}}$ above threshold.

Formula (2.38) bears the name of Gailitis. It is a useful relation for evaluating excitation rates, which correspond to a cross section averaged over the electron energy distribution (usually, a Maxwell distribution) (see (2.5)).

2.4 The distorted-wave approximation (DW)

In the DW approximation, the trial radial functions are calculated while neglecting the coupling between channels. They are solutions of the equation

$$\left\{ \frac{d^2}{dr^2} - \frac{\ell_i(\ell_i+1)}{r^2} + V_i(r) + k_i^2 \right\} \mathcal{F}_i(r) = 0 \tag{2.39}$$

There are many ways to define $V_i(r)$. HENRY (1981) gives an extensive review on the subject. Among all the methods available, the one developed at University College London, DWUC, has been used widely for astrophysical applications. It may serve here as a typical example to illustrate some basic ideas.

In this method, potential $V_i(r)$ is chosen to be a modified
Thomas-Fermi-Dirac potential which contains a scaling parameter λ_i
for each set of radial orbitals i with a given angular momentum 1_i.
This potential satisfies the two boundary conditions

$$r\, V_i(r) \xrightarrow[r \to 0]{} Z \qquad (2.40)$$

$$r\, V_i(r) \xrightarrow[r \to \infty]{} z$$

For $k_i^2 > 0$, the asymptotic form of $\mathscr{F}_i(r)$ can be chosen so that

$$\mathscr{F}_i(0) = 0 \qquad (2.41)$$

$$\mathscr{F}_i(r) \underset{r \to \infty}{\sim} \frac{1}{\sqrt{k_i}} \sin(\eta_i + \tau_i)$$

The constant phase shift τ_i is a function of k_i^2 and 1_i. It arises
from the fact that the potential $V_i(r)$ differs from $z\,r^{-1}$ for r is
small.

By suitable matrix transformations, the set of solutions to
the Schrödinger equation (2.14) can be expressed in terms of sine
and cosine of phase ($\eta_i + \tau_i$):

$$F_{ij}(\underline{\rho}|r) \underset{r \to \infty}{\sim} \frac{1}{\sqrt{k_i}} \left(\sin(\eta_i + \tau_i)\, \delta_{ij} + \cos(\eta_i + \tau_i)\, \rho_{ij} \right) \quad (2.42)$$

The relation with the formalism based on matrices $\underset{\sim}{R}$ and $\underset{\sim}{S}$ is summarized in the following expressions:

$$\left.\begin{array}{l} \underset{\sim}{R} = \left(\sin\tau + \cos\tau\,\underline{\rho} \right)\left(\cos\tau - \sin\tau\,\underline{\rho} \right)^{-1} \\[2mm] F(\underline{\rho}|r) = F(\underset{\sim}{R}|r)\left(\cos\tau - \sin\tau\,\underline{\rho} \right) \end{array}\right\} \quad (2.43)$$

$$\underset{\sim}{S} = e^{i\tau}\, S_\rho\, e^{i\tau} \qquad (2.44)$$

with $\qquad S_\rho = (1 + i\underline{\rho})(1 - i\underline{\rho})^{-1}$

Note that for $i \neq j$ (which corresponds to an excitation process),
$|S_{ij}| = |(S_\rho)_{ij}|$. The Kohn-Hulthen variational principle can be
applied to $\underline{\rho}_{ij}$ in the same way as to $\underset{\sim}{R}$ (see (2.26)) :

$$\delta\left\{ <\psi_i|H - E|\psi_j> - \frac{1}{2}\rho_{ij} \right\} = 0 \qquad (2.45)$$

When the trial function is "uncoupled", we have :

$$F_{ij}^t (r) = \mathscr{F}_i (r) \, \delta_{ij}$$

$$P_{ij}^t = 0$$

$$\psi_j^t = \sum_i \mathscr{A} \, \Theta_i \, \frac{1}{r} \, F_{ij}^t \qquad (2.46)$$

To the first order, formula (2.45) can be read as:

$$\langle \psi_i \mid H-E \mid \psi_j \rangle - \langle \psi_i^t \mid H-E \mid \psi_j^t \rangle = \frac{1}{2} \left(P_{ij} - P_{ij}^t \right) \quad (2.47)$$

hence

$$P_{ij} = -2 \, \langle \psi_i^t \mid H-E \mid \psi_j^t \rangle \qquad (2.48)$$

Similarly to the close-coupling approximation, orthogonality conditions imposed on the radial functions $\mathscr{F}_i (r)$ and on the target radial functions $P_{nl}(r)$ are welcome, in particular, because they greatly simplify the algebraic formulation (note that Lagrange multipliers are still required in the equations).

Besides algebraic coefficients, estimating P_{ij} from (2.48) requires the evaluation of two different types of radial integrals:

$$I \, (n\ell, \, k\ell) = \int_0^\infty dr \, P_{n\ell} (r) \left\{ -\frac{d^2}{dr^2} + \frac{\ell(\ell+1)}{r^2} - \frac{2Z}{r} \right\} \mathscr{F}_{k\ell} (r) \qquad (2.49)$$

$$R_\lambda (n_1\ell_1, k_2\ell_2; \, n_1'\ell_1', k_2'\ell_2') = \int_0^\infty dr_1 \int_0^\infty dr_2 \, P_{n_1\ell_1} (r_1) \mathscr{F}_{k_2\ell_2} (r_2) \, \frac{r_<^\lambda}{r_>^{\lambda+1}} \, P_{n_1'\ell_1'} (r_2) \mathscr{F}_{k_2'\ell_2'} (r_2)$$

where $r_< = \min(r_1, r_2)$ and $r_> = \max(r_1, r_2)$.

It is easy to show that $I(nl, kl) = I(kl, nl)$. The DW approximation gives very good results from medium to highly ionised atoms, on condition that resonances are not important.

2.5 The Coulomb-Born approximation (CB)

The CB approximation is very similar to the DW approximation, but instead of the potential defined in (2.40), the choice made is

$$V_i (r) = \frac{z}{r} \qquad (2.50)$$

Consequently, the solutions are analytical :

$$\mathscr{F}_{k_i\ell_i} (r) = f_{k_i\ell_i} (r) \qquad (2.51)$$

where $f_{k_i l_i}$ is the function defined in (2.10). Note that potentials (2.40) and (2.50) are the same for r large and that the differences appear in the short range region. Exchange is neglected in the strict CB approximation. When exchange is included, the approximation is called Coulomb-Born-Oppenheimer (CBO). In fact, the difference

consists in inserting (CBO) or not (CB) the antisymmetrisation operator in the expression for Ψ_j^t in (2.46).

Let us remark that the complexity of DW or CBO calculations is
comparable when numerical methods are used to calculate radial func-
tions. In the CBO formalism, $\underset{\sim}{\text{P}}$ is identical to $\underset{\sim}{\text{R}}$ because $\tau_i = 0$.
For a lowly-ionised target, the DW method gives much better results
because the potential "felt" by the colliding electron is very different, for r small, from the asymptotic potential. The average scattering potential due to the ion is present, in great part, in the
phase τ_i. Therefore, $\underset{\sim}{\text{P}}$ remains small, which is the main condition
for the validity of the Kohn-Hulthen principle. CBO gives its best
results for highly ionised atoms.

2.6 The Hydrogenic approximation

For very highly-ionised atoms, the ionic electrons are almost
hydrogenic. therefore, the Nucleus Coulomb potential, Zr^{-1}, can be
used for both ionic and colliding electrons. All the radial integrals
can then be solved analytically. Interestingly, they can be calculated once and used in many different collision problems.

2.7 The Coulomb-Bethe approximation (CBe)

For an angular momentum l_i large, the colliding electron radial
function $\mathcal{F}_{k_i l_i}(r)$ is very small inside the target and begins to
oscillate outside the target. This allows one to simplify the R_λ
integrals (2.49) by setting $r_2 > r_1$ everywhere :

$$R_\lambda\left(n_1 \ell_1 k_2 \ell_2; n_1' \ell_1' k_2' \ell_2'\right) \simeq \int_0^\infty dr_1 \, P_{n_1 \ell_1}(r_1) r_1^\lambda P_{n_1' \ell_1'}(r_1) \int_0^\infty dr_2 \, \mathcal{F}_{k_2 \ell_2}(r_2) \frac{1}{r_2^{\lambda+1}} \mathcal{F}_{k_2' \ell_2'}(r_2)$$

$$(2.52)$$

This is valid only for λ small because $r_2^{-\lambda-1}$, in the second
integral, is larger for small values of r_2 than for large values of
r_2. On the other hand, for $\lambda = 0$, the first integral vanishes for
excitation processes $(n_1 l_1 \neq n_1' l_1')$. Consequently, the only practical
case of interest is $\lambda = 1$, which corresponds to electric dipole
transitions in the target. This "non-penetrating" electron approximation is called the CBe approximation, when functions $\mathcal{F}_{k_2 l_2}$ are
taken to be coulombic functions of charge z.

Inserting integral R_λ in the calculation of matrix $\underset{\sim}{\text{P}}$ and
assuming

$$|T_{ij}| \simeq |2 P_{ij}|$$

$$(2.53)$$

one can relate the collision strength Ω (2.21) directly to the
radiative line strength

$$S_{ij} = |<\Theta_i| \vec{r} |\Theta_j>|^2 \tag{2.54}$$

which is linked to the first integral in (2.52), while the second one is of the type $|\int_0^\infty dr\, f_i(r) \frac{1}{r^2} f_j(r)|^2$

2.8 The $\bar{g}$ approximation

From the CBe approximation, BURGESS (1961), SEATON (1962) and VAN REGEMORTER (1962) derived a very simple approximation for the collision strength, valid within a factor of 2 :

$$\Omega_{ij} = \frac{8\pi}{3\sqrt{3}} S_{ij} \bar{g}_{ij} \tag{2.55}$$

where $\bar{g}_{ij} \simeq 0.2$ most usually.

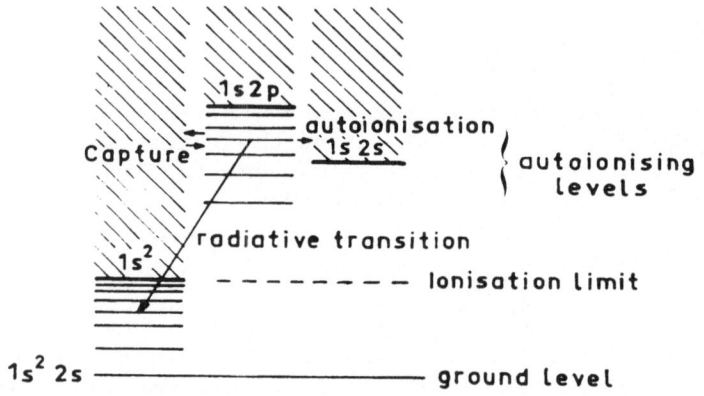

Fig. 1 : Energy diagram of Li-like and He-like ions

3. Resonances

3.1 Basic ideas

In an electron-ion collision, the physical origin of a resonance lies in the interaction between a nearly bound state and a scattering continuum. Such nearly bound states are known under the

name of autoionising states. Figure 1 presents the example of a
3-electron ion system. The ground level is $1s^2 2s\ ^2S_{1/2}$. The excited
levels correspond to the configurations $1s^2 nl$ where the valence
electron $2s$ is excited to quantum numbers nl. The first ionisation
limit is $1s^2\ ^1S_o$. Now, if we consider the excited inner-shell electron 1s, we must include among the possible configurations 1s2s2p
and, more generally, 1s2sn'l' or 1s2pn"l". The excitation energy of
levels corresponding to such configurations is larger than the energy required to ionize the 2s valence electron. These excited levels
are therefore above the first ionisation limit; they lie in the continuum. These states are generally unstable and autoionisation takes
place. This is a consequence of the impossibility of representing
such states with bound configuration only : for example, if we use a
configuration expansion method to represent the exact wavefunction,
we must include Slater determinant wavefunctions containing at least
one free electron. Inversely, in a collision formalism which contains
such Slater wavefunctions, one has to include bound configurations.
This can be done, as we have already seen in Section 2.2 by adding
bound channels or correlation functions in the expansion of the
wavefunction. It is a common feature in CC calculations, but it is
not standard practice in DW or CBO calculations. EISSNER and SEATON
(1972) have developed a method in which the trial functions defined
in (2.46) are modified to allow for the inclusion of correlation
functions Φ_m taken, for example, as bound configurations :

$$\psi_j^t = \psi_j^{DW} + \sum_m c_m^\dagger \Phi_m \qquad (3.1)$$

where
$$\psi_i^{DW} = \sum_i \mathcal{A}\,\Theta_i\,\frac{1}{r}\,\mathcal{F}_i(r)\,\delta_{ij}$$

To illustrate this point, we can go back to the example of Figure 1:
Φ_m would then be "1s2pnl" and ψ_1^{DW} and ψ_2^{DW}, "$1s^2 + e$" and
"1s2s + e'," respectively.

Parameter $c_m^\dagger$ can be obtained variationally, in the way followed
for the radial functions in the CC method :

$$\delta\psi_j^t = \sum_m \delta c_m^\dagger \Phi_m \qquad (3.2)$$

The variational principle again reads

$$\langle \delta\psi_i^t \mid H-E \mid \psi_j^t \rangle = 0 \qquad (3.3)$$

which implies, in the present case :

$$\langle \Phi_m \mid H-E \mid \psi_j^{DW} \rangle + \sum_{m'} c_{m'}^\dagger \langle \Phi_m \mid H-E \mid \Phi_{m'} \rangle = 0 \qquad (3.4)$$

For the sake of simplicity, it is very convenient to diagonalise the functions ϕ_m separately. We obtain functions $\bar{\phi}_m$, such that

$$< \bar{\phi}_m | H | \bar{\phi}_{m'} > = \bar{E}_m \, \delta_{mm'} \tag{3.5}$$

It is always possible to follow this procedure, because the matrix $< \phi_m | H | \phi_{m'} >$ is hermitian (or symmetric if the ϕ_m's have real radial functions). The relation between the two types of functions is of the form :

$$| \phi_m > = \sum_{m'} | \bar{\phi}_{m'} > < \bar{\phi}_{m'} | \phi_m > \tag{3.6}$$

Equation (3.4) becomes

$$< \bar{\phi}_m | H - E | \psi_j^{Dw} > + \bar{c}_m^j \left(\bar{E}_m - E \right) = 0 \tag{3.7}$$

where

$$\bar{c}_m^j = \sum_{m'} c_{m'}^j < \bar{\phi}_m | \phi_{m'} > \tag{3.8}$$

Therefore, we obtain for coefficient $\bar{c}_m^j$ a familiar variational expression

$$\bar{c}_m^j = \frac{< \bar{\phi}_m | H - E | \psi_j^{Dw} >}{E - \bar{E}_m} \tag{3.9}$$

In terms of functions $\bar{\phi}_m$, the trial wave function is written thus:

$$\psi_j^t = \psi_j^{Dw} + \sum_m \bar{c}_m^j \bar{\phi}_m \tag{3.10}$$

Once the best trial function has been determined, the Kohn-Hulthen principle (2.26) can be applied to determine ρ_{ij} :

$$\rho_{ij} = -2 < \psi_i^t | H - E | \psi_j^t > \tag{3.11}$$

i.e., in the present case :

$$\rho_{ij} = \rho_{ij}^{Dw} - 2 \sum_m \frac{< \psi_i^{Dw} | H - E | \bar{\phi}_m > < \bar{\phi}_m | H - E | \psi_j^{Dw} >}{E - \bar{E}_m} \tag{3.12}$$

Note that expression (3.12) contains a pole for $E = \bar{E}_m$, which is the energy of the autoionising state m whose bound part is described by the function $\bar{\phi}_m$.

We now want to derive an approximation expression for matrices $\underset{\sim}{S}_\rho$ and $\underset{\sim}{S}$ (see (2.44)). We use the notation

$$\ell_{mj} = \sqrt{2} < \bar{\phi}_m | H - E | \psi_j^{Dw} >$$

$$(\ell^+)_{im} = \sqrt{2} < \psi_i^{Dw} | H - E | \bar{\phi}_m > \tag{3.13}$$

$$a_{ij} = - \rho_{ij}^{DW} \tag{3.13}$$

$$\bar{E}_{mm'} = \bar{E}_m \, \delta_{mm'}$$

We assume that a_{ij} and b_{mj} are very small matrix elements. In matrix notation, (3.12) reads:

$$\underset{\sim}{\rho} = - \left(\underset{\sim}{a} + \underset{\sim}{b}^+ (E - \bar{E})^{-1} \underset{\sim}{b} \right) \tag{3.14}$$

Inserting (3.14) in (2.44), we find that

$$\underset{\sim}{S}_\rho = \left(1 - i \underset{\sim}{a} - i \underset{\sim}{b}^+ (E - \bar{E})^{-1} \underset{\sim}{b} \right) \left(1 + i \underset{\sim}{a} + i \underset{\sim}{b}^+ (E - \bar{E})^{-1} \underset{\sim}{b} \right)^{-1} \tag{3.15}$$

Neglecting $\underset{\sim}{a}^2$, $\underset{\sim}{b}^2$, $\underset{\sim}{a} \cdot \underset{\sim}{b}$, $\underset{\sim}{b} \cdot \underset{\sim}{a}$ as compared to the first order terms $\underset{\sim}{a}$ and $\underset{\sim}{b}$, and using the identity :

$$\underset{\sim}{b} \left(1 + i \underset{\sim}{b}^+ (E - \bar{E})^{-1} \underset{\sim}{b} \right)^{-1} \equiv \left(1 + i \underset{\sim}{b} \underset{\sim}{b}^+ (E - \bar{E})^{-1} \right)^{-1} \underset{\sim}{b} \tag{3.16}$$

one finally obtains that

$$\underset{\sim}{S}_\rho \simeq 1 - 2 i \underset{\sim}{a} - 2 i \underset{\sim}{b}^+ \left((E - \bar{E}) + i \underset{\sim}{b} \underset{\sim}{b}^+ \right)^{-1} \underset{\sim}{b} \tag{3.17}$$

Summing over open channels i,

$$\left(\underset{\sim}{b} \underset{\sim}{b}^+ \right)_{mm'} = \sum_i b_{mi} \, b_{im'}^* \tag{3.18}$$

We define a square matrix $\underset{\sim}{W}$ such that

$$\underset{\sim}{W} = \bar{E} - i \underset{\sim}{b} \underset{\sim}{b}^+ \tag{3.19}$$

$\underset{\sim}{W}$ is complex and symmetric. It can be diagonalised with the result that the elements of the diagonal matrix $\underset{\sim}{\bar{W}}$ are

$$\bar{W}_m = \mathcal{E}_m - \frac{i}{2} \gamma_m \tag{3.20}$$

The relation between $\underset{\sim}{W}$ and $\bar{W}$ is of the form

$$\underset{\sim}{W} = \underset{\sim}{X}^t \, \bar{W} \, \underset{\sim}{X} \tag{3.21}$$

where $\underset{\sim}{X}$ is orthogonal matrix ($X X^t = X^t X = 1$). Defining

$$\underset{\sim}{\tilde{b}} = \underset{\sim}{X} \cdot \underset{\sim}{b} \tag{3.22}$$

we can then rewrite formula (3.17), for $i \neq j$:

$$(S_\rho)_{ij} \simeq -2 i \, a_{ij} - 2 i \sum_m \frac{\tilde{b}_{im}^* \, \tilde{b}_{mj}}{(E - \mathcal{E}_m) + \frac{i}{2} \gamma_m} \tag{3.23}$$

The second part on the right-hand side of (3.23) is known as a Lorentz profile.

Assuming that the resonance described by (3.23) is well isolated from any other resonance, for $i \neq j$, we have

$$|T_{ij}|^2 = |(S_p)_{ij}|^2 \simeq 4 \left| a_{ij} + \frac{\tilde{\ell}_{im}^* \tilde{\ell}_{mj}}{(E - \mathcal{E}_m) + \frac{1}{2}\gamma_m} \right|^2 \quad (3.24)$$

The distance ΔE between two resonances can be defined as

$$\Delta E \simeq \left(E_p - \frac{z^2}{2(n+\frac{1}{2})^2} \right) - \left(E_p - \frac{z^2}{2(n-\frac{1}{2})^2} \right) \simeq \frac{z^2}{n^3} \quad (3.25)$$

where E_p is the energy of the threshold towards which the resonance series converges.

In the case where $\Delta E \gg \gamma_m$, we average $|T_{ij}|^2$ over the resonance to get

$$\frac{1}{\Delta E} \int_{\Delta E} |T_{ij}|^2 dE = 4|a_{ij}|^2 + \frac{2\pi}{\Delta E} \cdot \frac{4|\tilde{\ell}_{im}|^2 |\tilde{\ell}_{mj}|^2}{2\sum_{i'} |\tilde{\ell}_{mi'}|^2} \quad (3.26)$$

where we have assumed that

$$\gamma_m \simeq 2 \sum_{i'} |\ell_{mi'}|^2 \quad (3.27)$$

which is valid because $\underset{\sim}{X}$ is almost real, due to $\underset{\sim}{b}$ being very small.

Now, the time-dependent perturbation theory tells us that the autoionisation probability for the pseudo (or nearly) bound state m to decay into channel i is

$$A_{mi}^a = 2|\ell_{mi}|^2 \quad (3.28)$$

Therefore, from (3.27) and (3.28), γ_m is interpreted as the sum of all the autoionisation probabilities.

Comparing (3.26) with the Gaïlitis formula (2.38), for $i \neq j$, we infer that

$$|\mathcal{S}_{ij}|^2 = 4|a_{ij}|^2 \quad ; \quad |\mathcal{S}_{ip}|^2 = \frac{2\pi}{\Delta E} A_{im}^a \quad (3.29)$$

and consequently, from (3.25),

$$A_{im}^a \simeq \frac{z^2}{2\pi n_3} |\mathcal{S}_{ip}|^2 \quad (3.30)$$

Formula (3.30) is very useful for extrapolating the autoionisation probability for $k^2_p < 0$, from transition matrix elements calculated for $k^2_p > 0$. It is useful to remember that

$$A^a_{im} \propto \frac{1}{n^3} \tag{3.31}$$

3.2 Radiative effects : Dielectronic Recombination

So far, no attention was paid to the radiation field. The Schrödinger equation yields stable excited states, while those are, in fact, unstable, due to radiative emission. The autoionising states we have described in Section 3.1 were supposed to decay through auto-ionisation only. This is not always the case. In particular, for very highly ionised atoms, the theory must be extended in order to get results which are physically meaningful. In that case, the radiative probabilities dominate autoionisation probabilities :

$$A^r \gg A^a$$

and the main decay process to be considered is the radiative emission.

Figure 2 shows the different processes which must be included in the representation of a "real" situation

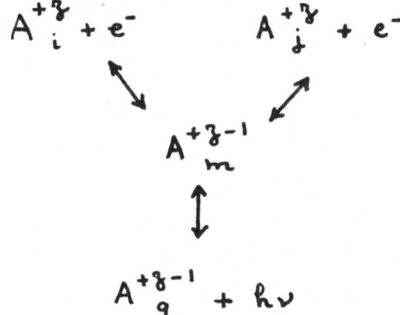

Fig. 2.   Transition from/to an autoionising state.

Starting from A^{+Z}_i, the capture of one electron allows for the formation of $A^{+(Z-1)}_m$, which can either autoionize to give A^{+Z}_j, or decay to $A^{+(Z-1)}_q$ by emitting a photon.

Taken together, the electron capture followed by the photon decay are the two processes which constitute the dielectronic recombination, provided that $A^{+(Z-1)}_q$ is a normal bound state.

It is possible to represent theoretically such processes by introducing in the collision problem "open radiative channels", in a formalism similar to the DW approximation.

The photon spherical waves are the magnetic and electric multi-poles. For example, the electric dipole behaves in the following way: (SHORE and MENZEL, 1967)

$$\mathcal{E}^d(k_q u_q \mid \vec{r}\,\sigma) = \left\{ \frac{1}{\sqrt{3}} \sum_{ab} C^{211}_{a\ell u_q} Y^a_2(\hat{r}) \delta(\ell|\sigma) \frac{f_{k_q 2}}{r} - \sqrt{\frac{2}{3}} \frac{1}{\sqrt{4x}} \delta(u_q|\sigma) \frac{f_{k_q 0}}{r} \right\} \quad (3.32)$$

In (3.32), $f_{k1}(r)$ is related to the regular spherical Bessel function (see formula (2.10) but with $\eta_i = k_i r - 1_i \pi/2$). From the asymptotic development of $f_{k1}(r)$, we find that

$$\mathcal{E}^d(k_q u_q \mid \vec{r}\,\sigma) \underset{r \to \infty}{\sim} U(u_q \mid \hat{r}\sigma) \frac{1}{\sqrt{c}} \frac{\sin(k_q r)}{r} \quad (3.33)$$

where $\hat{r}$, σ are the angular and spin (=1) variables of the photon, and k_q, u_q are the wave number and spin component of the same photon. c is the speed of light ($c = 1/\alpha \simeq 137$ in atomic units).

For r small,

$$\mathcal{E}^d(k_q u_q \mid \vec{r}\,\sigma) \underset{r \to 0}{\sim} -\sqrt{\frac{\pi}{6c}} k \, \delta(u_q|\sigma) \quad (3.34)$$

Just as in the case of the collision problem, it is useful to play on the conservation of the $LS\pi$ of the total system during the radiative process. We use the notation :

$$\gamma_q = (\alpha_q S_q L_q M_{S_q} M_{L_q} \mid u_q) \quad ; \quad \Gamma_q = (\alpha_q S_q L_q \mid LS\pi M_L M_S) (3.35)$$

A modified DW expansion reads

$$\Psi^t_\alpha = \Psi^{DW}_\alpha + \sum_m \bar{c}^\alpha_m \bar{\Phi}_m + \sum_q \left(\bar{\Phi}_q \frac{f_{k_q 2}}{r} + \bar{\Xi}_q \frac{f_{k_q 0}}{r} \right)$$
$$= \Psi^{DW}_\alpha + \sum_m \bar{c}^\alpha_m \bar{\Phi}_m + \Psi^{ph}_\alpha \quad (3.36)$$

where $\bar{\Phi}_q$ and $\bar{\Xi}_q$ are obtained by coupling $\bar{\Phi}_q$ with the first and the second parts of the R.H.S. of (3.32). Of course, we make use of a relation equivalent to (2.13) :

$$(\gamma_q \mid \Gamma_q) = C^{L_q \, 1 \, L}_{M_{L_q} u_q M_L} \delta_{S_q S} \quad (3.37)$$

Symbol α is used in (3.36) as an extension of j when electron and photon channels are present in the problem. The "exact" asymptotic wavefunction, solution of the equation including a radiation field part in the Hamiltonian can be written thus :

$$F_{q\alpha}(\rho \mid r) \underset{r_{N+2} \to \infty}{\sim} \frac{1}{\sqrt{c}} \left(\sin(k_q r) \delta_{q q} + \cos(k_q r) P_{q\alpha} \right) \quad (3.38)$$

Of course, the $\tilde{P}$ matrix now contains the electron and photon channels. It can be shown that a perturbation treatment in the continuum (DIRAC, 1958; FANO, 1961) leads to a formula similar to (3.12)

$$\rho_{qj} = -2\left(\Psi_q^{ph} \mid H\text{-}E \mid \Psi_j^{Dw}\right) - 2\sum_m \frac{\left(\Psi_q^{ph}\mid H\text{-}E\mid \bar{\Phi}_m\right)\left(\bar{\Phi}_m\mid H\text{-}E\mid\Psi_j^{Dw}\right)}{E - \bar{E}_m} \quad (3.39)$$

Term $\left(\Psi_q^{ph} \mid H\text{-}E \mid \Psi_j^{Dw}\right)$ is related to the "radiative recombination"

$$A_j^{+z} + e^- \longrightarrow A_q^{+(z-1)} + h\nu \quad (3.40)$$

and term $\left(\Psi_q^{ph}\mid H\text{-}E\mid\bar{\Phi}_m\right)$ is related to the "spontaneous emission"

$$A_m^{+(z-1)} \longrightarrow A_q^{+(z-1)} + h\nu \quad (3.41)$$

The resonance analysis of section (3.1) can be followed here and a formula similar to (3.26) can be derived, on condition that radiative probabilities be included.

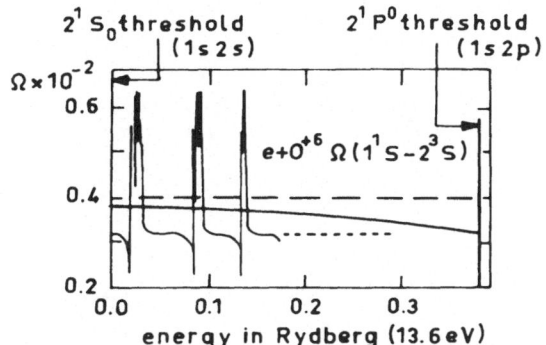

Fig 3: $1s^2\,{}^1S_0 - 1s2s\,{}^3S$, collision strength for O VII energy range between the $1s2s\,{}^1S_0$ and $1s2p\,{}^1P_1^0$ threshold. ---- background; $\langle\Omega_d\rangle$ ——— ; $\langle\Omega\rangle$ - - - -

Figure 1 is convenient to discuss the resonance effects : the higher autoionising states $1s2p\,nl$ lie between the $1s^2$ and $1s2s$ ionisation limits of the 3-electron system. They can be populated by capture of an electron $(1s^2 + e^-)$ and then decay into $(1s2s + e^-)$. The $1s2s\,{}^2S$ term can therefore be excited directly from $1s^2$ or through the decay of a $1s2pnl$ resonance populated by $1s^2 + e^-$ captures. However, the $1s2pnl$ autoionizing state can also decay via a radiative transition to $1s^2nl$. The efficiency of the $1s2s$ indirect excitation process through a resonance is therefore reduced by the radiative decay process. Figure 3 displays numerical results by

PRADHAN (1981) concerning $1s^2$ ^1S $-$ $1s2s$ ^3S collision strengths in
the case of O^{+6}. The lower curve is the background of the resonances;
it corresponds to the direct excitation process alone. $\langle \Omega \rangle$ includes
the average resonance effect, but in a theory excluding radiative
decay, and $\langle \Omega_d \rangle$ includes both the average resonance effect and
the radiation field effect. From the discussion above, it appears
natural to find $\langle \Omega_d \rangle$ somewhere in between the background and $\langle \Omega \rangle$.

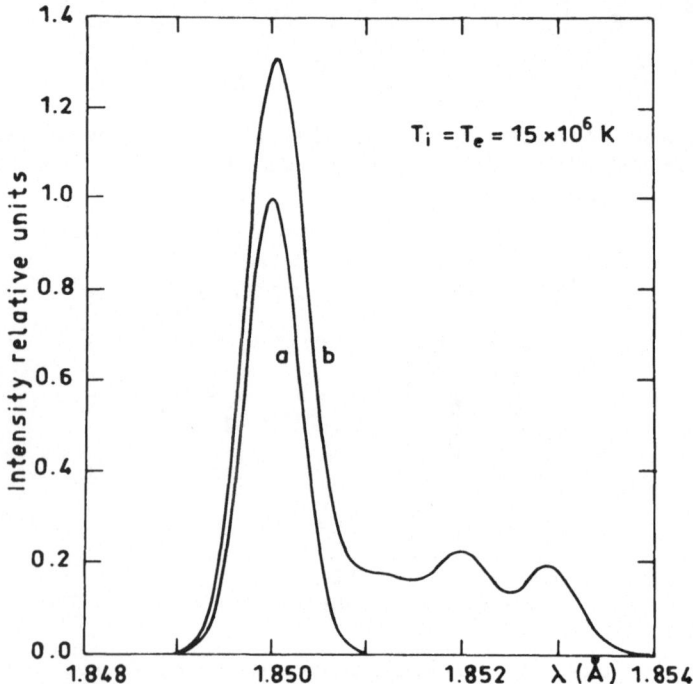

Fig 4 : Fe XXV resonance line intensity. a) pure resonance
line. b) apparent resonance line (including unresolved
satellites)

When the $1s2pnl$ autoionizing levels decay through radiative
transitions, they emit photons whose wavelengths are close to the
$1s2p$ $-$ $1s^2$ He-like resonance lines. The result of such emissions
appears either in the profile of the resonance lines or well separated from them (BELY-DUBAU et al., 1979)(see Fig 4). The extra lines
are therefore unresolved or resolved; they are called the satellite
lines of the $1s2p$ $-$ $1s^2$ transition. The radiative probability asso-

ciated to these lines increases as z^4 along the isoelectronic sequence, while the autoionisation probability associated to the same
lines remains almost constant. Satellite lines are consequently more
intense for highly-ionized atoms than for lowly ionized atoms.
(see lectures by A.H. GABRIEL in this volume).

In the case of the recombination of an ion other than H-like,
He-like or rare gas-like (Ne, Ar ...), dielectronic recombination
takes place for n = 10 - 100, and 1 $\geqslant$ 5. This process is very sensitive to the electron density in high-density plasmas.

Now, we have seen that, for 1 large, the CBe approximation gives
good results. This is also true, to some extent, for the $\overline{g}$ approximation. This fact was exploited elegantly by BURGESS (1965) who derived
a well-known general formula which shows essentially that dielectronic recombination depends only on the line strength.

Note about the ionisation process

We do not have enough time to consider the theory of ionisation
in any detail. However, we want to stress the importance of the excitation of autoionising states, which, via autoionisation, contribute
to ionisation. This contribution is included in all "modern" ionis-
tion equilibrium calculations (JORDAN, 1970; SUMMERS, 1972; JACOBS et
al., 1977).

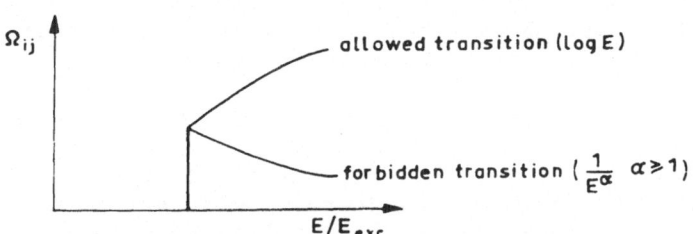

Fig 5 : allowed and forbidden transition behaviour of collision
strength Ω_{ij} as a function of energy.

4. Relativistic effects

For very highly-ionised atoms, the structure within the target
"terms" α_i S_i L_i π_i shows more and more. The separation between
the different "levels" α_i S_i L_i J_i π_i increases considerably as
the ion charge goes up. The radiative transition probabilities associated with these levels do not follow statistical rules any more.
This is all due to the relativistic effects.

The following example should serve to illustrate the importance of such effects : in a relativistic formalism, the transition

$$1s2p \ ^3P_1 \ \rightarrow \ 1s^2 \ ^1S_o + h\nu$$

can take place as an electric dipole transition, which is an allowed
transition, because we can represent the $^3P_1$ state as

$$"1s2p \ ^3P_1" = a \ 1s2p \ ^3P_1 + \sqrt{1 - a^2} \ 1s2p \ ^1P_1.$$

$J\pi$ and not $LS\pi$ are now "good" quantum numbers

In this volume, GABRIEL discusses the behaviour of a collision
strength Ω_{ij} as a function of the energy E (Fig 5). For an allowed
transition, Ω_{ij} behaves as logE, while it behaves like $1/E^\alpha$,
with $\alpha \geqslant 1$, for a forbidden transition. As a consequence, Ω for
the transition

$$1s^2 + e^- \ \rightarrow \ 1s2p \ ^3P_1 + e^-$$

displays an "allowed" behaviour when a relativistic formalism is
required ($\eta \gg$) and a "forbidden" behaviour when relativistic
effects can be neglected ($\eta \ll$).

4.1 Non-relativistic wavefunctions

For moderately charged ions, it appears justified to use a non-
relativistic method to build up zero-order wavefunctions and to include the relativistic Breit-Pauli contribution as a perturbation.
The Hamiltonian for the total system becomes

$$H = H_{NR} + H_{BP} \tag{4.1}$$

where H_{BP} contains the Breit-Pauli terms one chooses to include in
the formalism. There are one-body terms : mass variation, Darwin
and spin-orbit; and two-body terms : i) fine-structure: mutual spin
orbit, spin-other orbit and spin-spin interactions; ii) non-fine-
structure: Darwin, contact spin-spin and orbit-orbit interactions.
The atomic structure code developed at University College London by
EISSNER et al., (1974) is based on the (NR + BP) hamiltonian. The
program calculates non-relativistic, multiconfigurational expansion
wavefunctions and energy terms in a modified Thomas-Fermi statistical
model potential. Then, relativistic corrections are introduced in
the diagonalisation procedure and the non-relativistic and "relativistic" wavefunctions are related to one another by a matrix transformation of the form :

$$\overline{\omega}\left(\Delta_i J_i M_{J_i}\right) = \sum_{S_i L_i} f_{J_i}\left(\Delta_i \alpha_i S_i L_i\right) \chi\left(\alpha_i S_i L_i J_i M_{J_i}\right) \tag{4.2}$$

where

$$\chi(d_i S_i L_i J_i M_{J_i}) = \sum_{M_{S_i} M_{L_i}} C^{L_i S_i J_i}_{M_{L_i} M_{S_i} M_{J_i}} \bigodot (d_i S_i L_i M_{S_i} M_{L_i}) \tag{4.3}$$

The wavefunction for an entrance channel j is now

$$\overline{\Psi}_j \sim \frac{1}{\sqrt{N+1}} \sum_{i=1}^{\overline{NCHOP}} \overline{\bigodot}_i \frac{\overline{F}_{ij}(r_{N+1})}{r_{N+1}} \tag{4.4}$$

where

$$\overline{\Gamma}_i \doteq (\beta_i J_i \, l_i \, J \pi M_J) \tag{4.5}$$

Note that $\overline{NCHOP} \gg NCHOP$ because the fine structure increases the number of open exit channels : many different LS π contribute to one J π .

As far as collisions are concerned; JONES (1975) gave a detailed description of a method for building a DW computer program using such wavefunctions, when the colliding electron is "relativistic".

A simpler approach was described by SARAPH (1972) and SAMPSON and GOLDEN (1978). The colliding electron is taken to be non-relativistic. Therefore, l_i is a "good" quantum number and the pair coupling scheme :

$$\vec{J}_i + \vec{l}_i = \vec{K} \qquad\qquad \vec{K} + \vec{1/2} = \vec{J} \tag{4.6}$$

can be used.

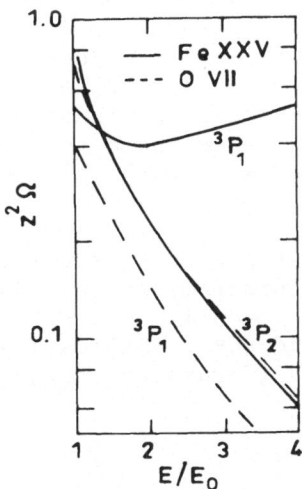

Fig 6 : Reduced collision strength for $1s^2 \, {}^1S_0 - 1s2p \, {}^3P_1, \, {}^3P_2$ transitions for O VII and Fe XXV

Playing with matrix transformations like (4.2) and employing Racah recoupling coefficients W, SARAPH (1972) established the relation between matrix R in the LS π representation and matrix $\underline{R}$ in the J π representation. A computer program, called JJOM allows for the derivation of the fine-structure R from the non-relativistic $\underline{R}$, which can be calculated in the CC or DW approximation. SCOTT and BURKE (1980) have described recently a method to implement the Breit-Pauli Hamiltonian within the RMATRX framework. They applied their method to Fe XXIII. This approach is the most refined way to calculate cross-sections with the Breit-Pauli Hamiltonian.

Figure 6 illustrates the importance of relativistic effects. It displays the behaviour of the reduced collision strength $z^2 \Omega$ as a function of the reduced energy E/E_{exc} for transitions $1s^2\ ^1S_0$ to $1s2p\ ^3P_1$ and $^3P_2$ in the case of O VII and Fe XXV. The $^1S_0 - ^3P_2\ z^2\ \Omega$ happens to be the same for O VII and Fe XXV, as expected, while the marked difference seen for the $^1S_0 - ^3P_2\ z^2\Omega$ is easily explained by the relativistic effects, important in Fe XXV, which can change a "forbidden" tendency into an "allowed" tendency, as explained above.

4.2 Relativistic wavefunctions

To account for relativistic effects, the "best" approach is to build a fully relativistic equation including the Dirac monoelectronic equation and the Breit two-electron interaction. This approach is used at present to calculate bound-state wavefunctions (see the lecture by DESCLAUX in this volume). It is tempting to generalize the technique to the collisional problem, but the work is still in progress. The possibility to include the Breit-Dirac Hamiltonian into the CC approximation has been considered (CHANG (1975)). However, as far as we know very few calculations have been performed (CHANG (1977)). The first relativistic collisional calculations published are those by WALKER (1974) who extended the CB method to allow for relativistic effects in the case of the electron impact excitation of hydrogenic ions.

We wish to mention that progress is being made in the direction towards efficient computer programs for building relativistic wavefunctions in a collisional formalism. For the CC approximation, the groups of GRANT (Oxford) and BURKE (Belfast) are much involved, while the group of KLAPISCH (Jerusalem) concerns itself with the DW approximation.

REFERENCES

Bely O., Moores D.L., Seaton M.J., 1963, Atomic Collision Processes,
      ed. M.R.C. McDowell (Amsterdam, North Holland) p 304
Bely-Dubau F., Gabriel A.H., Volonté S., 1979, M.N.R.A.S., 189, 801
Berrington K.A., Burke P.G., Le Dourneuf M., Robb W.D., Taylor K.T.,
      Vo Ky Lan, 1978, Comput. Phys. Commun., 14, 367
Burgess A., 1961, Mém. Soc. Roy. Sci. Liège, 4, 299
Burgess A., 1965, Astrophys. J. Lett., 141, 1588
Burke P.G., Schey H.M., 1962, Phys. Rev., 126, 147
Chang J.J., 1975, J. Phys. B, 8, 2327
Chang J.J., 1977, J. Phys. B, 10, 3335
Crees M.A., Seaton M.J., Wilson P.M.H., 1978, Comput. Phys. Commun.,
      15, 23
Dirac P.A.M., 1958, The Principles of Quantum Mechanics (Oxford,
      Clarendon Press) 4th Ed., Chap 8
Eissner W., Jones M., Nussbaumer H., 1974, Comput. Phys. Commun.,
      8, 270
Eissner W., Seaton M.J., 1972, J. Phys. B, 5, 2187
Fano U., 1961, Phys. Rev., 124, 1866
Gailitis M., 1963, Sov. Phys.-JETP, 17, 1328
Ham F.S., 1955, Solid State Physics, ed. F. Sertz and D. Turnbull,
      (New York, Academic Press) vol. 1 p 127
Henry R.J.W., 1981, Physics Reports, 68, 1
Jacobs V.L., Davis J., Kepple P.C., Blaha M., 1977, Astrophys. J.,
      215, 690
Jones M., 1975, Phil. Trans. Roy. Soc. London, A277, 587
Jordan C., 1970, M.N.R.A.S., 148, 17
Pradhan A.K., 1981, Phys. Rev. Lett., 47, 79
Sampson D.H., Golden L.B., 1978, J. Phys. B, 11, 54
Saraph H.E., 1972, Comput. Phys. Commun., 3, 256
Scott N.S., Burke P.G., 1980, J. Phys. B, 13, 4299
Seaton M.J., 1955, Comptes Rendus, 240, 1317
Seaton M.J., 1962, Atomic and Molecular Processes, ed. D.R. Bates
      (New York, Academic Press) p 374
Seaton M.J., 1969, J. Phys. B, 2, 5
Shore B.W., Menzel D.H., 1967, Principles of Atomic Spectra
      (New York, John Wiley) Chap 6
Summers H.P., 1972, M.N.R.A.S., 158, 225
Van Regemorter H., 1962, Astrophys. J., 136, 906
Walker D.W., 1974, J. Phys. B, 7, 97

CONTRIBUTORS

ANDERSEN, J.U.   Institute of Physics, University of Aarhus
                       DK-8000 Aarhus C, Denmark

BACKE, H.   Institut fur Physik der Universitat Mainz
                       Postfach 3980, D-6500 Mainz, West Germany

BARAT, M.   Laboratoire des Collisions Atomiques et
                       Moleculaires, Universite Paris-Sud, Bat. 351
                       91405 Orsay Cedex, France

BOSCH, F.   Gesellschaft fur Schwerionenforschung mbH.
                       D-6100 Darmstadt, Fed. Rep. of Germany

COCKE, C.L.   Department of Physics, Kansas State University
                       Manhattan, Kansas 66506, U.S.A.

CRANDALL, D.H.   Oak Ridge National Laboratory
                       Oak Ridge, Tennessee 37830, U.S.A.

DESCLAUX, J.P.   Centre d'Etudes Nucleaires de Grenoble
                       DRF/Laboratoire d'Interactions Hyperfines
                       85X-38041 Grenoble Cedex, France

DRAKE, G.W.F.   Department of Physics, University of Windsor
                       Windsor, Ontario N9B 3P4, Canada

DUBAU, J.   Observatoire de Meudon, 92190 Meudon, France

HINNOV, E.   Princeton University, Plasma Physics Laboratory
                       Princeton, New Jersey, U.S.A.

MARTINSON, I.   Department of Physics, University of Lund
                       S-223 62 Lund, Sweden

SOFF, G.   Gesellschaft fur Schwerionenforschung mbH
                  D-6100 Daemstadt, West Germany

WINTER, H.   Institut fur Allgemeine Physik, Technische
                  Universitat Wien, Karlsplatz 13, A-1040 Wien
                  Austria

ZEIPPEN, C.J.   Observatoire de Paris, 92190 Meudon, France